Papermaking Science and Technology

a series of 19 books covering the latest technology and future trends

Book 5

Mechanical pulping

Series editors
Johan Gullichsen, Helsinki University of Technology
Hannu Paulapuro, Helsinki University of Technology

Book editor
Jan Sundholm, Finnish Pulp and Paper Research Institute

Series reviewer
Brian Attwood, St. Anne's Paper and Paperboard Developments Ltd.

Book reviewer
Joseph A. Kurdin

Published in cooperation with the Finnish Paper Engineers' Association and TAPPI

Cover photo by Martin MacLeod

ISBN 952-5216-00-4 (the series)
ISBN 952-5216-05-5 (book 5)

Published by Fapet Oy
(Fapet Oy, PO BOX 146, FIN-00171 HELSINKI, FINLAND)

Copyright © 1999 by Fapet Oy. All rights reserved.

Printed by Gummerus Printing, Jyväskylä, Finland 1999

Printed on LumiMatt 100 g/m^2, Enso Fine Papers Oy, Imatra Mills

Certain figures in this publication have been reprinted by permission of TAPPI.

Foreword

Johan Gullichsen and Hannu Paulapuro

PAPERMAKING SCIENCE AND TECHNOLOGY

Papermaking is a vast, multidisciplinary technology that has expanded tremendously in recent years. Significant advances have been made in all areas of papermaking, including raw materials, production technology, process control and end products. The complexity of the processes, the scale of operation and production speeds leave little room for error or malfunction. Modern papermaking would not be possible without a proper command of a great variety of technologies, in particular advanced process control and diagnostic methods. Not only has the technology progressed and new technology emerged, but our understanding of the fundamentals of unit processes, raw materials and product properties has also deepened considerably. The variations in the industry's heterogeneous raw materials, and the sophistication of pulping and papermaking processes require a profound understanding of the mechanisms involved. Paper and board products are complex in structure and contain many different components. The requirements placed on the way these products perform are wide, varied and often conflicting. Those involved in product development will continue to need a profound understanding of the chemistry and physics of both raw materials and product structures.

Paper has played a vital role in the cultural development of mankind. It still has a key role in communication and is needed in many other areas of our society. There is no doubt that it will continue to have an important place in the future. Paper must, however, maintain its competitiveness through continuous product development in order to meet the ever-increasing demands on its performance. It must also be produced economically by environment-friendly processes with the minimum use of resources. To meet these challenges, everyone working in this field must seek solutions by applying the basic sciences of engineering and economics in an integrated, multidisciplinary way.

The Finnish Paper Engineers' Association has previously published textbooks and handbooks on pulping and papermaking. The last edition appeared in the early 80's. There is now a clear need for a new series of books. It was felt that the new series should provide more comprehensive coverage of all aspects of papermaking science and technology. Also, that it should meet the need for an academic-level textbook and at the same time serve as a handbook for production and management people working in this field. The result is this series of 19 volumes, which is also available as a CD-ROM.

When the decision was made to publish the series in English, it was natural to seek the assistance of an international organization in this field. TAPPI was the obvious partner as it is very active in publishing books and other educational material on pulping and papermaking. TAPPI immediately understood the significance of the suggested new series, and readily agreed to assist. As most of the contributors to the series are Finnish, TAPPI provided North American reviewers for each volume in the series. Mr. Brian Attwood was appointed overall reviewer for the series as a whole. His input is gratefully acknowledged. We thank TAPPI and its representatives for their valuable contribution throughout the project. Thanks are also due to all TAPPI-appointed reviewers, whose work has been invaluable in finalizing the text and in maintaining a high standard throughout the series.

A project like this could never have succeeded without contributors of the very highest standard. Their motivation, enthusiasm and the ability to produce the necessary material in a reasonable time has made our work both easy and enjoyable. We have also learnt a lot in our "own field" by reading the excellent manuscripts for these books.

We also wish to thank FAPET (Finnish American Paper Engineers' Textbook), which is handling the entire project. We are especially obliged to Ms. Mari Barck, the project coordinator. Her devotion, patience and hard work have been instrumental in getting the project completed on schedule.

Finally, we wish to thank the following companies for their financial support:

A. Ahlstrom Corporation
Enso Oyj
Kemira Oy
Metsä-Serla Corporation
Rauma Corporation
Raisio Chemicals Ltd
Tamfelt Corporation
UPM-Kymmene Corporation

We are confident that this series of books will find its way into the hands of numerous students, paper engineers, production and mill managers and even professors. For those who prefer the use of electronic media, the CD-ROM form will provide all that is contained in the printed version. We anticipate they will soon make paper copies of most of the material.

List of Contributors

Artamo Arvi – M.Sc. (Eng.), Managing Director, Rinheat Oy, E-mail: arvi.artamo@rinheat.fi – main contribution: heat recovery

Haikkala Pekka – B.Sc. (Eng.), Product and RTD Manager, Valmet Corp., Stock Preparation, E-mail: pekka.haikkala@valmet.com – main contributions: grinders, grinding processes

Hautala Jouko – B.Sc. (Eng.), Research and Development Manager, Valmet Corp., Stock Preparation, E-Mail: jouko.hautala@valmet.fi – main contribution: screening

Heikkurinen Annikki – Lic. Tech., Senior Research Scientist, KCL, E-Mail: Annikki.Heikkurinen@kcl.fi – main contribution: characterization of pulps

Holmbom Bjarne – D. Tech., Professor, Åbo Akademi University, E-mail: bholmbom@abo.fi – main contribution: environmental impacts

Hourula Ismo – M.Sc. (Eng.), Technology Manager, Valmet Corp., Stock Preparation, E-mail: Ismo.Hourula@valmet.com – main contribution: screening

Huusari Erkki – M.Sc. (Eng.), Manager Development, Sunds Defibrator, E-mail: erkki.huusari@sundsdefibrator.com – main contributions: TMP refiners, refining of shives and coarse fibers

Härkönen Esko – D. Tech., Research Manager, UPM-Kymmene Oyj, E-mail: esko.harkonen@upm2.upm-kymmene.com – main contribution: fundamental mechanisms in refining

Jussila Tero – M.Sc. (Eng.), Manager Technology, Valmet Corp., Stock Preparation, E-mail: Tero.Jussila@valmet.com – main contributions: screening systems, control of screening, flow sheets for various paper grades

Karojärvi Risto – M.Sc., Research Manager, Valmet Corp., Stock Preparation, E-mail: risto.karojarvi@valmet.com – main contributions: pulpstones, groundwood wood feeding systems

Kortelainen Juha – M.Sc. (Eng.), Research Scientist, Tampere University of Technology, E-mail: Juha.Kortelainen@mit.tut.fi – main contribution: TMP control

Kurdin Joseph A. – Consultant, E-mail: ja_kurdin@prodigy.net – main contributions: APMP process, zero effluent mill

Leskelä Leena – Customer service assistant, KCL, E-mail: Leena.Leskela@kcl.fi – main contribution: characterization of pulps

Liimatainen Heikki – B.Sc. (Eng.), General Sales Manager, Groundwood Systems, Valmet Corp., Stock Preparation, E-mail: heikki.liimatainen@valmet.com – main contributions: co-editor of grinding and screening chapters, PGW pulp family

Lindholm Carl-Anders – D. Tech., Senior Assistant, Helsinki University of Technology, Pulping Technology, E-mail: Carl-Anders.Lindholm@hut.fi – main contributions: chemimechanical pulping and bleaching

Lucander Mikael – M.Sc., Research Scientist, KCL, E-mail: Mikael.Lucander@kcl.fi – main contributions: fundamental mechanisms in grinding, TGW

Lumiainen Jorma – B.Sc. (Eng.), Vice President, Technology, Sunds Defibrator Valkeakoski Oy, E-mail: jorma lumiainen@sundsdefibrator.com – main contribution: post refining

Manner Hannu – Lic.Tech., Professor, Lappeenranta University of Technology, E-mail: hannu.manner@enso.com – main contribution: environmental impacts

Mannström Markus – M.Sc. (Eng.), Project Manager, Pilot Plant, KCL, E-mail: Markus.Mannstrom@kcl.fi – main contribution: pulp storage

Nystedt Harri – M.Sc. (Eng.), Research Scientist, Tampere University of Technology, E-mail: Harri.Nystedt@mit.tut.fi – main contribution: TMP control

Pitkänen Markku – M.Sc. (Eng.), Managing Director, GL&V/Celleco , E-mail: Markku.Pit@pp.kolumbus.fi – main contributions: cleaning, thickening

Reponen Pekka – M.Sc. (Eng.), Development Manager, Enso Publication Paper Ltd, E-mail: pekka.reponen@enso.com – main contribution: environmental impacts

Salmén Lennart – Ph. D., Manager Wood and Fiber Physics Research, STFI, E-mail: Lennart.Salmen@stfi.se – main contribution: rheological behaviour of wood

Sundholm Jan – M.Sc. (Eng.), Senior Research Scientist, KCL, E-mail: Jan.Sundholm@kcl.fi – main contributions: introduction, what is mechanical pulping, history, reasons for difference in energy consumption between grinding and refining, introduction to TMP, main TMP process types, how TMP refiner type and process conditions affect pulp properties, flow sheets for various paper grades, future outlook

Tienvieri Taisto – M.Sc.(Eng.), Research Manager, Fibre Processing, UPM-Kymmene Corp., Paper Divisions, E-mail: Taisto.Tienvieri@upm-kymmene.com – main contributions: chip handling, flow sheets for various paper grades

Tuovinen Olli – M.Sc. (Eng.), Research Manager, Valmet Corp., Stock Preparation, E-mail: olli.tuovinen@valmet.com – main contributions: raw materials, grinding process control

Varhimo Antero – M.Sc. (Eng.), Senior Research Scientist, KCL, E-mail: Antero.Varhimo@kcl.fi – main contribution: raw materials

Vuorio Petteri – M.Sc. (Eng.), Sales Manager, Sunds Defibrator Fiber Applications, E-mail: petteri.vuorio@sundsdefibrator.com – main contribution: refiner segments

Preface
Jan Sundholm

This book is the successor of a chapter on mechanical pulping by H. Paulapuro, J. Vaarasalo, and B. Mannström in a Finnish textbook on pulping technology (Chapter 11 in the textbook *Puumassan Valmistus,* in Finnish, edited by N.-E. Virkola, Teknillisten tieteiden akatemia, Turku, 1983). The themes from the predecessor are extended to screening, reject refining, and other pulp treatments to offer a more complete approach to mechanical pulping.

In editing a textbook, which also should be used as a handbook and reference book, it is a problem to judge how much emphasis should be put on fundamentals and how much on the practices of technology. One might think that fundamentals should be stressed, as these change slowly with time. Casey, on the other hand, stated in the preface to the second edition of *Pulp and Paper* that "ideas are changing faster than practice in the paper industry," and that the first edition because of this was more out of date than if it had been merely a presentation of the technology (Casey, J.P., *Pulp and Paper,* 2nd edition, Interscience Publishers, New York, 1960). This book handles both fundamentals and practical technology and is thus applicable for both scientists and process engineers, but as editor I have chosen to put somewhat more stress on the practice of mechanical pulping than on the fundamental viewpoints. The goal is to include material that will be meaningful and useful to a spectrum of readers from students to mill managers, and from scientists to engineers.

In order to get expert opinions, I have tried to select recognized experts in each subprocess and field of mechanical pulping. It is my belief that this team of 28 expert authors is the best guarantee for opinions and facts that could be relied on. Writers have their own styles and opinions, which I have not tried to suppress (not very much anyway). On the contrary, in my opinion, their sometimes differing opinions help the reader to get both a diversified and a deeper understanding of the world of mechanical pulping. I offer my warmest thanks to the large team of authors who have managed to write their contributions in addition to their normal workload. Of the authors, I especially want to acknowledge the coordination efforts by Heikki Liimatainen of Valmet Corporation regarding the grinding, screening, and pulp treatment chapters. I also thank I. Forsskåhl, I. Nurminen, and M. Ranua of KCL and K. Repo of CTS Engineering for giving useful comments on some of the texts. Thank you to the Finnish Pulp and Paper

Research Institute (KCL) which made it possible for me to coordinate the efforts put into this book.

Finally I wish to express my great gratitude to the main editors of this textbook series: Hannu Paulapuro and Johan Gullichsen for having faith in me as editor all along the process of producing this book and for giving valuable remarks on the manuscript, and to the TAPPI reviewer Joseph A. Kurdin for giving valuable comments and especially for pointing out special fields of interest for North America and, in the end, writing some paragraphs himself!

I wish to dedicate this book to two pioneers of the science of mechanical pulping and former colleagues to me at KCL: Georg von Alfthan (retired) and Bo Mannström (†), whose work was essential in forming the scientific basis for the Finnish paper industry's expansion in mechanical printing papers during the 1980s and '90s.

September 1998
Jan Sundholm

Table of Contents

1. Introduction 12
2. What is mechanical pulping? 17
3. History of mechanical pulping 23
4. Fundamentals of mechanical pulping 35
5. Raw materials 67
6. Grinding and pressure grinding 107
7. Thermomechanical pulping 159
8. Chemimechanical pulping 223
9. Screening and cleaning 251
10. Refining of shives and coarse fibers 289
11. Bleaching 313
12. Thickening, storage, and post refining 345
13. Flow sheets for various paper grades 365
14. Environmental impacts of mechanical pulping 375
15. The character and properties of mechanical pulps 395
16. Future outlook 415
    Conversion factors 422
    Index 425

CHAPTER 1

Introduction

1   On the contents of this book 12
2   How to interpret the numbers 13
3   Abbreviations used 13

CHAPTER 1

Jan Sundholm

Introduction

In this introductory chapter, the reader will find information on how to read this book, on what is included and what is not included in the book, and on how to interpret the data in text, figures, and tables. The chapters of the book are quite independent of each other, which means the reader can choose to read only those chapters of personal interest.

1 On the contents of this book

What is mechanical pulping? provides basic facts on wood and how to mechanically produce paper fibers (mechanical pulping). The chapter describes some of the characteristic features of mechanical pulps as opposed to chemical pulps. It also includes a nomenclature with short process descriptions and explains the main end uses of mechanical pulps.

History of mechanical pulping gives a short description of the invention of grinding and the evolution of today´s processes.

Fundamentals of mechanical pulping explores the fundamental mechanisms of defibration and fiber development in grinding and refining starting from the reological behavior of wood and ending with some remarks on the reasons for the high and varying energy consumption in mechanical pulping.

The chapter **Raw materials** explains how the most important variable in mechanical pulping is the wood raw material. The characteristics of the produced pulp is highly dependent on the wood species used and its quality. The book Chemical Pulping describes wood handling in more detail.

The **Grinding and pressure grinding, Thermomechanical pulping,** and **Chemimechanical pulping** chapters tell about the different mechanical pulping processes: wood feeding systems, chemical impregnation systems, grinder and refiner constructions, pulpstones and refiner plates, process control, as well as how process conditions affect pulp properties.

The chapters **Screening and cleaning, Refining of screen rejects and coarse fibers,** and **Thickening, storage, and post refining** discuss the increasingly important pulp processing steps that follow grinding or refining.

In **Flow sheets of pulp processing for various paper grades,** typical state-of-the-art process layouts provide data for board, newsprint, and magazine papers.

The **Bleaching** chapter is an introduction to the chemical operations necessary for the lignin-retaining bleaching (brightening) of mechanical pulps.

The chapter **Environmental impacts** introduces the various environmental problems associated with mechanical pulping.

The chapter **Properties** discusses both the fundamental aspects of characterization of fiber fractions and the standards used in practice for testing the suitability of mechanical pulps for papermaking and various paper products.

The final **Future outlook** chapter tries to look ahead at what is to come. How will the paper grades continue to develop? What kind of improvements will we get in today´s processes? Will we get new low-energy processes?

As this book concentrates on the technical aspects of mechanical pulping, there are certain areas of mechanical pulping that fall outside the scope of the book. Such areas are the economics of mechanical pulping as well as production and trade statistics.

2 How to interpret the numbers

The text contains a lot of numerical information mainly from the testing of various pulps. Unless otherwise mentioned, ISO standards are probably used in deriving the data. (See chapter "Properties" for more information on standards.) However, the reader should be aware that all references do not give exact information on the standard test methods used. Also, in most cases, only values in the same table or figure can be compared to each other. One reason for this is that different standards might have been used for different tables or figures. Other reasons include the use of varying process conditions, different technology, varying wood raw material, etc.

If not otherwise stated, production numbers are given on the basis of air dry metric tons ($t_{90\%}$). Energy consumption is given as kWh/metric ton or MWh/metric ton (per air dry metric ton). Other favored units are bar and kPa (pressure), °C (temperature), l/s (flow rate), and m/s (speed). As in most technical texts, pressure values are normally given as over pressure values, bar(e). If not otherwise stated, the raw material is Norway spruce *(Picea abies)*.

3 Abbreviations used

SC	Supercalendered paper, uncoated magazine paper
LWC	Lightweight coated paper, coated magazine paper
SEC	Specific energy consumption
CSF	Freeness
DIP	Deinked recovered paper pulp
GW (SGW)	Stone groundwood
PGW	Pressure groundwood

CHAPTER 1

PGW70	PGW with shower water temperature 70°C
PGW95	PGW with shower water temperature 95°C
PGW-S	Super pressure groundwood
PGW-S95	PGW-S with shower water temperature 95°C
PGW-S120	PGW-S with shower water temperature 120°C
TGW	Thermo groundwood
RMP	Refiner mechanical pulp
PRMP	Pressure refiner mechanical pulp
TMP	Thermomechanical pulp
WTMP	TMP with water impregnation
SD	Single disc
DD	Double disc
CD	Conical disc
SC	Single disc conical
LC	Low consistency
HC	High consistency
Twin refiner	Refiner with two refining gaps
CMP	Chemimechanical pulp
CTMP	Chemithermomechanical pulp
BCTMP	Bleached CTMP
BCMP	Bleached CMP
CRMP	Chemi-refiner mechanical pulp
DWS	Dilution water sulfonation
OPCO	Interstage sulfonation process developed by Ontario Paper Co.
TMCP	Thermomechanical chemical pulp
LFCMP	Long fiber CMP
CTLF	Chemically treated long fibers
SLF	Sulfonated long fibers
G-CTMP	Groundwood CTMP
$CTMP_R$	Reject CTMP
APTMP	Alkaline peroxide thermomechanical pulp
APMP	Alkaline peroxide mechanical pulp
AP/BCTMP	Alkaline peroxide bleached CTMP
SCMP	Sulfite or sulfonated CMP
UHY	Ultrahigh-yield pulp
UHYS	Ultrahigh-yield sulfite
UHYSP	Ultrahigh-yield sulfite pulp
VHY	Very high yield
VHYS	Very high-yield sulfite
VHYSP	Very high-yield sulfite pulp
Na_2SO_3	Sodium sulfite
H_2O_2	Hydrogen peroxide
NaOH	Sodium hydroxide

COD	Chemical oxygen demand
TOC	Total organic carbon
BOD	Biological oxygen demand
TDS	Total dissolved solids

CHAPTER 2

What is mechanical pulping?

1 Mechanical pulping and mechanical pulps 17
2 Nomenclature and short description of processes 18
3 Characteristic features of mechanical pulps 19
4 End use 21

CHAPTER 2

Jan Sundholm

What is mechanical pulping?

1 Mechanical pulping and mechanical pulps

Wood is a natural composite material and is far more complex than any man-made composite. Wood consists of fibers with lignin in between, and thus can be compared to, for instance, a carbon fiber composite material, which consists of carbon fibers held together by a synthetic resin. What makes the structure so complex is that the wood fiber is in itself a composite; it consists of fibrils with lignin and hemicellulose holding it together. And, furthermore, the fibrils consist of microfibrils which also in themselves are composites.

    The purpose of all pulping processes is to separate the fibers from the wood and to make the fibers suitable for papermaking. The fibers can be separated from each other without being too damaged only if the lignin and, to a large extent, also the hemicellulose is either dissolved and removed as in chemical pulping, or more or less softened as in mechanical pulping. Both chemical and mechanical pulp fibers must be mechanically treated (refined, beaten) before the fibers become suitable for papermaking. The difference is that for chemical pulp this is a separate process in the paper mill, while for mechanical pulps this happens during the mechanical pulping process itself. This loosening up of the fiber structure is even more important in mechanical pulping because a separated but otherwise "native" mechanical fiber is far too stiff to produce a smooth and strong paper. In the mechanical pulping process, the wood and the fibers are fatigued by vibrational forces caused by the same stone or plate pattern that separates the fibers until the structure loosens up in the desired way .

    The pulp from processes where only small amounts of chemicals are used (and thus only a small amount of lignin is dissolved) also are considered to belong to the mechanical pulp family. These are generally called chemimechanical pulps.

    An ideal mechanical pulp produces a sheet of paper with high opacity, brightness, bulk and smoothness, and a suitable pore structure at a low basis weight and without an excessive use of reinforcement pulp, with its negative influence on opacity and bulk. The advantage of mechanical pulp compared to chemical pulp is that, by using mechanical pulp, it is possible to produce low basis weight paper with sufficient opacity and bulk. Therefore an ideal mechanical pulp also must have sufficient strength properties.

CHAPTER 2

One can say that in an ideal mechanical pulping process:
- The fibers must be separated from the wood.
- Fiber length must be retained.
- The fibers must be delaminated (internal fibrillation of the fiber wall).
- Abundant fines must be peeled off from the middle lamella and the primary and secondary layers of the fiber wall.
- The remaining secondary wall must be fibrillated (external fibrillation).

In practice, there are two major ways to produce mechanical pulp on an industrial scale:
- By pressing wood logs against a revolving pulpstone (grinding).
- By disintegrating wood chips in a disc refiner (refining).

It must perhaps be pointed out that neither of these processes is what above is called an ideal mechanical pulping process. Grinding tends to produce a pulp with excellent optical properties but not so good strength properties. Chip refining generally produces pulps with good strength properties but poorer optical properties. Depending on the end product demands, both processes can be adjusted toward better optical and bulk properies on one hand and better strength properties on the other hand.

2 Nomenclature and short description of processes

The major mechanical pulps and processes are listed in Table 1. The table also gives short descriptions of the processes. The numerous subprocesses are not listed here but are discussed in the chapters on grinding, thermomechanical pulping, and chemimechanical pulping. Many of the process names have a commercial background (for example, PGW is a Valmet process), and total consensus for all the names listed in Table 1 has not yet been reached.

What is mechanical pulping?

Table 1. Nomenclature for mechanical pups. The main processes of today are the GW, PGW, TMP, and CTMP processes. The yield values are for fresh debarked Norway spruce wood and modern closed processes with a water usage 10–20 m^3/metric ton.

GW (SGW)	**Stone Groundwood**
	Atmospheric grinding of logs using a pulpstone. The temperature of the shower water normally 70°C–75°C. Yield 98.5%.
PGW	**Pressure Groundwood**
	Logs are ground in pressurized conditions (2.5 bar) with a Valmet grinder at shower water temperatures lower than 100°C. Yield 98.5%.
PGW-S	**Super Pressure Groundwood**
	Logs are ground in pressurized conditions (4.5 bar) with a Valmet grinder at shower water temperatures higher than 100°C. Yield 98%.
TGW	**Thermo Groundwood**
	Atmospheric grinding of logs with a Voith chain grinder and a special system that allows shower water temperatures of 80°C or more. Yield 98.5%.
RMP	**Refiner Mechanical Pulp**
	Atmospheric refining of chips using a disc refiner. No other treatment than washing of chips and possibly some atmospheric presteaming before refining. Yield 97.5%.
PRMP	**Pressure Refiner Mechanical Pulp**
	As RMP, but pressurized refining at an elevated temperature. Yield 97.5%.
TMP	**Thermomechanical Pulp**
	Chips are preheated with (pressurized) steam and refined under pressure at elevated temperatures. The steam (over)pressure is normally 3–5 bar and the steam temperature is correspondingly 140°C–155°C. Yield 97.5%.
CMP	**Chemimechanical Pulp**
	1. General name for all chemimechanical pulps, produced by refining or grinding of chemically pretreated wood. Yield 80%–95%.
	2. Pulp that is manufactured from chemically and normally also thermally pretreated chips by refining under atmospheric or pressurized conditions. Relatively strong chemical treatments are used. Yield typically lower than 90%.
CTMP	**Chemithermomechanical Pulp**
	Pressurized refining of chemically pretreated chips or coarse pulp. Relatively mild treatments. Yield typically over 90%. Several subprocesses, see the chapter on chemimechanical pulping.

3 Characteristic features of mechanical pulps

Typical of mechanical pulping is the high yield, normally 97%–98% for Norway spruce, while that of chemical pulping normally is 45%–50%. In ecological and national economy terms, this means that more paper can be produced out of limited wood resources. As far as corporate economy is concerned, the raw material costs decrease with higher yield. Because of the high yield, the mechanical pulp contains all components of the wood raw material in about the same ratios as in wood.

CHAPTER 2

Compared to the chemical pulping processes, mechanical pulping offers further advantages: low capital costs and an uncomplicated process. Mechanical pulps have certain good paper properties such as a high light-scattering power, a fairly high brightness, a high smoothness, a good formation, and a high bulk.

As a general rule, mechanical pulps are produced at the paper mill and tailor-made for the end use in a lot of varieties, whereas the chemical pulps are produced only in a few major grades. Chemical pulps must be refined at the paper or board mill before end use.

Naturally the mechanical pulping processes have their drawbacks. The mechanical processes require high-quality wood raw materials, and the consumption of electric energy is very high. The electric energy consumption can be as high as 3.5 MWh/metric ton for the mechanical pulp in top quality magazine paper.

Because of their very nature, the bonding capacity of mechanical fibers is lower than that of chemical pulp fibers, which leads to lower overall strength properties. Mechanical pulps also often contain impurities. Mechanical pulps cannot be bleached to the same level as chemical pulps, and their brightness stability is limited. Figures 1 and 2 show the differences in strength and light scattering between some softwood mechanical pulps on one hand, and kraft and standard deinked pulp on the other hand. Of the mechanical pulps, the refiner pulps (TMP and CTMP) have the best tear strength but the lowest light scattering. The opposite goes for the groundwood pulps (GW and PGW), which have the best optical properties but the lowest strength properties.

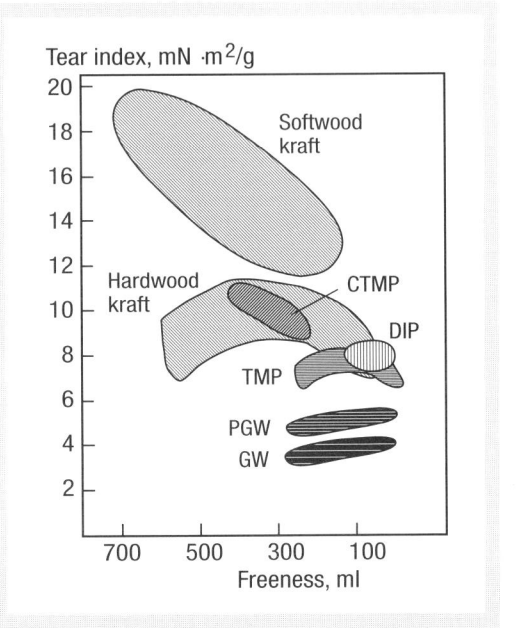

Figure 1. The tear strength of mechanical and chemical pulps. Norway spruce.

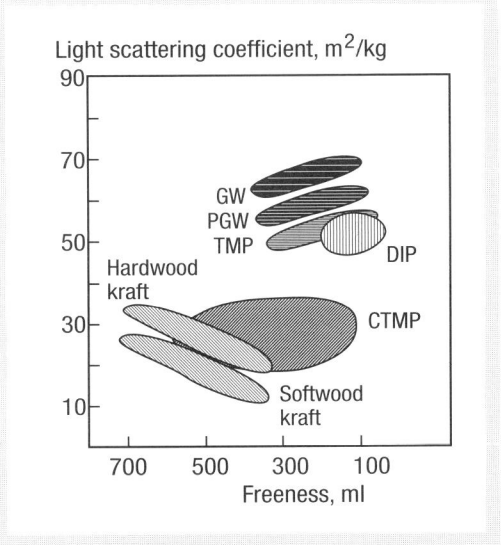

Figure 2. The light scattering ability of mechanical and chemical pulps. Norway spruce.

What is mechanical pulping?

Due to the quite harsh process conditions in mechanical pulping, all mechanical pulps contain more or less intact fibers, fragmented fibers, and fine material. These pulp fractions have different properties, and the properties of a certain mechanical pulp are dependent on both the quantity and quality of these different fiber fractions. For instance, the fines fraction of a groundwood pulp can be completely different in quality from the fines fraction of a thermomechanical pulp.

4 End use

Mechanical pulps are mainly used for the production of mechanical printing papers (also called wood-containing printing papers). These are printing papers which have good opacity and printability at low basis weight, but only a limited strength and durability, and which furthermore need to be comparatively cheap. The major grades are newsprint as well as uncoated (SC) and coated (LWC) magazine papers. The main requirements for these grades are good runnability and good printability. All main pure mechanical pulps (GW, PGW, and TMP) are used in all mechanical printing paper grades, but the economy of manufacturing and the quality of the final product depend on which mechanical pulp type is used. Typically it is easier to achieve the required printability level by using groundwood (GW or PGW), while it is easier to achieve a good runnability level by using TMP. Newsprint can contain up to 100% mechanical pulp, but the trend is to use more and more DIP in newsprint manufacture. SC paper usually contains 55%–65%, and LWC paper usually 35%–40% mechanical pulp.

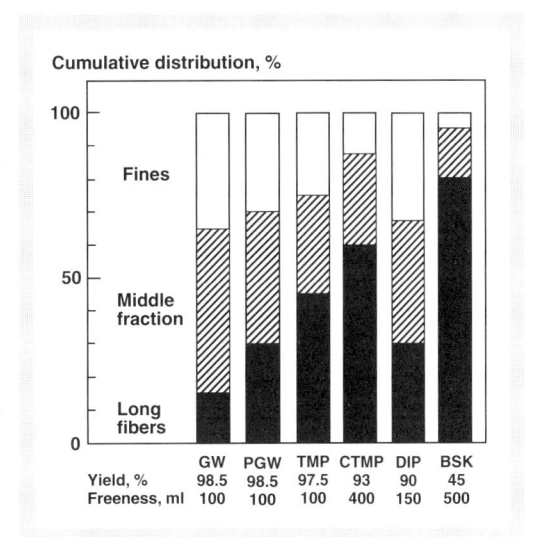

Figure 3. Pulp fraction distributions for various pulps. Note that the properties of the fractions differ for the various pulps. Norway spruce.

Other applications for mechanical pulps are various board grades, wallpaper, fine papers, soft tissue, and absorbent and molded products. Folding boxboard with its three-layer composition has traditionally been manufactured with groundwood in the middle layer. A fast growing board application for CTMP has been liquid packaging board, which originally was manufactured as solid kraft board, but now more and more is made in the same way as folding boxboard, but with bleached and washed CTMP in the middle layer. CTMP also is used as fluff pulp in absorbent products.

Some wallpaper grades contain up to 70% mechanical pulp. Although often named "wood-free," some fine papers are manufactured with an addition of 5%–20% mechanical pulp. Those fine papers with around 20% mechanical pulp are often called slightly mechanical or part-mechanical papers. Many soft tissue and molded products used to be mechanical pulp based, but currently are made primarily from recovered paper pulp.

CHAPTER 3

History of mechanical pulping

1	**The invention of the grinding process**	**23**
2	**The development of the grinding process**	**24**
2.1	The first commercial grinders	24
2.2	The first reject refiner	27
2.3	The brown groundwood and the chemigroundwood processes	27
2.4	The hot grinding process	27
2.5	Pulpstone development	27
2.6	The development of the main grinder types	28
2.7	Pressure grinding (PGW)	28
3	**Thermomechanical pulping (TMP)**	**29**
3.1	Groundwood reject refining	29
3.2	The original thermomechanical pulping process for hardboards	29
3.3	The development of RMP for printing papers	30
3.4	The development of the modern TMP	31
4	**Novel methods to produce mechanical pulps**	**32**
5	**Development of testing methods**	**32**
	References	33

CHAPTER 3

Jan Sundholm

History of mechanical pulping

This chapter contains a short description of the "milestones" in the technical development of mechanical pulping. This information does not deal with geographical and commercial aspects of the history of mechanical pulp manufacture. The reference list at the end of the chapter is also a list of recommended books and articles for those who want to learn more of what has happened during the now more than 150 years of mechanical pulping.

1 The invention of the grinding process

Already in the 18th century, J. Ch. Shäffer (Germany) described how sawdust and other residuals from sawing of timber could be mechanically defibrated and used as papermaking pulp[1]. His ideas, however, did not lead to the development of a commercial process. The time was not yet ready for such a development, and it would take almost another century before the invention of the grinding process in 1843–44.

Friedrich Gottlob Keller (1816–95) and Charles Fenerty (1821–92) are credited independently with the discovery of wood grinding and the first mechanical pulp suitable for papermaking.

Charles Fenerty lived in Novia Scotia where his family farmed some land and operated a sawmill. He was a well educated man who, because of the sawmill, knew about wood and wood properties. Because of a nearby rag paper mill, he was also acquainted with the papermaking process. According to a letter he wrote to the Halifax newspaper in 1844, he already by1835 had begun experimenting with pressing billets of wood against the surface of a revolving grindstone. Fenerty seems to have been ingenious and visionary but lacking in practicality, for he failed to pursue his discovery[2].

Figure 1. Keller's grinder according to Voelter[14].

CHAPTER 3

Friedrich Gottlob Keller, a technically interested weaver and bookbinder, first started to think about possibilities to produce papermaking fibers out of wood after reading a technical article in 1840 about the difficulties in obtaining enough old rags for the increasing paper production. He derived the idea to use wood from the way wasps build their nests. The problem was how to separate the fibers from the wood. He finally thought of using friction after remembering how he as a child used to make chains out of cherrystones. The cherrystones were put into a depression in a board and were ground to flat rings by rubbing the board against a wet rough stone surface. It was obvious during this process that the surface of the board also was ground, and Keller remembered that after the grinding he would find a flat whitish cover of dried fibers on the surface of the stone. This cover resembled paper[3].

Keller then made his first groundwood pulp in 1843 simply by turning the handle of a small grindstone partly immersed in water while also pressing a piece of wood against the stone surface. The grindstone was a typical sandstone used for sharpening tools. In 1844 he built an improved, but still hand-driven grinder with a production capacity of about 2 kg per hour. With this grinder, he and his wife produced about 100 kg of groundwood which, when mixed with rag fibers, was used to produce some copies of the Frankenberg weekly newspaper in October 1845. That same year Keller got a patent for his invention.

2 The development of the grinding process

As we look back on the history of technical developments, we can recognize four periods of rapid development:

- From 1852 until 1867, Heinrich Voelter and Johann Matthäus Voith developed not only the grinders but also screening, reject handling, and dewatering techniques. In 1867, Voelter showed a complete grinder room at the World's Fair in Paris.

- In the 1880s, the hot grinding and high production grinding were developed in the United States.

- Immediately after the First World War, the main grinder types of today were developed. The most important of these are the chain grinder (Voith) and the two pocket hydraulic grinder (Valmet).

- In the late 1970s and the early 1980s, the modern pressure groundwood process was developed by the Tampella Company (now Valmet).

2.1 The first commercial grinders

In 1846, Keller sold the rights to further patent his invention to Heinrich Voelter. Voelter developed the grinding process in cooperation with the machine shop J. M. Voith in Heidenheim. The first two mill scale grinders were built by Voith and installed in Voelters paper mill in 1852. These grinders had several presses, each of which could be reloaded independently of the others, while grinding in the others continued. In 1854, Voelter and Voith sold their first commercial grinders to Papierfabrik a. d. Sihl in Switzerland. This

History of mechanical pulping

grinder type won a gold medal at the first German Industrial Fair in Munich in 1854 and also received a lot of attention at the 1855 World's Fair in Paris. Between 1852 and 1860, more than 20 grinders were built and delivered to several countries in Europe. The first Voith grinders in North America were installed in 1866 in Quebec. During the rest of the 19th century, there was a lot of experimenting with different types of grinders such as ones with vertical axes and others with longitudinal grinding.

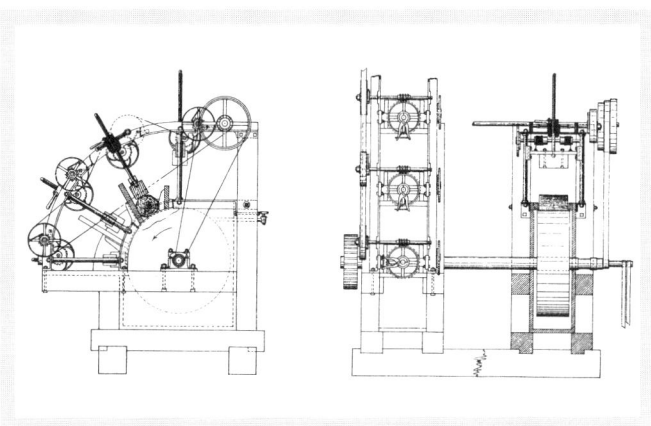

Figure 2. Voelter and Voith's mill scale grinder, the "Defibreure" (1852).

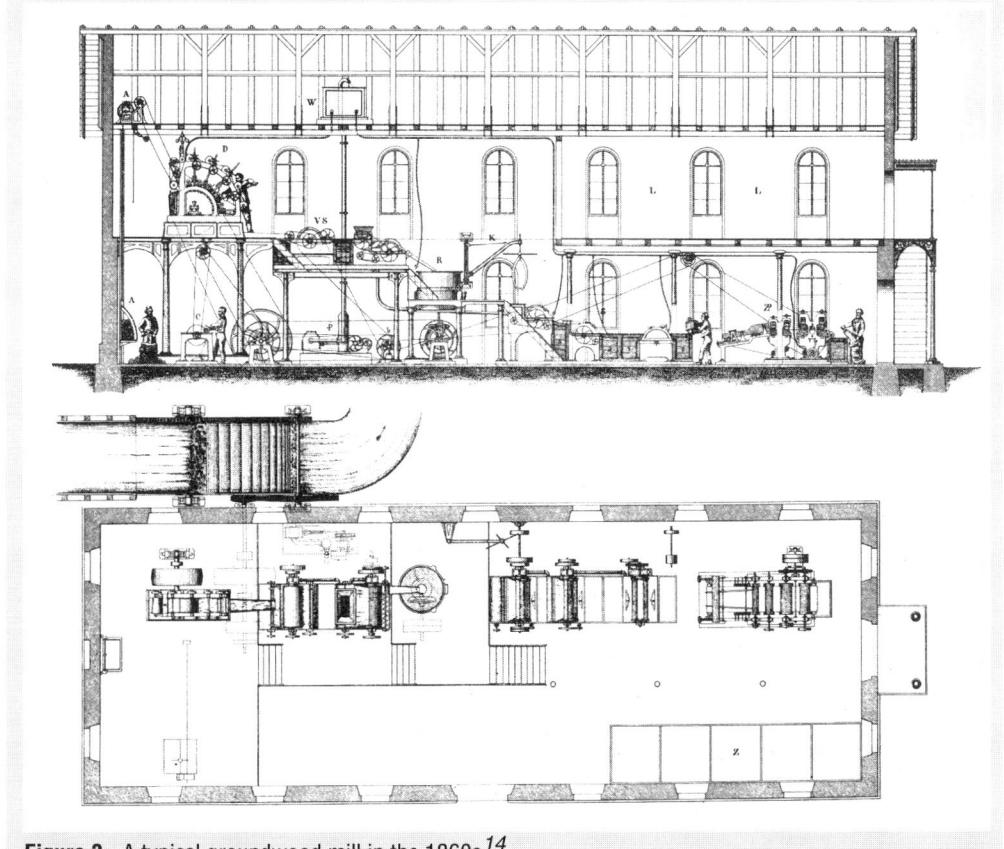

Figure 3. A typical groundwood mill in the 1860s[14].

Figure 4. Brochure on groundwood mill equipment showing the complete grinder room shown at the World´s Fair in Paris 1867. At the bottom is a note stating that the brochure was printed on paper containing 50% spruce groundwood[14].

2.2 The first reject refiner

The groundwood from the first Voith grinder contained too many shives to be used in any high-quality papers. In order to mill down the shives, Voith constructed a "Raffineur" in 1859. This refiner had two facing grindstones, between which the shives were ground to usable pulp.

2.3 The brown groundwood and the chemigroundwood processes

The brown groundwood process was developed in Germany around 1868. In this process, the wood logs are first steamed or cooked in pressurized digesters for several hours after which they are ground. This method produced groundwood with very good strength properties and a brown color. It was used for "natural brown" wrapping papers, high strength papers, and paperboards until it was displaced by kraft pulp at the beginning of the 20th century. Another development of the brown groundwood process, the chemigroundwood process, came some 50 years later, but failed to make any significant impact on the development of grinding.

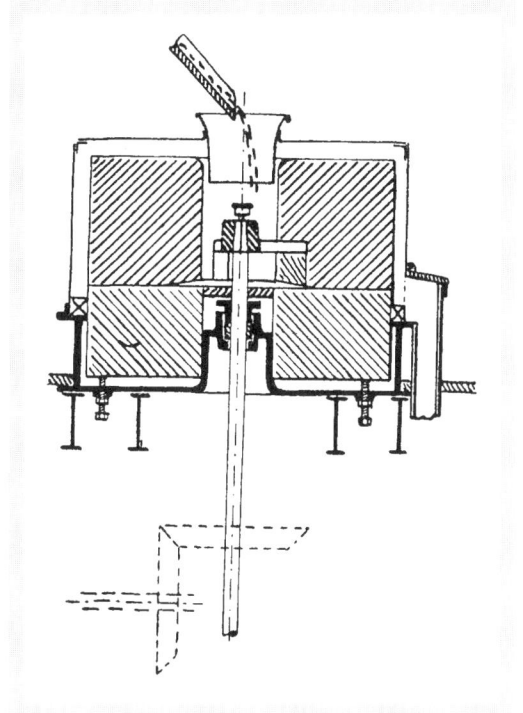

Figure 5. Voith's "Raffineur" (1859).

2.4 The hot grinding process

During the 1880s and '90s, the essential part of the groundwood process development took place in the United States. It was here that the introduction of the hot grinding method and the high production units occurred. The hot grinding was a result of the development of the "high-speed" Francis water turbine, the modern arrangement with direct horizontal drive, and finally the hydraulic loading. The new hydraulic loading enabled much higher pressloads and, because of the higher power usage per grinding area, the temperature in the grinding zone raised substantially and the pulp got much longer fibers[4].

2.5 Pulpstone development

The early pulpstones were cut from sandstone in the same way that small grindstones for the sharpening of tools were made. These sandstones were not of very high or even quality and were soon replaced by man-made pulpstones, which had been manufactured since the 1870s. These pulpstones were made by cold casting from crushed and screened quartz sand and cement. Early manufacturers were Hercules (Germany) and Norrøna (Norway).

The Norton and Carborundum Companies developed the synthetic ceramic stones of today in North America in the 1920s. The first ceramic stones had abrasive ceramic segments attached to a heavy iron spider, but later the iron spider was replaced by a cheaper strongly reinforced concrete core.

2.6 The development of the main grinder types

Development of the two dominating grinder types of today took place in the 1920s. Installation of the first large Voith chain grinder occurred in the Schongau mill in 1922. The design of this grinder was for one meter long wood and a pulpstone diameter of 1.5 meter. This was the first truly continuous and automatically operated grinder, and this grinder type has remained principally unchanged through the years. In 1984, Voith installed the first Thermogrinding (TGW) process based on this grinder type.

The Great Northern Company developed the other main grinder type, the Great Northern-Waterous-Tampella-Valmet grinder. This grinder has two large opposite pockets. Over each pocket, a low magazine holds a single charge of wood. Installation of the first Great Northern grinder occurred in the East Millinocket mill in 1926. The constructor was the Montague Machine Company. Because this grinder had many similarities to the magazine grinder produced by the Waterous Company, disputes broke out between the two companies. The two companies finally agreed to join forces in producing the Great Northern-Waterous grinder. In Finland, the Tampella Company secured a license to construct grinders of this design, and the Anjala mill of Tampella received the first delivery of such a grinder in 1937. After the temination of the license agreement in 1955, Tampella continued to build this type of grinder, identified as the two pocket Tampella hydraulic grinder. Today this grinder is known as the Valmet grinder. The main grinder types are described in detail in the chapter on grinding.

2.7 Pressure grinding (PGW)

At the beginning of the 1960s, F.G. Powell, F. Luhde, and K. C. Logan performed tests with a batchwise operating pressurized laboratory grinder[5]. They showed that it would be possible to produce a groundwood type pulp with much better strength properties by grinding at elevated temperatures. But, because of the low brightness and because of difficulties in feeding logs to a pressurized grinder, they felt these results could only be commercially applicated in the disc refining process.

Following the development of the TMP process, A. Lindahl (Sweden) became convinced that a sort of thermomechanical process also would be possible during grinding. He presented his ideas to the Tampella Company in Finland in 1976. At Tampella, M. Aario and P. Haikkala led the technical development of the new process, which was named pressure grinding (PGW). In 1977, Tampella modified a grinder at the Bure mill of MoDo into a pressurized grinder. Installation of the first PGW-grinders of a completely new design occurred in 1979 at Bure and at Anjala in Finland[12]. A. Kärnä provided final optimization of the PGW process with the grinder at Anjala[6]. By the middle of the 1980s, the process had developed into a major mechanical pulping method with worldwide industrial applications. Since 1993, the Valmet Corporation has manufactured PGW grinders. A description of the PGW process can be found in the chapter on grinding.

3 Thermomechanical pulping (TMP)

The modern TMP process has its origins in the "raffineurs" used for groundwood reject refining, the neutral-sulfite semichemical pulping processes, the Asplund hardboard process, and in the pressurized grinding trials mentioned earlier. But there were also early attempts to produce a "refiner mechanical pulp." Already in 1881–83, Rasch and Kirchner developed a method for making brown groundwood without a grinder. In this process, logs were steamed and then cut into chips and crushed in a Kollergang before being beaten in a Hollander.

3.1 Groundwood reject refining

As stated earlier, Voith built the first refiner using sandstone "discs" in 1859. This machine developed in 1893 into the Nacke´s refiner, with a horizontal drive. The stone surface of these refiners often had a grooved pattern to some extent similar to the disc patterns of today. At the end of the 19th century, metal discs appeared but, at first, only for the manufacture of brown groundwood. These machines equipped with metal discs can be considered to be the first disc refiners.

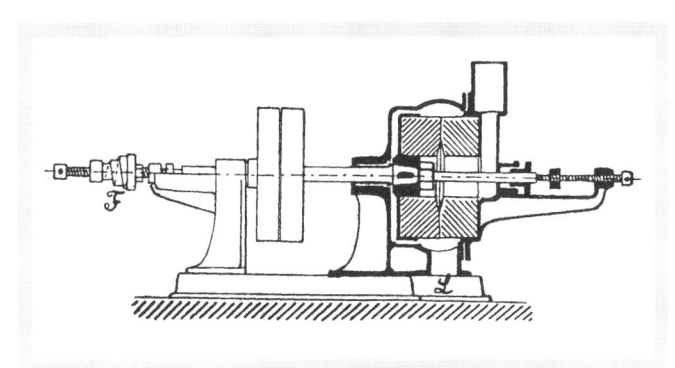

Figure 6. The Nacke refiner (Voith).

3.2 The original thermomechanical pulping process for hardboards

Early in this century, an attrition mill with two opposite counter-rotating discs was developed in the United States. This refiner, originally used for breaking down cottonseed and peanuts, developed into the Bauer Brothers Company double disc refiner, one of which was installed in 1925 at a plant making Insulite boards.

The inventor of the thermomechanical pulping process was A. Asplund (Sweden), who in the 1920s worked in the United States as a young engineer for W.H. Mason, the inventor of the masonite process. In 1927, Asplund supervised the building of a masonite mill in Sweden and thus began to think of ways to improve the masonite explosion process. In 1931, Asplund constructed a sort of pressurized refiner, which was the basis of his thermomechanical pulping process. KMW manufactured the first real continous Asplund Defibrators for the Ljusne hardboard mill in Sweden in 1934. Asplund´s continuous pulping method consti-

Figure 7. Refiner stone surface pattern.

tuted the basis for modern TMP. The KMW team also used the technical solutions regarding continuous chip feeding and pulp discharge in developing the Kamyr continous cooking method[7, 13].

Figure 8. An early Asplund Defibrator unit (1940s).

Asplund originally intended his invention also for the production of paper pulp, but it soon became clear that the dark lignin-covered fibers produced by this high steam pressure process could not be refined further to make a good papermaking pulp.

3.3 The development of RMP for printing papers

In 1929, the Bauer double disc refiner was used to defiber spent chestnut chips, which were used to manufacture a semichemical corrugating board. In 1938 at Blandin Paper Company (United States), a Bauer refiner was used to defiber aspen chips, soaked in chemicals, to produce a pulp for groundwood printing papers. Several other mills followed suit, and the semimechanical pulping processes developed in the direction of the high-yield sulfite processes and in the direction of the CTMP processes of today.

But it was not until the 1950s that real interest arose in producing pure mechanical pulp using refiners only. According to L. Eberhart of the Bauer Company[8], the original idea came up in 1948 during discussions between himself, K. Kirkpatrick, and J. Perry of the Norton Company. After running abrasive coated plates delivered by Norton, comparisons with metal plates at the Bauer pilot plant proved the metal plates to be the best. In 1955, Bauer announced the development of a "new and revolutionary method for production of mechanical pulp – groundwood from wood chips"[9]. In this process, wood chips were first defibered in a screw press before being refined in double disc refiners. At about the same time, the Sprout Waldron Company announced a similar process using single disc refiners. Also the Defibrator Company developed refiner mechanical pulping processes in the 1950s. The first trials with softwood chips were performed in Kvarnsveden

(Sweden) in 1957 using an open discharge defibrator. The first commercial mill RMP installation was a Bauer plant at Crown Zellerbach (United States) in 1960.

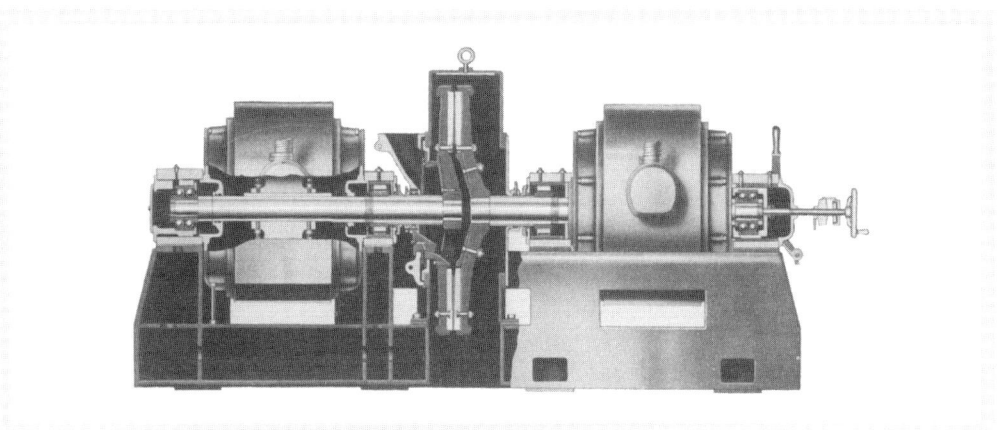

Figure 9. Bauer double disc refiner from the early 1950s.

This method at first had many names such as "refinedwood pulp," "chip groundwood," and "supergroundwood," but soon the abbreviation RMP (Refiner Mechanical Pulp) became accepted. The first RMP refiners operated at quite low consistency, until it was found in the early 1960s that high consistency refining produced a pulp of superior quality. RMP refiners and systems were videly sold in the 1960s and in the beginning of the '70s by Bauer, Sprout Waldron, Defibrator, Sunds (Bauer license), Enso (Bauer license), and Jylhävaara.

3.4 The development of the modern TMP

A pilot plant at an Anglo-Canadian mill installed the first pressurized Bauer refiner in 1963 in Quebec City, Canada. During that same year, a pilot plant of the Billerud Jössefors mill in Sweden installed the first pressurized Asplund refiner intended for paper pulp. The efforts of the Defibrator company led to the first commercial TMP mill at Rockhammar, Sweden, in 1968.

The real breakthrough for the TMP process came after the 1973 International Mechanical Pulping Conference in Stockholm[10,11]. At that conference, papers from both the Defibrator and the Bauer Companies argued that presoftening of the wood chips prior to refining at elevated temperatures resulted in a pulp with better properties as well as lower energy consumption than RMP. It soon became evident that the energy consumption of TMP in fact was much higher than for RMP but, because of TMP´s much improved strength properties and less shives, the RMP process lost the competition to TMP, which became a major process success. Later development determined that it was not the preheating of chips, but the pressurized refining, that was the important factor in the TMP process. Modern TMP systems have little or no preheating (and should actually in the latter case be called PRMP systems).

The breakthrough installations for TMP came in 1974–76. The first large Defibrator TMP mill was built at the Hallstavik mill (Sweden) of the Holmen company in 1974 for newsprint. The first application for LWC came already that same year at Bowater Carolina (United States) with a three-stage Bauer refiner plant. In 1976, the Steilacoom mill of Boise Cascade (United States) started a 400 metric tons/day Sprout Waldron TMP system with Twin 50 refiners. Jylhävaara's first TMP plant delivery was in 1975 to Kaipola of United Paper Mills (Finland). In 1977, Jylhävaara pioneered a two-stage all pressurized "Tandem" TMP process at a pilot plant in the Kaipola mill. This enabled the recovery of pressurized steam for direct use to dry paper. This heat recovery has since proved itself most important for the economics of thermomechanical pulping, and today up to 70% of the energy can be recovered.

The high energy demand of the TMP process led to intensive research into possibilities for energy savings. Starting in 1984 at KCL (Finnish Pulp and Paper Research Institute), pilot plant trials with high disc speed led to the development of the Andritz high-speed TMP process, later called the RTS process. The first high-speed TMP installation was at Perlen (Switzerland) in 1994. Also in 1994 and based on the ideas of H. Höglund, the Thermopulp process was developed by Sunds Defibrator with the first installation at Ortviken (Sweden). The chapter on thermomechanical pulping describes these processes in more detail.

4 Novel methods to produce mechanical pulps

Through the years, there have been numerous attempts to produce mechanical pulp without normal grinders or refiners as we know them today. Several of these new processes have reached the mill stage, but in the end they all have failed. Such processes are the Bersano grinder, the chip grinder of the Oji Paper mill, and the Centrigrinder of Koehring Canada. Other attempts are the Bi-Vis double screw process developed by CTP (France) and the Stake explosion process.

During the 1990s, the high energy consumption of mechanical pulping has evolved into a threat to the future of mechanical printing papers, and it seems that the time is now at hand for new "low energy" mechanical pulping processes. Even if the attempts so far have not been successful, it is quite probable that such processes eventually will be developed.

5 Development of testing methods

During the first decades of groundwood production, the only way to test the pulp was the so-called "scoop test." In this test, diluted pulp was poured from one scoop into another, during which operation the pulp was looked at visually. In the 1890s, this test developed into the blue glass test, which uses a piece of blue glass with a wooden frame. Because of the contrast between the blue glass and the yellowish pulp, diluted pulp poured onto the glass allowed easy identification of shives and other impurities. The blue glass inspection also gave some indication of the fineness of the pulp.

The first real pulp measurement method came in 1913 with the invention of the Schopper Riegler dewatering test in Germany. The Schopper test was later modified in North America to become the Canadian Standard Freeness test, which today remains the most important test for mechanical pulps. For further information on testing methods, please see the chapter on pulp properties.

References

1. Kirchner, E., Die Holzschleiferei oder Holzstoff-Fabrikation, Verlag von Güntter-Staib, Biberach, 1912.

2. Carpenter, C. H., "The History of Mechanical Pulping," TAPPI Mechanical Pulping Committee, Atlanta, 1989.

3. Kunz, W.,"Friedrich Gottlob Keller. Gedanken zum 100. Todestag," PTS-TUD-Symposium Papierzellstoff und Holzstofftechnik ´95, PTS Verlag, München, 1995.

4. Sourander, I. and Solitander, E., Finska träsliperiföreningen 1892–1942, Frenckellska Tryckeri Aktiebolag, Helsingfors, 1943.

5. Powell, F. G., Luhde, F., Logan, K. C., Pulp Paper Mag. Can. 66(8):T399 (1965).

6. Kärnä, A., "Studies of pressurised grinding," dissertation, Anjalankoski, 1984.

7. Antoine, A., Svensk Trävaru- och Pappersmassetid. (4):247 (1978).

8. Eberhardt, L., Paper Age 101(11):38 (1985).

9. Eberhardt, L., Paper Trade J. 139(37):26 (1955).

10. Asplund, A., "Development of the thermo-mechanical pulping method," 1973 International Mechanical Pulping Conference Preprints, p. 15:1, SPCI, Stockholm, 1973.

11. Charters, M. T. and Ward, R. O., "Elevated temperature refining for high quality mechanical pulp," 1973 International Mechanical Pulping Conference Preprints, p.16:1, SPCI, Stockholm, 1973.

12. Aario, M., Haikkala, P., Lindahl, A., Tappi 63(2):139 (1980).

13. Richter, J., "The history of Kamyr continous cooking." Industry historical publication No. 7, edited by the Swedish Pulp and Paper Association, Stockholm, 1981.

14. Benedello, A., "Keller-Voelter. Die Einführung des Holzschliffs in der Papierindustrie," Papierfabrik Kabel Aktiengesellschaft, Hagen-Kabel, 1957.

CHAPTER 4

Fundamentals of mechanical pulping

1	**Rheological behavior of wood**	**35**
2	**Fundamental mechanisms in mechanical pulping and especially grinding**	**39**
2.1	The principles of defibrating wood by grinding	39
	2.1.1 The breakdown of the fiber structure by fatigue	40
	2.1.2 Removal of fibers from	43
2.2	Wood structure parameters affecting the breakdown process	44
	2.2.1 Thin walled/thick walled fibers	44
	2.2.2 The layer structure of the fiber	45
	2.2.3 Wood: a composite polymer with a viscoelastic nature	45
2.3	The main physical parameters contributing to the structural breakdown of wood in grinding	46
	2.3.1 Wood in a cyclic stress field	46
	2.3.2 Influence of amplitude, frequency, and temperature	47
2.4	Energy consumption in grinding	51
3	**Fundamental mechanisms in refining**	**51**
3.1	The development of fiber properties during refining	52
3.2	Traditional ways of describing refining	52
3.3	Theoretical approaches to understanding refining	53
3.4	Measured data from the plate gap	54
3.5	The present state of knowledge and understanding of fundamental mechanisms in refining	56
4	**Reasons for the difference in energy consumption between grinding and refining**	**57**
4.1	Energy consumption in grinding and refining	57
4.2	Reasons for the high energy consumption	57
4.3	The reason for higher energy consumption in chip refining	59
	References	61

CHAPTER 4

Lennart Salmén, Mikael Lucander, Esko Härkönen, Jan Sundholm

Fundamentals of mechanical pulping

1 Rheological behavior of wood

In mechanical pulping, the wood material is subjected to forces of various magnitude and duration which deform the fibers with the purpose of obtaining a suitable papermaking pulp. Since wood is a viscoelastic, natural polymeric material, its response to mechanical treatment is greatly affected by temperature, moisture, and time under load. In this respect, the transition temperatures of the wood polymers, i.e., the temperatures characterizing for each polymer the rather abrupt change from a stiff to a soft material (softening), are of great importance. In a viscoelastic material, the properties are neither elastic like a spring, nor viscous like a liquid, but are a combination of these states. The deformation of the material depends not only on the force applied but also on its duration. Under a specific load, the material is increasingly deformed with time or, if the material is kept with a specific deformation, the force successively diminishes with time[13].

Under mechanical pulping conditions, the wood material always contains more water than that which corresponds to full saturation of the fiber wall, roughly a moisture ratio of 0.33 based on dry weight[26, 21]. Under such conditions, both the hemicelluloses and the amorphous cellulose are softened at 20°C so that only the lignin softening plays a critical role in the mechanical pulping processes[3]. The softening of watersaturated spruce wood occurs at about 90°C at a frequency of 0.5 Hz[17, 4] as shown in Fig. 1. With increasing frequency, the lignin softening is shifted to higher temperatures (Fig. 1) so that the lignin transition temperature is determined by the frequency of each particular operation in the mechanical pulping process. In partic-

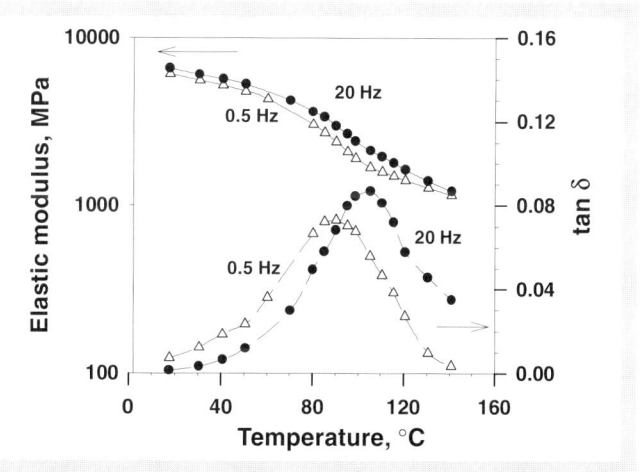

Figure 1. The elastic modulus and tanδ for wet spruce wood tested in the direction along the fiber axis as a function of temperature. The maximum in the tanδ curve is taken as the transition (softening) temperature for the wood[19].

CHAPTER 4

ular, two operations are of importance, the fiber separation and the further treatment to develop flexible fibers[8].

The separation of fibers is essential as it sets the character of the obtained fibers. For example, after separation at too high a temperature (Asplund pulp), it might be impossible to further treat the fibers to give a suitable papermaking furnish. The position at which the fracture takes place between fibers depends on the different characters of the various cellwall layers building up the fiber (see Fig. 2)[8, 5]. At low temperatures, when the lignin is stiff, the fracture occurs in an uncontrollable manner leading to greater amount of broken fibers and a high fines content. This is typical of refiner mechanical pulps (RMP) where the temperature at fiber separation is close to 100°C at frequencies that can be estimated to be in the region of some 1000 Hz[4]. For processes operating at higher temperatures, such as the pressurized systems for thermomechanical pulps (TMP), the fracture zone is moved further outward toward the secondary S_1 and the primary wall of the fiber, and a higher long fiber content is obtained[5]. With chemical treatment, as in chemithermomechanical pulps (CTMP), the lignin properties are altered so that its transition/softening temperature is lowered, and the fiber separation can be further improved, substantially reducing the shives (unseparated fiber bundles) content[7]. In this case, although fracture occurs in the middle lamella region, the chemical changes in the lignin still make it possible to further treat the fibers mechanically to a suitable pulp.

Mechanical treatment of the separated fibers is essential in order to make the fibers more flexible. At the same time, fines are produced. The creation of lamellar cracks in the fiber wall as well as a reduction of their outer perimeter by peeling off of the outer layers achieves flexibilization of fibers[9, 10, 12]. In both cases, some sort of fracture process occurs which is dependent on the physical characteristics of the fiber walls and is thus also affected by the frequency dependence of the lignin softening. In refining operations, so many processes are occurring at the same time (for instance, transpor-

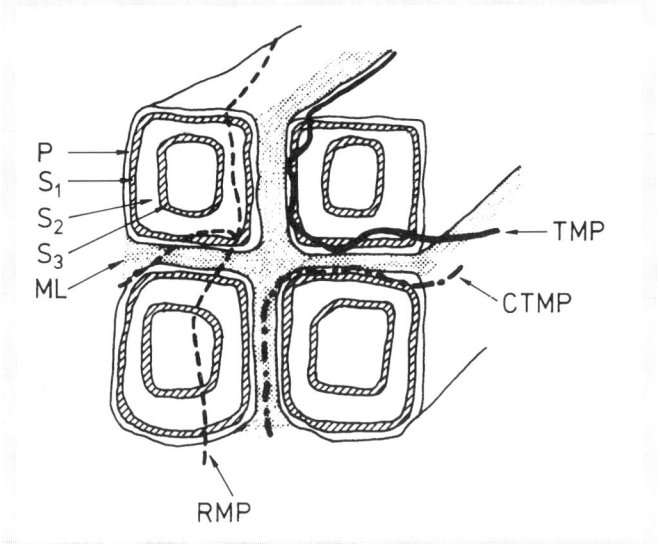

Figure 2. Schematic diagram of fracture zones in softwood as affected by different mechanical processes, RMP (refiner mechanical pulp), TMP (thermomechanical pulp), and CTMP (chemithermomechanical pulp) adapted after Franzén[5]. The cellwall layers are indicated as: P (primary wall); S_1, S_2, S_3 (secondary walls); and ML (middle lamella).

tation of material out of the refining zone) that the dependence on frequency is difficult to relate to material characteristics. For a fatigue process in a polymeric material and also in wood at temperatures above the brittle ductile transition, lower frequencies are more effective in producing fractures[18, 24]. This is also observed in grinding operations[11] where the different process operations of fiber separation and fiber treatment are more easily distinguished.

It is thus clear that the viscoelastic characteristics of the lignin polymer in wood have a large influence on the behavior of the wood in refining operations. This lignin softening is a reflection of the cooperative motion of the lignin polymer chain occurring when the energy in the form of temperature is sufficient[13]. With increasing frequency of the externally applied forces, still more energy has to be applied to overcome this rate of deformation, so that the wood appears to be stiffer at higher frequencies, especially in the region of lignin softening, as shown in Fig. 3[17]. For water saturated wood, it has been demonstrated that the elastic modulus follows the general time-temperature superposition principle (the WLF equation[29]) which makes it possible to calculate the properties of the wood over a large frequency-temperature interval[17]. Thus the properties of the wood can be estimated at refiner frequencies as exemplified in Fig. 3.

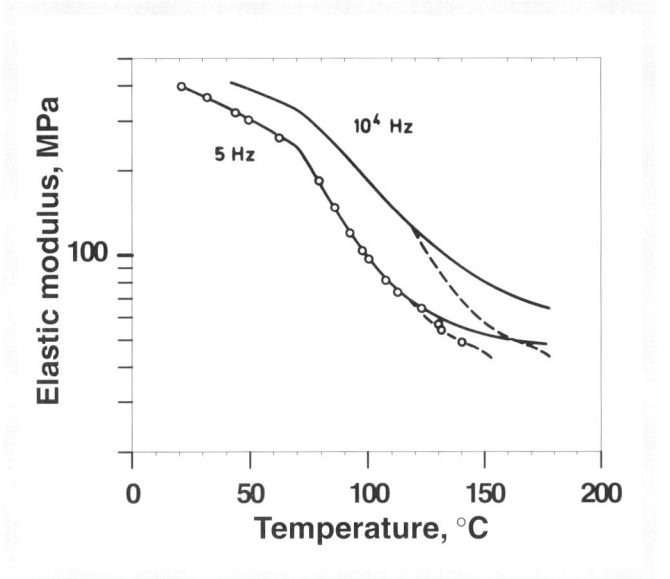

Figure 3. The elastic modulus of wet wood as a function of frequency at 5 and 10^4 Hz. At 5 Hz measured and calculated data are compared, whereas at 10^4 Hz only calculated data are given based on WLF and Arrhenius predictions made from experimental frequency-temperature measurements. Dotted lines indicate the deviation of the true behavior from the WLF-relation at higher temperatures[17].

An increase in pressure generally leads to a higher transition temperature for a polymer. The pressures used in refiner operations, however, are too small to have any major effect on the lignin transition/softening temperature; it shifts only a few degrees with a decade increase of pressure[1].

The lignin transition/softening temperature, like that of any polymer, is related to the polymer structure and, in the case of lignin, this varies with origin. There is a clear difference in structure between lignin from hardwoods and softwoods; the former contains a substantial amount of syringyl units (two methoxyl groups), while the latter is composed almost entirely of guaiacyl units (one methoxyl group)[23]. With a larger

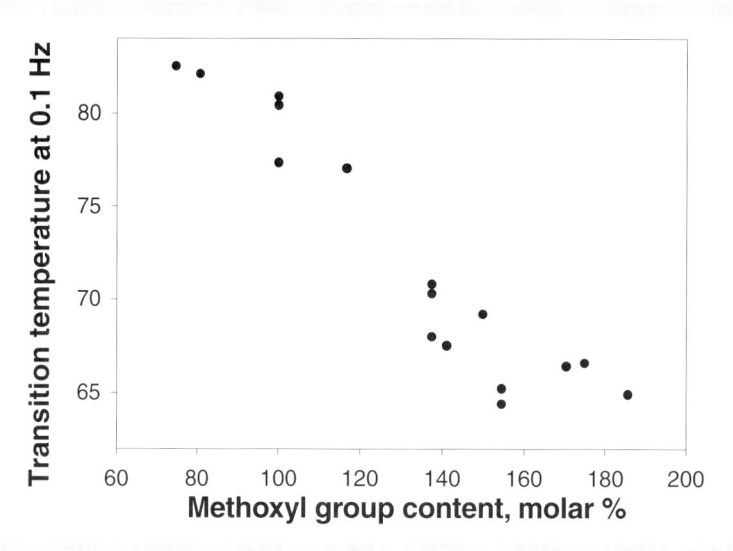

Figure 4. The transition/softening temperature of wet wood at a stress frequency of 0.1 Hz versus methoxyl group content[16]. 100 molar % represents normal softwood, while lower values relate to compression wood and higher values to various hardwood species.

number of methoxyl groups, it is more difficult for the lignin to crosslink and the transition temperature of the lignin is lower as shown in Fig. 4[16]. Compression wood lignin contains parahydroxyphenylpropane units with no methoxyl groups and is consequently more crosslinked[14] and thus has a higher transition temperature[15]. In normal softwood, the lignin structure also varies somewhat with the location in the cell wall, and the middle lamella lignin is somewhat more crosslinked than the secondary wall lignin[25]. No major difference in transition/softening temperature, however, has been noticed between these layers[8]. Instead the primary wall has been found to be different with a lower transition/softening temperature than the other walls, due to a blending effect of about 15% protein within the lignin in this cellwall layer[22].

Chemical changes in the structure of the lignin naturally also affect its softening behavior. Mechanical pulps include chemithermomechanical pulps (CTMP) and chemimechanical pulps (CMP) where either sulfonic acid groups (sulfonation) or carboxylic acid groups (peroxide treatment) are incorporated into the lignin. In both cases, the introduction of these ionic groups leads to a lowering of the lignin transition/softening temperature to a degree dependent on the amount of ionic groups introduced[2, 20]. With the introduction of ionic groups, the properties also become dependent on the type of counterion present, and sodium gives a lower transition temperature than calcium. In the undissociated proton form, the transition is even lower in the case of sulfonic acid groups but is highest in the case of carboxylic acid groups as is seen in Fig. 5[20].

The chemical treatment also affects the swelling of the wood, which increases with increasing content of ionic groups in the lignin[6]. The swelling properties in the native form also are affected somewhat by the counterion because the swelling in the

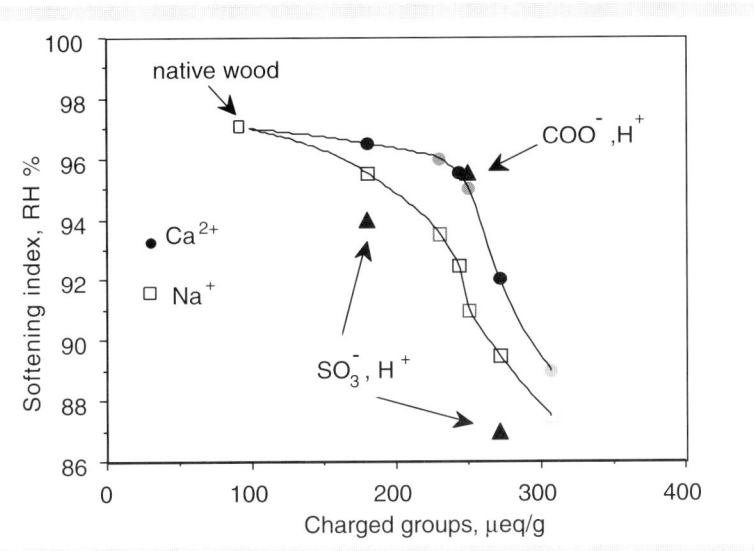

Figure 5. The softening index for wood having different counterions to the charged groups (taken from the damping curve in moisture scans at 90°C as a measure of the softening point) as a function of the content of charged groups in wood subjected to sulfonation or peroxide treatment[20].

sodium form is greater than in the calcium form. This is due to the presence of carboxylic acid groups attached to the xylan of the hemicelluloses in wood[6].

Since wood is a viscoelastic material, energy is lost and converted to heat in the deformation process[13]. For small reversible deformations, this energy loss has a maximum at the transition/softening temperature. In fact, the peak is used in mechanical measurements to define the point of softening. At higher deformations, when permanent changes are also occurring in the wood, much higher amounts of energy are taken up by the wood during the deformation[28]; this energy is mostly lost as heat. This generation of heat is the cause of the heating up of wood close to the surface in grinding operations and to the production of steam in refiner operations. At higher temperatures, the energy needed to deform the wood is reduced and also the energy that is lost in the material as a cause of the deformation[27]. Measurements show that the permanent changes per number of deformation are greater the higher the temperature is, i.e., fatigue of wood is more rapid at higher temperatures. The higher the temperature is, the smaller is the amount of energy needed to achieve a permanent mechanical change to the wood structure.

2 Fundamental mechanisms in mechanical pulping and especially grinding

2.1 The principles of defibrating wood by grinding

The main principle in grinding as well as in all mechanical defibration processes is to bring the wood raw material into a cyclic oscillating stress field whereby the absorbed mechanical energy breaks down the structure of the fibrous raw material. Successively

CHAPTER 4

the fibers are separated from each other and, in a more or less controlled operation, both uncut and flexible fibers are yielded to a certain extent depending on, e.g., the wood moisture content and temperature.

In grinding, wood logs are pressed or forced against a rotating pulpstone together with a large amount of water which acts both as a cooling and a lubricating agent. The surface layer of the pulpstone is a hard and porous ceramic composite. The logs, with a typical length of 1–1.5 m, are oriented in a pocket or a magazine parallel with the shaft of the stone and fed against the stone with a typical speed of 1–2 mm/s. The "Grinding" chapter describes more in detail the technical aspects of grinding.

In the early 1960s, D. Atack's work at Paprican provided the first more explicit description of the grinding. Atack suggested that two distinct actions ruled fiber removal in the groundwood process[30, 31]:

- The preliminary breakdown, or loosening, of the wood structure into what Atack called an embryonic pulp – this phase consumes most of the applied energy.

- Removal of the loosened wood structure by some mechanism, similar to a combing action, which is controlled by the height of the grit in the pulpstone. This phase consumes relatively little energy.

Atack suggested that the liberated fibers undergo further breakdown sequences during passage through the grinding zone. However, because of the complexity of the fiber entanglement, the nature of this post grinding could not be determined at the time Atack stated his principles, nor has it in later times been clearly proven to take place. The late 1980s and the 1990s have seen a newly born interest to explain the mechanisms in grinding[80, 81, 82]. Also in these studies the working approaches are based on the same hypothesis as in Atack´s earlier work.

2.1.1 The breakdown of the fiber structure by fatigue

In all mechanical defibration processes for paper and board, the ultimate goal is to produce both fines and flexible fibers with good bonding characteristics. In grinding, the fiber structure in the wood matrix is stressed cyclically in a series of shear as well as compression and decompression stages. Because wood has a complex multifiber structure (Fig. 6) and is a composite polymer with viscoelastic behavior, certain conditions must be fulfilled:

1) Softening of the polymeric wood material prior to the separation of the fibers is essential in order to produce uncut fibers. The more water is absorbed by the wood polymer components, the larger is the plastizing effect at a specific temperature. Under dry conditions lignin and hemicellulose become plastic (rubbery) in the temperature interval between 180 and 220°C[79, 83]. The more hydrophilic polymers, hemicellulose and amorphous cellulose, absorb up to 4–5 times more water than lignin and have glass-rubber transition at water saturated conditions at relatively low temperatures close to 20°C[84]. In lignin the plasticizing effect of water has a limit at water contents as low as 5%. For water saturated isolated lignin the softening takes place at

Fundamentals of mechanical pulping

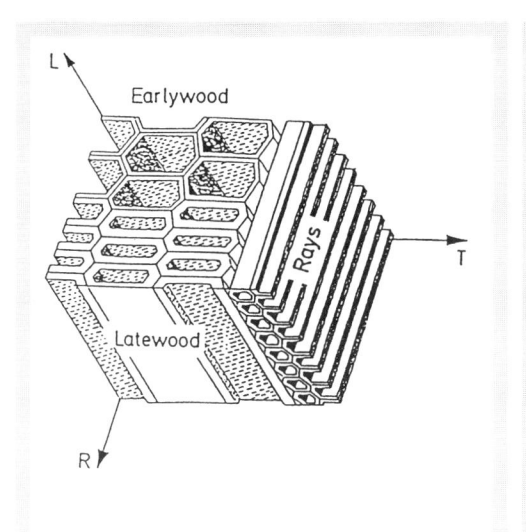

Figure 6. The multifiber structure of softwood[32].

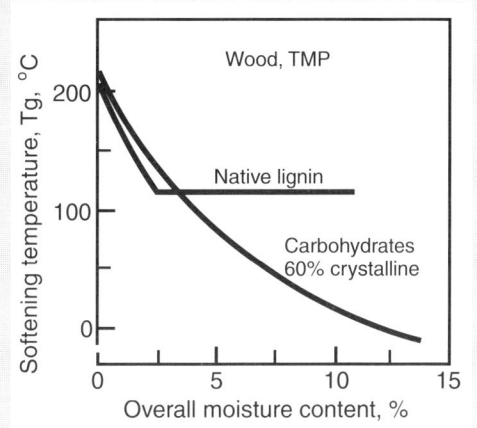

Figure 7. Influence of water on the softening temperature for wood and TMP. The calculations are based on a lignin content of 29% and on the assumption that the lignin absorbs 20% and the carbohydrates 80% of the total water absorbed[79].

80°C–90°C. Additional water does not result in a considerable further softening of the lignin. At conditions typical for mechanical pulping processes, the softening temperature for *in situ* lignin (as part of the wood fiber matrix) is higher, typically in the range of 100°C–130°C. (Fig. 7). Thereby the lignin polymer, being the stiffest wood component at conditions prevailing in mechanical pulping, plays the most important part for the thermal softening of the wood. In practice, on a production scale, wood for mechanical pulping should have a minimum moisture/wood ratio of 0.3–0.4 but preferably over 0.5.

2) In grinding with a ceramic pulpstone, the active grits protruding above the mean level of the pulpstone surface generate the cyclic shear and compression pulses (Fig. 8). The grits have to be of a certain dimension (preferably having a particle size in the range of 0.2–0.5 mm) in order to achieve the

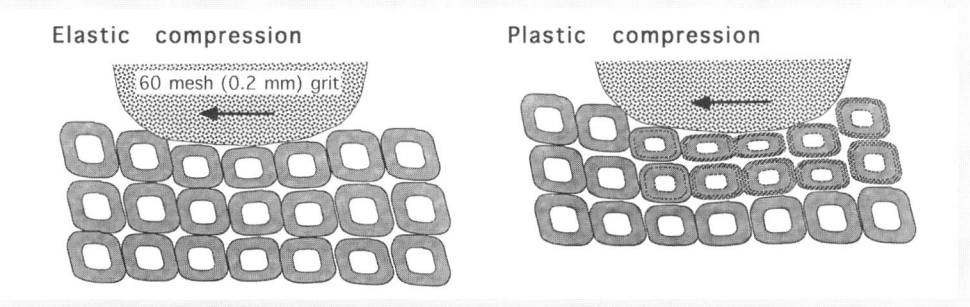

Figure 8. The elastic (reversible) and the plastic (irreversible) deformation of the tracheid cell in the wood matrix due to the stress pulses caused by the pulpstone grits in grinding. In plastic deformation, the cell is compressed sufficiently (preferably 10%-50%) and thereby causes fatigue changes of the wood structure[33].

desirable structural breakdown of the wood matrix. In order to achieve plastic deformations in the fiber structure within a reasonable time, the applied wood feeding load on the active grits in contact with the wood has to exceed a certain level. If this level is too low, the wood and the fiber matrix is fatigued too slowly and/or to an inappropriate level, and the cyclic elastic deformations of the wood just heat up the wood resulting in high specific energy consumption. On the other hand, if the pulpstone grits are too small in size, excessive fiber cutting follows and results in poor quality pulp.

3) The presence of shower water in the groundwood process in essential for several reasons. Firstly it acts as a cooling medium of the pulpstone as the energy consumed in the process is converted nearly totally to heat (on an average 1300 kWh/t at CSF = 100 ml). Secondly the shower water acts as a lubricating film between the grits and the wood and thereby substantially reduces the power required to drive the pulpstone at constant speed by eliminating the surface interaction component of the sliding friction between the grit and the wood[31]. The water film also prevents burning of the wood surface by eliminating direct surface contact between the grit and the wood.

4) The temperature plays a central role when breaking down the wood structure into unbroken fibers. To a large extent, the shower water temperature and the amount of water (grinding consistency) govern the temperature level in grinding (or more specifically in the grinding zone). If the shower water temperature is low in relation to the lignin softening temperature, the softening of the fibers (caused by the energy uptake as heat in the compression/decompression fiber layer close to grinding zone) is reversed when the softened fibers enter the grinding zone (Fig. 9). The too low shower water temperature stiffens the fibers which, in turn, alters the amount of fiber breakage as the fibers are ripped off from the wood matrix. A too high shower-temperature, on the other hand, (with a sufficient casing over pressure to prevent boiling) causes undesirable darkening of the groundwood pulp.

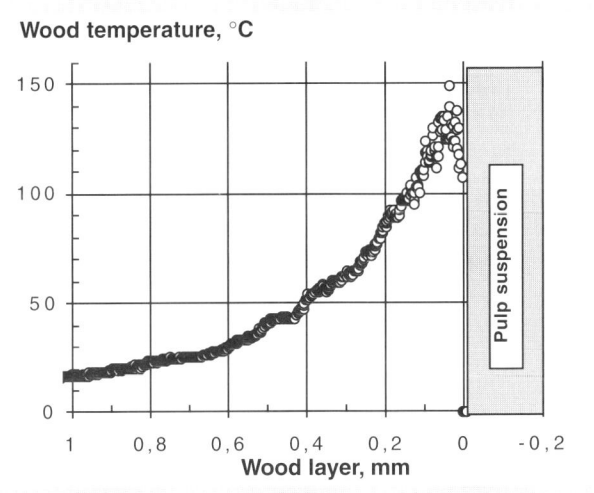

Figure 9. Temperature distribution in the hot wood zone in the 1 mm wood layer preceding the grinding zone. The grinding trial was carried out at the KCL pilot pressure grinder with the following process conditions: PGW–S (450 kPa/75°C), stone 38A601, peripheral speed 20 m/s, and wood feed 0.75 mm/s[34].

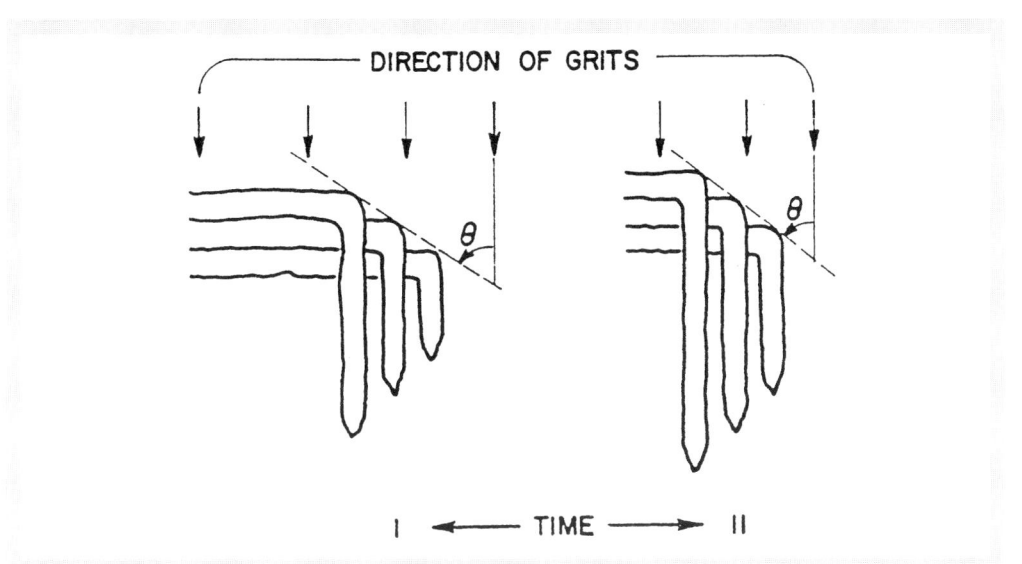

Figure 10. In the grinding of wood to produce papermaking pulp, the wood structure is first loosened by the conditioned grits and the loosened fibers are subsequently peeled away. This schematic diagram shows two stages in the peeling of loosened fiber and, as indicated, fibers are always removed in a strict sequence[31].

2.1.2 Removal of fibers from the wood by a peeling action

As the wood is fed by the force of the piston in the wood magazine against the revolving pulpstone, the fibers closest to the grits protruding from the stone surface are ready to be loosened by a peeling action. The peeling generally starts from one end of the fiber (occasionally from both). Fibers peel in a strict sequence and in the same direction (Figs. 10 and 11). Fibers contracted first by the advancing grits always peel a little in advance of the next ones, and so on. The loose ends of the fibers that have been separated from the parent wood (wood matrix) and from the neighboring fibers are combed into the direction of the rotating stone (motion of the grits). Incipient peeling occurs over a front that is

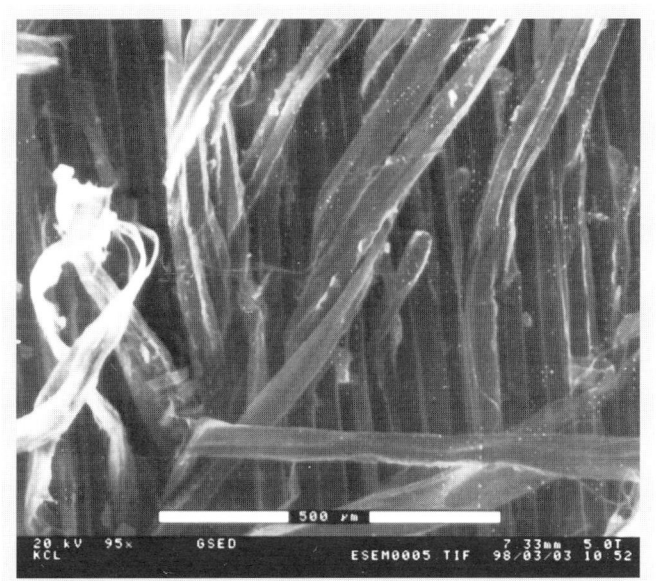

Figure 11. Surface of ground wood rejected quickly from the wood pocket of a laboratory grinder. The picture has been taken in an ESEM microscope using a 95x magnification. The partly peeled off fibers and fiber bundles can be seen clearly. The direction of the grits has been from the left to the right of the picture.

CHAPTER 4

at an angle generally close to 45° to the grit direction.

It is during this peeling action of the groundwood process that most of the fines material essential for pulp properties is produced. If the active grits in pulpstone surface are too small or sharp, as in freshly prepared (sharpened) stone, a large proportion of the fibers are cut into short particles with poor bonding properties.

2.2 Wood structure parameters affecting the breakdown process

Wood is built up into a rather complex structure, both on a macroscopic and on a microscopic plane. Each building component affects the wood structure in a specific way. To break down each component in an effective way to yield long and flexible fibers with good papermaking properties, the physical rules governing material properties of each component should be taken into consideration.

2.2.1 Thin walled/thick walled fibers

In softwoods. the shape and form of the tracheids have distinct seasonal variations: earlywood and latewood (Fig. 12). As shown in the figure, the earlywood fibers are thin walled and have a larger fiber thickness in the radial direction. For Norway spruce, the

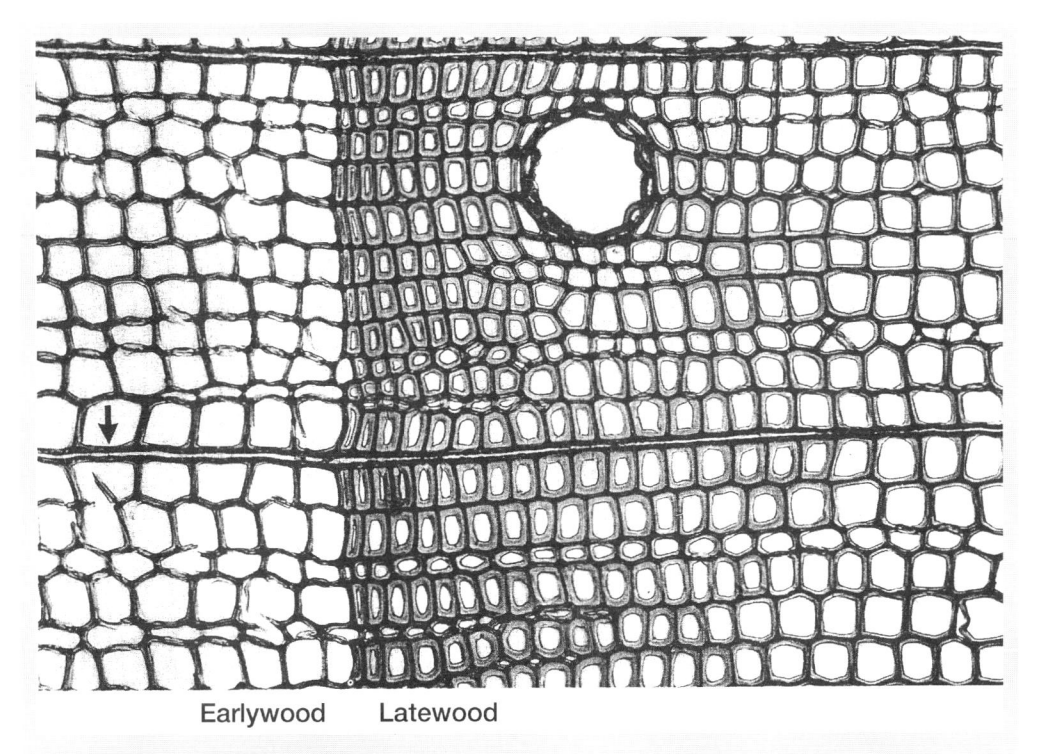

Figure 12. Cross-section of *Picea abies* at early-latewood boundary. From the right to the left, wide, thin-walled earlywood tracheids are gradually changed into narrow, thick-walled latewood tracheids. The arrow drawn in the SEM-picture indicates a ray. At the margin of the early- and latewood, the large opening is a vertical resin canal[35].

Fundamentals of mechanical pulping

ratio of the early- and latewood width is around 2. We can easily understand that, when compared to late wood fibers, earlywood fibers are far more accessible to compressive (across to grain) deformation.

2.2.2 The layer structure of the fiber

Softwood fibers are built up to four sets layers or the P, S1, S2, and S3 -walls, that is, the primary (P) wall and the three secondary (S1, S2, S3) (Fig. 13). Furthermore, a lignin rich middle lamella (M) glues the fibers together. To be able to mechanically separate the tracheids in the wood matrix from each other into both long uncut and well bonding fibers, one must understand the physical and chemical properties of each wall and take into account each specific property when tailoring the process parameters to get the most desirable pulp characteristics.

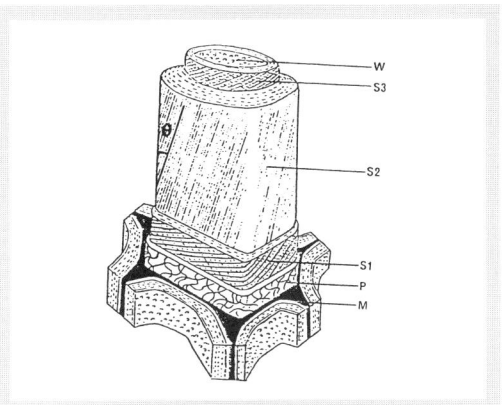

Figure 13. A schematic picture of softwood tracheid structure composed of the middle lamella (M) and the four fiber walls (P, S1, S2, and S3)[36].

2.2.3 Wood: a composite polymer with a viscoelastic nature

Wood is built up of basically three polymers: (1) cellulose which is a linear macro molecule (to a greater part crystalline), (2) linear amorphous polyoses molecules jointly called hemicellulose, (3) lignin, a crosslinked amorphous polymer. These three polymer groups also have distinctly different chemical and physical characteristics as well as different distributions in the different cell walls (Fig. 14).

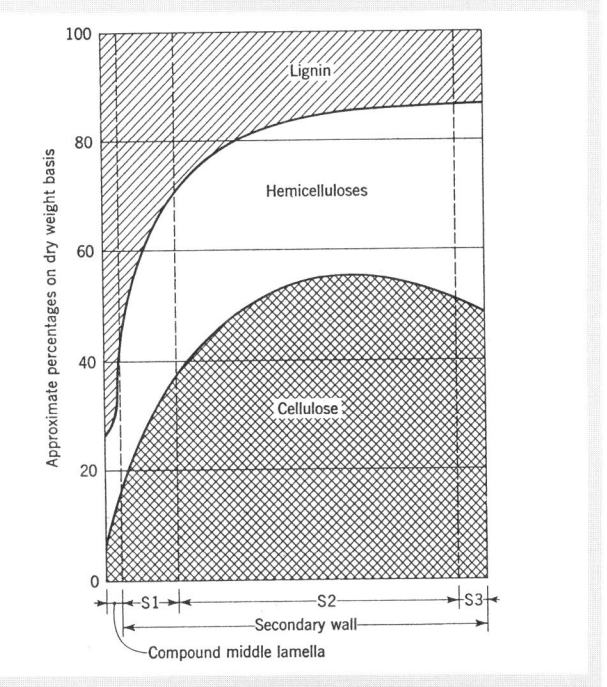

Figure 14. An approximate percentual distribution of the most important chemical components in the different layers of a softwood tracheid[37].

CHAPTER 4

2.3 The main physical parameters contributing to the structural breakdown of wood in grinding

2.3.1 Wood in a cyclic stress field

During grinding, a cyclic stress mechanism defiberizes wood. In the cyclic load sequence, the pressing and shearing forces of the pulpstone grits passing transversally over the fibers in the grinding zone sequentially compress (stress) and decompress (relax) the wood fibers (Fig. 15). The grits in a newsprint pulpstone are on an average approximately 0.6 mm apart and cyclically load the fibers at frequencies in the range of 40–50 kHz. The compression/relaxation pulse is transported into the wood on a right angle from the grinding zone plan where most of the kinetic energy is transformed to heat due to the viscoelastic nature of the wood. The damping effect of the polymeric wood material reaches to a depth of approximately 1 mm, resulting in a rapid heating of this layer. Deeper in the wood, a sharp drop in temperature can be detected (Fig. 9). This damping effect in a viscoelasic material is also called hysteresis loss. The temperature in the so-called hot zone reaches a relatively high level due to the fact that damping happens very fast in the direction perpendicularly away from the grinding zone. Because the cell thickness of e.g. Norway spruce fibers, are approximately 20–40 µm, there are some 10 fiber layers in the hot zone and the rapid rise in temperature starts at a depth of some 20–50 fiber layers.

The temperature in the hot zone reaches its maximum in the wood matrix close to the grinding zone after which it declines to a level governed by the circumstances prevailing in the grinding zone. A maximum develops because, as the hot wood zone

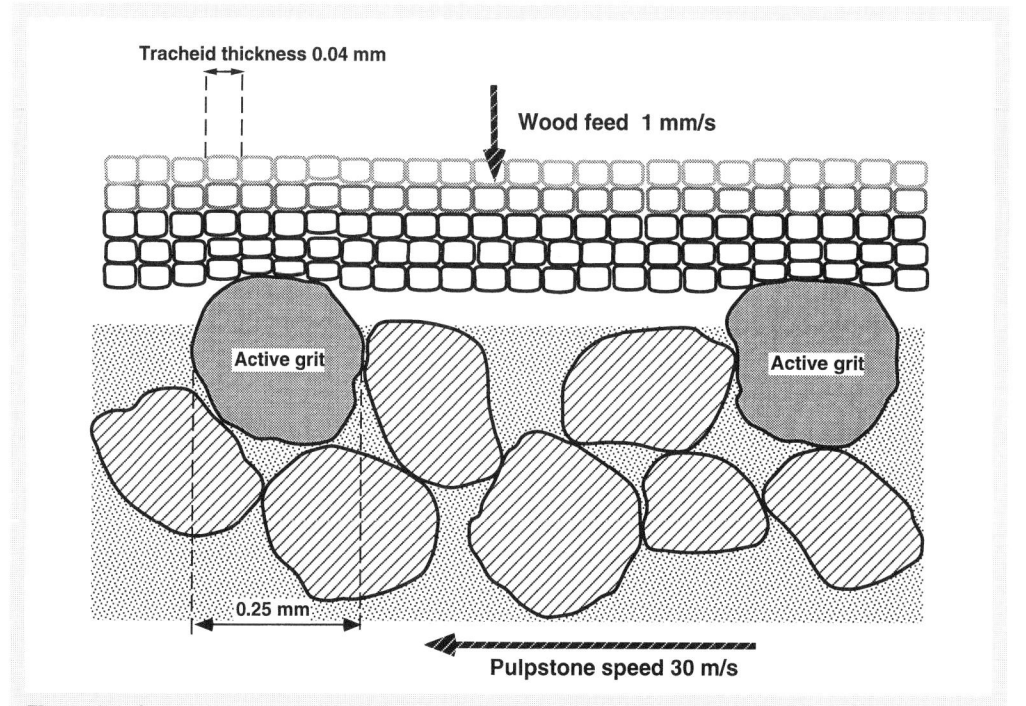

Figure 15. Schematic presentation of the grinding zone where the pulpstone grits (mesh 60) and tracheids are scaled to correct dimensions[38].

approaches the colder stone surface, it is cooled at the same time hysteresis loss deeper in the wood generates heat. From its maximum deeper in the wood, the temperature declines steeply to the temperature level of the showered stone surface.

2.3.2 Influence of amplitude, frequency, and temperature

To weaken the structure of wood by fatigue, the most important parameters are strain amplitude, frequency, and temperature [18, 39]. Also the structure and the rheologial behavior of wood tracheid cells play a major part. A condition which enables the fibers to deform by compression and shear forces is the hollow structure of the fiber. The hollow space or lumen involves frequently over 2/3 of the total cross section area of the fiber (Fig. 12). Recent work by Salmén and Uhmeier [28, 40] shows that, to get irreversible structural changes in wood by means of radial compression, the wood sample has to be stressed beyond the elastic region to the plateau region where a preferably plastic strain takes part (Fig. 16–18).

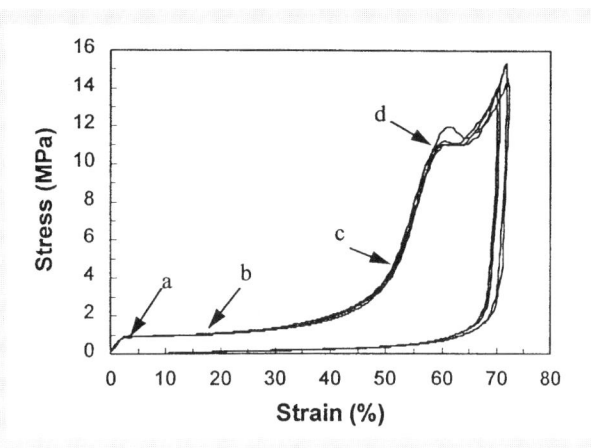

Figure 16. Stress versus strain curves for several specimens in a radial compression at 98°C of wet wood through the elastic (a), the plateau (b), and the densification region (c) of the earlywood fibers to the point (d) where densification of the latewood fraction starts [40].

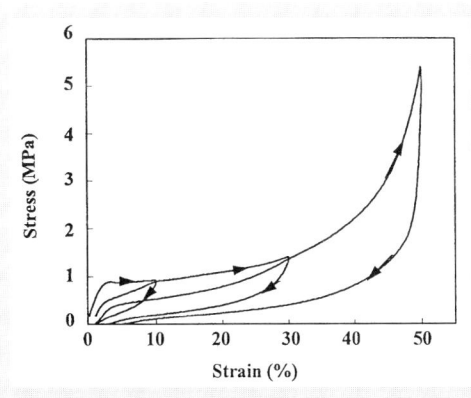

Figure 17. Stress versus strain for three consecutive compression cycles to 10, 30, and 50% strain at 98°C. The specimens were relaxed 10 minutes between the compressions [40].

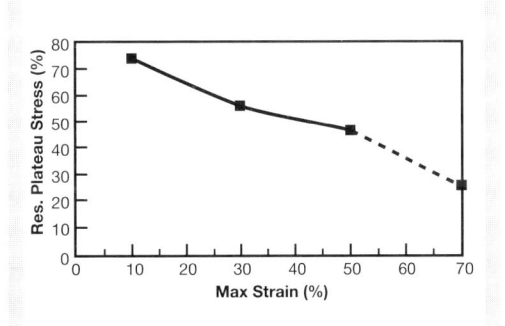

Figure 18. The resulting plateau stress versus the maximum strain in the previous compression. The specimens compressed to 70% strain have been fractured [28].

CHAPTER 4

In grinding, it is most probable that the applied stress perpendicular to the grinding zone on the wood is distributed as compression pulses over a fairly thick fiber layer (20–50 fibers as stated earlier) before it is damped. This implies that most of the deformation takes place in the elastic region of the stress-strain curve with very little fatigue but due to a large degree on hysteresis loss[40]. However, by repeated compression and shear pulses on the wood, the probability for plastic deformation or fatigue on the fiber structure grows (Fig. 19). This is the main reason for poor efficiency in mechanical pulping and, consequently, the high energy consumption connected to the process.

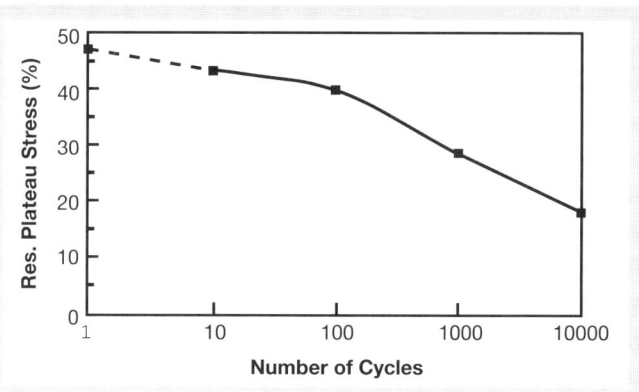

Figure 19. The resulting plateau stress versus the number of sinusoidal compressions (10%–40% strain) at 10 Hz and 98°C. The value for one compression corresponds to an initial slow ramp to 50% strain[28].

Amplitude

Work done by Salmén and others[28, 39, 40] has shown that increasing the amplitude compressing in cyclic loading in a straightforward way reduces the amount of compression cycles needed to achieve a certain amount of structural breakdown (fatigue or change in the elastic modulus), as shown in Fig. 20. As a consequence, the amount of energy needed in defibration can be expected to be reduced. The larger the strain or the more compression work is carried out in the densification region of the stress strain cycle curve, the larger is the irreversible structural change in the wood. In grinding, it can be expected that a major part of the large amplitudes (compressions in the plastic region) are absorbed by the early wood fibers, whereas the latewood fibers needing higher stress levels are plastically deformed to a lesser extent.

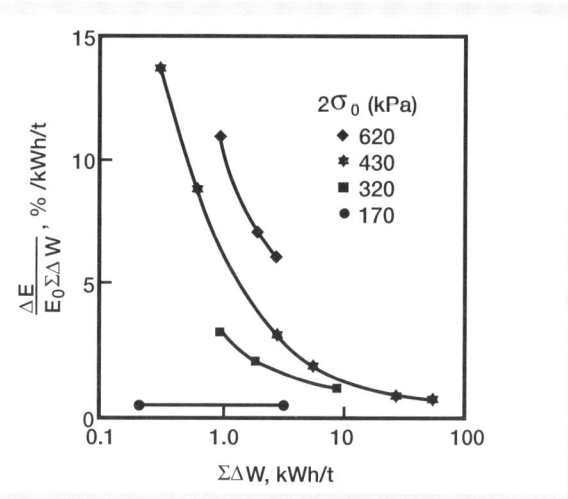

Figure 20. Efficiency of the structural breakdown process as a function of energy absorbed for tests in compression across the grain at 100°C. The tests were carried out in amplitude (stress) ranges: 620, 430, 320 and 170 kPa[39].

Fundamentals of mechanical pulping

Temperature

To break down the structure of wood and at the same time avoid excessive fiber cutting, it is essential that separation of fibers takes place at a temperature when the material is not brittle. As stated earlier in this chapter, both the amorphous cellulose and hemicellulose components in the fiber layers soften at temperatures close to 20°C when the fiber is saturated with water. Evidently the lignin component in the fiber structure dominates the fiber softening in mechanical pulping (grinding). The temperature should not be lower than the glass transition (Tg) temperature. At low frequencies, this temperature is close to 90°C for water-saturated spruce wood (Fig. 1). With increasing grit frequencies, the softening temperature wood is higher, some 6°C–8°C per decade of increase in frequency[19]. A commercial scale pulpstone 1.8 m in diameter running, e.g., 300 rpm, and a having a grit structure with active grits separated approximately 0.6 mm apart, generates frequencies in the vicinity of 50 kHz. The grinding temperature should be much higher than 90°C; in fact, it should be past 120°C to assure minimum fiber cutting at fiber separation. Very high temperatures way above Tg produces long and uncut fibers but unfortunately ones with unsuitable bonding properties. To achieve an acceptable combination of flexible well bonding and long fibers in grinding, one should defibrate at a temperature close to the glass trasition of lignin.

Trials at laboratory scale[41–44] and at mill scale[45–47] clearly demonstrate the effect of high process temperatures when grinding both spruce and fir softwood species. In the temperature range of 70°C–140°C, fiber length and strength properties improved at higher temperatures most commonly achieved by elevating the shower water temperature. The higher shower water temperatures used in the PGW process (70°C–95°C) and in the PGW-S process (120°C) not only yield higher long fiber fractions but also more flexible fibers with more developed bonding properties[48].

Frequency

Work by Salmén and others[18, 39] shows that the structural changes occurring in wood fibers subjected to mechanical fatigue in laboratory tests are similar to changes occurring in the long fibers of mechanical pulp. Due to the viscoelastic nature of wood, its mechanical properties are dependent on the frequency of the mechanical action. Also the softening temperature of the wood increases with increasing frequency. Still, for fatigue properties of poly-

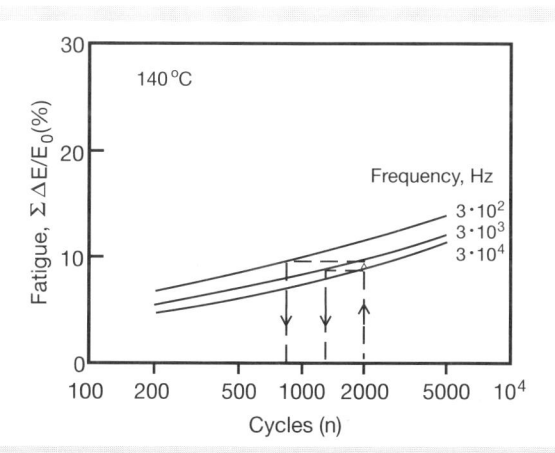

Figure 21. Calculated fatigue of wood across the grain at a temperature of 140°C for the frequencies 300, 3000, and 30 000 Hz. The reduction in energy consumption (i.e., in the number of cycles) when decreasing the frequency one magnitude at 2000 cycles is indicated by dotted lines[18].

CHAPTER 4

mers, knowledge is at present quite limited as to the effects of temperature and frequency and especially for the fatigue in the vicinity of the glass transition temperature of the material.

Recent investigations have contributed to more facts on the behavior of wood material at different strain frequencies. Experimental data with a material testing apparatus (MTS) on the fatigue of wood[18, 28] show that structural degradation due to mechanical deformation occurs faster at lower frequencies (Fig. 21). It is to be noted that the differences between frequencies diminishes as the frequency level increases. However, a major decrease in frequency from, e.g., 30 kHz to 3 kHz, results in a clearly noticeable energy reduction. Trials carried out on a laboratory grinder[11] confirm these findings. Energy consumption is lower and pulp strength properties are better when grinding at lower stone peripheral speeds (Fig. 22).

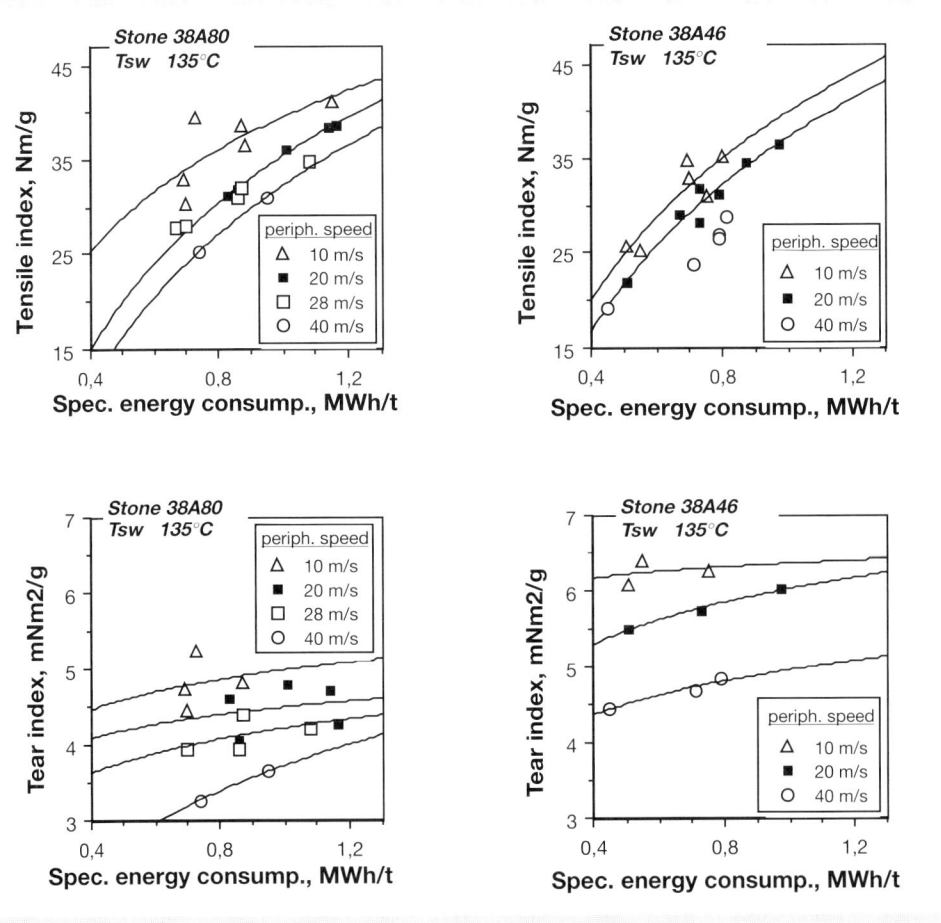

Figure 22. The tensile and tear indices versus energy consumption of PGW-S spruce *(Picea abies)* pulps produced at three and four pulpstone peripheral speed levels. The pulps were ground at KCL's semi-pilot grinder as two series using a fine (38A80) and a coarse (38A46) ceramic stone[49].

2.4 Energy consumption in grinding

Generally it can be said that, at the same grinding process conditions, better pulp strength properties demand a larger energy input for the defibration of the fibers. However, energy consumption can be reduced to a certain degree by switching some process parameters in a less energy consuming direction:

- Higher process temperature (PGW, PGW-S)
- Lower deformation frequency (lower pulpstone peripheral speed)
- High strain amplitudes.

3 Fundamental mechanisms in refining

In the refining process, wood is fed in chip form into a narrow gap and, during their passage, the chips are defibrated and fibrillated. This can take place in one refiner or it can continue in subsequent refiners. The most common process type is thermomechanical pulping (TMP), in which the refiner is pressurized and the steam pressure corresponds to the temperature of saturated steam.

The TMP process has mainly been developed during the construction and startup of new and larger process units; most of the knowledge is gained via intensive testing and design changes introduced on these occasions. The knowledge gained is therefore of a qualitative rather than quantitative nature.

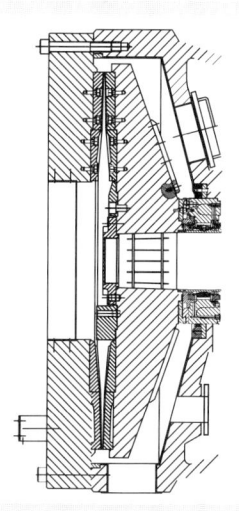

Figure 23. Housing, rotor, rotor and stator segments, and breaker bar segment of a TMP-refiner with a rotor diameter of 65 in.

The following text describes the refining phenomena in qualitative terms for a typical refiner as illustrated in Fig. 23. The chips come to the center of the refiner and strike the edges of the breaker bar. Immediately the chip is broken down into small pieces[50], and the refining of these pieces begins when the pieces strike each other as well as the refiner rotor and stator edges. Centrifugal force drives this coarse wood and pulp mixture outward in radial direction and the plate gap becomes smaller. The interaction between rotor, stator, and fiber defibrates and fibrillates the fiber material to the final freeness level. The collisions between fiber and rotor bar edges and collisions between fibers, friction between fiber and segment surfaces, and internal friction in the fiber phase consume a considerable amount of energy. This energy is transformed into heat which increases the temperature of the water and fiber and thus evaporates the water into steam. This steam has a strong influence on the fiber flow in the plate gap. Depending on the pressure conditions before and after the refiner, some of the steam flows toward the chip feed (thus called flowback steam) and some flows forward with the fiber flow. The steam flow is restricted due

CHAPTER 4

to fiber in the narrow plate gap and, because of this, the pressure and the temperature in the plate gap can be noticeably higher than those in the refiner housing. The heat in the refining process changes the rheological properties of wood and fiber, and this heating has an important influence on the final pulp quality in refining.

The main mechanisms in refining are the defibration and fibrillation of fibers. Defibration and fibrillation are determined partly by rheological properties of wood and fiber and partly by flow conditions in the refiner. The earlier part of this chapter discusses the former. Refiner geometry and the basic laws of physics determine the flow of steam and fiber and steam generation in the refiner. The quantitative concepts which are based on these physical laws are the velocities, volume fractions, pressures and temperatures of steam and fiber, and the power dissipation by different mechanisms. The power dissipation mechanisms are fibers striking the rotor and stator edges, fibers gliding against each other and against segment surfaces, and collision between fibers as a result of turbulence. The creation of new surfaces probably binds some energy, but the amount is very small compared to the energy bound to heat flows. Power dissipation also occurs in the steam phase, but it is negligible compared to the power dissipation in the combined fiber and water phase. The values of these variables vary according to the position in the plate gap and, to control the refining, these values should be known during the entire passage of a fiber through the refiner.

3.1 The development of fiber properties during refining

This chapter discusses the forces applied on wood and fiber in the refiner. Chapter 15 "The character and properties of mechanical pulps" explains the mechanisms that tell us about the fiber development, or how the fiber responds to the refining forces.

3.2 Traditional ways of describing refining

Where refining occurs, mechanical conditions are very harsh. Temperature and pressure, up to 180°C–200°C and 1–1.5 MPa, respectively, are relatively high; mechanical abrasion is also substantial. Rotor and stator are located in a housing with thick steel walls. This means that the refining process has been described in terms which are not necessarily accurate.

The most important factor in describing the refining process is the specific energy consumption (SEC). This is derived by calculating the energy consumed per produced pulp ton. The distance between the segments in the plate gap controls the energy consumption in refining; in operational conditions, energy consumption increases as the plate gap narrows.

The concepts of gentle refining and harsh refining were introduced in the earlier stages of development of the TMP process. Gentle refining produces well fibrillated, long mechanical pulp, while harsh refining produces shorter and less fibrillated pulp. The specific characteristics of gentle pulping include high SEC, high dry-content of pulp and long residence time of pulp in the refiner. Correspondingly, the specific characteristics of harsh pulping include low SEC, low dry-content of pulp and short residence time of fiber in the refiner. An increase in rotational speed produces harsher refining.

The geometry of the refiner segments is also used to control the refining phenomenon. Typical refiner segments are shown in Fig. 24. The expressions "coarse" and "dense" are used to describe the segments: A coarse segment is one characterized by wide bars and grooves, while a dense segment is one with narrower bars and grooves.

Important factors in controlling the radial speed of fiber are the amount and height of dams and the open cross sectional area available for the pulp and steam flow. The depth of grooves and the plate gap determine the open cross sectional area. The expression plate gap usually means the distance between the rotor and stator bars at the outer periphery of the refiner. Several radial conical sections form the segment surfaces; thus, the plate gap is wide at the inner edge of the segment and becomes narrower on the way outward along the radius. The aim is for the rotor and stator surfaces to be parallel at the outermost section when the refiner is in operation.

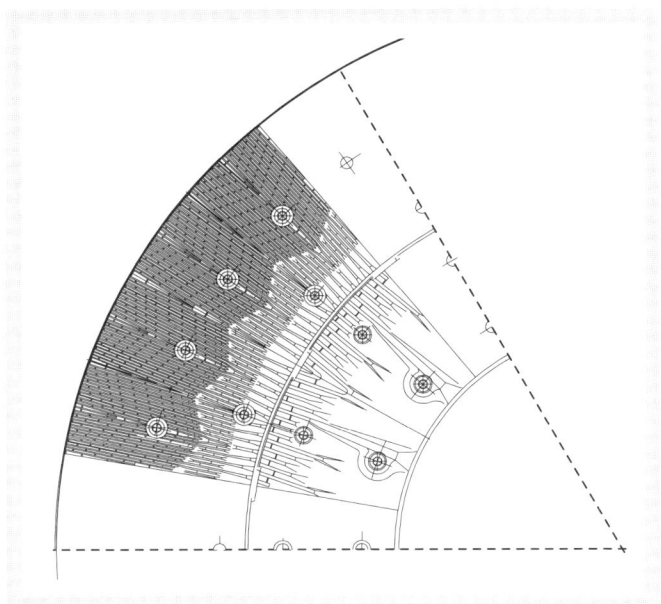

Figure 24. Inner and outer segments in a 65-in. SD refiner

The common refiner designs are single disc refiner (SD), double disc refiner (DD), and conical disc refiner (CD). Flow conditions and power dissipation mechanisms, even if not precisely known, are specific to each type of refiner. Each refiner type produces a specific type of pulp. The CD refining process is the gentlest, the DD process the harshest, and the SD refiners are somewhere in between. The refiner type therefore is also used to describe the pulp quality.

3.3 Theoretical approaches to understanding refining

Theoretical models always contain functions, variables, or constants the value of which must be determined experimentally. The theoretical model is, however, the only reasonable way to combine the influences of several factors and to find out their combined influence on refining. The key point for a theoretical model is to predict the fiber and steam velocities and the power dissipation by different mechanisms in the plate gap. When radial fiber velocity is known, it gives the volume fraction of fiber and also the residence time of fiber in the plate gap. This knowledge enables us to understand different mechanisms of refining.

Miles, Lunan, and May[51] were the first to attempt the calculation of steam speed in the plate gap. Their model contains only the rotor and is one-dimensional, describing

CHAPTER 4

the motion in radial direction. Miles and May[52] present a model to calculate the motion of fiber in the plate gap. These models give steam and fiber velocities as well as steam pressure distribution in the plate gap.

Miles and May[52] introduced the concept of refining intensity, which describes whether the refining is gentle or harsh. This is arrived at by dividing the total specific energy by the total number of bar impacts during a fiber passage. It requires the residence time of the fiber and the radial power dissipation distribution to be known. To measure the residence time of fiber in the refiner requires special arrangements and special techniques, and it is usually unknown in a regular refiner. The power dissipation distribution is also unknown and usually is assumed to be evenly distributed over the segment surface. These factors limit the applicability of the refining intensity as a quantitative concept.

Härkönen *et al.*[53] published a calculation model to calculate the flow phenomena in a refiner plate gap. The model gives the fiber and steam velocities, fiber and steam pressures, fiber and steam temperatures, dry content of pulp, and power dissipation in the plate gap. The model is rotationally symmetric and includes rotor, stator, and plate gap. Also included are the feed zone in the center of the refiner and the refiner housing. The model contains all physical factors to describe the steam and fiber motion as well as different power dissipation mechanisms. However, this model contains several parameters which must be determined by testing.

When fiber and steam velocities and power dissipation by different mechanisms are known, it is possible to simulate the fiber refining phenomena. Currently the knowledge of these variables is limited and there are few studies on this topic. Strand and Mokvist[54] described the refiner performance by applying the communition theory. DiRuscio and Balchen[55] presented a dynamic model for fiber size distribution between refiner discs. Both models give a qualitative view on the refining phenomena in the plate gap. These studies lead to a large set of coefficients. The connections between the value of these coefficients and, e.g., segment geometry, temperature, wood quality, or energy consumption, etc., are not known.

3.4 Measured data from the plate gap

Several researchers have measured temperature distribution in the plate gap. The temperatures measured in the gap are higher than those in the housing and at the inlet. The level of temperature depends on the power level in refining. To control the axial force balance in the refiner, it is necessary to know the steam pressure in the plate gap. It is possible to measure the pressure or, if we assume that steam is saturated and we know its temperature, it is possible to obtain steam pressure values from handbooks. Technically it seems to be a difficult task to measure pressure and temperature at the same point; no published results are available. There have been several trials to measure the superheating of steam. According to Strand, superheating is less than $5°C$[56] and, according to Engstrand and Karlström, superheating can be $3–5°C$[57].

Fundamentals of mechanical pulping

Härkönen et al. also concluded in their trials that superheating is less than 5°C[58]. In practical studies, steam is assumed to be saturated and, on these assumptions, it is possible to transform temperature values into pressure values. Usually, axial thrust in the rotor bearing is measured and, when pressure distribution round the rotor is known, it is possible to calculate the compression force directed at the fiber pad in the plate gap[59]. When steam is assumed to be saturated, the position of the maximum temperature indicates the position of the steam flow turning point. The position of the peak temperature divides the steam flow into flow-back steam and forward flow steam to the blow pipe.

The following temperature measurements are given as examples. Engstrand et al. measured the temperature distribution in a 60-in. twin refiner, in both the first and second stages[60] (Fig. 25). Härkönen et al. measured temperature distribution in a first-stage 65-in.[4] and in a second-stage 68-in.[61] single disc refiner (Fig.26).

The measurement of residence time requires the utilization of some type of tracer material. The tracer can be a dye or radioactive material. The tracer is fed

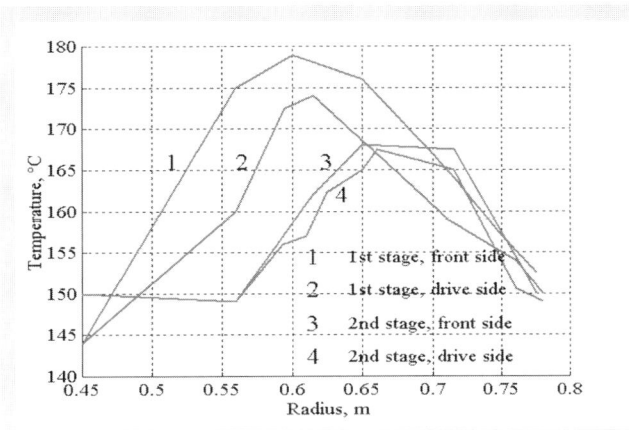

Figure 25. Measured temperature distribution in the plate gap in a first-stage and in a second-stage 60-in. twin-refiner.

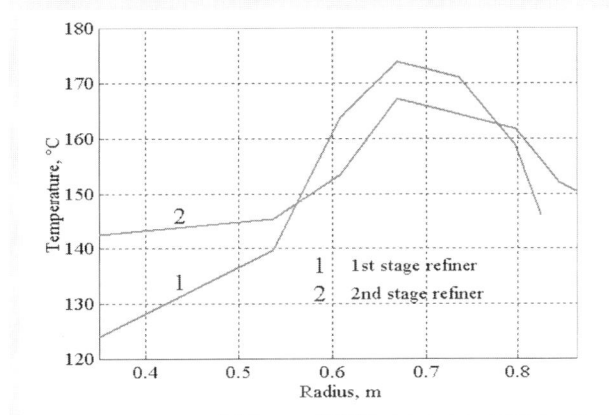

Figure 26. Measured temperature distribution in the plate gap in a 65-in. first-stage and in a 68-in. second-stage SD refiner.

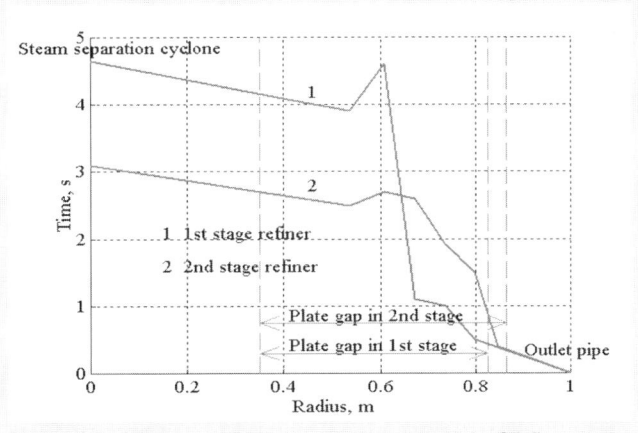

Figure 27. Average residence time of fiber in a 65-in. SD first-stage refiner and in a 68-in. SD second stage refiner measured from feed point to outlet pipe.

CHAPTER 4

into the process before the refiner and its passage through the refiner is followed by sensors. At present, there is only one set of residence time measurement data available for a modern mill in a production scale refiner. Härkönen et al. measured, with a radioactive tracer, the residence time in a 65-in. SD refiner in the first stage[53], and in a 68-in. refiner in the second stage[62] (Fig.27). Results indicate that there is intensive mixing in the inner section of the plate gap. After the flow has passed the point of maximum temperature, it passes through the gap very rapidly in the first-stage refiner and, in the second-stage refiner, the residence time is a little higher. In the outer section, the radial speed of pulp is <1.0 m/s. May et al.[63] estimated pulp residence time by calculating the amount of compressed pulp in the plate gap. Their theoretical study was made for a 185-kw double disc atmospheric refiner. The residence time in the plate gap according to their study is 0.07 s in the first-stage refiner and 0.023 s in the second-stage refiner. Quellet et al.[64] measured residence time in a laboratory refiner using dye as a tracer. They performed their study using a 37-kW laboratory refiner. The residence time of fibers in the plate gap varied for different measurements and conditions between 0.5 and 3 seconds.

There are several reports on visual observations of the phenomena in the plate gap. Atack et al.[65] report on high-speed photography of fiber flow in the plate gap. This study was carried out in a 1.87 MW 42-in. refiner. Stationwala et al.[66] report further trials using high-speed photography in the same refiner. Atack et al. describe the high-speed photography of pulp flow patterns in a 5 MW 60-in. refiner[50]. These studies clearly show the tangential orientation of fibers on the stator bars and the flow pattern in stator grooves. Depending on the amount of dams, pulp is stagnant in the stator or it flows backward in the stator grooves in the inner segment section together with flow-back steam and outward in the outer segment section. This filming does not show the motion in the rotor grooves, but it is reasonable to assume that there is an extensive vortex motion in the inner section of the plate gap. Alahautala et al.[67] developed a technique to visualize the plate gap phenomena by means of endoscope and video filming in a 12 MW 65-in. SD refiner.

3.5 The present state of knowledge and understanding of fundamental mechanisms in refining

As mentioned above, the TMP process has been developed experimentally. This means that many phenomena in the refining process are described in qualitative rather than quantitative terms. In many studies and explanations, the refiner is considered as a "black box"; the process has some controlled input values, and output values are observed without any deeper analysis of refiner construction and refining phenomena. Experience shows that refiners do not work as equal "black boxes." It follows that the transfer of experimental results and data from one refiner to another does not necessarily lead to the results observed earlier.

However, we must remember that this type of developmental work has given the refiner mechanical pulping processes the leading position in mechanical pulping. Experimental work will continue to play an essential role in development in the future. Basic studies conducted into wood material and fiber as well as into the phenomena inside the refiner might well shorten the development path or limit it to some defined directions.

Fundamentals of mechanical pulping

4 Reasons for the difference in energy consumption between grinding and refining

Atack presented one very clear review on the fundamental aspects of grinding and refining at the 1981 International Mechanical Pulping Conference[68]. He pointed out clear differences in the mechanisms of grinding and refining, and he based his own explanation of the great difference in energy consumption between grinding and refining on these differences. He also highlighted the physical reasons for the high energy consumption in mechanical pulping – what Pöyry[69] some years earlier called "the Achilles' heel of mechanical pulp."

4.1 Energy consumption in grinding and refining

Figure 28 shows the total electric energy usage per air dry metric ton in producing groundwood and refiner mechanical pulp for different end uses. The raw material is Norway spruce. Of the total energy consumption, normally some 200–300 kWh/t is used for pumping, etc., which means that in most cases about 90% of the energy is used for grinding or refining (reject refining included).

4.2 Reasons for the high energy consumption

In 1934, Campbell[70] published calculations showing that the theoretical efficiency of grinding for a certain specific surface was only 0.012% (only 0.22 kWh/t out of 1770 kWh/t). Similar calculations presented on several occasions since then have slightly improved this theoretical efficiency, although not to any decisive

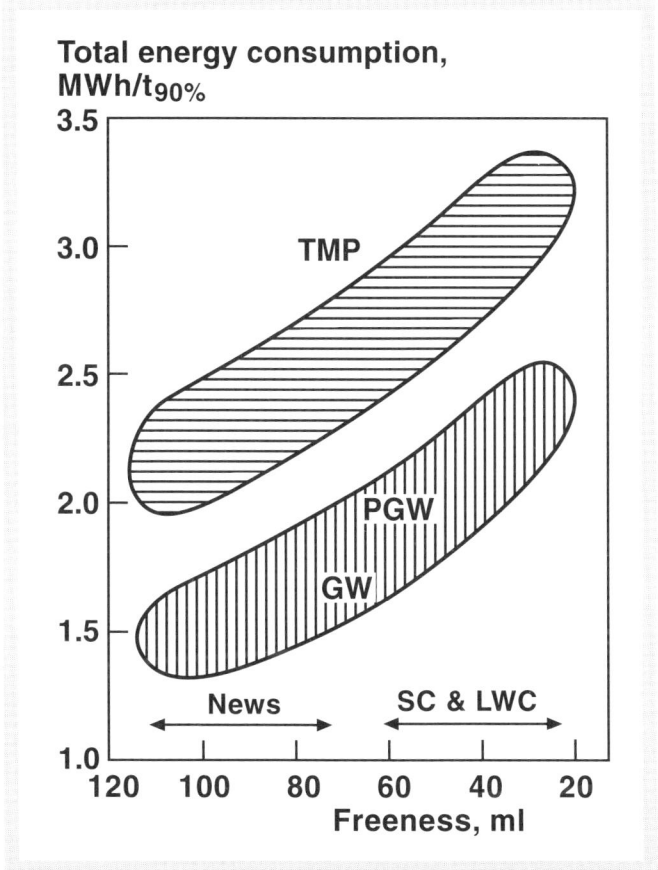

Figure 28. The total electric energy consumption for GW, PGW, and TMP as a function of freeness and end use. Norway spruce, Finnish mills.

degree. This approach would suggest that some 90%–99% of the energy used for the production of, say, PGW or TMP is lost. But, as Atack pointed out, mechanical pulping is a "highly specialized attrition process in which the objective is to produce 'debris' with

CHAPTER 4

certain physical characteristics." That is why energy consumption is so much higher for mechanical pulping than for merely grinding the wood into particles with a certain specific surface.

In an ideal mechanical pulping process:

- The wood fibers must be separated from each other.
- Fiber length must be retained.
- The fibers must be delaminated (internal fibrillation).
- Large amounts of fines with suitable bonding and light scattering characteristics must be produced from the middle lamella and the primary and secondary layers of the fiber wall.
- The remaining secondary wall must be fibrillated (external fibrillation).

For all this to take place simultaneously or at least in something resembling the right order, the disintegration of the wood must be directed at certain specific areas in the fiber wall. This is achieved by altering the softness and toughness of the fiber wall components in relation to each other. As the various polymer components of the fiber wall have different softening temperatures, this can be achieved by raising or lowering the temperature of the fibers during defibration and further processing. The differences in the components of the fibers are, however, so slight in this respect that the only way the disintegration can be directed to a specific area is to use a fatiguing process. After a certain, extremely high number of quite weak but still sufficiently strong impacts, the structure will usually break in a particular place because of fatigue. This also means that a large proportion of the impacts which do not directly cause the breaking down of the wood are still necessary for the process, as they cause the weakening of the wood and fiber structure at the right places preceding the actual breaking down. Thus, the conclusion is that the efficiency of the mechanical pulping processes is higher than generally thought.

What then would be a realistic approximation of the efficiency? According to results from pilot trials with normal spruce wood, but using extremely low peripheral stone speed and a special "stone," it seems to be possible to produce 100 ml freeness GW using only about 700 kWh/t at the grinder, when normal GW of the same strength requires 1200 kWh/t[71]. This indicates that energy reductions of at least 40% are possible. A fair guess is that the efficiency of the process is somewhere in the range of 40%–60%. The minimum amount of energy for producing a newsprint-type groundwood pulp then is 500–700 kWh/t (grinding and reject refining energy only). Assuming the same ratio for the high strength refiner pulps, the minimum energy requirements for a TMP-like pulp of normal news quality then is 800–1200 kWh/t (refining only). We have thus somewhat improved the bad reputation of mechanical pulp in terms of theoretical efficiency, but there is still much room for large practical improvements in energy efficiency.

4.3 The reason for higher energy consumption in chip refining

The production of TMP for a certain end use consumes about 40% more electric energy than does production of GW or PGW. Why is this so? Atack's hypothesis[68] is that "chip refining is an almost idealized form of longitudinal grinding." He bases this theory on results of Montmorency[72], which he interprets as meaning that longitudinal grinding would require considerably higher amounts of energy than normal transversal grinding. Kirchner[73] published similar results back in 1912. In Kirchner's tests, output was about 80% lower with longitudinal grinding, and energy consumption was six times higher than with normal transversal grinding.

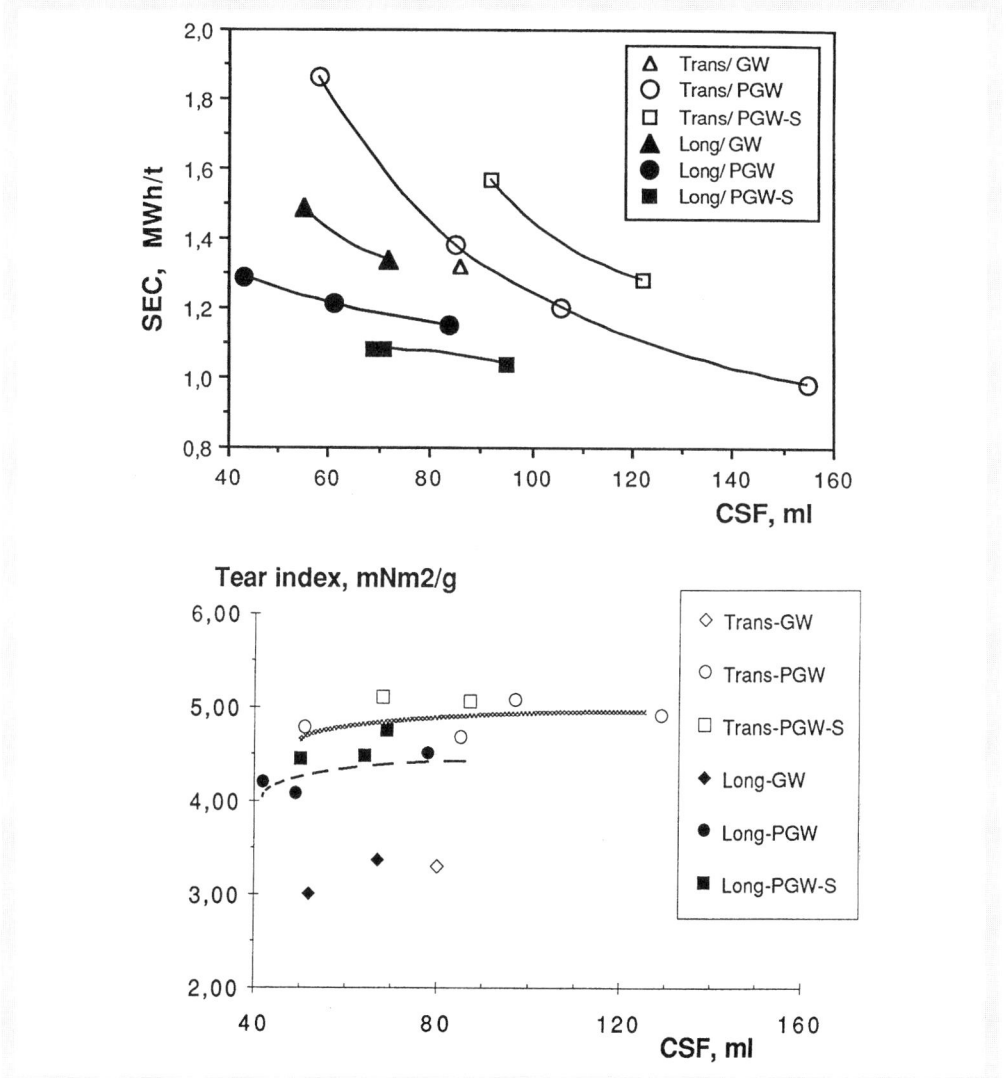

Figure 29. Energy consumption and tear strength for pilot PGW using normal and longitudinal grinding[74].

CHAPTER 4

What would have happened if Kirchner had increased output by pressing the wood harder against the stone? According to Lucander[74, 75], longitudinal grinding has to be run with considerably higher output than transversal grinding in order to reach the same freeness. Figure 29 shows results from tests with a small pilot facility at KCL[74]. The results of the tests unambiguously indicate a lower energy consumption and the same or a slightly lower tear resistance for longitudinal groundwood, pressure groundwood, and super pressure groundwood. The conclusion is that the difference in energy consumption between grinding and refining cannot be explained by comparing refining with longitudinal grinding. The comparison with longitudinal grinding might explain certain differences in pulp properties but not the difference in energy requirements.

How then can the high energy consumption of refiner mechanical pulp be explained? The author tends to support the hypothesis, first put forward by Luhde[76] and later supported by Brecht[77] and presented here in a slightly developed form, that fibers and fiber bundles are not as firmly fixed during refining as they are during grinding, and that's why grinding is more efficient in breaking down wood structure. During grinding, the fibers are firmly positioned at least at one end of the fiber the whole time they are treated by the grits. The passing grits hit the fibers both before and during the peeling action. We can imagine a fiber, still attached to the wood, which is hit too hard by a passing grit. The fiber cannot move out of the way, and as a result it will most likely be damaged or cut. Now imagine the same fiber placed over the edge of the refiner bar and imagine that it receives the same blow from one of the bars of the opposite disc. Since the fiber is held over the bar edge only by relatively weak friction forces, the fiber, which is light in itself, can slide off the bar and thus avoid being cut. Therefore the refining process is much "gentler" and is able to avoid fiber cutting much more successfully than in grinding. Yet, for the same reason, the refining process is also more inefficient at breaking down the wood into the smallest particles possible.

It is worth noting that the above hypothesis, if proved correct, disproves Atack's theory concerning regrinding in grinding. According to Atack[68], the actual processing of the fibers takes place after they have been removed from the wood, and not, as in the hypothesis presented here, mostly while the fibers are attached to the wood.

The difference pointed out above between the mechanisms of refining and grinding can explain much of the difference in energy consumption. There are, however, also a number of other possible reasons. Steenberg pointed out that refining is a cooperative process in which a large part of the energy is consumed in mixing, that is, in creating new configurations[78]. Steenberg's model is not in conflict with the hypothesis presented above. Although it can be asserted that the fibers can assume new positions and configurations even while they are partly attached to the wood as in grinding, this actually happens more easily and to a greater extent during refining. Thus some of the differences between grinding and refining can also be explained using Steenberg's model.

Finally, there are differences in the properties of the processing material, in the processing frequencies, in the processing speed, and in the consistency that influence energy consumption. However, the fundamental difference between grinding and refining is that the wood particles are not immobilized during disc refining as they are during grinding.

References

1. Atack, D., Philosophical Magazine A. 43(3):619 (1981).
2. Atack, D. and Heitner, C., Trans. Tech. Sect. (Can. Pulp Pap. Assoc.) 5(4):99 (1979).
3. Back, E. L. and Salmén, N. L., Tappi 65(7):107 (1982).
4. Becker, H., Höglund, H., Tistad, G., Paperi ja Puu 59(3):123 (1977).
5. Franzén, R., Nordic Pulp Paper Res. J. 1(3):4 (1986).
6. Hammar, L. Å., Htun, M., Ottestam, C., Salmén, L., Sjögren, B., J. Pulp Paper Sci. 22(6):J219 (1996).
7. Heitner, C. and Atack, D., Svensk Papperstid. 85(12):R78 (1982).
8. Htun, M., Salmén, L., Eriksson, L., "A better understanding of wood as a material (a way to increased energy efficiency when making mechanical pulps?" in Energy Efficiency In Process Technology (P.A. Pilavachi, Ed.) Elsevier Appl. Sci., London, 1993, pp. 1086–1095.
9. Kure, K.A., "The alteration of the wood fibers in refining," 1997 International Mechanical Pulping Conference Proceedings, SPCI, Stockholm, p.137.
10. Kärenlampi, P., Paperi jaa Puu 74(8):650 (1992).
11. Lucander, M., Lönnberg, B., Haikkala, P., J. Pulp Pap. Sci. 11(2):J35 (1985).
12. Mohlin, U. B., "Fiber development in mechanical pulp refining," 1995 International Mechanical Pulping Conference Proceedings (Ottawa), CPPA, Montreal, p. 71.
13. Nielsen, L. E. and Landel, R. F., Mechanical Properties of Polymers and Composites, Marcel Dekker, New York, 1994.
14. Nimz, H. H., Robert, D., Faix, O., Nemr, M., Holzforschung 35(1):16 (1981).
15. Olsson, A. M. and Salmén, L., "Mechanical spectroscopy (a tool for lignin structure studies," Cellulosics: Chemical, Biochemcal and Material Aspect (J.F. Kennedy, G. O. Phillips, and P. A. Williams, Eds.) Ellis Horwood, Chichester, 1993, pp. 255–262 .
16. Olsson, A. M. and Salmén, L., Nordic Pulp Pap. Res. J. 12(3):140 (1997).
17. Salmén, L., J. Materials. Sci. 19(9):3090 (1984).
18. Salmén, L., J. Pulp Paper Sci. 13(1):J23 (1987).
19. Salmén, L., "Directional viscoelastic properties of the fiber composite wood," Progress and Trends in Rheology II (H. Giesekus, Ed.) Steinkopff Verlag, Darmstadt, 1988, pp. 234–235.

20. Salmén, L., J. Pulp Paper Sci. 21(9):J310 (1995).

21. Salmén, L. and Berthold, J., "The swelling ability of pulp fibers" in The Fundamentals of Papermaking Materials (C. F. Baker, Ed.) Pira International, 1997, pp. 683–701.

22. Salmén, L. and Pettersson, B., Cellulose Chem. Technol. 29(3):331 (1995).

23. Sjöström, E., Wood Chemistry, Fundamentals and Applications, Academic Press, London, 1993.

24. Skibo, M. D., "The effect of frequency, temperature and materials structure on fatigue crack propagation in polymers," Ph. D. dissertation, Lehigh University, Lehigh, 1977.

25. Sorvari, J., Sjöström, E., Klemola, A., Laine, J. E., Wood Sci. Technol. 20(1):35 (1986).

26. Stamm, A. J., Wood and Cellulose Science, Ronald Press, New York, 1964.

27. Uhmeier, A. and Salmén, L., J. Eng. Mat. Technol. 118(3):289 (1996).

28. Uhmeier, A. and Salmén, L., Nordic Pupl Pap. Res. J. 11(3):171 (1996).

29. Williams, M. L., Landel, R. L., Ferry, J. D., J. Am. Chem. Soc. 77:3701 (1955).

30. Atack, D. and May, W. D., Pulp Paper Mag. Can. 63(1):T10 (1962).

31. Atack, D., "Mechanics of Wood Grinding Trend," The Activities of the Pulp and Paper Research Institute of Canada, Report No. 19, 1971, pp. 6–11.

32. Koponen. S., Toratti, T., Kanerva, P., Wood Sci. Technol. 25(1):25 (1991).

33. Lucander, M., KCL internal report.

34. Lucander, M., KCL internal report.

35. Ilvessalo-Pfäffli, M-S., Fiber Atlas, Identification of Papermaking Fibers, Springer-Verlag, Berlin, Heidelberg, 1995.

36. Ilvessalo-Pfäffli, M-S. and Laamanen, J., KCL internal report.

37. Panshin, A. J. and DeZeeuw, C., Textbook of wood technology. Vol. I: Structure, identification, uses and properties of the commercial woods of the United States and Canada. 3rd edition, McGraw-Hill, New York, 1970.

38. Lucander, M., KCL internal report.

39. Salmén, L. and Fellers, C., Trans. Tech. Sec., Can. Pulp Pap. Ass. 9(4):TR93 (1982).

40. Salmén, L., Dumail, J. F., Uhmeier, A., "Compression behavior of wood in relation to mechanical pulping," 1997 International Mechanical Pulping Conference Proceedings, SPCI, Stockholm, p. 207.

41. Powell, F. G., Luhde, F., Logan K. C., Pulp Paper Mag. Can. 66(8):T399 (1965).

42. Atack, D., Fontebasso, J., Stationwala, M. I., Tappi J. 66(7):75 (1983).

43. Lucander, M., "Einfluss der Prozessvariabeln auf die Qualität des Druckschliffs," PTS Holzstoff Symposium München 1985, TVW Papiermaschinen GmbH, Drucksliffvorträge.

44. Lucander, M., "Recent investigations of pressure grinding at elevated grinding pressure," PTS-TUD-Symposium Papierzellstoff und Holzstofftechnik '95, PTS Verlag, München, p.7.

45. Haikkala, P., Liimatainen, H., Manner, H., Tuominen, R., "Pressure groundwood (PGW), super pressure groundwood (PGW-S) and thermomechanical pulp (TMP) in wood-containing printing papers," 1989 International Mechanical Pulping Conference Proceedings, KCL, Helsinki, p. 36.

46. Pasanen, K., Peltonen, E. Haikkala, P., Liimatainen H., Tappi J. 74(12):63 (1991).

47. Honkanen, K. and Yrjövuori, R., "The first years of the first super pressure groundwood (PGW-S) plant," 1993 International Mechanical Pulping Conference Proceedings, PTF, Oslo, p. 44.

48. Tuovinen, O. and Liimatainen, H., "Fibers, fibrils and fractions (an analysis of various mechanical pulps, 1993 International Mechanical Pulping Conference Proceedings, PTF, Oslo, p. 324.

49. Lucander, M., KCL internal report.

50. Atack, D., Stationwala, M. I., Huusari, E., Ahlqvist, P., Fontebasso, J., Perkola, M., Paperi ja Puu 71(4):689 (1989).

51. Miles, K. B., Dana, H. R., May, W. D., "The Flow of Steam in Chip Refiner," 1980 International Symposium on Fundamental Concepts of Refining, The Institute of Paper Chemistry, Appleton, WI, p. 30.

52. Miles, K. B. and May, W. D., J. Pulp Paper Sci. 16(2):J63 (1990).

53. Härkönen, E., Ruottu, S., Ruottu, A., Johansson, O., "A theoretical model for a TMP-refiner," 1997 International Mechanical Pulping Conference Proceedings, SPCI, Stockholm, p. 95.

54. Strand, B. C. and Mokvist, A., "The Application of Communition Theory to Describe Refiner Performance," 74th CPPA Annual Meeting Notes, CPPA, Technical Section, Montreal, PQ, 1988, p. 115.

55. Di Ruscio, D. and Blachen, J. G., "A state space model for the TMP process," 1992 International Control Systems Conference Proceedings, CPPA, Montreal, p. 107.

56. Strand, B., private conversation, 1993.

57. Engstrand, A., private conversation, 1997.

58. Härkönen, E., UPM-Kymmene, internal report,1994.

59. Härkönen, E., Tienvieri, T., Viljakainen, E., Huusari, E., Mäkivaara, J., Teoreettinen ja kokeellinen tutkimus perusilmiöistä TMP-jauhatuksessa. Kestävä Paperi - tutkimusohjelman raportti 8, 1995, Oy Keskuslaboratorio-Centrallaboratorium Ab, Espoo, Finland (in Finnish).

CHAPTER 4

60. Engstrand, P., Karlström, A., Nilsson, L., "The Impact of Chemical Addition on Refining Parameters," 1995 International Mechanical Pulping Conference Proceedings, CPPA, Montreal, p. 281.

61. Parta, J., UPM-Kymmene, internal report,1997.

62. Parta, J., UPM-Kymmene, internal report,1997.

63. May, W. D., McRae, M. R., Miles, K. B., Lunan, W. E., J. Pulp Paper Sci. 14(3):J47 (1988).

64. Quellet, D., Bennington, C. P. J., Senger, J. J., Borisoff, J. F., Martiskainen, J. M., J. Pulp Paper Sci. 22(8):301 (1996).

65. Atack, D., Clayton, D. L., Quinn, A. E., Stationwala, M. I., "High speed photography of wood pulping in a disc refiner," 1984 International Congress on High Speed Photography and Photonics Conference Proceedings, SPIE, Strasbourg, p. 348.

66. Stationwala, M. I., Miller, C., Atack, D., Karnis, A., "Application of high speed photography to chip refining," 1990 International Congress on High-Speed Photography and Photonics Conference Proceedings, SPIE, Strasbourg, p. 237.

67. Alahautala, T., Vattulainen, J., Hernberg, R., "Visualization of pulp refining in a rotating disk refiner," 1997 IMEKO World Congress Proceedings, p.60.

68. Atack, D., "Fundamental differences in energy requirement between the mechanical pulping processes," 1981International Mechanical Pulping Conference Proceedings, PTF, Oslo, p. VI:5.

69. Pöyry, J., "Are energy costs the Achilles' heel of mechanical pulp?" 1977 International Mechanical Pulping Conference Proceedings, The Finnish Paper Engineers´ Association, Helsinki, 1977, p. 13.

70. Campbell,W. B., Pulp Paper Can. 35(4):218 (1934).

71. Lucander, M., internal reports, KCL,1990.

72. Montmorency, W. H., Pulp Paper Mag. Can. 49(3):115 (1948).

73. Kirchner, E., Die Holzschleiferei Oder Holzstoff-Fabrikation, Verlag von Güntter-Staib, Biberach, 1912.

74. Lucander, M., "Longitudinal grinding," unpublished work for the Kuitu program, 1993.

75. Lönnberg, B. and Lucander, M., "Heat distribution in the grinding zone (Theoretical aspects and the effect upon pulp properties," 1981 International Mechanical Pulping Conference Proceedings, PTF, Oslo, p. I:3.

76. Luhde, F., Das Papier 16(11):655 (1962).

77. Brecht, W., Wochenblatt für Papierfabrikation 107(11/12):394 (1979).

78. Steenberg, B., "A model of refining as a special case of milling," 1980 International Symposium on Fundamental Concepts of Refining Preprints, Institute of Paper Chemistry, Appleton, 1980, p. 107.

79. Back, E. and Salmén, L., Tappi 65(7):107 (1982).

80. Lönnberg, B., Paperi ja. Puu 79 (4):249 (1997).

81. Björkqvist, T., Lautala, P., Finell, M., Lönnberg, B., Saharinen, E., Paulapuro, H., "Development and verification of a grinding process mode," 1995 International Mechanical Pulping Conference (Ottawa) Preprints, CPPA, Montreal, p. 65.

82. Björkqvist, T., Lautala, P., Saharinen, E , Paulapuro, H., Koskenhely, K., Lönnberg, B., "Behavior of spruce sapwood in mechanical loading," 1997 International Mechanical Pulping Conference Preprints, SPCI, Stockholm, pp 199.

83. Sakata, I., Senju, R., J. Appl. Polym. Sci. 19(10):2799 (1975).

84. Cousins, W. J., Wood Sci. Technol. 12(1):161 (1978).

CHAPTER 5

Raw materials

1	**General properties**	**67**
1.1	Basic density	68
1.2	Moisture content	68
1.3	Juvenile wood	70
1.4	Heartwood	71
1.5	Unsuitable parts of raw material	71
2	**Wood species**	**75**
2.1	Characteristics	75
2.2	TMP	78
2.3	Groundwood processes	80
3	**Within-tree property variation**	**85**
3.1	Fiber length	85
3.2	Fiber width	86
3.3	Cell wall thickness	88
3.4	Fibril angle	89
4	**Property variation between trees**	**89**
5	**Different wood assortments**	**91**
6	**Wood procurement**	**97**
7	**Wood handling**	**98**
8	**Seasonal variation**	**100**
	References	101

CHAPTER 5

Antero Varhimo, Olli Tuovinen

Raw materials

Wood raw material is an essential parameter in the manufacture of mechanical pulps. An ideal wood raw material for mechanical pulping can be defiberized at a low cost, i.e., with a low energy consumption without process disturbances, and the pulp will give a suitable sheet quality for the end product in question.

Because pulp and paper products are largely made of fibers, the fiber morphology of the original wood raw material can be expected to create the basis for final sheet properties and energy consumption in mechanical defiberizing. Chemical composition of the wood will directly affect pulp properties, especially brightness and the stablility of the process through possible pitch problems. Mechanical quality of the original wood raw material (defects, knot content, trunk shape, etc.), wood procurement, and wood handling at the mill are additional factors in determining the final properties of the raw material.

Wood is far away from being a homogenous material. Large natural variation exists in the fiber and chemical properties between various wood species and within a specified species. The genetic inheritance and environmental conditions determine the wood and fiber properties in a tree stem at a certain age. Wood and fiber properties found in different parts of a trunk vary, resulting in a wide range of property variation in the raw material entering the pulping processes.

Measuring of the important wood and fiber properties is often time-consuming and thus traditionally seldom practiced in mill conditions. The wood species is normally known and sometimes logs originating from different wood stands (thinning, clear-cutting) are processed separately. Sorting of wood logs according to diameter, age, or growth rate is practiced in some cases. Sawmill chips originating from the sapwood of mature trees are a raw material entering the mill as its own stream having properties different from those of roundwood. Little, in fact is done to minimize the wood property variation at the mill although better control and sorting of the wood raw material would no doubt lead to a more stable process and give better opportunities to match the various demands set for the mechanical pulp from the end product.

1 General properties

Many properties have been used in describing the quality of wood mechanical pulping and properties of the pulp. Those reported having significance are listed below[1-7].

- Wood species
- Basic density
- Moisture content
- Fiber properties (length, width, cell wall thickness, microfibril angle, etc.)

CHAPTER 5

- Content of earlywood and latewood
- Content of heartwood and sapwood
- Content of mature and juvenile wood
- Content of compression wood
- Knotwood content
- Defects (decay)
- Chemical composition of wood (lignin, extractives)
- Effect of wood procurement and handling (bark content, chip quality)
- Effect of impurities (sand, dirt, etc.).

It is difficult to figure out the relative importance of the various parameters because many of the properties are more or less strongly interrelated. In general, all the properties can be derived from fiber properties, chemical composition of the wood, and technical wood quality. In the following text, emphasis has been put on these properties and the comparison of various wood species and the others have had more general attention.

1.1 Basic density

Basic density is one of the most studied wood properties, and it varies substantially between and within various wood species. Basic density is not, however, an independent property but is determined by several characteristics of wood. Since the density of cell wall material is generally considered to be almost constant (between 1.5 and 1.6 g/cm^3), basic density is determined mainly by the proportion of cell wall material, i.e., fiber cross sectional dimensions. For example, latewood percentage has been reported to correlate strongly with the basic density of Norway spruce, when the juvenile wood zone is excluded[8].

Since poor general correlation between wood basic density and TMP properties as well as the energy consumption in refining has been reported[5, 9], the variation patterns for basic density are not reviewed here. However, it must be pointed out that the economic significance of basic density is very large, because wood is mainly purchased by a unit volume to the mill. High-density wood thus gives a better pulp yield per unit volume of wood than low-density wood. Furthermore, *variations* in wood basic density result in quality variations for mechanical pulps.

The chip flow usually is metered by volume, and wood density variations will cause fluctuations in the production rate. If the control system of the refiner cannot compensate for the changing wood density, the specific energy applied will fluctuate and pulp quality will be reduced by nonuniform refining.

1.2 Moisture content

It is commonly acknowledged that a sufficient moisture content is essential for the quality of mechanical pulps. Softening of the lignin in wood prior to the separation of the fibers is a prerequisite for producing uncut fibers and further development of fiber properties. Temperature and wood moisture content have a tremendous effect on the thermal softening of lignin in the wood matrix.

Several factors influence the moisture content of the wood arriving at the fiberizing zone:

- Basic density
- Heartwood and resin content
- Wood procurement and storage
- Remoistening in the process.

In mechanical pulping, there is thus always present at least small-scale moisture variation. In many cases when the moisture content of wood changes, other changes occur simultaneously. For example, drying during storage and transportation is accompanied with fungal activities in wood resulting in the loss of brightness. With increased heartwood content, not only the moisture content but also other properties such as fiber dimensions change because of their pith to cambium variation pattern.

Uncontrolled variations in the wood moisture content cause variations in the consistency during defiberization. Too high a refining consistency, for example, is known to affect pulp strength properties dramatically. Pulp of uneven quality is obtained, because the process control systems are not fully able to even out the changes caused by the wood moisture content variations. In a controlled process, the effect of wood moisture content decreases in the order GW>PGW>TMP.

Increasing dry matter content of the wood impairs the properties of groundwood pulps, especially in the 70%–80% range where the fiber saturation point lies. The long fiber content of pulp decreases and strength properties start to deteriorate even after moderate drying (Fig. 1).

The lowered bonding ability and increased shives content cause linting problems in offset printing. Rewetting dry logs before grinding results in similar pulp properties to those obtained with original moist wood logs[3].

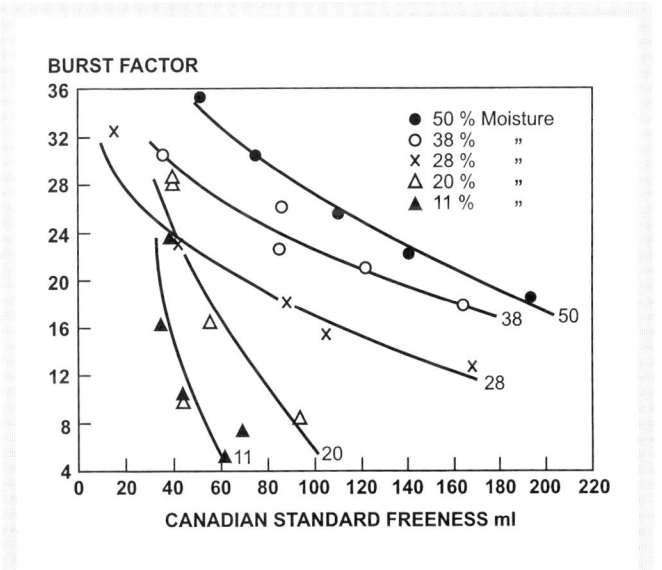

Figure 1. Effect of dry matter content on the burst index versus freeness of Black spruce groundwood pulps[3].

CHAPTER 5

The effect of moisture content in thermomechanical pulping is not as drastic as in groundwood processes because of the effect of chip washing and preheating. It is widely believed that the energy consumption for a given pulp freeness will increase along with increasing dry matter content of the chips. For example, Miles[10] found out that the specific energy consumption in refiner mechanical pulping of Black spruce increases in the moisture content range 23%–58% (dry matter content 77%–42%). It has also been stated, however, that the initial chip moisture content has no significant effect on the energy consumption and the pulp properties except for shives content, provided the moisture content is kept above the fiber saturation point[2,11].

Uniformity of moisture content in chips will promote the quality of TMP. With proper impregnation, pulp quality can be improved (Table 1).

Table 1. Properties of TMP from softwood chips having different dry matter contents. Values at 100 ml CSF[12].

Chips	"Dry"	Normal impregnation	Water impregnation
Dry matter content, %	54.4	45.3	35.7
SEC, kWh/t	1750	1885	2060
Bauer McNett, +16 mesh	11.8	25.0	32.7
Bauer McNett, +30 mesh	25.0	20.5	17.7
Bauer McNett, −200 mesh	21.4	25.0	23.6
Shives, %	1.17	1.22	0.78
Tensile index, Nm/g	42.5	49.0	51.3
Burst index, kPam^2/g	2.11	2.80	3.00
Tear index, mNm^2/g	7.8	9.0	10.3
Wet strength, N/m	60	75	91
Scattering coeff, m^2/kg	44.6	42.5	43.0

In these experiments, the moisture contents have all been within fairly normal range. With thorough impregnation, the long fiber content of TMP has been clearly increased to result in better strength properties (tensile, burst, and tear). The energy consumption for a given pulp freeness has been lowest for the driest chips, but the situation would look different when compared at a given tensile index.

1.3 Juvenile wood

In softwoods, juvenile wood is situated around the pith where the cells have not fully matured. In Norway spruce, the first 20–30 growth rings are usually juvenile wood. It is not a single property, but a zone where many wood properties change from pith to the cambium, very rapidly nearest the pith. Properties having the variation pattern with significance for mechanical pulping include:

- Generally increasing: basic density, fiber properties (length, width, cell wall thickness, latewood percent), cellulose content
- Generally decreasing: fiber wall microfibril angle, lignin content.

Juvenile wood exists in all trees; the maximum content can be found in young trees and in tops of older trees. The later sections "Wood property within-tree variation" and "Mechanical pulps from different wood assortments" provide a more detailed review of the variation of fiber properties and the effect of juvenile wood on mechanical pulping and pulp properties

1.4 Heartwood

Differences between heartwood and sapwood exist in many wood properties, for instance:

- Dry matter content
- Presence of juvenile wood (varying basic density, fiber length, width, cell wall thickness, fibril angle)
- Wood color and extractives content.

Heartwood thus by no means has a single property but is even more complex than juvenile wood. Different wood species deviate markedly from each other as regards their heartwood characteristics. For example, spruces *(Picea spp.)* have no colored heartwood as the genera *Larix, Pinus,* and *Pseudozuga,* but instead have a clearly higher dry matter content in the heartwood than in the sapwood.

In spruces, the pit membrane in bordered pits seals the aperture during heartwood formation, stopping the flow of liquids through the pit. The permeability of such wood is thus dramatically reduced. Transformation of sapwood to heartwood starts in Norway spruce grown in southern Finland at about 30 years of age or less.

In most other conifers, the formation of heartwood is accompanied by accumulation of extractives.

For example, the (acetone soluble) extractives content in Scots pine is normally about 3%, whereas that of heartwood is typically about 5%[61]. Heartwood in the butt section of old pine trees usually has an even higher extractives content.

The effect of heartwood on mechanical pulping arises from the many property differences from sapwood. The main effects of heartwood formation arise from the differences in moisture and exractives contents. These effects are largely dependent on wood species and changes in the pulping process and will not be reviewed further here. The effect of heartwood formation alone would, moreover, be difficult to differentiate from the effects of the differences in the fiber properties of heartwood and sapwood (juvenile wood and mature wood).

1.5 Unsuitable parts of raw material

All pulps meant for paper manufacture consist of fibers or particles formed from the fibers during the manufacturing process. Therefore all other parts of a tree except defectless wood in the trunk can be regarded as unsuitable or less suitable raw material sources for pulp. Thus bark (both inner and outer bark), branches and knotwood, needles and leaves, roots, and also the top log of a tree below a practical diameter must be excluded or minimized when selecting the raw material for mechanical pulping.

CHAPTER 5

The bark in conifers is morphologically and chemically very heterogenous. It consists of inner bark (bast), which contains all the living cells in the bark and outer bark (cork), which protects the cells under it from mechanical damage, drying, and temperature changes. Pulpwood logs are normally debarked before the pulping process, but efficient bark removal may be difficult if the logs are dry or frozen. High bark contents can be found in sawmill chips during the wintertime because they consist mainly of chipped slabwood residues.

The most striking effect of increased bark content is the decrease in pulp brightness. In spruce TMP, the brightness reduction caused by bark (inner + outer bark) is 2–3 brightness units/% of bark in the chips, Fig. 2[2,5,13]. The color of inner bark is originally almost the same as that of wood, but becomes darker during storage. The effect of inner bark on pulp brightness is thus not as drastic as that of the whole bark. Bark has a significant effect on dithionite or peroxide bleaching of mechanical pulp. The brightness loss caused by bark remains after bleaching if the chemical charges are not increased. High amounts of bark will inevitably make the pulp unsuitable for the highest quality papers because sufficient brightness cannot be reached.

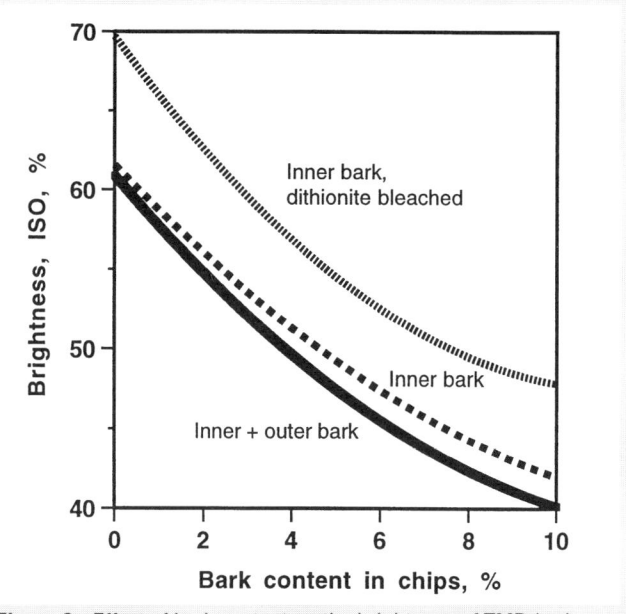

Figure 2. Effect of bark content on the brightness of TMP (redrawn from Refs. 2, 5, and 13).

In addition to reduced brightness, bark has a negative effect on other properties of TMP (Table 2).

Table 2. Effect of bark on TMP properties at CSF 100 ml[5].

Chip blend	Bark free	5% inner bark	5% outer bark
SEC, kWh/b.d.t	1900	1540	1770
Mini shives, PFI, %	1.0	1.3	1.5
BMcNett +30 mesh, %	39	36	34
BMcNett −200 mesh, %	27	28	27
Tear index, mNm^2/g	7.5	7.15	6.35
Tensile index, Nm/g	36	36	33
Density, kg/m^3	420	420	420
Scattering coeff., m^2/kg	54	52	54
Brightness, ISO, %	63	50	49

At constant CSF, increased bark content results in a lower energy consumption, a higher shives content, a lower long fiber content (BMcNett +30 mesh), and a lower tear index. Inner bark has a fibrous nature and thus has a smaller effect on strength properties than nonfibrous outer bark.

At practical bark contents, the deterioration of properties other than brightness are of minor importance with the exeption of the risk for impurities in the final product.

Knots are formed from tree branches and thus cannot be avoided in the wood arriving at the mechanical pulping plant. The amount of knotwood depends on the number and size of the branches they originate from, the age at which the branches die, and the length of time dead branch stubs remain on the tree[6]. The average knotwood volume in Norway spruce grown in Finland varies between 0.5 and 2.0%, corresponding to a weight range of 1.1%–6.7%[14].

The properties of knotwood differ markedly from those of normal stemwood (Table 3).

Knots have a substantially higher basic density and dry matter content than stemwood. They are composed mostly of compression wood, the fibers of which are shorter and narrower than those of stemwood. The microfibril angle in the S_2-layer, which influences fiber stiffness, is larger in compression-wood fibers than in normal wood fibers. Knotwood is also chemically different from stemwood with a clearly higher lignin and extractives content and lower cellulose content.

Table 3. Wood and fiber properties of knotwood and stemwood in Norway spruce (*Picea abies* (L.) Karst)[14–17].

Wood property	Stemwood	Knotwood	Stemwood	Knotwood
Basic density, kg/m^3	328–450	800–900		
Dry matter content, %	41/75[a]	76–81		
Fiber length, mm	1.9–3.7	1.0–1.1		
Fiber diameter, µm	46.3	36.0		
Cell wall thickness, µm	4.0–6.0	3.0–5.5		
Fiber coarseness, mg/m	0.205	0.080		
Lignin content, %	28.6	33.0	27.6	34.9
Extractives, %	0.5–2.0[b]	10.8–13.4[b]	1.0[c]	11.0[c]
Cellulose, %			43.6	27.9
Glucomannan, %			12.0	7.7
Xylan, %			5.5	5.8
Other carbohydrates, %			3.5	11.7

[a] sapwood/heartwood, [b] acetone, [c] dichloromethane

The presence of knots in the chips decreases the specific energy consumption to a given pulp freeness value compared with a knot-free pulp. Pulps containing knots also have a lower fiber length and long-fiber content than knot-free pulps resulting in reduced strength properties, especially tear index (Fig. 3).

CHAPTER 5

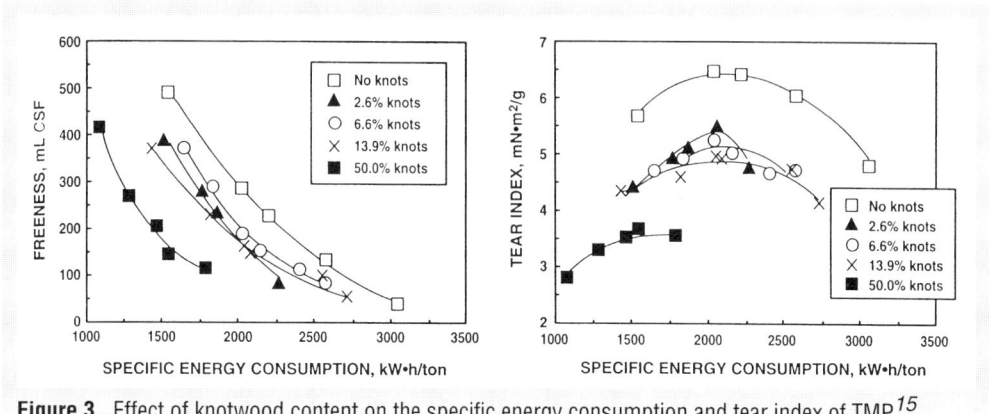

Figure 3. Effect of knotwood content on the specific energy consumption and tear index of TMP[15].

The presence of even small amounts of knots degrades pulp properties. In addition to strength properties, brightness and surface properties are also impaired. The negative effect of knotwood on TMP properties is probably caused by insufficient liquor impregnation, the chemical composition of knots, the morphology of knot fibers, and the interactions between knots and knot-free chips during refining. Knotwood also contains large amounts of metal ions that increase the consumption of chemicals, especially in peroxide bleaching[15].

The energy consumption in thermomechanical pulping of branchwood, which is closely related to knotwood, has been found to be larger than that of stemwood (Table 4).

Table 4. Thermomechanical pulping (CSF 100) of stemwood and branchwood of Norway spruce [*Picea abies* (L.) Karst][5].

Wood raw material	Stemwood	Branchwood
SEC, kWh/bdmt	1900	2040
Tensile index, Nm/g	40	14
Tear index, mNm^2/g	7.2	2.5
Density, kg/m^3	440	350
Scattering coeff., m^2/kg	64	64
Brightness, ISO, %	64	55

Similar to knotwood, branchwood gives TMP with low strength and brightness. The addition of small amounts of branchwood to normal chips shows only marginal negative effects on pulp quality.

Knots are a part of wood arriving at the mill and thus cannot be avoided. Most of the knots, however, are found in the over-thick fraction of chips. Thickness screening of chips provides thus a means not only to improve the stability of the refining process but also to improve pulp quality.

2 Wood species

Wood species is probably the most significant wood parameter for mechanical pulping and pulp quality. Several wood quality variations, i.e., genetic origin, environmental factors, tree age, and quality variation within a tree stem also have a strong effect, which tends to make the comparison of different species difficult.

It is generally recognized that species of the spruce family, especially Norway spruce (*Picea abies*), are the most favorable raw materials for mechanical pulping and provide pulp properties ideal for various paper products. The superiority of the Spruce family has been attributed to the favorable fiber properties (length/wall thickness, fibril structure), low extractives content, and high initial brightness of the wood. The spruces thus give simultaneously good strength, optical, and smoothness properties.

Hardwoods differ from softwoods in terms of a more complex fiber morphology and chemical composition. Mechanical pulps from hardwoods exhibit good light scattering and sheet surface properties, whereas the strength properties usually are poor. Especially in refiner pulping, hardwood chips therefore are pretreated chemically to obtain a compromise between the strength and optical properties. Hardwoods usually are used in small quantities mixed with softwoods.

2.1 Characteristics

Softwood species used for mechanical pulping[18–23] include:

Spruces

- Norway Spruce [*Picea abies* (L.) Karst], Black spruce [*Picea mariana* (Mill) B.S.P]
- White spruce [*Picea glauca* (Moench) Voss], Red spruce (*Picea rubens* Sarg.)
- Sitka spruce [*Picea sitchensis* (Bong.) Carr.],
- Engelmann spruce [*Picea Engelmanni* (Parry) Engelm.]

Firs

- Balsam fir [*Abies balsamea* (L.) Mill.]
- Silver fir (*Abies alba* Mill.)

Pines

- Radiata pine [*Pinus radiata* D. Don], Scots pine (*Pinus sylvestris* L.)
- White pine (*Pinus strobus* L.), Red pine (*Pinus resinosa* Ait.)
- Jack pine (*Pinus banksiana* Lamb.)
- Lodgepole pine (*Pinus contorta* Dougl.)
- Ponderosa pine (*Pinus ponderosa* Dougl. ex Laws)
- Caribbean pine (*Pinus caribaea* Morelet)
- Patula pine (*Pinus patula* Schl. Et Cham.)
- Virginia pine (*Pinus virginiana* Mill.)

CHAPTER 5

Southern pines

- Loblolly pine (*Pinus taeda* L.), Slash pine (*Pinus elliottii* Engelm.)
- Longleaf pine (*Pinus palustris* Mill.), Shortleaf pine (*Pinus echinata* Mill.)
- Pitch pine (*Pinus rigida* Mill.)

Hemlocks

- Western hemlock [*Tsuga heterophylla* (Raf.) Sarg.]
- Eastern hemlock [*Tsuga canadensis* (L.) Carr.]

Other softwood species

- Douglas-fir [*Pseudotsuga menziesii* (Mirb.) Franco]
- Tamarack (Eastern Larch) [*Larix laricina* (Du roi) K. Koch]

Wood species like the cedars (*Thuja occidentalis* L. and *Thuja plicata* Donn ex D. Don) and many larches, for example, European larch (*Larix decidua* Mill.) are considered to have too low of a wood brightness for mechanical pulp.

Wood and fiber properties of various pine species are compared with those of White spruce, Black spruce, and Balsam fir, which are the preferred Canadian softwood species for papermaking (Table 5).

Table 5. Fiber characteristics and chemical composition of various softwood species[21].

Species	Black spruce	White spruce	Balsam fir	Jack pine	Lodgepole pine	Ponderosa pine	Loblolly pine	Slash pine	Radiata pine	Caribbean pine
	Picea marana	*Picea glauca*	*Abies balsmea*	*Pinus banksiana*	*Pinus contorta*	*Pinus ponderosa*	*Pinus taeda*	*Pinus elliottii*	*Pinus radiata*	*Pinus caribaea*
Fiber length, mm	3.5	3.3	3.5	3.5	3.1	3.6	3.6	4.2	4.0	2.6-3.9
Fiber width, µm	25-30	25-30	30-40	28-40	35-45	35-45	35-45	-	35-45	40-50
Cell wall thickness, µm	2.2	2.4	2.5	2.5-2.9	3.0	2.4	3.3	4.2	3.0	6.0-7.1
Coarseness, mg/100m	16-29	-	25	27-40	23	26-46	30-56	29-67	-	-
Specific gravity, o.d g/cm³	0.45	0.42	0.37	0.46	0.43	0.42	0.54	0.66	0.43	0.33-0.68
Lignin, %	27.6	29.4	29.4	28.3	27.7	25.6	28.6	26.8	28.9	26.2-31.2
Extractives (EtOH-Benz), %	2.2	2.0	2.5	4.0-4.2	3.5	4.4-5.0	3.2-5.4	3.4-6.0	5.4	4.2

The tracheids of pine have an average length of 3.6 mm and a diameter of 35–45 µm, which are comparable with those of spruces and balsam fir. The cell wall thickness of pine, on the other hand, is greater for pine than for spruce or fir tracheids. Pine fibers are therefore more rigid and require more chemical and/or mechanical treatment to collapse and confer adequate strength to the sheet[21]. Southern pines also have a high fiber coarseness and a high ratio of latewood fibers to earlywood fibers. Pines also have a high level of extractives soluble in organic solvents, which often leads to pitch problems in pulp and paper mills.

The differences in wood and fiber properties between Scandinavian Norway spruce and Scots pine are similar, but the difference in fiber coarseness is relatively small. One of the main reasons for not using pine for mechanical pulp is the high amount and character of extractives it contains.

Various poplar species are best suited hardwoods for mechanical pulping, other species such as *Gmelina arborea* have also been studied[24-27]. Poplar species with potential for mechanical/chemimechanical pulping include:

> European species: European aspen (*Populus tremula*)
>
> North American species: Trembling aspen (*Populus tremuloides*), Cottonwood (*Populus deltoides*), Balsam poplar (*Populus balsamifera*), Bigtooth aspen (*Populus grandidentata*)
>
> Hybrid poplars: P. tremula x tremula, P. tremula x tremuloides, P. deltoides x nigra, P. deltoides x trichocarpa, P. trichocarpa x deltoides, P. maximowiczii x trichocarpa

The fiber morphology of hardwood differs greatly from that of softwoods. Softwoods contain essentially only one type of fibers (tracheids), whereas hardwoods contain normal libriform fibers, large-diameter vessel elements, and a larger proportion of parenchyma cells than softwoods.

Hardwoods contain generally significantly lower amounts of lignin and correspondingly higher amounts of cellulose and hemicellulose than softwoods. Wood and fiber properties of selected poplar species are given in Table 6.

Table 6. Wood and fiber properties of selected poplar species[25].

Scientific name	Location	Fiber length mm	Coarseness mg/m	Lumen diameter μm	Double cell wall thickness μm	Vessel cells %	Lignin %
P. tremula x tremula	Finland	0.90	0.121	16.3	6.1	27.6	20.4
P. tremula x tremuloides	Finland	0.96	0.132	15.8	7.1	24.8	19.9
P. tremula x tremuloides	Finland	0.93	0.103	14.7	5.3	23.3	20.3
P. tremuloides	Ohio, USA	1.02	0.132	14.0	7.0	26.9	21.2
P. balsamifera	Ohio, USA	0.88	0.112	16.4	6.2	29.3	24.1
P. grandidentata	Ohio, USA	1.08	0.136	18.4	5.7	29.2	20.9
P. deltoides x nigra	Surray, UK	1.00	0.135	19.1	6.8	29.6	26.0
P. trichocarpa	Surray, UK	0.95	0.111	17.1	4.8	20.8	24.5
P. maximowiczii x trichocarpa	Eberswalde, Germany	0.99	0.138	16.0	6.1	26.0	21.5
P. maximowiczii x trichocarpa	Eberswalde, Germany	0.99	0.143	16.0	7.6	26.7	21.9

CHAPTER 5

The following information provides a brief comparison of pulp properties and energy consumption of the most commonly used species for TMP and groundwood processes.

2.2 TMP

As mentioned above, spruces are considered to be the most suitable wood species for thermomechanical pulping. Especially Norway spruce (*Picea abies*) grown in Northern and Central Europe has been used as a reference for the best quality wood species for TMP. North American spruces, especially Black spruce (*Picea mariana*) and White spruce (*Picea glauca*) are comparable with Norway spruce. Firs are considered to be relatively close to spruces; for example, Balsam fir is commonly used mixed with spruces in Canada. Silver fir (*Abies alba*) is also used mixed with spruce in France although certain pulp properties (tensile index, brightness) at a given freeness are slightly inferior to spruce TMP[59].

Table 7 gives the properties of various pine species compared with Black spruce/Balsam fir mixture as raw material for TMP[21,27].

Table 7. Thermomechanical pulping of various wood species (CSF 100 ml)[21,27].

Wood species	Spruce/Fir P.mariana/ A.balsamea	Jack pine Pinus banksiana	Southern pine Pinus taeda	Caribbean pine Pinus caribea	Aspen Populus tremuloides
Refining energy, MWh/t	1.92	2.20	2.14	2.56	
Bauer McNett					
+14, %	12.2	7.0	9.7	19.7	
−200, %	30.2	32.5	33.2	–	
Fiber length, mm	1.29	1.00	1.19	1.32	
Rejects, %	0.45	0.10	0.30	0.50	0.50
Density, kg/m^3	391	341	303	297	365
Tensile index, Nm/g	40.8	34.9	31.9	26,7	23
Tear index, mNm^2/g	9.4	8.5	9.2	10,2	3.1
Scattering coeff., m^2/kg	63.5	59.4	49.5	53,6	68.0
Brightness, %	58.0	47.8	53.3	54,1	58.0

Rejects: [a]Somerville, [b]Pulmac (0.15 mm)

The pines consume 10%–30% more refining energy to reach the same pulp freeness. The spruce/fir pulp has not only greater strength properties than the pine pulps but also a higher light-scattering coefficient and brightness. In general, pine pulps must have lower freeness to produce adequately developed fibers to generate a sheet with properties comparable to spruce/fir.

Regional shortages of spruce in Scandinavia have encouraged the use of Scots pine and Lodgepole pine for TMP. The differences of pine compared with spruce[20,28,29] can be summarized as follows:

- Higher energy consumption to given pulp freeness (10%–20%)
- Lower tensile index (10%–25%)
- Lower tear index (10%–20%)
- Higher light-scattering coefficient (about 5%).

Lodgepole pine consumed even more energy than Scots pine but produced pulp with strength properties close to spruce[20]. In practice, pitch problems caused by pine extractives limit the use of pine more than the inferior strength properties. Pitch deposits on wires, felts, rolls, and other process equipment cause web breaks and increase the need for equipment washing.

Radiata pine grown in New Zealand, Australia, and South America is an exception among pines.

It requires more refining energy than spruce to produce TMP of a given freeness. The pulps have, however, less shives and a higher long fiber content than spruce pulps produced under the same pulping conditions[23]. When compared at a given tensile index, the radiata pine pulps have a superior tear index and a lower light-scattering coefficient than spruce pulps (Fig. 4).

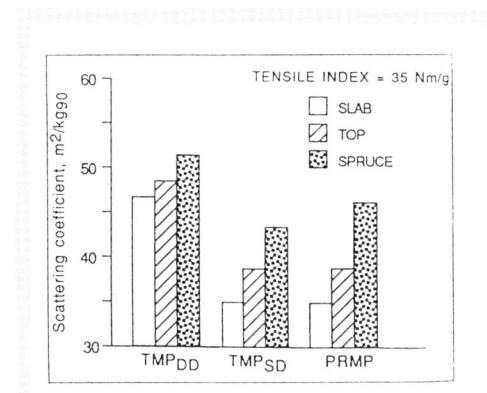

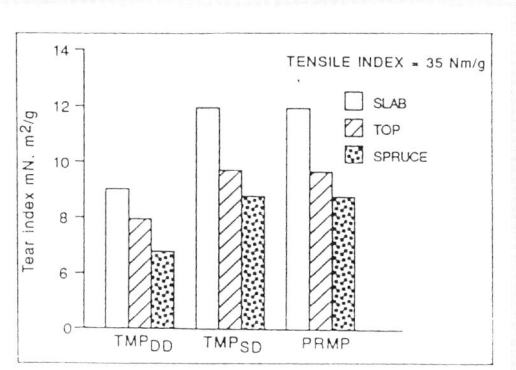

Figure 4. Tear index and scattering coefficient of PRMP and TMP pulps from spruce (*Picea abies*) and *Pinus radiata* (topwood/slabwood) at a given tensile index[23].

The disadvantages of pine species, especially the southern pines, for mechanical pulping such as large thick-walled fibers, high extractives content, and low brightness can be reduced by proper raw material selection. On the basis of natural variation in fiber properties caused by tree age and position in the stem, it is clear that choosing juvenile southern pine wood (for example, thinnings) will clearly reduce the problems caused by high fiber coarseness. Brightness also may be improved and pitch problems eased by selecting younger trees with lower proportion of colored, resinous heartwood.

Table 6 shows properties of TMP from trembling aspen. The pulp has a good light-scattering coefficient and brightness but has poor bonding properties, resulting in poorer strength properties than those of spruce GW. Thus, in order to achieve a suitable

CHAPTER 5

compromise between strength and optical properties, chemical pretreatment of chips is a prerequisite if furnish costs are to be maintained within reasonable limits[27]. Other hardwood species than those from the genus *Populus* usually are unsuitable as raw material for TMP (Table 8)[60]. From the species studied, white birch and maple clearly give lower quality TMP than aspen and poplar, and red oak and eucalypt were clearly the poorest species.

Table 8. Thermomechanical pulping of various hardwood species (CSF 90 ml)[60].

Wood species	Aspen	Poplar	Eucalypt	White birch	Red oak	Sugar maple
	Populus tremuloides	*Populus deltoides*	*Eucalyptus viminalis*	*Betula papyrifera*	*Quercus rubra*	*Acer saccharum*
Wood density, kg/m³	350	413	465	472	548	607
Refining energy, MWh/t	2.3	2.6	2.3	3.3	3.1	4.0
Density, kg/m³	435	457	321	404	286	390
Tensile index, Nm/g	25.5	27.8	12.6	21.2	10.2	20.0
Tear index, mNm²/g	2.1	2.3	1.4	2.5	1.7	1.6
Scattering coeff., m²/kg	77.0	87.9	57.6	67.0	48.3	66.5
Brightness, %	60	55	35	39	31	46

2.3 Groundwood processes

It is common knowledge that wood species of the Spruce family perform very favorably in grinding processes. This is especially true for Norway spruce (*Picea abies*). Also some fast growing woods of the *Pinus* family have turned out to be very applicable in the pressurized grinding process. On the other hand, slow growing pines like *Pinus sylvestris* are not considered equally suitable for grinding processes because of their big extractive content, low brightness yield, and stiff and bulky fibers. However, *Pinus sylvestris* is to some extent ground in the PGW process mixed with spruce in Middle-Europe.

The applicability of hardwood is mostly controlled by density, fiber cell wall thickness, and average fiber length, which are important factors when considering the ability to absorb energy and defiberize in grinding.

Generally, the proportional response of various wood species in pressurized grinding processes increases in the following order: pine > spruce > aspen. There is a substantial increase in strength of aspen pulps when comparing GW and PGW, but PGW-S does not give any further increase. Spruces respond equally favorably both in PGW and PGW-S. Pines have the best proportional response in pressurized grinding processes of all commercial wood species.

This chapter introduces pit pulp properties from various wood species that have economical significance in large scale.

Raw materials

Softwoods

Spruce

Different spruces are the most common wood species in grinding processes. They have a favorable density, fiber morphology, and good initial brightness. It is also typical that they have a good response in pressurized grinding process too.

Spruces have a lot of commercial significance as raw material for groundwood processes, especially in Europe and North-America.

Spruce PGW pulp exhibits high long fiber content, high tear, and tensile strength with good brightness and light-scattering ability. This combination makes them ideal pulp for value added papers. This is very true for PGW70, which combines good optical properties of GW and good strength and printability properties of PGW[30]. Spruce PGW-S pulps exhibit high strength, but lower brightness, which makes them most suitable for newsprint type paper grades. Table 9 shows typical qualities for Norway spruce (*Picea abies*)[30].

Tables 10 and 11 show examples of Black spruce (*Picea mariana*) and Balsam fir (*Abies balsamea*) pit pulp properties.

Table 9. Groundwood pit pulps from Norway spruce (*Picea abies*) at 80 ml CSF[30].

Process	GW	PGW70	PGW	PGW-S
Specific energy consumption, MWh/t	1.4	1.4	1.4	1.4
Bauer McNett +28 fraction,%	18	26	30	32
Bauer McNett −200 fraction,%	34	28	28	28
Density, kg/m^3	360	370	360	360
Tensile index, Nm/g	25	36	36	40
Tear index, mNm^2/g	4.1	5.8	6.1	7.1
Scattering coefficient, m^2/kg	63	62	61	57
Brightness (ISO), %	64	64	63	61

Table 10. Groundwood pit pulps from Black spruce (*Picea mariana*) at 80 ml CSF.

Process	GW	PGW	PGW-S
Specific energy consumption, MWh/t	1.3	1.3	1.3
Bauer McNett +28 fraction, %	14	26	34
Bauer McNett −200 fraction, %	38	33	30
Density, kg/m^3	370	360	360
Tensile index, Nm/g	31	38	44
Tear index, mNm^2/g	3.5	5.0	6.0
Scattering coefficient, m^2/kg	62	62	58
Brightness (ISO), %	61	61	59

Table 11. Groundwood pit pulps from Balsam fir (*Abies balsamea*) at 80 ml CSF.

Process	GW	PGW	PGW-S
Specific energy consumption, MWh/t	1.5	1.5	1.5
Bauer McNett +28 fraction, %	12	25	33
Bauer McNett −200 fraction, %	40	30	30
Density, kg/m^3	350	350	350
Tensile index, Nm/g	27	34	38
Tear index, mNm^2/g	3.0	4.5	5.5
Scattering coefficient, m^2/kg	69	68	67
Brightness (ISO), %	58	59	56

In general, all wood species of spruce perform basically alike in grinding processes. Balsam fir differs from spruces. The basic difference in the pit pulp properties of spruce and fir pulp is that fir has lower strength properties, but a clearly higher light-scattering coefficient. Fir is usually ground together with various other spruce species.

Pines

Slow growing pines like Scots pine (*Pinus sylvestris*) or Lodgepole pine (*Pinus contorta*) are not commonly used in grinding. Some attempts have tried to utilize this raw material. Table 12 shows an example of these furnishes[37].

In terms of pit pulp qualities, Scots pine performs very good in PGW and PGW-S processes. With moderate SEC, pulp qualities that are comparable to spruce can be achieved. On the other hand, these numbers do not reflect the problems that these furnishes possess. The pulps made from these pines exhibit a high content of rough, thick walled fibers that need to be handled in the pulping process. Even if this problem could be solved by intensive screening and a reject handling procedure, another even more limiting factor still exists. The high content extractives of pine trees generate pitch problems in the papermaking processes.

Table 13 provides another example of slowly growing pines. *Pinus contorta* performs even better than Scots pine in terms of pit quality. Its use in groundwood processes is limited for the same reasons found with *Pinus sylvestris*.

Table 12. PGW and TMP unscreened pulps from Scots pine (*Pinus sylvestris*) at 100 ml CSF[37].

Process	PGW	PGW-S	TMP
Specific energy consumption, MWh/t	1.0	1.0	2.0
Bauer McNett +28 fraction, %	19	27	31
Bauer McNett −200 fraction, %	31	29	29
Tensile index, Nm/g	20	24	24
Tear index, mNm^2/g	3.4	4.1	4.5
Scattering coefficient, m^2/kg	61	60	55
Brightness (ISO), %	66	64	63

Table 13. PGW pulps pit pulps from Lodgepole Pine (*Pinus contorta*) at 100 ml CSF[37].

Process	PGW	PGW-S
Specific energy consumption, MWh/t	1.1	1.1
Bauer McNett +28 fraction, %	26	35
Bauer McNett −200 fraction, %	32	30
Tensile index, Nm/g	23	27
Tear index, mNm^2/g	4.1	5.1
Scattering coefficient, m^2/kg	63	62
Brightness (ISO), %	65	62

Table 14. Groundwood pit pulps from Radiata Pine (*Pinus radiata*) at 80 ml CSF[32].

Process	GW	PGW
Specific energy consumption, MWh/t	1.4	1.4
Bauer McNett +28 fraction, %	14	25
Bauer McNett −200 fraction, %	34	28
Density, kg/m^3	390	390
Tensile index, Nm/g	39	43
Tear index, mNm^2/g	5.1	6.0
Scattering coefficient, m^2/kg	62	59
Brightness (ISO), %	60	60

Unlike slow growing North-European species, fast growing pines are widely used for groundwood production. The main regions where pines are used for groundwood are the southeastern region of the United States (*Pinus taeda*), Chile (*Pinus araucaria, Pinus radiata*), South Africa (*Pinus patula*), and New Zealand (*Pinus radiata*).

Radiata pine *(Pinus radiata)*

Radiata pine performs in many ways like spruce in grinding. They both have similar energy requirement but, due to their fiber characteristics, Radiata pine pulps show better strength and lower light-scattering characteristics. An example is given in Table 14.

Southern Pine *(Loblolly pine, Pinus taeda)*

Southern pine (*Pinus taeda*) typically has fibers that are bulky and have a thick cell wall. In atmospheric grinding, it yields low strength and poor printing characteristics. These handicaps can be compensated by grinding down to lower freeness than spruce[34]. Southern pine also responds very favorably in the PGW-S process, yielding a high long fiber content and tear index. Table 15 shows typical pit pulp qualities.

Table 15. Groundwood pit pulps from Southern Pine (Loblolly pine, *Pinus taeda*) at 100 ml CSF.

Process	GW	PGW	PGW-S
Specific energy consumption, MWh/t	1.5	1.5	1.5
Bauer McNett +28 fraction, %	9	21	31
Bauer McNett −200 fraction, %	40	32	29
Density, kg/m^3	370	370	350
Tensile index, Nm/g	17	22	29
Tear index, mNm^2/g	2.5	4.5	5.9
Scattering coefficient, m^2/kg	60	63	58
Brightness (ISO), %	55	58	54

Hardwoods

Poplars

The woods of the *Populus* family have quite a substantial economical significance in North America as a raw material for groundwood-based value added papers. Typical properties for American aspen (*Populus tremuloides*), cottonwood (*Populus deltoides*), and European aspen (*Populus tremula*) both in GW and PGW processes are listed in Tables 16, 17, and 18, respectively.

Common characteristics for the pulps from various wood species of the *Populus* family are low long fiber yield, high bulk, and relatively low strength values. On the other hand, these pulps show very good optical properties. This quality profile makes them very suitable for coated paper grades, in which they promote good printability, stiffness, smoothness and very good paper formation, and opacity[31]. Moderate strength values can be promoted by alkaline peroxide treatment. Peroxide bleaching not only improves brightness and strength values, but also improves important surface characteristics like linting and fiber roughening[35].

In North America, various aspen species are commonly used for groundwood production. The main reason for this is that, in many areas, aspen is the only available wood resource. In Europe, aspen is still considered a secondary wood resource. It is used in large scale for groundwood pulping only in Italy and Finland. Recent implementation has shown, however, that this is about to change. It has been found that with proper treatment aspen PGW makes an excellent raw material in coated wood-free papers[36].

Table 16. Groundwood pit pulps from American aspen (*Populus tremuloides*) at 200 ml CSF[31].

Process	GW	PGW
Specific energy consumption, MWh/t	1.2	1.2
Bauer McNett +28 fraction, %	13.3	14.6
Bauer McNett –200 fraction, %	32.0	27.6
Density, kg/m^3	356	368
Tensile index, Nm/g	16.0	19.6
Tear index, mNm^2/g	2.7	3.1
Scattering coefficient, m^2/kg	73.5	72.1
Brightness (ISO), %	63.0	61.8

Table 17. Groundwood pit pulps from cottonwood (*Populus deltoides*) at 100 ml CSF[31].

Process	GW	PGW
Specific energy consumption, MWh/t	1.4	1.5
Bauer McNett +28 fraction, %	10.7	12.7
Bauer McNett –200 fraction, %	37.0	32.5
Density, kg/m^3	340	340
Tensile index, Nm/g	11.5	19.0
Tear index, mNm^2/g	2.3	3.2
Scattering coefficient, m^2/kg	68.0	69.0
Brightness (ISO), %	63.0	61.0

Table 18. Groundwood pit pulps from European aspen (*Populus tremula*) at 200 ml CSF[31].

Process	GW	PGW
Specific energy consumption, MWh/t	1.2	1.5
Bauer McNett +28 fraction, %	10.5	15.3
Bauer McNett –200 fraction, %	33.1	24.7
Density, kg/m^3	363	368
Tensile index, Nm/g	14.8	18.5
Tear index, mNm^2/g	2.4	2.9
Scattering coefficient, m^2/kg	71.7	70.8
Brightness (ISO), %	69.0	68.4

3 Within-tree property variation

Because fibers form the basic structure of various paper grades, fiber properties in the original wood raw material can be expected to have a substantial effect on sheet properties. Fiber properties were already seen to vary greatly between different wood species, resulting in variation of the energy consumption in mechanical defiberizing and pulp properties. Fiber properties vary not only between wood species but, also within and among trees of the same species. The following text provides a brief description of the variation of fiber properties within wood species as well as the significance of fiber property variation for mechanical pulp and paper properties. Norway spruce (*Picea abies*) grown in Scandinavia is mostly used as the reference species.

3.1 Fiber length

Within-species variation in fiber properties exist between trees, within the trunk of a tree, and within a selected growth ring. Fiber dimensions vary within a tree trunk both radially (from pith to cambium) and vertically (from butt to top). The variation pattern is similar in different tree trunks. Fiber (tracheid) length has the greatest variation among fiber properties.

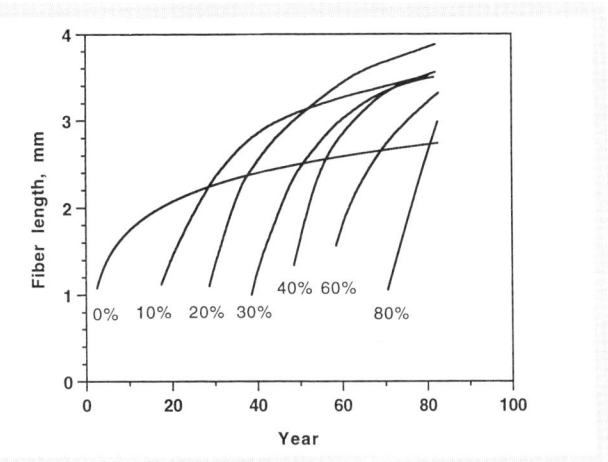

Figure 5. Average fiber length in the stem of a 96-year old Norway spruce stem at various heights of the stem. The fiber lengths are plotted against the age of the tree. Redrawn from Ref. 38.

Variation in *tracheid length* follows the variation in the length of cambial cells, which in turn depends on cambium age. Thus the fiber length increases from pith to bark at all heights of the stem[38–42]. Fig. 5 illustrates this for a 96-year old Norway spruce stem[38]. The average fiber length for the stem was 2.88 mm, the minimum value was 0.93 mm, and the maximum value 3.88 mm respectively.

The fiber length increases rapidly near the pith during the juvenile period, but slows down gradually and almost evens out when the cambium gets very old. The increment is slowest at the stump level. The longest fibers in this tree can be found at the outer parts of the stem at a stem height of 10%–40%. Table 19 summarizes the average fiber lengths in various radial sections of Norway spruce at breast height[42].

The fiber length varies in the vertical direction of the stem within a selected growth ring. The fiber length increases from the butt to the middle part of the stem and then decreases again toward the top of the stem (Fig. 6)[38]. For many conifers, maximum average tracheid length occurs at a height level of 30%–40% of the stem[43].

CHAPTER 5

Table 19. Fiber length in various radial locations in the stem. Average values and ranges of variation for 30 Norway spruce trees from two stands. Measured at breast height[42].

Radial location in the stem	Average fiber length, mm	Range of fiber length variation, mm
Inner heartwood	1.9	1.28–2.70
Middle part	3.0	1.69–3.88
Outer sapwood	3.7	2.80–4.29

In addition to the fiber length variation in radial and vertical directions within a stem, variation occurs also within a selected growth ring. Fiber length increases from earlywood to latewood and reaches its maximum value at the end of the growth season. In conifers, the variation in fiber length within a growth ring is on an average 12%–25%[40].

Wood handling at the mill has a significant effect on fiber length in the wood raw material for TMP and CTMP. The original fiber length in the pulpwood arriving at the mill is reduced in chipping to an extent determined by the target chip length and the formation of fines.

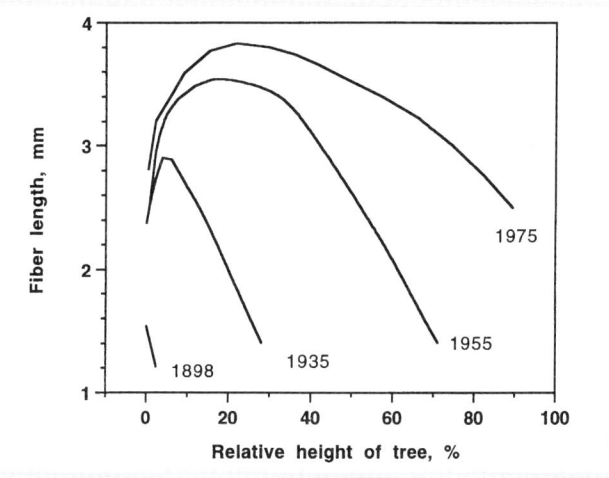

Figure 6. Variation of fiber length in selected growth rings of a 96-year old Norway spruce stem as a function of relative height of the tree. The growth rings are named after the year they were formed. Redrawn from Ref 38.

3.2 Fiber width

The average *tracheid width* shows the same kind of variation pattern within the stem as tracheid length. The fiber width is smallest near the pith increasing rapidly at the juvenile period, but evens out more rapidly than fiber length and may even start to decrease when the cambium gets very old (Fig. 7)[38]. The minimum value for fiber width

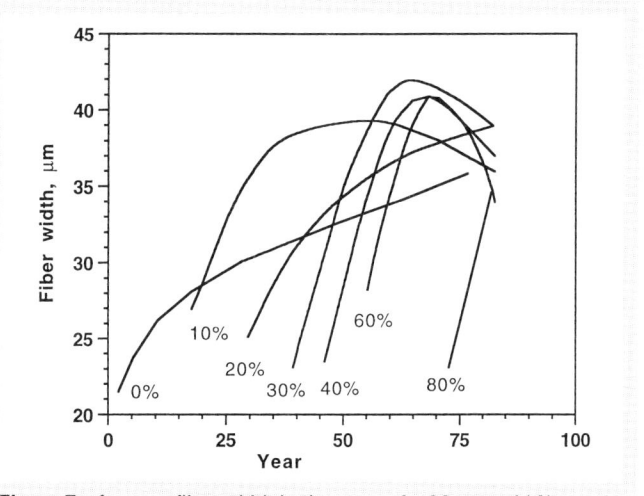

Figure 7. Average fiber width in the stem of a 96-year old Norway spruce stem at various heights of the stem. The fiber widths are plotted against the age of the tree. Redrawn from Ref. 38.

for a single spruce stem studied was 21 μm and the maximum value was 42 μm. The widest fibers in this tree were found at the outer parts of the stem at a stem height of 30%–60%[36].

The tracheid width measured in tangential direction is about the same for earlywood and latewood tracheids of conifers and it does not change much along the cambium age, whereas that measured in the radial direction is substantially larger for earlywood than for latewood. The increase in the average tracheid width from pith to cambium as shown in Fig. 6 is probably due to the increase in the radial width of earlywood since the radial diameter of latewood fibers does not show marked changes (Fig. 8)[44].

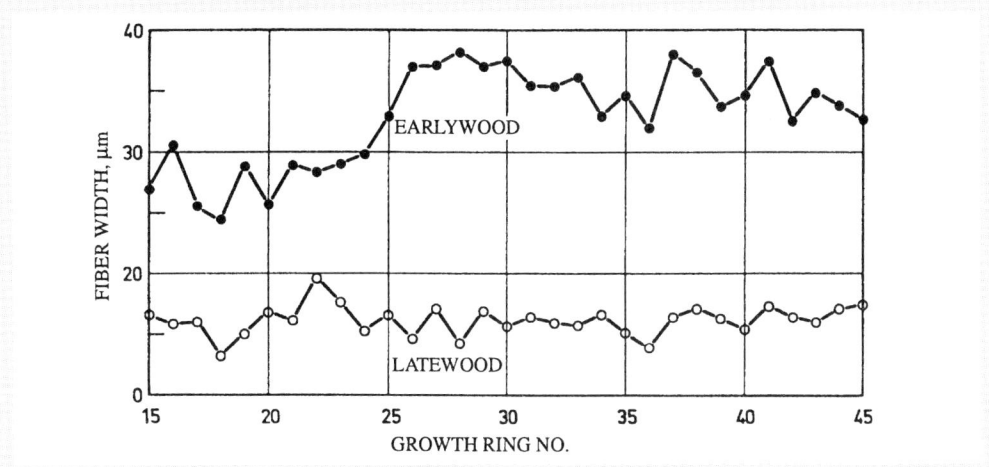

Figure 8. Variation of earlywood and latewood tracheid width in radial direction at the stump level of a pine tree (*Pinus sylvestris*)[44].

The fiber width varies in the vertical direction of the stem within a selected growth ring. The fiber width increases from the butt to the middle part of the stem and then decreases again toward the top of the stem. Table 20 illustrates this for Lodgepole pine grown in Finland[41].

Table 20. Variation of tracheid diameter in the earlywood and latewood of Lodgepole pine (*Pinus contorta*) grown in Finland. Measurements were made at the third growth ring from the pith at various height levels of the stem. Average values of eight stems from two stands[41].

Height in stem m	Earlywood Radial	Earlywood Tangential	Latewood Radial	Latewood Tangential
1.3	27.3	30.6	22.0	30.3
4	29.1	31.9	22.2	29.4
8	29.3	31.2	20.9	28.0
12	27.8	29.8	22.2	30,0
16	26.6	30.2	21.6	29.2

CHAPTER 5

The earlywood tracheid diameter is at its largest in the middle part of the stem, whereas the pattern for latewood diameter variation seems to be more vague.

3.3 Cell wall thickness

The variation in *cell wall thickness* has been investigated less than that of fiber length and width.

The average cell wall thickness has been found to increase from pith to cambium for many conifers[40,41,44]. The cell wall thickness depends greatly on the annual growth conditions, and the increase from pith to cambium is more pronounced in latewood than in earlywood (Figs. 9 and 10). Controversial results concerning the variation in vertical direction of the stem exist, but the average cell wall thickness can be expected to decrease toward the top of the stem due to the radial variation pattern of the cell wall thickness.

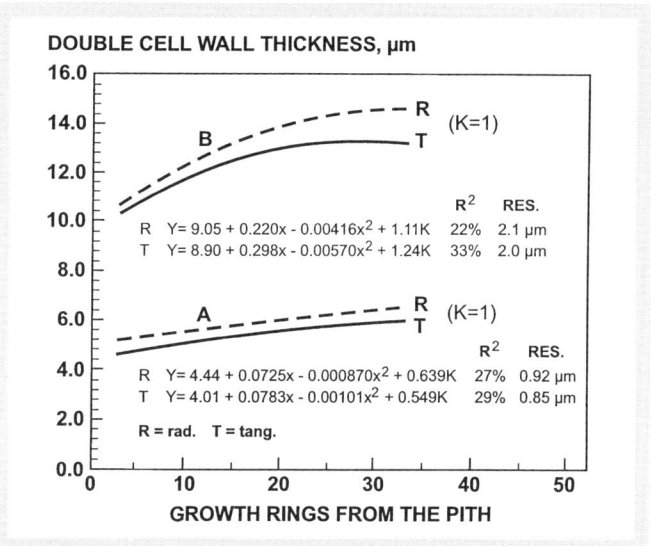

Figure 9. Variation in earlywood and latewood cell wall thickness from pith to the cambium. Lodgepole pine (*Pinus contorta*) grown in Finland. Average values from eight trees from two stands[41].

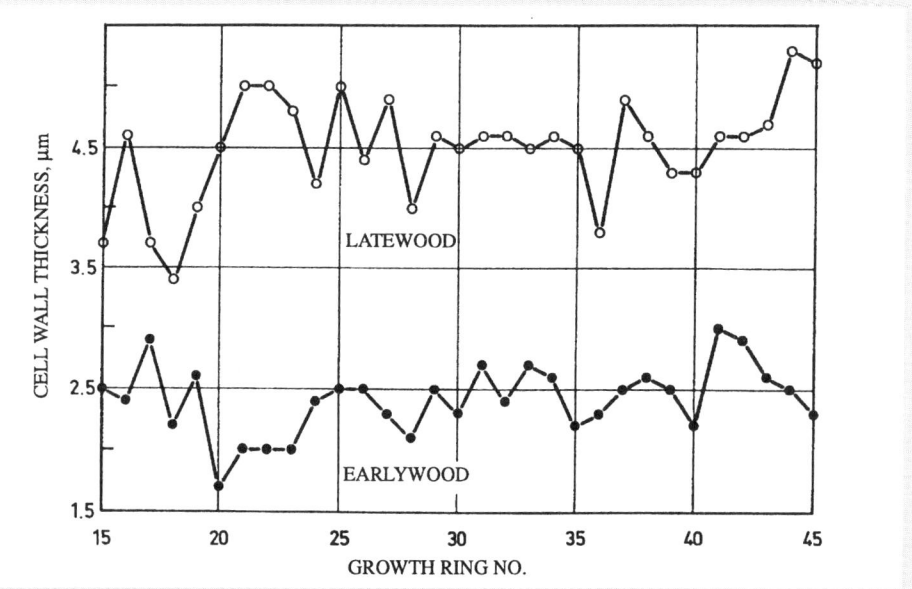

Figure 10. Variation in the cell wall thickness in the radial direction of earlywood and latewood tracheids at the stump level of a pine tree (*Pinus sylvestris*)[44].

Since both the tracheid diameter and cell wall thickness of earlywood and latewood differ greatly from each other within the same tree, the average variations in these dimensions are usually well correlated with the earlywood-latewood ratio. As shown above, the variation in tracheid diameter is associated especially with earlywood, whereas the variation in cell wall thickness is more associated with latewood.

3.4 Fibril angle

The average *fibril angle* in the S2-layer of softwood tracheids varies as follows[45]:

- It decreases from pith to the cambium at all heights of the stem. The angle is highest near the pith where the fiber length is smallest. It decreases rapidly in radial direction toward the cambium and thereafter remains approximately constant.
- It is smaller in the latewood than in the earlywood of a selected growth ring.
- It increases from butt to top inside a selected growth ring.
- It is smaller in narrow growth rings than in wide growth rings.

The variation of fibril angle is negatively correlated with fiber length and positively correlated with fiber width within a tree trunk. The variation of fibril angle along with the width of a growth ring and between earlywood and latewood visualize the effect of growth rate on the fibril angle; the fibril angle of a slowly grown fiber is smaller than that of a fast grown fiber[45].

4 Property variation between trees

In addition to within-tree variation, within-species variation includes geographical variation, stand-to-stand variation, and variation among individual trees[46].

Geographical variation results from the action of natural selection, favoring individuals that are well adapted to the local environmental conditions (climate, soil). Geographical variation, as for all phenotypic variation, has two components: genetic and environmental.

Considerable *stand-to-stand differences* can occur within one geographical region. Most of these are due to differing site conditions, although genetic differences also exist.

The most remarkable variability is represented by the differences *among individual trees within a stand*. Genetic causes have varying importance in causing the differences. For example, tree-to-tree competition can greatly exaggerate small initial differences resulting from genetics or from microsite variation. The combined effects of environment and competition can often mask genetic differences[46].

Tree age also has a very significant effect on the variation of the average properties between individual trees.

The amount and distribution of genetic and environmental variation among wood properties differs greatly from species to species. It is determined by the species range, the species distribution (the species can be either continuous or broken up into smaller populations) or by the evolutionary history of the species. Some species such as Red

CHAPTER 5

pine (*Pinus resinosa*) are extremely uniform, while other species such as Loblolly pine, Ponderosa pine, and Douglas-fir are highly variable[46].

Genetic control is measured by analyzing and subdividing the total variation into its components. The total phenotypic variance (V_P) can be divided into two main components: the genetic variance (V_G) and the environmental variation (V_E).

The genetic variance, in turn, can be separated into additive genetic variance (V_A) and dominance variance (V_D). Thus:

$$V_P = V_A + V_D + V_E \tag{1}$$

The degree of genetic control is usually expressed by the term heritability, which is the ratio between the additive genetic variance and the total phenotypic variance (narrow sense heritability, $h^2 = V_A/V_P$) or the ratio between the genetic variance divided by the phenotypic variance [broad sense heritability or gross heritability, $H^2 = (V_A + V_D)/V_P$].

Genetic control of wood properties

Fiber length differences among stands can be considerable and are nearly always large among trees within stands (Fig. 11).

Tracheid length of softwood species is strongly inherited. For hardwoods, there is a moderate genetic control of tracheid length. In conifers, the evidence for inheritance of tracheid length is more complete, with narrow sense heritabilities ranging from 0.28 to 0.97 and broad sense heritabilities ranging from 0.56 to 0.86[46].

There is definite evidence for genetic control of cell wall thickness, radial tracheid diameter, and radial and tangential cell width. This control does not seem to be as strong as the control of tracheid length. Particularly in conifers, indirect evidence from the inheritance of wood basic density and latewood content strongly supports the strong inheritance of fiber dimensions.

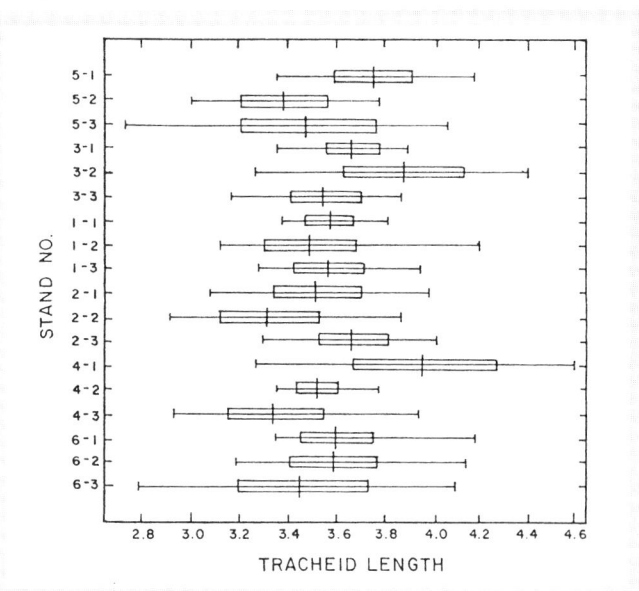

Figure 11. Variation of softwood fiber length from stand-to-stand and between trees within a stand[46].

For wood chemical composition, it seems that lignin content is under strong genetic control, but the range of variation is small. Cellulose content also has a strong genetic component, but its inheritance pattern is such that selection is ineffective. However, vegetative propagation can utilize the differences among clones, and meaningful genetic gains have been obtained in the eucalypts.

Most wood properties are moderately to strongly inherited and can be modified readily by breeding. The breeding objectives for wood quality have to be defined in terms of the proposed end use. The wood properties required for different end uses (structural lumber, energy use, pulp and paper manufacture) differ from each other. In the temperate zones, the lag between the initiation of a breeding program and the harvest of the resulting mature tree can be 50–75 years. The product requirements will probably change considerably during such a long period[46].

5 Different wood assortments

The natural wood property variation between and within various wood species has a substantial effect on mechanical pulping and pulp properties. The quality variation patterns and practical limitations in real life prevent the sorting of wood into fully homogenous fractions ideal for the manufacturing of pulps with even quality and desired properties for the end use. In the following, a selected review of the property variation of mechanical pulps attainable through the selection and sorting of the wood within a single species is given.

Table 21 provides properties of TMP from logs cut from various positions of the stem[47].

Table 21. Properties of thermomechanical pulps from logs taken in various positions of the stem. *Pinus radiata* trees with small branches, values interpolated to CSF 100 ml[47].

Log position	Butt		Second	Top	
Wood density	low	high	high	low	high
SEC, MWh/bdt	2.15	2.30	2.05	2.20	2.20
Bauer McNett, +30 mesh, %	37.0	40.5	35.0	28.0	28.5
Bauer McNett, −200 mesh, %	33.5	31.5	28.8	30.0	33.0
Tensile index, Nm/g	42.0	42.0	40.5	35.0	36.0
Tear index, mNm^2/g	10.0	11.5	9.2	8.0	7.9
Sheet density, kg/m^3	380	370	370	355	360
Air resistance, s/100 ml	67	60	42	51	43
Scattering coeff., m^2/kg	56.3	55.8	59.9	63.9	65.3
Brightness (ISO), %	57.5	59.5	59.0	59.8	60.5

Butt logs produced TMP with the highest and the top logs with the lowest long fiber content. The fines contents showed much less variation. Regardless to whether the comparison is made as a function of freeness, sheet density, or energy consump-

CHAPTER 5

tion, the tensile index of the butt log pulps is greater than that of the second and top log pulps. Although the differences are even more pronounced, the same result is obtained in the comparison of tear index. The order of the different samples is reversed when the comparison is made with pulp optical properties (Fig. 12).

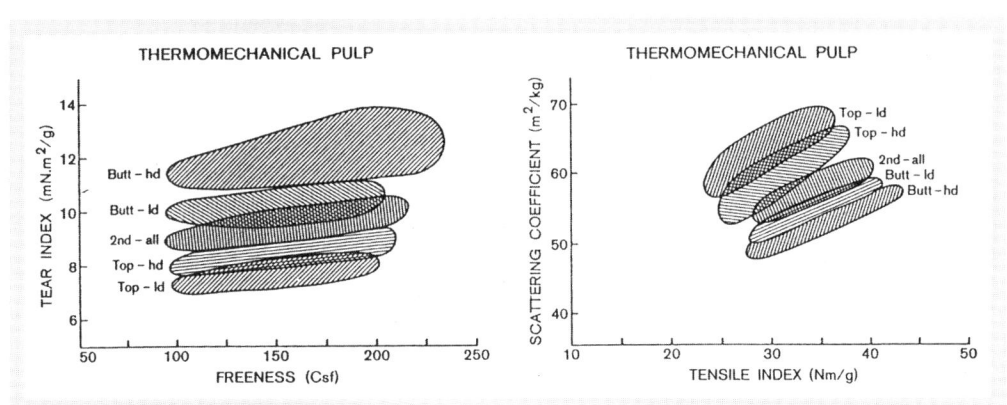

Figure 12. Properties of thermomechanical pulps produced of logs from different locations in the stem. (Pinus radiata)[47].

Log position has a more pronounced effect on the inverse relationship between the strength and optical properties of TMP than on wood basic density. Log position as discussed here is an indication of wood age. As both wood density and fiber properties are positively correlated with wood age, fiber properties must have a major influence on pulp properties.

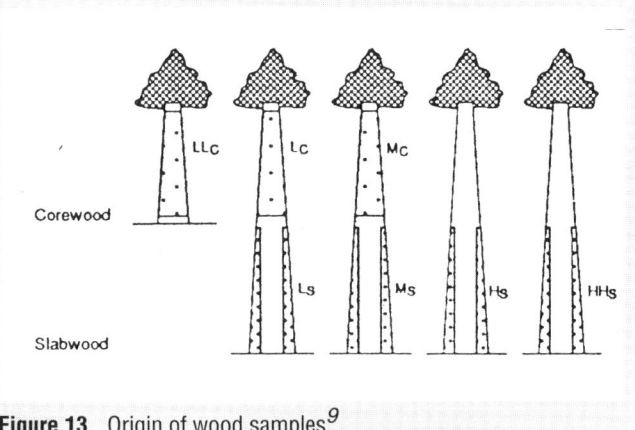

Figure 13. Origin of wood samples[9].

Further examination of the influences of wood basic density and fiber length on TMP quality included manufacturing thermomechanical pulps from seven *Pinus radiata* samples with varying ages and basic densities (Fig. 13, Table 22[9]).

Table 22. Effect of wood characteristics on TMP quality, pulp properties at CSF 100 ml, calculated from regression equations[9].

Sample code Wood type	LLC Core-wood	LC Core-wood	MC Core-wood	LS Slab-wood	MS Slab-wood	HS Slab-wood	HHS Slab-wood
Age, a*	11	11	13	26	27	44	46
Original fiber length, mm	2.2	2.9	2.5	3.3	3.0	3.6	3.5
Basic density, kg/m^3	345	375	425	390	465	515	565
Pulp properties							
Fiber length, mm	1.8	1.7	1.8	1.9	2.1	2.3	2.4
Bauer McNett, +30 mesh, %	23.3	25.4	29.6	34.5	35.9	39.8	39.5
Bauer McNett, +100/200 mesh, %	7.2	9.4	7.1	8.0	5.9	6.1	6.0
Bauer McNett, −200 mesh, %	33.6	33.6	31.2	31.3	30.9	31.1	30.1
Tensile index, Nm/g	34.3	32.6	34.6	39.8	38.2	37.8	37.3
Tear index, mNm^2/g	8.9	7.5	8.5	9.5	10.8	11.2	11.6
Sheet density, kg/m^3	368	352	351	373	358	356	348
Air resistance, s/100 ml	42	38	32	65	41	50	51
Scattering coeff., m^2/kg	52.9	56.7	50.5	51.7	49.5	48.2	49.8

*measured at base of the sample logs 6 m in length

The results showed that:

- There were no direct trends between either wood basic density or fiber length and energy consumption

- A high original fiber length generally resulted in a high pulp long fiber content and tear index, possibly also in high tensile index and low sheet density.

- The nature of wood and TMP property relationship is complex; there probably exist compensating mechanisms which balance out the relative contributions of thin and thick-walled fibers and their coarse fines, in the development of these properties.

Earlier, the properties of juvenile wood showed a deviation from those of mature wood. Table 23 shows the effect of the differences for refiner mechanical pulps of Canadian wood species[48]. The juvenile and mature wood samples were separated from logs taken from the butt section of the sample trees. As shown in the table, the juvenile wood required about 8% more energy than mature wood to refine pulps to 100 ml CSF. Furthermore, pulps from juvenile wood pulps had:

- A lower fiber length and long fiber content and thus lower strength properties (tear, tensile) than mature wood pulps

- A higher fines content and scattering coefficient than mature wood pulps.

CHAPTER 5

Table 23. Refiner mechanical pulping of juvenile and mature wood of managed second-growth Douglas-fir, Jack pine, and Lodgepole pine[48].

Wood species	Douglas-fir		Jack pine		Lodgepole pine	
Wood type	Juvenile	Mature	Juvenile	Mature	Juvenile	Mature
SEC, MJ/kg	11.8	11.0	12.4	11.5	13.4	12.1
Fiber length, mm	1.95	2.16	1.76	2.02	1.83	2.28
Bauer McNett, +48 mesh, %	53.7	57.0	45.5	50.9	52.5	58.3
Bauer McNett, −200 mesh, %	27.9	26.3	31.2	28.5	27.3	24.1
Tensile index, Nm/g	33	36	27	29	35	40
Tear index, mNm^2/g	7.2	8.4	5.9	6.8	6.0	7.1
Scattering coeff., m^2/kg	55.1	49.1	59.8	54.2	65.1	58.9

The sheet properties of juvenile and mature wood thus had the same kind of difference as thermomechanical pulps manufactured from the top and butt section of the stem. In this case, the origin of the difference probably also lies in the fiber property variation between the samples.

The properties of TMP made from different wood assortments of Norway spruce show analogous variation as observed for juvenile/mature wood and top/butt logs (Table 24)[49].

Table 24. Properties of TMP from different wood assortments. Norway spruce grown in Norway, specific energy consumption 2000 kWh/bdt[49].

Wood assortment	Thinnings	Top logs	Butt/middle logs	Sawmill chips (slabwood)
CSF, ml	124	117	128	102
Bauer McNett, +30 mesh, %	32.0	36.2	40.1	n.a.
Fiber length, mm	1.51	1.74	1.92	1.88
PFI mini shives, %	0.68	0.34	0.31	0.08
Tensile index, %	38.7	39.2	39.4	39.2
Stretch, %	2.2	2.2	2.3	2.4
Tear index, mNm^2/g	6.4	7.0	7.6	7.5
Density, kg/m^3	427	413	432	435
Air permeability, Gurley, μm/Pas	3.5	1.7	2.2	2.5
Scattering coeff., m^2/kg	57.9	55.3	54.0	53.4
Brightness (ISO), %	65.5	64.0	63.4	64.9

Pulp properties affected by wood assortment are fiber length, Bauer McNett fractions, shives content, tear index, air permeability, brightness, and scattering coefficient. Shives content, tear index, and scattering coefficient show the most striking variation. Thinnings and slabwood (sawmill chips) represent the extremes, with thinnings giving the highest shives content, lowest tear index, and highest scattering coefficient, while the slabwood gives the opposite values.

Observations reveal modest differences in the energy consumption. On the other hand, young trees of *Pinus radiata* (thinnings) require more refining energy than mature wood. On an average, thinnings require about 10% more refining energy than the wood of mature trees, taken from either the top or the outer part of the base as slabwood[50,51]. The thinnings also give TMP with higher sheet density, tensile index, tear index, and light scattering coefficient than top logs.

In addition to laboratory or pilot scale experiments, most of the raw material effects described above have been verified in mill scale trials with Norway spruce (Table 25)[52].

Table 25. Properties of second stage and newsgrade end-use thermomechanical pulps manufactured from different spruce wood (*Picea abies*) assortments[52].

Wood assortment	First thinnings		Regeneration cuttings		Sawmill chips	
Properties in the wood raw material						
Fiber length, mm	1.91		2.39		3.02	
Fiber diameter, µm	22.7		31.8		39.8	
Cell wall thickness, µm	1.8		3.3		3.8	
Pulp properties	2nd stage	End-use	2nd stage	End-use	2nd stage	End-use
CSF, ml	120	88	120	74	120	80
SEC, kWh/metric ton	1765	1967	1772	2036	1780	2007
Bauer McNett, +28 mesh, %	31.0	29.5	42.1	37.6	44.1	40.1
Bauer McNett, -200 mesh, %	28.7	30.0	23.6	27.0	23.5	26.8
Fiber length, mm	1.46	1.32	1.72	1.63	1.81	1.72
Fiber diameter (+14 fraction), µm	–	24.3	–	29.4	-	34.1
Cell wall thickness (14 fraction), µm	–	2.0	–	2.6	-	3.1
Tensile index, %	41.9	40.8	43.8	47.2	44.1	45.9
Stretch, %	3.3	3.4	3.4	3.8	3.0	3.1
Tear index, mNm^2/g	7.60	7.65	8.54	8.39	9.03	8.60
Sheet density, kg/m^3	392	405	386	412	391	414
Air resistance, Bendtsen, ml/min	315	170	460	130	420	120
Scattering coeff., m^2/kg	60.7	61.4	54.9	56.4	53.2	54.1
Brightness (ISO), %	62.9	61.7	58.8	59.6	57.8	58.2

Here again, sawmill chips give TMP with the highest long fiber fraction, smallest proportion of fines, highest tear index, and lowest scattering coefficient. TMP from thinnings clearly have the best optical properties, whereas the strength properties are the poorest. TMP from old-growth forest (regeneration cuttings) fall in between the other two pulps in most properties. No differences are found in pulp specific energy consumption.

CHAPTER 5

All the results on the effects of different wood assortments are not totally consistent. Differences arise because the magnitude of wood property variation varies from species to species and variation can also exist in the origin of the samples between the studies. Furthermore, the combined effects of wood raw material and process conditions (for example, refining intensity) have not been adequately studied.

The wood property variation is a major source of pulp quality variation, and there is always variation in the wood and fiber properties. To improve pulp homogenity, the following measures can be taken:

- Different wood species are not mixed.
- Pulpwood from different stand origins are processed separately.
- The different wood assortments are not mixed but processed separately.
- If the stand origin is unknown, pulpwood is sorted by age, diameter, or growth rate.
- Only fresh wood is used.
- Proper operation of the wood handling is instilled at the mill.

Most of the measures above are obvious on the basis of results presented. The stand origin affects Norway spruce quality for mechanical pulping[49,53]. For example, wood grown on an agricultural stand had shorter and wider fibers with thinner cell walls than wood grown on a forest stand. Thus TMP from the agricultural stand had inferior strength properties but a higher light-scattering ability and a higher specific energy consumption to reach a certain tensile index than that from the forest stand[53].

The use of fiber dimensions for sorting criteria would certainly be useful. Due to measuring difficulty, it is not practical. The relationship between growth ring pattern and fiber dimension, however, is known; growth ring measurements, thus, can be used as log sorting parameters for Norway spruce leading to more stable quality at the end of the production line[54]. At Hallsta Paper Mill in Sweden, pulpwood is classified in three categories for TMP production[55]. The classification is based on growth ring width and the classes are:

I Wood resulting in high brightness (large growth rings and much earlywood)

II Intermediate class

III Wood resulting in good strength properties (narrow growth rings).

The different spruce wood classes are used for the manufacture of tailored pulps for improved and superimproved newsprint, SC magazine, and office paper.

The following sections review the importance of using fresh wood and the effect of wood handling.

6 Wood procurement

The first decisions concerning the quality of wood are made at the forest, when the wood is divided into assortments or quality classes. To operate in an optimal way, wood procurement should:

- Provide the customers (various mills with different processes) with the best possible raw material in sufficient quantities for continuous mill operation
- Deliver the wood mechanically undamaged and free from impurities
- Deliver the wood fresh, i.e., as fast as possible
- Not lose the knowledge of wood origin during transportation
- Operate cost effectively.

The time between felling the tree and processing it cannot always be minimized in real life, but the logs must be stored for some time. The wood quality during transportation and storage suffers impairment through mechanical damage (drying, introduction of impurities such as sand and soot), biological effects (fungi, bacteria, insects), and chemical effects (tannin damage, extractives reactions).

The negative effects encountered in mechanical pulping are[56]:

- Loss of wood substance
- Loss of moisture content
- Increased fiber losses in debarking and chipping through drying and decay
- Increase of impurities in the logs or chips
- Increased costs for wood handling
- Reduction in pulp brightness
- Reduction in pulp strength.

The loss in pulp brightness is the most serious adverse consequence of prolonged time between felling and processing. The loss in the brightness of thermomechanical pulp and PGW pulp from Finnish sprucewood is nearly independent of the type of transportation and storage when the wood felled during winter is used before midsummer. The brightness of TMP made in August from floated and irrigated pulpwood is 2 to 3 and the brightness of PGW pulps 2 to 4 units lower than those made from fresh pulpwood during the same period (Fig. 14)[57].

The study made at pilot scale revealed no significant changes in the strength properties of the pulps. The linting and printing characteristics of newsprint made from the thermomechanical pulps studied were not affected by the type of storage or transportation.

CHAPTER 5

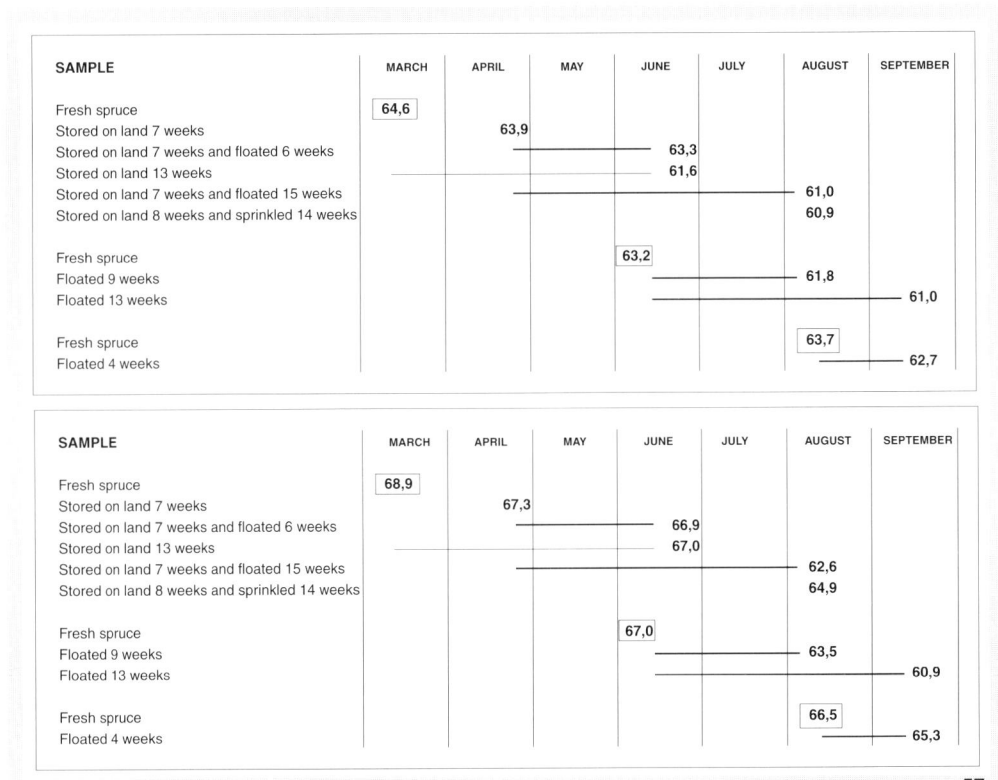

SAMPLE	MARCH	APRIL	MAY	JUNE	JULY	AUGUST	SEPTEMBER
Fresh spruce	64,6						
Stored on land 7 weeks		63,9					
Stored on land 7 weeks and floated 6 weeks				63,3			
Stored on land 13 weeks				61,6			
Stored on land 7 weeks and floated 15 weeks						61,0	
Stored on land 8 weeks and sprinkled 14 weeks						60,9	
Fresh spruce				63,2			
Floated 9 weeks						61,8	
Floated 13 weeks							61,0
Fresh spruce						63,7	
Floated 4 weeks							62,7

SAMPLE	MARCH	APRIL	MAY	JUNE	JULY	AUGUST	SEPTEMBER
Fresh spruce	68,9						
Stored on land 7 weeks		67,3					
Stored on land 7 weeks and floated 6 weeks				66,9			
Stored on land 13 weeks				67,0			
Stored on land 7 weeks and floated 15 weeks						62,6	
Stored on land 8 weeks and sprinkled 14 weeks						64,9	
Fresh spruce				67,0			
Floated 9 weeks						63,5	
Floated 13 weeks							60,9
Fresh spruce						66,5	
Floated 4 weeks							65,3

Figure 14. The effect of transportation and storage of spruce wood on the brightness of TMP and PGW pulp[57].

7 Wood handling

In northern districts, the modern roundwood handling consists of deicing, drum debarking, disc chipping, effective chip screening, and chip storage. Bark processing equipment and a separate effluent treatment plant is included as well. Purchased chips are usually stored separately from roundwood chips. These operations are reviewed in detail in Book 6 of this textbook series.

Mechanical pulping demands the following special requirements on the wood raw material handling at the mill:

- There must be minimal bark content in the chips or groundwood logs, which are best obtained with fresh roundwood.
- Efficient deicing must be provided during the wintertime.
- At the same time, the formation of breakage in the drum must be minimized to avoid wood losses and reduced chip quality.
- Groundwood processes require straight logs which have to be cut into fixed length; an even diameter distribution is advantageous for the filling of the grinder pocket.
- Chip quality must match the demands set by the refining process.

Raw materials

The uniformity of chip size distribution, bulk density, and wood source are probably the most important factors determining the chip quality for mechanical pulping[58]. Chip properties which influence their effective bulk density also affect the applied specific energy. Such properties include wood basic density and chip dimensions. The effects of chip size distribution independent of wood quality are not yet totally established.

The average chip length for TMP chips (about 22 mm) is smaller than that of chips for kraft pulping. The shorter the chip is along the grain, the more fibers have been cut. Thus, short chips should theoretically result in low strength properties, especially the tear index. TMP properties were only to a small extent dependent on the chip length within the length range of 16 to 23 mm[11]. Only the McNett long fiber fraction (+30 mesh) and the tear index suffered some impairment with decreasing chip length (Fig. 15). Mixing fines or overlarge chips into homogenous chips also gave a somewhat poorer pulp quality.

Chip dimensions did not affect the relationship between CSF and energy consumption, but other studies[62,63] indicate increasing specific energy requirement with increasing chip size.

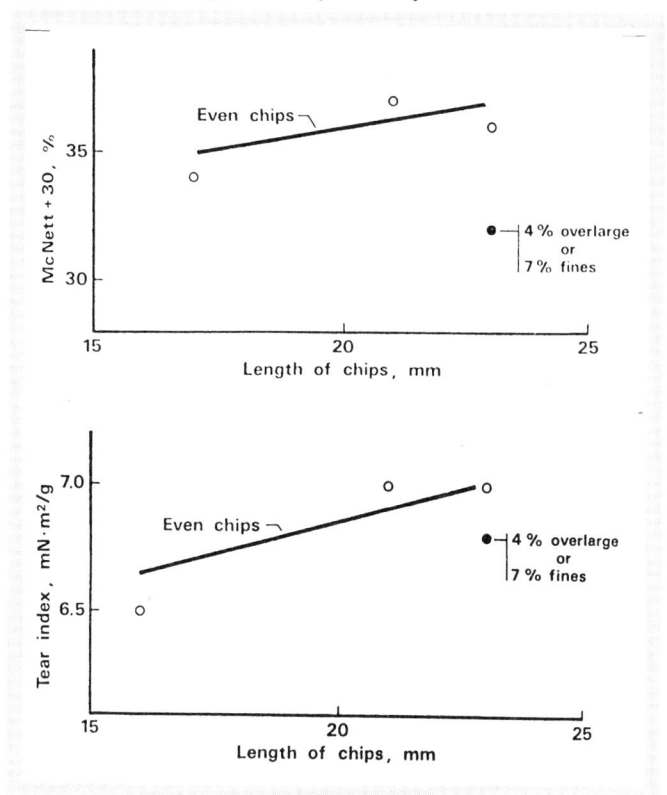

Figure 15. Effect of chip length on the McNett long fiber fraction and tear index of TMP, SEC 1800 kWh/bdt[11].

Extreme chip fractions have the following impact on TMP:

The addition of *over-large chips* to accept chips causes an uneven feed to the refiner and thus reduces pulp quality. The fraction often contains large amounts of knotwood, the effect of which was reviewed earlier.

The *over-thick fraction* contains most of the knotwood present in roundwood logs. The increase in the over-thick fraction usually decreases pulp fiber length and long fiber fraction and impairs strength properties and brightness. Thick chips also cause instable refining and increase energy consumption.

The *fines fraction* is known to lower the energy consumption and decrease pulp strength properties, sheet density, brightness and scattering coefficient, and increase the shives content[5, 11]. It also increases linting problems. The smallest fines fraction (< Ø 3 mm) is considered to be the most harmful one.

In conclusion, in spite of the lack of clear evidence of favoring a particular chip size, signs of decreasing pulp strength along with smaller sizes exist. There is, however, a demonstrated advantage in removing the fines fraction and over-size fraction. Uniformity, both in chip size and wood properties, leads to improving product quality and machine efficiency[58].

8 Seasonal variation

Properties of the wood raw material entering the pulping process show a seasonal variation pattern. This is caused by annual changes in the growing trees, climatic conditions, and the history of the raw material between felling and processing. Annual cycles exist in the following raw material properties[57, 64-67]:

- dry matter content
- chip size distribution
- bark content
- amount of lipophilic extractives and inorganic constituents
- brightness.

In conifers growing at the temperate zone, the moisture content of freshly cut trees is highest during the wintertime. Both moisture content and brightness of logs to be stored are also well preserved. Freezing of the logs, however, increases the bond strength between wood and bark, making the debarking more difficult. Increased bark content of the chips is accompanied by an increase in the extractives content of the chips. Chipping of frozen logs yields a smaller chip size than obtained during the warm season. Because of these variations one would expect the brightness of mechanical pulps to be lowest in the summertime due to the brightness loss of wood during transportation and storage, provided the bark content during the wintertime can be controlled. As a consequence of the brightness drop, the consumption of bleaching chemicals would increase and in case of limited bleaching capacity, lead to lower brightness in the final paper. The reduced chip size during the cold season can be expected to yield lower strength properties during the wintertime. Grinding frozen logs would be expected to reduce pulp strength. Reduced pulp strength often leads to problems in pressroom runnability[64]. Several studies have confirmed these patterns[57, 64-67]. There are, however, casual changes in the wood origin (stand origin, wood species mix, ratio of sawmill chips to roundwood) which may disturb the variation patterns.

References

1. Giertz, H. W., "Basic wood raw material properties and their significance in mechanical pulping," 1977 International Mechanical Pulping Conference Proceedings, The Finnish Paper Engineers' Association, Helsinki, p. 1.1.

2. Hartler, N., Nordic Pulp Pap. Res. J. 1(1):4 (1986).

3. de Montmorency, W. H., Pulp Paper Mag. Can. 65(6):T235 (1964).

4. de Montmorency, W. H., Pulp Paper Mag. Can. 66(6):T325 (1965).

5. Brill, J. W., "Effects of wood and chip quality on TMP properties," 1985 International Mechanical Pulping Conference. Proceedings, SPCI, Stockholm, p. 153.

6. Tyrväinen, J., "Wood and fiber properties of Norway spruce and its suitability for thermomechanical pulping," (Diss.) Acta forestalia fennica 249, 1995, 155 pp.

7. Rudie, A., "Wood and how it relates to the paper products," TAPPI 1995 Pulping Conference Proceedings, TAPPI PRESS, Atlanta, p. 807.

8. Hakkila, P., "Investigations on the basic density of Finnish pine, spruce and birch wood," (Diss.) Commun. Inst. For. Fenn. 61.4, 1966, 98 pp.

9. Corson, S., "Wood characteristics influence pine TMP quality," 1991 International Mechanical Pulping Conference (Minneapolis) Proceedings, TAPPI PRESS, Atlanta, p. 243.

10. Miles, K. B., Paperi ja Puu 72(5):509 (1990).

11. Eriksen, J. T., Hauan, S., Gaure, K., Mattans, A. L., "Consequences of chip quality for process and pulp quality in TMP production," 1981 International Mechanical Pulping Conference Proceedings, PTF, Oslo, p. 19.

12. Barbe, M. C., Janknecht, S., Sauriol, J. F., "The importance of chip impregnation on refiner pulp quality," 1993 International Mechanical Pulping Conference Proceedings, PTF, Oslo, p. 17.

13. Varhimo, A., unpublished results, KCL (The Finnish Pulp and Paper Research Institute), 1994.

14. Lehtonen, I. "Knots in Scots pine and Norway spruce and their effect on the basic density of stemwood," Commun. Inst. For. Fenn. 95.1, 1978, 34 pp.

15. Sahlberg, U., Tappi J. 78(5):162 (1995).

16. Boutelje, J., Svensk Papperstid. 69(1):1 (1966).

17. Eskilsson, S., Svensk Papperstid. 75(10):397 (1972).

18. Hatton, J. V. and Johal, S. S., Pulp Paper Can. 97(12):T420 (1996).

CHAPTER 5

19. Harris, G., Tappi J. 76(6):55 (1993).

20. Härkönen, E. J., Heikkurinen, A., Nederström, R., "Comparison between different species of softwood as TMP raw material," 1989 International Mechanical Pulping Conference Preprints, KCL, Helsinki, p. 390.

21. Barbe, M. C., Dessureault, S., Janknecht, S., Paper Southern Africa 13(1):10 (1993).

22. Isenberg, I. H., "Pulpwoods of the United States and Canada, Part 1," Conifers, 3rd edn., Institute of Paper Chemistry, Appleton, WI, 1980, 219 pp.

23. Richardson, J. D., Corson, S. R., Foster, R. S., Appita J. 45(1):33 (1992).

24. Petit-Conil, M., de Choudens, C., Chantre, G., "Selection of poplar clones for thermomechanical pulping," 1995 International Mechanical Pulping Conference(Ottawa) Proceedings, CPPA, Montreal, p. 1.

25. Lehto, J., "Various poplar species as raw material for paper grade mechanical pulps," 1995 International Mechanical Pulping Conference (Ottawa) Proceedings, CPPA, Montreal, p. 9.

26. Baker, D. L. and Siedlak, E. R., "TMP – key to development of mechanical book papers from a tropical hardwood – Gmelina arborea." 1977 International Mechanical Pulping Conference Proceedings, The Finnish Paper Engineers' Association, Helsinki, p. 6B.1.

27. Jackson, M., Åkerlund, G., Falk, B., "High yield pulp from North American aspen (Populus tremuloides)," 1985 International Mechanical Pulping Proceedings, SPCI, Stockholm, p. 170.

28. Lindström , C., Bovin, A., Falk, B., Lindahl, A., "Scots pine as raw material in thermomechanical pulping," 1977 International Mechancal Pulping Conference Proceedings, The Finnish Paper Engineers' Association, Helsinki, p. 4.1.

29. Sundholm, J., unpublished results, KCL (The Finnish Pulp and Paper Research Institute).

30. Liimatainen, H. and Tuovinen, O., "Fibers, fibrils and fractions – an analysis of various mechanical pulps," 1993 International Mechanical Pulping Conference Proceedings, PTF, Oslo, p. 324.

31. Liimatainen, H., Pylkkö, J., Tuovinen, O., "Pressurized grinding improves the potential of the wood species of the Populus family as a raw material for value added papers," Proceedings of 1992 SPCI-ATICELCA European Pulp and Paper Week, Bologna, Italy, ATICELCA, Milan, p. 312.

32. Anthony, A., Jensen, C., Chambers, D., Moller K., "The influence of furnish on newsprint quality using pinus radiata," 1989 Appita Annual General Conference Proceedings, Appita, Brisbane, p. A28.1.

33. Salakari, H., Pulp and Paper 59(9):190 (1985).

34. Aario, M., "Use of PGW to improve southern pine mechanical pulp for wood-containing papers," TAPPI 1985 Pulping Conference Proceedings, TAPPI PRESS, Atlanta, p. 119.

35. Haikkala, P., Liimatainen, H., Tuovinen, O., "Effect of alkaline peroxide in aspen PGW process," TAPPI 1989 Pulping Conference Proceedings, TAPPI PRESS, Atlanta, p. 103.

36. "Kirkniemen menestyksen takana luova hulluus ja kehitystyössä onnistuminen" (in Finnish). Mesäliiton viesti 3/97, p. 12-13.

37. Haikkala, P., Liimatainen, H., Tuominen, R., "Impact of superheated shower waters on pressure groundwood pulp properties," 1988 Appita Annual General Conference Proceedings, Appita, Hobarth, p. A24.1.

38. Atmer, B. and Thörnqvist, T., "The properties of tracheids in spruce (Picea abies Karst.) and pine (Pinus silvestris L.)" (in Swedish), The Swedish University of Agricultural Sciences, Department of Forest Products, Report No. 134, Uppsala, 1982, 59 p.

39. Kucera, B., Wood Fiber Sci. 26(1):152 (1994).

40. Panschin, A. J. and deZeeuw, C., Textbook of Wood Technology – Vol. I: Structure, Identification, Uses and Properties of Commercial Woods in the United States and Canada, 3rd edn., McGraw-Hill, New York, 1970, 705 p.

41. Saranpää, P., Silva Fennica 19(1):21 (1985).

42. Saarnijoki, S., unpublished report, The Finnish Forest Research Institute, 1966, 29 p.

43. Dinwoodie, J. M., Forestry 34(12):125 (1961).

44. Ilvessalo-Pfäffli, M-S. and Laamanen, J., unpublished reports, KCL (The Finnish Pulp and Paper Research Institute), 1964, 1975.

45. Ilvessalo-Pfäffli, M-S., unpublished report, KCL (The Finnish Pulp and Paper Research Institute), 1981.

46. Zobel, B. J. and van Bujtenen, J. P., Wood Variation, Its Causes and Control, Springer-Verlag, Berlin, Germany, 1989, 363 pp.

47. Corson, S. R., "Thermomechanical and refiner mechanical pulping of New Zealand grown Radiata pine," 1983 International Mechanical Pulping Conference Proceedings, PTF, Oslo, p. 1.

48. Hatton, J. V., "Pulping and papermaking properties of managed second-growth softwoods," TAPPI 1995 Pulping Conference Proceedings, TAPPI PRESS, Atlanta, p. 673.

49. Braaten, K. R., Palm, A., Omholt, K., "Wood classification leads to more uniform TMP," 1993 International Mechanical Pulping Conference Proceedings, TAPPI PRESS, Atlanta, p. 23.

50. Murton, K. D. and Corson, S. R., Appita J. 45(5):327 (1992).

CHAPTER 5

51. Corson, S., "Tree and fiber selection for optimal TMP quality," 1997 International Mechanical Pulping Conference Proceedings, SPCI, Stockholm, p. 231.

52. Tyrväinen, J., "The influence of wood properties on the quality of TMP made from Norway spruce (Picea abies) – Wood from old-growth forests, first-thinnings and sawmill chips," 1995 International Mechanical Pulping Conference (Ottawa) Proceedings, CPPA, Montreal, p. 23.

53. Brolin, A., Noren, A., Ståhl, E. G., Tappi J. 78(4):203 (1995).

54. Wang, X. and Braaten, K. R., Nordic Pulp Paper Res. J. 12(3):196 (1997).

55. Wahlgren, M. and Karlsson, L., "Experiences of classifying spruce for tailormade mechanical pulps for printing papers," 1997 International Mechanical Pulping Conference Proceedings, SPCI, Stockholm, p. 227.

56. Pulkki, R., "Effect of harvest systems on wood quality," 1990 CPPA Annual Meeting Preprints, CPPA, Montreal, p. 379.

57. Pennanen, O., Terävä, J., Laamanen, J., Lucander, M., Varhimo, A., Paperi ja Puu 75(4):182 (1993).

58. Wood, J. R., "Chip quality effects in mechanical pulping (a selected review," TAPPI 1996 Pulping Conference Proceedings, TAPPI PRESS, Atlanta, p. 491.

59. Internal reports, KCL (The Finnish Pulp and Paper Research Institute), UPM-Kymmene Corporation, 1989, 1994.

60. Marton, R., Goff, S., Brown, A. F., Granzow, S., Tappi 62(1):49 (1979).

61. Hakkila, P., "Geographical variation of some properties of pine and spruce pulpwood in Finland," Commun. Inst. For. Fenn. 66.8, 1968, 60 pp.

62. Hoekstra, P. L., Veal, M. A., Lee, P. F. and Sinkey, J. D., "The effects of chip size on mechanical pulp properties and energy consumption" 1993 International Mechanical Pulping Conference Proceedings, PTF, Oslo, p. 185.

63. Liukkonen, S., Unpublished results, KCL (The Finnish Pulp and Paper Research Institute), 1998.

64. Tyrväinen, J. "Influence of seasonal wood variation on TMP properties, bleach consumption and newsprint runnability," 1997 International Mechanical Pulping Conference Proceedings, SPCI, Stockholm, p. 213.

65. Tuominen, R., Haikkala, P. and Liimatainen, H. Paperi ja Puu 73(4):346 (1991).

66. Samson, A. and Schaffner, R. "Maximizing TMP strength by controlling chip quality," TAPPI 1993 Pulping Conference Proceedings, TAPPI PRESS, Atlanta, p. 999.

67. Fuhr, B. J., Henry, D., Leary, G. and Smith, G. Pulp & Paper Can. 99(2):T61 (1998).

CHAPTER 6

Grinding and pressure grinding

1	**Introduction**	**107**
1.1	Grinder productivity	108
2	**Two-pocket grinder for GW and PGW processes**	**109**
2.1	Tampella two-pocket atmospheric grinder	110
2.2	Modern Valmet grinder	110
2.3	Montague two-pocket grinders	112
2.4	Valmet pressure grinders	113
2.5	Pressure groundwood processes	116
3	**Chain grinders and thermogroundwood (TGW)**	**121**
4	**Pulpstones and pulpstone treatments**	**125**
4.1	Pulpstones	125
4.2	Abrasive specifications and their influence on pulpstone behavior	126
4.3	Pulpstone treatments	128
4.4	Burr treatments for pulpstones	128
	4.4.1 Pulpstone truing	129
	4.4.2 Pulpstone sharpening	129
	4.4.3 Dulling of pulpstone	131
	4.4.4 Pulpstone grooving	131
4.5	Pulpstone surface control with ultrahigh pressure water	132
	4.5.1 The waterjet equipment	133
	4.5.2 Pulpstone treatment with waterjet	134
	4.5.3 Impacts of waterjet conditioning on grinding process	134
5	**Grinder feeding systems**	**135**
5.1	Feeding systems for the chain grinders	136
5.2	Charging system for the pocket grinders	136
6	**Process control**	**139**
6.1	Introduction	139
6.2	Sources of variation	139
6.3	The classic grinding model	140
6.4	Load controls of a two-pocket grinder	142
6.5	Pros and cons of different control modes	146
6.6	Load controls of a chain grinder	147
6.7	Grinder group controls	147
6.8	Commercial products for grinder group controls	149
7	**The PGW pulp family and various paper grades**	**150**

7.1	Pulp properties	151
8	**Requirements of various paper grades**	**151**
8.1	Suitability for various grades	152
8.2	Other influencing factors	153
	References	154

CHAPTER 6

Heikki Liimatainen, Pekka Haikkala, Mikael Lucander, Risto Karojärvi, Olli Tuovinen

Grinding and pressure grinding

1 Introduction

The first commercial stone grinders were placed in production in Germany as early as 1852. They started a new era in pulp making by making it possible to use cheap wood logs as raw material for paper and board making in rapidly increasing quantities. "History of mechanical pulping" describes the historical development of the grinders and the grinding process in more detail[1].

Since those days, the stone grinders have been intensively developed to improve their productivity and level of automation as well as to continuously improve the groundwood pulp quality. This has been accomplished mainly through the implementing of inventions which other associated engineering techniques have developed through the years.

The main principle of today's stone grinding is still very much the same as in the early days; the logs are pressed against a rotating, properly conditioned or "sharpened" pulpstone, while the stone surface is simultaneously cleaned and cooled by using shower waters.

Regarding the grinder construction itself, there has been the development of many different ideas and pieces of machinery for pressing the logs against the rotating pulpstone – even in the commercial scale. These commercial and pilot grinders, including their brand names or major manufacturers, can be classified as follows:

- **Continuous grinders**
    - Chain grinder (Voith, Miag, Koehring)
    - Screw grinder (Füllner, Miag)
    - Hydraulic grinder (KMW)
- **Hydraulic pocket grinders**
    - Three- and four-pocket grinders (several manufacturers; historical)
    - Two-pocket grinder (Great-Northern, Tampella, Koehring, Montaque, KMW, Voith) -Two-pocket high magazine grinder (Koehring, Montaque)
- **Ring grinders** (Roberts, Tampella)
- **Pressure grinders** (Tampella, Valmet)

CHAPTER 6

- **Other grinders**
    - Berzano grinder (KMW; Pilot)
    - Centri grinder (Koehring; Pilot)
    - Chip grinder (Oji).

Today the major grinder designs are the atmospheric and pressurized two-pocket grinders as well as the chain grinders, which are described in more detail further on. For more information on the other grinder designs, refer to descriptions in the literature[1-3].

1.1 Grinder productivity

Many improvements have been made in grinder productivity. The first commercial grinders had a power demand of only 30 kW, but today the modern two-pocket grinders can have 9000 kW motors and use logs of up to 1.6 meters in length.

The production capacity of a stone grinder can be calculated when known information includes the motor power and the specific energy consumption (SEC) for each wood species to reach the desired groundwood pit freeness (Eq. 1).

$$GrinderProduction \text{ (t/h)} = k \times \frac{Motorpower \text{ (kW)}}{SEC \text{ (kWh/t)}} \qquad (1)$$

where k is coefficient of the grinder efficiency.

SEC proves to be fairly constant for each wood species, and it is not actually dependent on either the grinder design or the grinding process. For instance, Fig. 1 shows SEC for Norway spruce (*Picea abies*) and for corresponding softwood species, as used by Valmet.

[Graph: Specific Energy Consumption (kWh/t) vs Grinder Pit Freeness (ml). Curve equation: SEC = 3700 - 510 ln (CSF)]

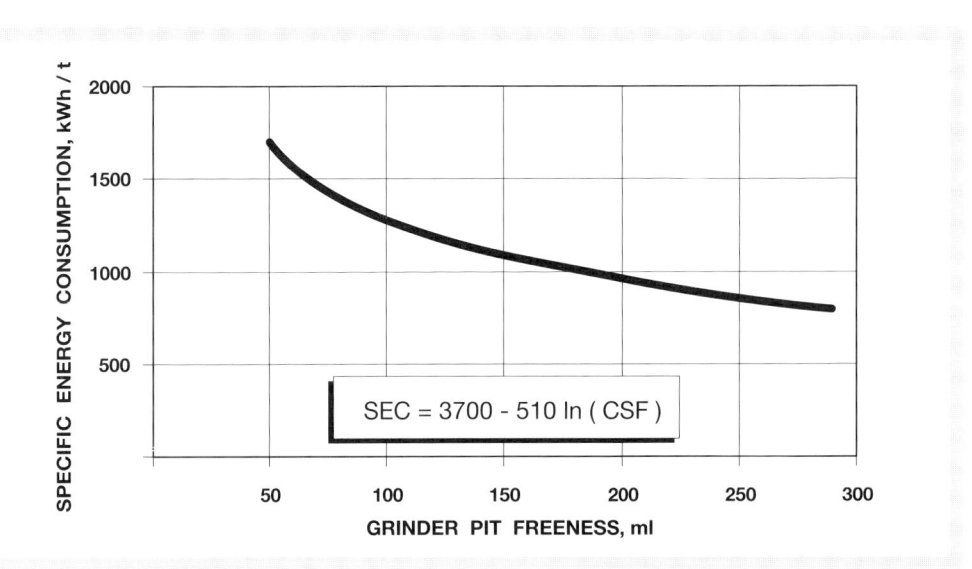

Figure 1. Specific Energy Consumption for GW and PGW pit pulps as a function of the groundwood pit freeness (*Picea abies*).

For an exact description of the grinder SEC and tonnage, information should be gathered as to whether the tons are metric or short tons as well as air dry (90%) or bone dry (100%) tons. This is an important factor because these terms still vary from country to country as well as from one mill to another. In this book, all tons are air dry metric tons (a.d. metric tons) if not otherwise stated.

For operational and pulp quality reasons, the grinders cannot continuously be operated at their maximum motor load. Thus, the formula requires the constant "k" for the grinder operational efficiency. The value of this constant "k" is in the range of 0.9 to 0.7, and it is affected by several factors like:

- The set limits for the grinder pit freeness
- Stone sharpening techniques and actual pulpstone sharpness
- Grinder control techniques
- Log charging time for the discontinuous two-pocket grinders
- Mechanical grinder efficiency.

The pulpstone pheripheral speed for modern grinders is normally selected to be in the range of 25 to 36 m/s. It is obvious that slightly better groundwood pulp quality is achieved at a lower pheripheral speed. However, as the grinder shaft torque, hydraulic load, and maximum motor power are almost linearly related to this speed, higher pheripheral speeds are favored in order to minimize the groundwood mill investment costs.

Maximum grinder power is mainly restricted today by the the concrete core pulpstone and the grinder shaft construction as the pulpstone is tightened by the shaft flanges. For maximum grinder load and groundwood pulp quality, dull stones and high shoe forces are required. To achieve the necessary defiberizing frequencies into the wood matrix and to transport the pulp out of the grinding zone, the stone surface has to be modified accordingly. In this respect, the pulpstone grooving and the high-pressure waterjet stone conditioning technique have already much improved the productivity of the highly loaded grinders (see "Pulpstones and pulpstone treatments"). Intensive developmental work regarding these issues makes it obvious that the maximum grinder power and productivity will be further increased in the near future.

2 Two-pocket grinder for GW and PGW processes

In accordance with its name, the two-pocket grinder has two separate grinder pockets, which are operated batchwise. The logs are dropped into the pockets through the feeding gates on top of the grinder (Fig. 2), the gates and pressure shoes of which are operated by water hydraulics (Fig. 6).

This two-pocket or hydraulic grinder is also known as the Great Northern grinder, which refers to the Great Northern paper mill in northern Maine in the United States. This grinder was first developed in 1926 to increase grinder loads and especially to increase the log length from three to four feet, as these longer and heavier logs were too difficult to feed manually into three- and four-pocket grinders.

CHAPTER 6

2.1 Tampella two-pocket atmospheric grinder

Tampella Inc. in Finland delivered its first stone grinder as early as 1869 and, since that time, more than 600 Tampella stone grinders have been delivered around the world[4]. Later on, Tampella Inc. acquired the license for Great Northern two-pocket grinders and, in 1937, a huge groundwood mill at the Tampella Anjala paper mill started up. This included 20 Tampella G-N grinders to produce GW pulp for two newsprint machines, which at that time were the biggest ones in Europe.

Since 1947, Tampella has delivered more than 250 atmospheric two-pocket grinders (Fig. 3) and, of these, some 160 grinders are still successfully in operation.

2.2 Modern Valmet grinder

Since 1992, Valmet Corp. has manufactured the Tampella grinders and are thus in this book referred to as Valmet grinders (Fig. 2).

The main body of the Valmet grinder is made of strong and corrosion-resistant cast iron to dampen any possible vibrations. Also stainless steel castings are available if so required, due to the closed or corrosive white water loops. Today the pulpstone diameter for all the Valmet grinders is 1.8 m and both grinding zones are 1.0 m long. Each grinder pocket is controlled individually even as the other pocket is in the pocket charging phase. (See "Process control.")

The width of the log feeding gate opening is 600 mm and the pocket charging time is approximately 30 s. The feed gates and the pressure shoes are operated by a water hydraulic system, which in principle is of the same type as in Valmet pressure grinders (Fig. 6). The hydraulic low pressure is rated for 12 bars and the high-pressure loading pressure for a max. 50 bars.

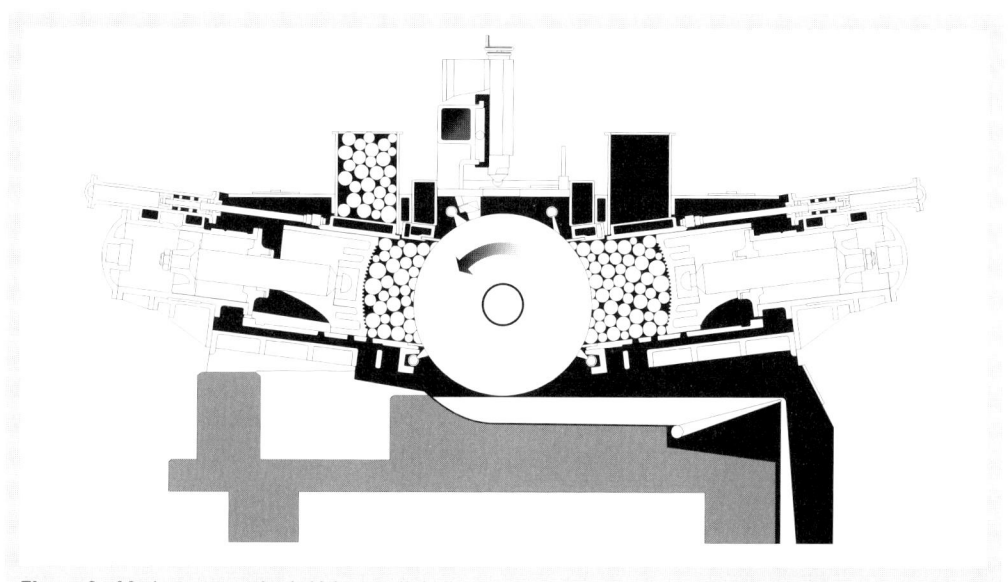

Figure 2. Modern atmospheric Valmet grinder.

Grinding and pressure grinding

Figure 3. Atmospheric Tampella grinders at Stora Fors board mill in Sweden.

There are four shower pipes to clean and cool the pulpstone. The net water pressure in the showers is 8–10 bars to effectively clean the stone surface; this is especially important for dull and highly loaded pulpstones. The amount of shower water flow is set so that the pit consistency will be approximately 1.5%–2.0%. On top of the pulpstone there is an adjustable plow arrangement to avoid the groundwood pulp from the first pocket from getting into the second pocket.

Today normally no brushes are installed into the shower pipes, but the shower water is pumped through a pressure screen. This shower water screen has a basket with small holes ($\varnothing$ = 0.8 mm) to screen the shower water out of any disturbing shives or fibers (Fig. 11). The flow of each shower pipe can be measured and adjusted individually; these flows also act as interlockings to prevent the pulpstone from burning.

The shower water temperature is normally in the range of 60 to 80°C, and today heat exchangers for cooling are often required due to more and more closed white water loops. Higher shower temperatures result in somewhat stronger groundwood pulp, but the higher temperatures can decrease the pulp brightness, increase the dissolved solids, and make the GW mill atmosphere steamy and uncomfortable.

Valmet pressurized shredders (Fig. 9), which are placed into the grinder pit pulp outlet, are today often favored in GW processes instead of the conventional vibrating bull screens. This has not only simplified the screen rooms, but has also greatly improved the working environment by better closing the atmospheric groundwood process.

CHAPTER 6

For the grinder layout, there are two major options – single or tandem drive. In the tandem drive, there are two grinders installed on one shaft with a torque sensor in between to measure the actual power for each grinder. Due to its lower investment cost, tandem drive is preferred – especially in bigger groundwood mills which have many grinders.

There are four models of the Valmet GW (and PGW) grinder according to the log length. Table 1 gives the maximum motor power ratings of this grinder model for both single and tandem drives at 330 rpm, which corresponds to a pulpstone peripheral speed of 31 m/s. If 300 rpm is preferred, then maximum motor power will also be 10% less to maintain an equal grinder shaft torque.

Table 1. The maximum motor power rates for Valmet grinders at 330 rpm stone speed (P = pressurized).

Grinder model	Log length	Max. motor size for single drive	Max. motor size for tandem drive
T1810 (P)	1.0 m	5.0 MW	9.0 MW
T1812 (P)	1.2 m	6.0 MW	10.5 MW
T1815 (P)	1.5 m	7.5 MW	13.5 MW
T1816 (P)	1.6 m	7.5 MW	13.5 MW

The low rpm grinder motors are normally direct-driven synchronous high-voltage motors. However, in some mills, more standardized 1500 rpm synchronous motors also have been installed and connected to a gearbox for the desired stone speed.

A modern Valmet grinder T1815 is thus capable of producing 80–100 tons/day of groundwood pulp at a low pit freeness (80–120 ml) typical for SC and LWC paper grades and as much as 170–200 metric tons/day of high freeness pulps (350–450 ml) used for folding boxboard grades.

2.3 Montague two-pocket grinders

The Montague Machine Company located in Massachusettes, USA, started manufacturing two-pocket grinders as early as 1920 in collaboration with the Great Northern Paper Company.

Later on, this grinder design was further improved for increased grinder through-put as well as to increase the log length from 1.2 m (4 ft) to 1.6 m (5 ft). This new design was named Great Northern Waterous Grinder, and some 90 Montague grinders of this type are still in operation in North America or abroad.

A modern Montague Waterous Grinder (Fig. 4) has a fabricated main frame and pressure shoes, which are clad with 316L stainlesss steel to avoid any corrosion inside the grinder. These grinders have electric stone lathes and special Posi-Clean shower pipes. The conventional T-stands, which were used to control the water hydraulics, have now been replaced with special gate and piston control valves.

Hydraulic low pressure is today typically rated to 14 bars (200 psi) and the higher load pressure to 41 bars (600 psi). In the latest delivery, Montague grinders for 1.2-m logs were equipped with 5.0-MW motors to produce 70 tons/day of aspen groundwood for making LWC papers.

In 1992, Montague acquired the grinder business from the Koehring-Waterous Company in Canada and, since then, Montaque has also been responsible for servicing the many existing Koehring-Waterous two-pocket and high magazine grinders operated in Canada and abroad.

Recently Montague Machine Co. delivered nine Montaque high magazine grinders of modern design to Canada to replace 18 old chain grinders.

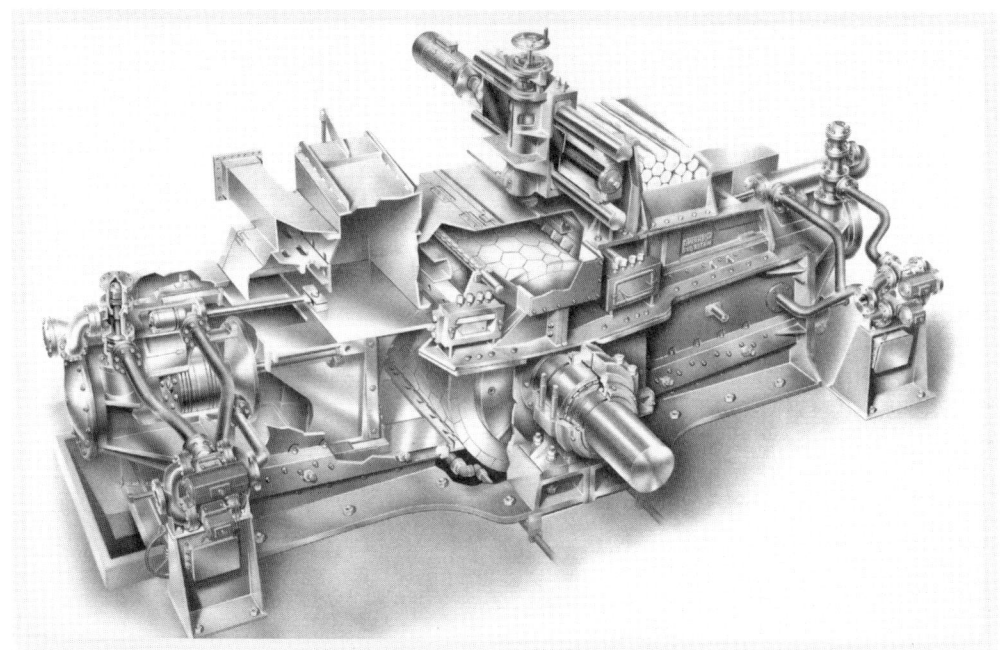

Figure 4. Modern Montague Waterous Grinder.

2.4 Valmet pressure grinders

In the 1970s, Tampella in Finland and MoDo in Sweden combined efforts to develop a new generation of two-pocket grinders: the PGW grinder design, where the grinding takes place under pressurized conditions and often at elevated temperatures – up to 140°C – to increase the fiber length and pulp strength properties.

The first production scale PGW trials were made in 1977 at MoDo's Bure market pulp mill in Sweden by modifying one of the atmospheric Tampella grinders into a discontinuously operated one-pocket pressurized grinder rated for 1 bar casing pressure[5,6].

The trial results obtained at Bure were so encouraging that both Tampella and MoDo decided to install a pressure grinder prototype into their mills. Thus the first continuously operating mill scale two-pocket pressure grinders were started at the Bure

CHAPTER 6

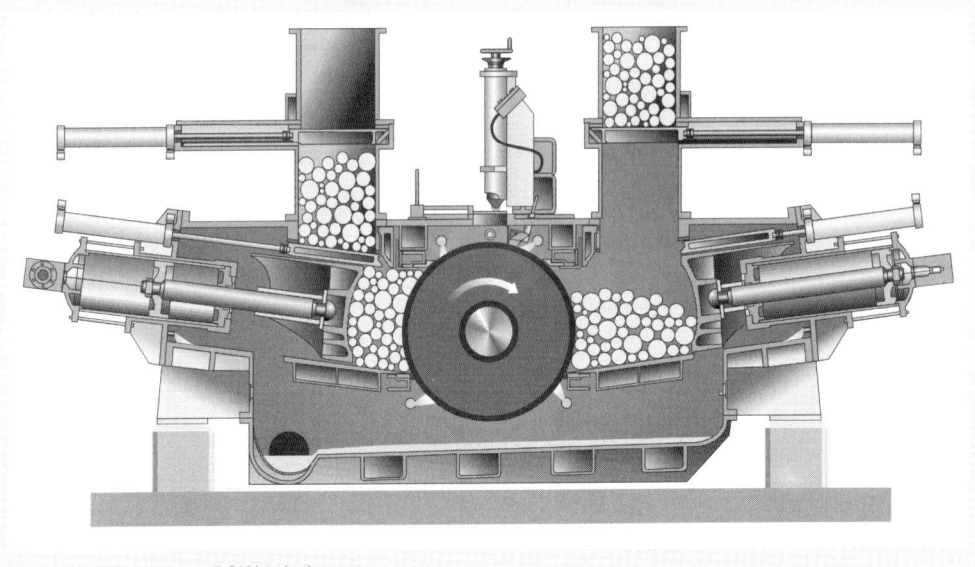

Figure 5. The Valmet PGW grinder.

groundwood mill in May 1979 and later in August at the Tampella paper mill in Anjalankoski[7, 8]. Both of these pressure grinders were rated for 3 bar casing pressure (Fig. 5).

The Valmet PGW grinder design is quite similar to the atmospheric two-pocket grinder (Fig.1). However, to feed the wood batches into the pressurized grinder, a pressure equalization chamber is a necessity and thus the upper log feeding gates are required. For the pressure grinders, the gate opening has been increased from 600 mm to 800 mm for feeding larger wood batches.

Naturally the earlier two-pocket grinder concept had to be redesigned for the pressurized grinding conditions. The grinder body therefore is now welded of heavy steel plates and all the surfaces in contact with the pulp are clad with stainless steel. The end blocks are made of stronger stainless steel castings.

The pressure inside the grinder casing is adjusted by compressed air, which is mainly required only when pressurizing the grinder for startups. The stone showering as well as the water hydraulic system are equal to those of the Valmet atmospheric grinders. Due to the elevated temperatures, cooling of the hydraulic water is now necessary and also a separate water filtering loop is included (Fig. 6). Two-sided mechanical seals are used for the grinder shaft to minimize the use of fresh water in the PGW process as well as to prevent the colder seal waters from bleeding onto the ceramic pulpstones.

The pressurized grinders were originally rated for 3 bar casing pressure, but soon it became evident that even stronger pulps could be produced at 5 bar grinder pressure and at elevated temperatures up to 140°C[9]. For this purpose, the design of a new series of pressure grinders has a stronger body rated for this 5 bar grinder pressure. These grinders are called super pressure (PGW-S) grinders (Fig. 7). The first PGW-S grinders were started at the Myllykoski paper mill in 1988 in Finland[10].

Grinding and pressure grinding

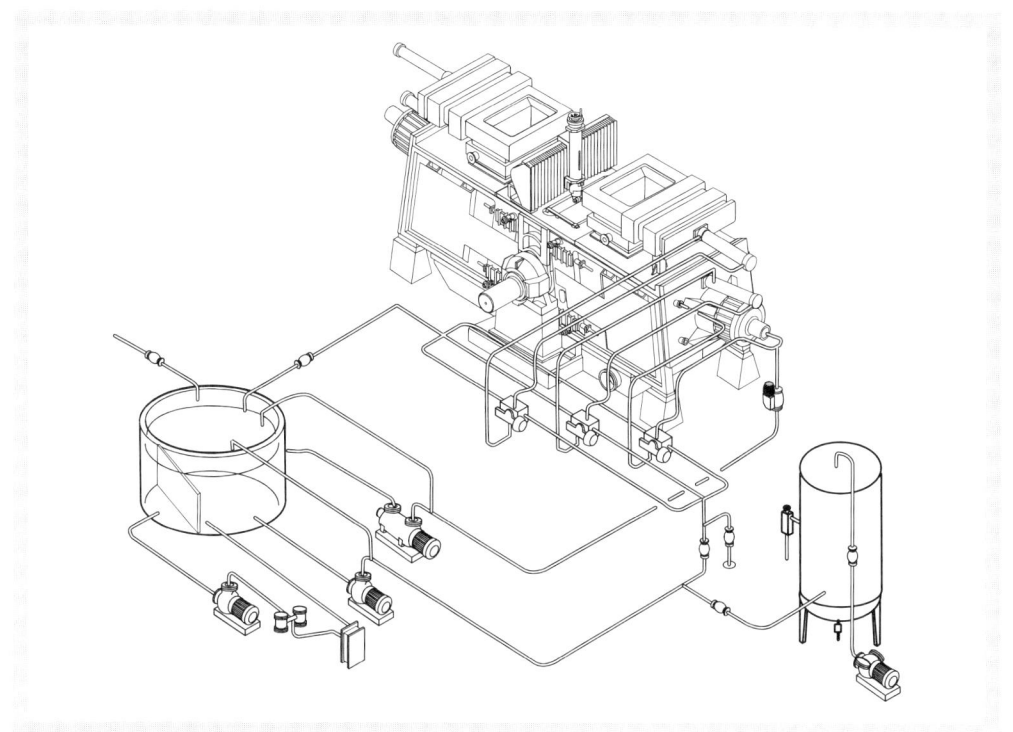

Figure 6. Water hydraulic system for Valmet pressure grinders.

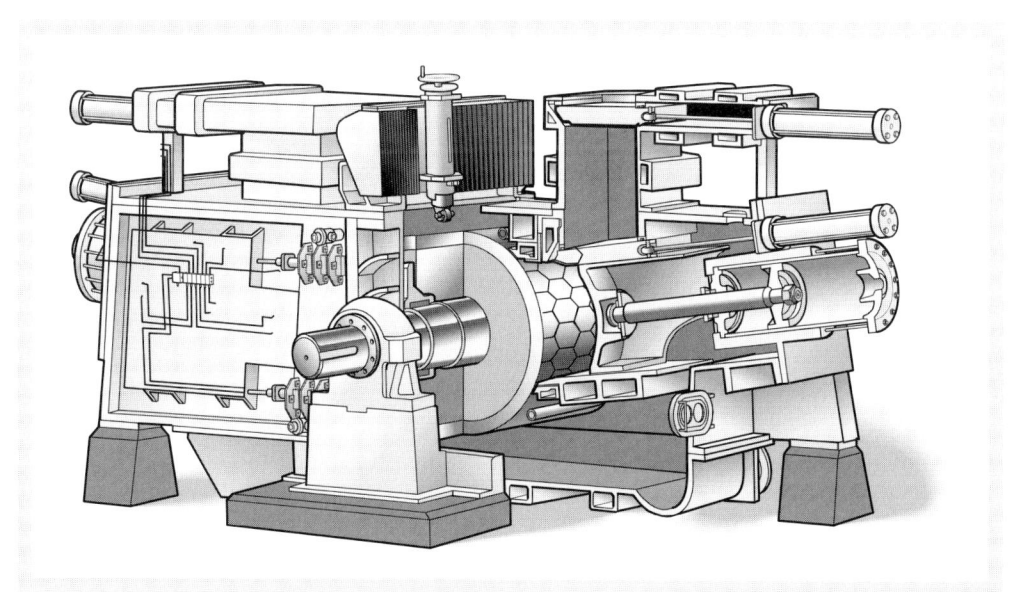

Figure 7. PGW-S grinder rated for 5 bar casing pressure.

CHAPTER 6

Since 1979, some 100 pressure grinders have been delivered worldwide for various paper grades and wood species (Table 2). The production capacity of these pressure grinders is more than 2.5 million tons annually.

Table 2. Pressurized grinders delivered in 1979–97 for different paper grades and wood species.

	Spruce, Fir	Pine	Aspen	Grinders, total
Newsprint	23	8	0	31
Uncoated mechanical	26	5	0	31
Coated mechanical	16	3	10	29
Fine papers	0	0	6	6
Folding boxboard	3	0	0	3
Total	68	16	16	100

2.5 Pressure groundwood processes

The elevated grinder casing pressure created by the compressed air simply increases the water boiling temperature (Fig. 8). An elevated casing pressure also makes it possible to use high shower water temperatures to increase the pulpstone surface temperature. Simultanously, the elevated grinder casing pressure prevents the evaporation of the water out of the the wood matrix and from the most critical grinding zone. The temperature of these are then related to that of saturated steam at the set grinder pressure.

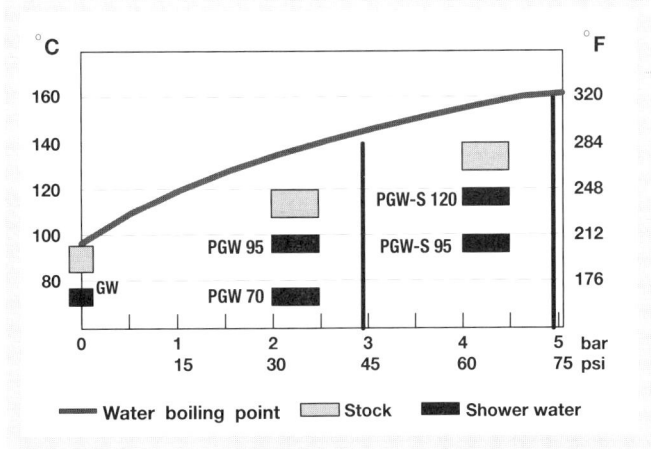

Figure 8. Process conditions for various grinding processes.

These for mechanical defibration more favorable process conditions contribute to the improved softening of the lignin polymer in the fiber. This results in a significantly increased fiber length in the PGW pulps when compared to GW pulp[11]. In the pressurized grinding, the grinder shower water temperature and flow are adjusted so that the pulp temperature is maintained well below the water boiling temperature at the set grinder pressure (Fig. 8). Unlike in the TMP refiners, the grinder pressures and temperatures can be adjusted separately according to the PGW pulp quality requirements. Thus there are various different combinations of the pressure groundwood process, which are named or coded as in the following examples:

PGW95: Grinder casing pressure is max. 3 bars and the shower water temperature is 95°C.

PGW-S120: Grinder casing pressure is max. 5 bars and the shower water temperature is 120°C.

To maintain the compressed air inside the pressure grinder, the blow valve controls the pulp level to be constant in the pulp outlet pipe prior to the pressurized shredder (Fig. 9). The shredder in the blow line is required to shred any wood particles in the pit pulp to avoid the plugging of the blow valve. The hole size of the shredder screen plate is normally set to 18 mm and the shredded wood particles, which typically amount to 1–2% of the pulp production, are shredded to the size of match sticks. Pulp can then be fed directly into the pressure screens, thus eliminating the conventional vibrating bull screens.

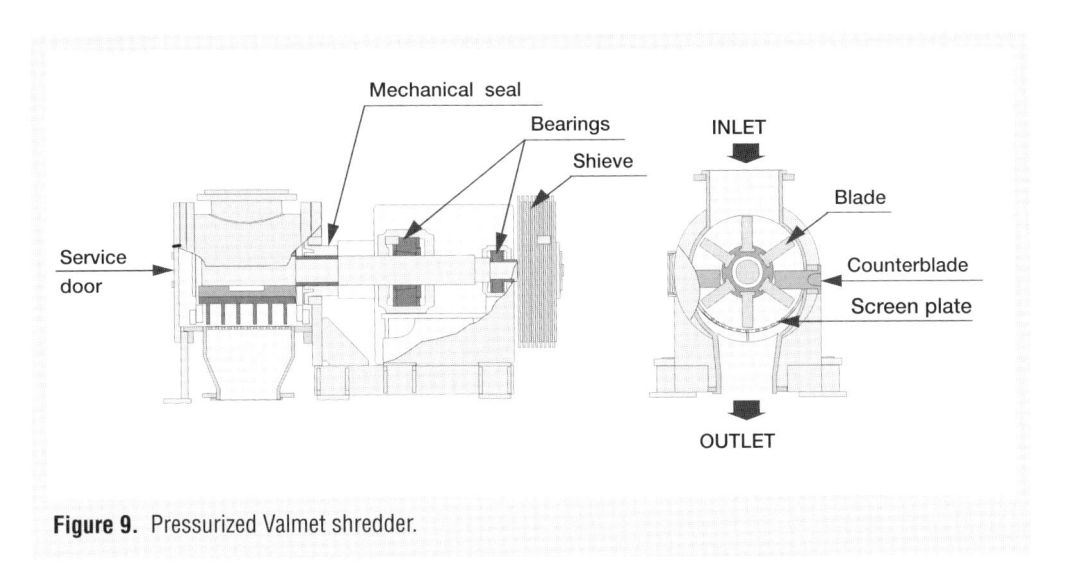

Figure 9. Pressurized Valmet shredder.

The early pressure grinding used the so-called PGW95 processes, which then included a "Hot Loop" or an intermediate thickening stage to return the hot filtrate back to the grinder showers prior to the pulp screening stage (Fig. 10). The pulp outlet temperature in the PGW95 process can be raised up to 110°C–120°C, and the excess heat is then recovered from the blow cyclone as an excess steam suitable for any heating purposes. Approximately 50% of the grinder motor power can be recovered, which corresponds to one ton of excess steam for each ton of PGW95 pulp produced. However, unlike in most TMP mills, this steam is atmospheric or its temperature is close to 100°C. For the intermediate thickening stage, Sunds pressurized drum thickeners were originally used, but recently pressurized Celleco Hedemora disc filters have been installed for the PGW Hot Loops.

CHAPTER 6

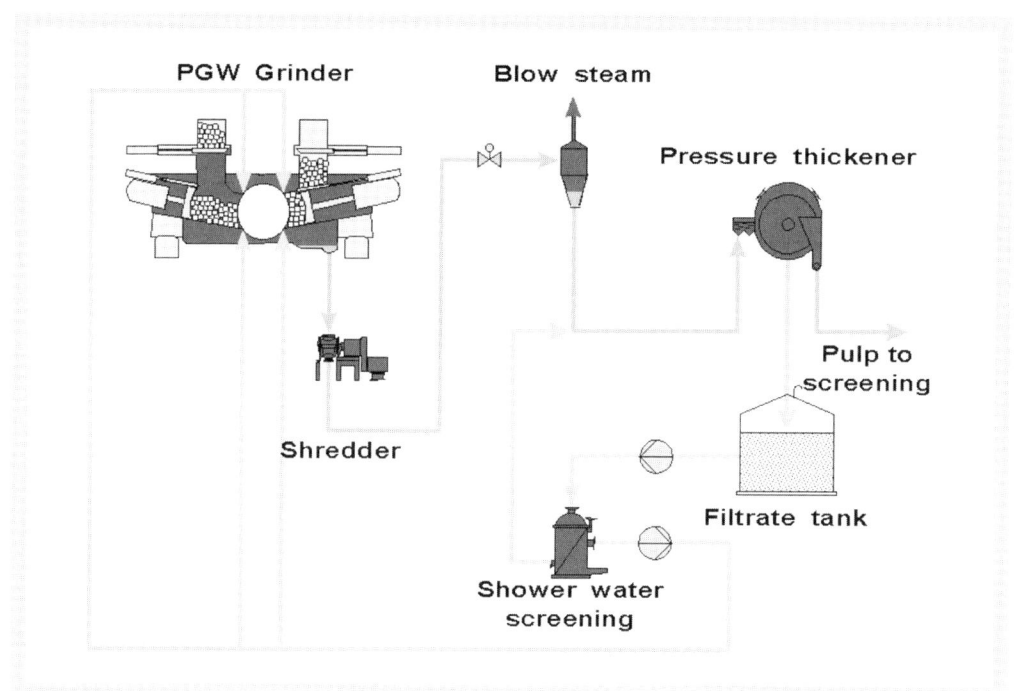

Figure 10. PGW95 process scheme including the Hot Loop.

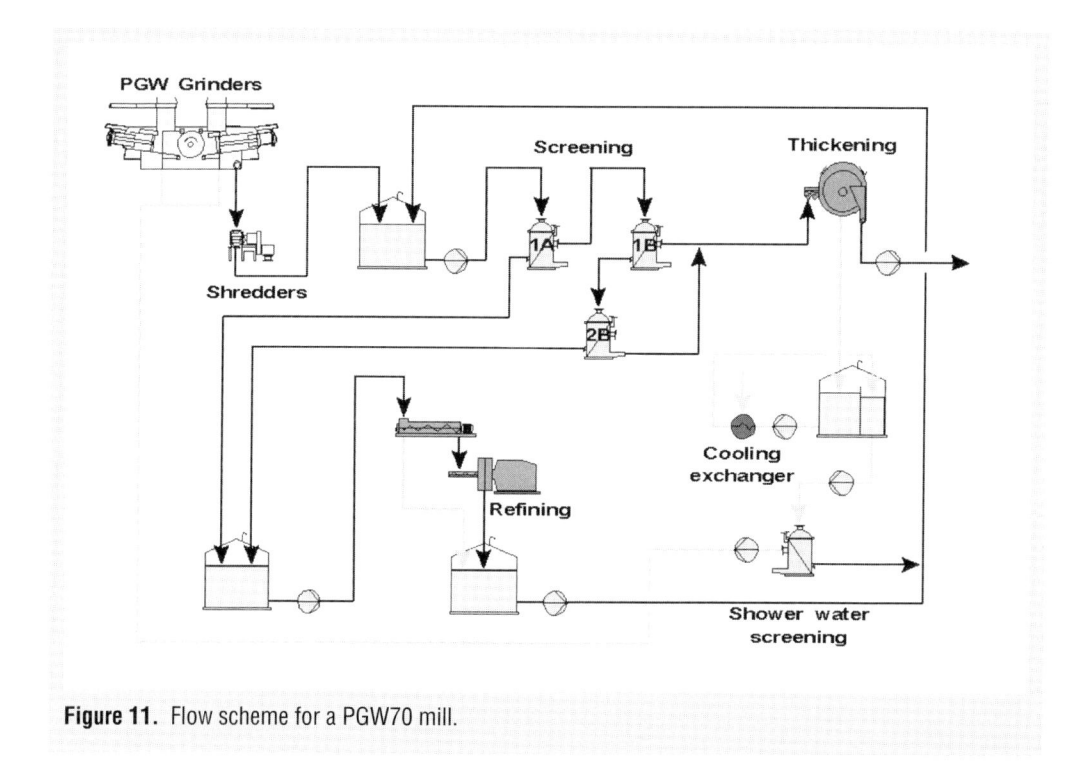

Figure 11. Flow scheme for a PGW70 mill.

The paper surface and printability requirements have become more and more demanding and especially so for the value added SC and LWC paper grades. For this purpose, excellent pulp brightness and light scattering are required and higher fines content and no coarse long fibers are favoured. Mill scale studies using Norway spruce proved that the PGW70 process with lower grinding temperatures has resulted in a better pulp initial brightness and paper printability for these value added paper grades. The pulp strength of these low freeness PGW70 pulps has decreased only slightly when compared to PGW95 pulps[12].

To reach the 70°C shower water temperature, the Hot Loop can now be abandoned, which simplifies the PGW process (Fig. 11). In the PGW70 mill, white water is cooled in a separate heat exchanger installed into the grinder shower water line. This cooling water is normally white water from the paper mill and, by this means, the PGW mill's excess energy can now be utilized to heat up the paper machine white water circuit.

For some paper grades – like newsprint – the pulp strength is of the most importance and the requirements for the paper printability and for brightness are not so demanding as for the value added papers. To maximize the fiber length and pulp strength, elevated shower water temperatures of up to 120°C should be used. In that case, PGW-S grinders, which are engineered for 5 bar casing pressure, are required because the pulp temperature can be as high as 140°C.

To reach these temperatures, a two-stage direct condenser-type heat exchanger system is added into the PGW-S Hot Loop (Figs. 12 and 13). In this heat exchanger, the hot PGW-S pulp is flashed in several stages and the flash steam is then used to heat up the incoming shower water up to 10°C at each stage. For efficient heating, a small vacuum pump is used to drain all the free air from the flash steam. Also for the adjustment of the shower water temperature, the amount of the free air in the flash steam can be controlled.

In several trials at Valmet Pulping Technology Center, it has been verified that – for spruce and especially for different pine species – PGW-S process conditions will result in much improved strength properties, but still the power consumption will stay at a level similar to that of the GW processes[13, 14]. The first commercial PGW-S120 mill having six T1815 PS grinders for a 490 tons/day pulping capacity was successfully started in 1991 at Voikkaa paper mill in Finland (Fig. 26)[15].

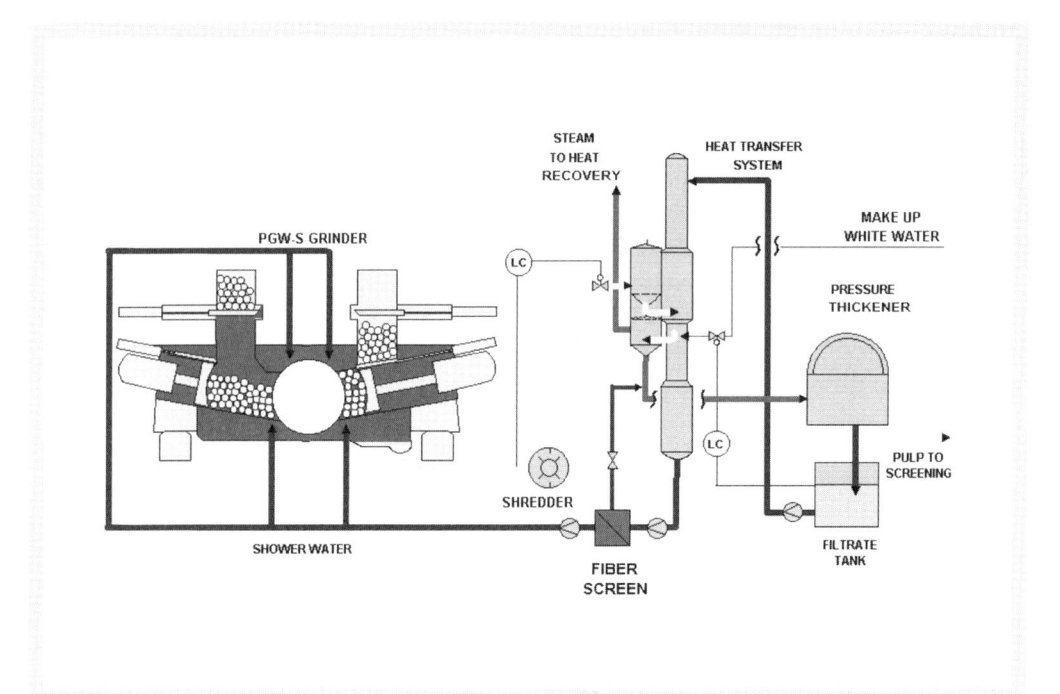

Figure 12. The Hot Loop for a PGW-S120 process.

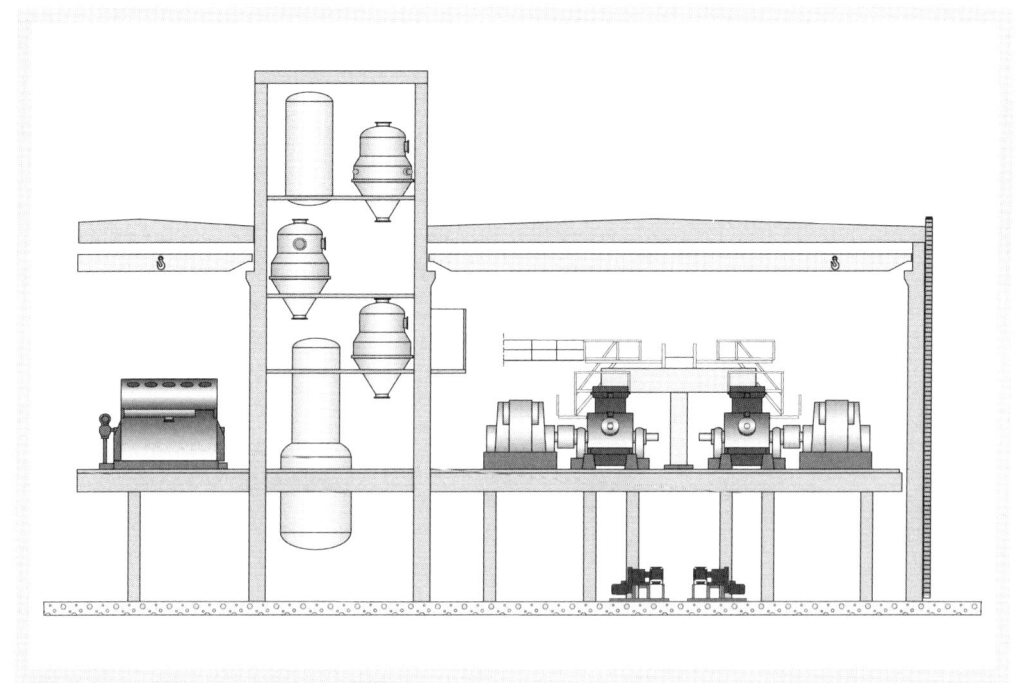

Figure 13. Layout for a PGW-S hot loop.

3 Chain grinders and thermogroundwood (TGW)

The only grinder type representing a truly continuously operating pulping device today is the atmospheric chain grinder[16]. The first of such design was built in 1915 by Voith. The first large production size chain grinder using 1-m long wood and 1.5-m stone diameter was installed in the Schongau mill in Germany in 1922 with a production capacity of 0.5 t/h.

The principle of the Voith grinder can be seen in Fig. 14. A single wood magazine is centered over the stone with large chains mounted on the two opposite sides of the magazine. The chains are moved downward, forcing the wood to be pressed against the rotating stone. The chain speed is controlled automatically in order to maintain a pre-set motor load, production rate, or specific energy value. Wood is charged automatically from the top of the magazine so that it is always full. In this sense, the chain grinder is simpler to operate than the pocket grinders.

There are also other advantages with the chain grinder as compared to pocket grinders. The stone is more accessible for cleaning and cooling. Also the motor load is not disrupted due to magazine changes as is the case with the pocket grinders. The wood magazine, together with the

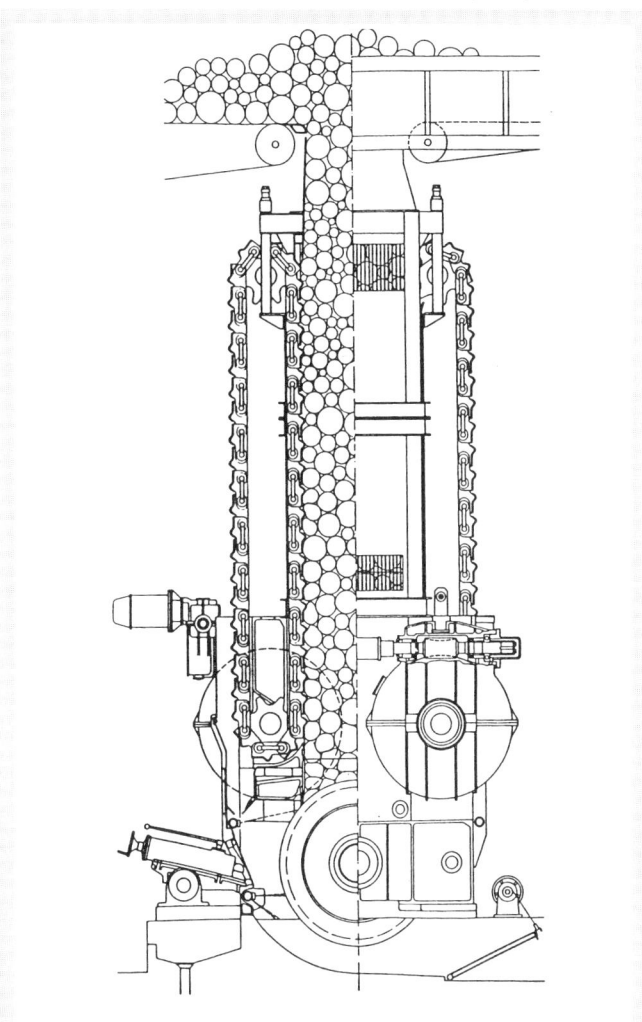

Figure 14. The principle of the Voith chain ginder.

chains and the drives, is supported by four screws by which the magazine can be lowered as the stone wears down. When the worn-out pulpstone has to be replaced, the magazine is lifted by the screws and the stone can be easily replaced by a new stone. On the other hand, the chain grinder has disadvantages and has not proven to be as acceptable to development as two-pocket grinders. The construction of the chain grinder does not enable magazines to take logs much longer than 1.3 m or to use motor

CHAPTER 6

loads exceeding 4000 kW because of the imbalance in loading (compared to the two-pocket grinder where loading takes part on two opposite sides of the pulpstone). In a continuous chain grinder, compared to a two-pocket grinder, a smaller part of the stone peripheral surface is used, 25% versus 40%. Because of the high magazine, installation of a chain grinder demands a two-story building. New chain grinders have not been built in the 1990s.

Numerous investigations have proven that raising the temperature in the grinder (either by altering the shower water temperature or by altering the pit temperature by means of raising the consistency, stone dulling, etc.) has a positive effect on pulp strength properties[11, 17–20]. Atmospheric grinding, however, has limitations concerning the maximum temperatures that can be used in the process. Higher shower water temperatures, together with high pit consistencies, generate pulp temperatures close to 100°C which in turn results in local overheating or boiling in the grinding zone. Too high grinding zone temperatures cause scorching and burning of the wood and the pulpstone, together with uneven (fluctuating) motor load conditions and deteriorated pulp quality.

Pressurizing the grinder in order to permit temperatures higher than 100°C demonstrated that the groundwood strength properties can be developed considerably. The PGW grinders have proved the benefits of grinding at higher temperatures. Due to the high magazine of the continuous chain grinder, the Voith grinder cannot (with reasonable constructional modifications) be pressurized. However in 1984–85, Meinecke[22, 23] presented Voith´s thermo grinding process (TGW) as an alternative to pressure grinding.

The idea of the TGW-process[23, 24] is to set the temperature at the beginning

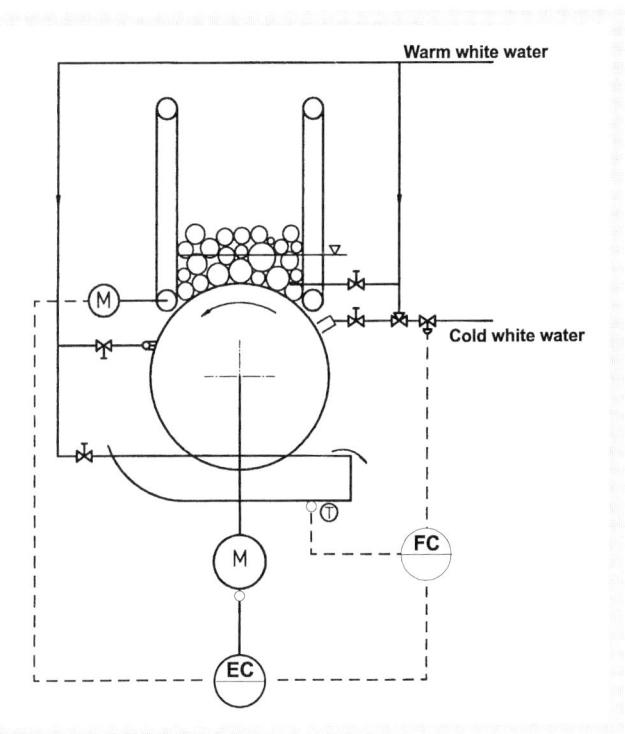

Figure 15. The TGW process.

of the grinding zone to such a level that the target pulp temperature of 100°C is reached just at the end of the zone. This can be reached by a shallow impoundment (the pond) in the wood magazine some 200–500 mm over the stone surface (Fig. 15). The impoundment is created as hot white water is added (showered) to the magazine; the pond over the stone is built up by special plates or seals. The seals replace the conventional finger bars and are adjusted to close tolerance to the stone surface. According to Voith, the

Grinding and pressure grinding

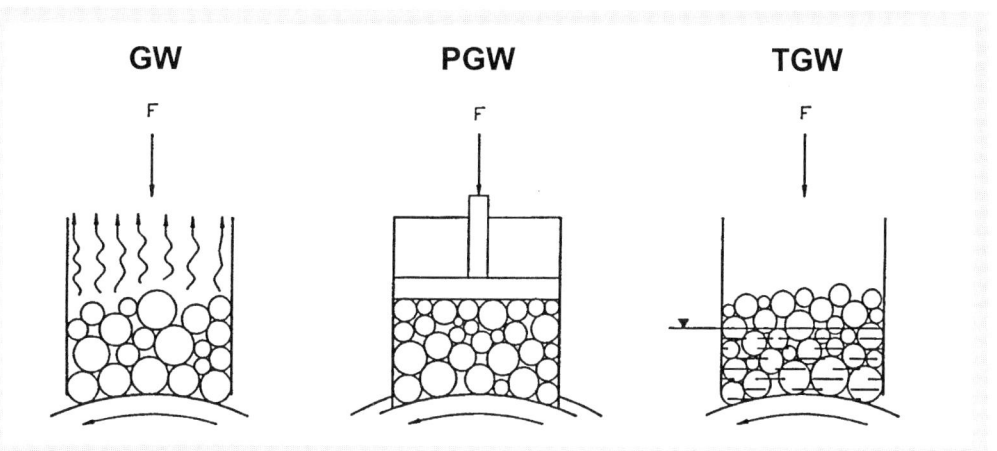

Figure 16. Principles of atmospheric grinding, pressure grinding and thermogrinding.
GW : atmospheric pressure, local overheating, evaporation of water steam
PGW: 2–5 bar overpressure, closed casing
TGW: 20–50 cm water column above the grinding zone.

white water pond in the wood magazine will condense the vapor developed in the grinding zone and thus prevent excess energy losses. In an open atmospheric process, the grinder can thus be operated closer to optimal defibration conditions (Fig. 16). However, it should be noted that, in a PGW grinder, the typical pressure of 2.5 bars corresponds to a 25-m water column.

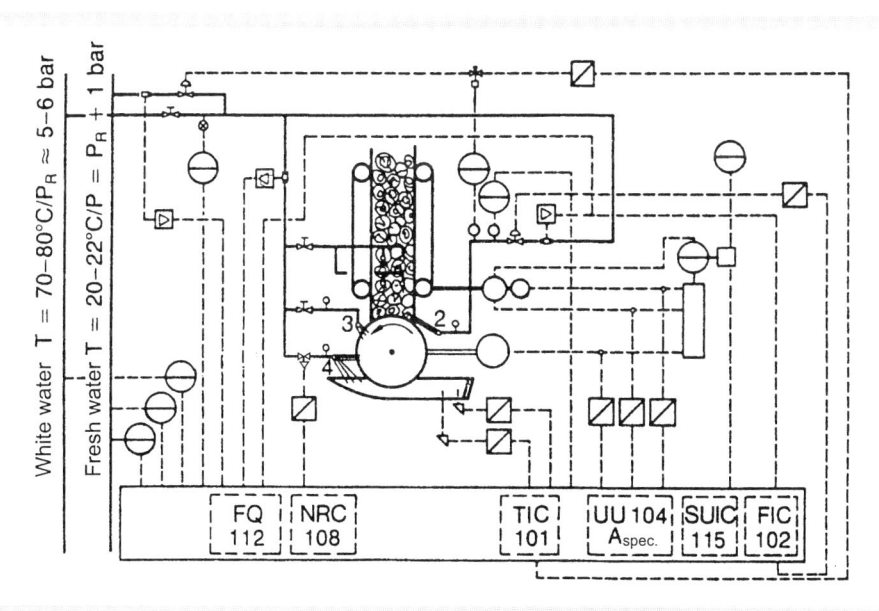

Figure 17. The control system of the TGW process.

CHAPTER 6

To reach a temperature not higher than 100°C at the end of the grinding zone, a very accurate control system is needed. The pulp temperature at the grinder pit and the specific energy consumption are used as input control signals. The controller is pre-set with boundary values for the pulp temperature and the specific energy. These values have proven to respond very promptly if the temperature in the zone tends to rise to a critically high level. This is due to the fact that the lubricating water film is disrupted as the temperature goes too high and boiling begins. When a pre-set limit is exceeded, the volume of white water to the uprunning part of the magazine is increased. Hereby the grinding zone temperature is lowered. The drop in the temperature avoids any excessive water evaporation in the grinding zone, and the specific energy consumption stabilizes again. The simultaneous recording of temperature and energy consumption makes it possible to differentiate whether the fluctuation has been caused by an excessive grinding zone temperature or by a nonuniform feeding of wood against the stone (Fig. 17). The use of the stability controllers in TGW also has a positive effect on stone sharpening; sharpening intervals have been reported to be considerably lengthened.

According to Voith and reported mill trials in Germany[16, 21, 24], thermogroundwood exhibits increased strengths over those of conventional groundwoods of comparable energy consumption (Fig. 18). The improvements are credited to two factors: (a) a slight increase in the grinding zone temperature and (b) more stable and more uniform grinding conditions, compared to those experienced in a conventional Voith grinder. By the end of the 1980s, some 60 chain grinders at 15 different paper mills had been upgraded to the Voith thermogroundwood pulping system[25]. The future of the TGW process is uncertain because chain grinders are not actively marketed any more.

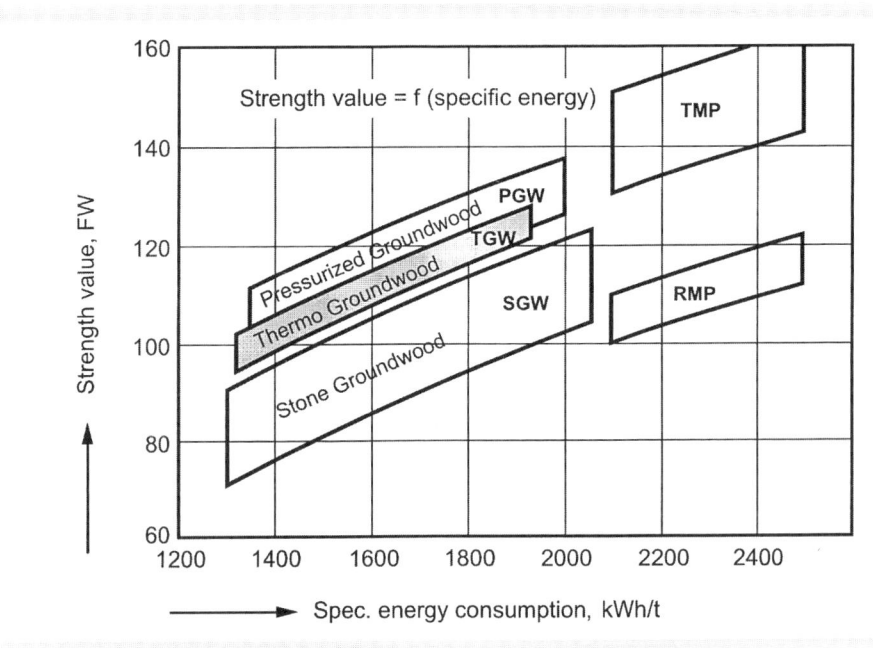

Figure 18. The strength properties of TGW pulp fall between those of SGW and PGW.

Grinding and pressure grinding

4 Pulpstones and pulpstone treatments

The grinder production rate, the specific energy consumption, and the quality of the produced pulp are dependent on many factors, e.g., grinding process, grinder type, wood species, wood quality, and freeness target. Also the types of pulpstones the mills are operating and the ways they treat and condition the pulpstones are most important in order to achieve the targets set by their paper or board quality and production requirements.

4.1 Pulpstones

Today the only used type of pulpstone is the so-called ceramic stone. All pulpstone manufacturers have their own detailed constructions which are based on the same basic structure – ceramic segments attached to a steel-reinforced concrete core. In most cases, each segment is fastened to the core using four steel studs or rods. The bolts can be preloaded before pouring the concrete to get stronger segment fastening. The spaces between the segments have elastic joint material filling.

There is also on the market a construction in which the segment fastening bolts go through the segment. In this construction, the bolt holes are usually kept open. This way the pulpstone surface has more pulp transportation capacity and better segment cooling than the other pulpstone types. Of course, the other pulpstone types can be equipped with axial or helical grooves or blind holes to achieve similar effects.

The pulpstone is fastened on the grinder shaft by pressing it between flanges (Fig. 19). The fastening torque is generally twice the nominal torque of the grinder.

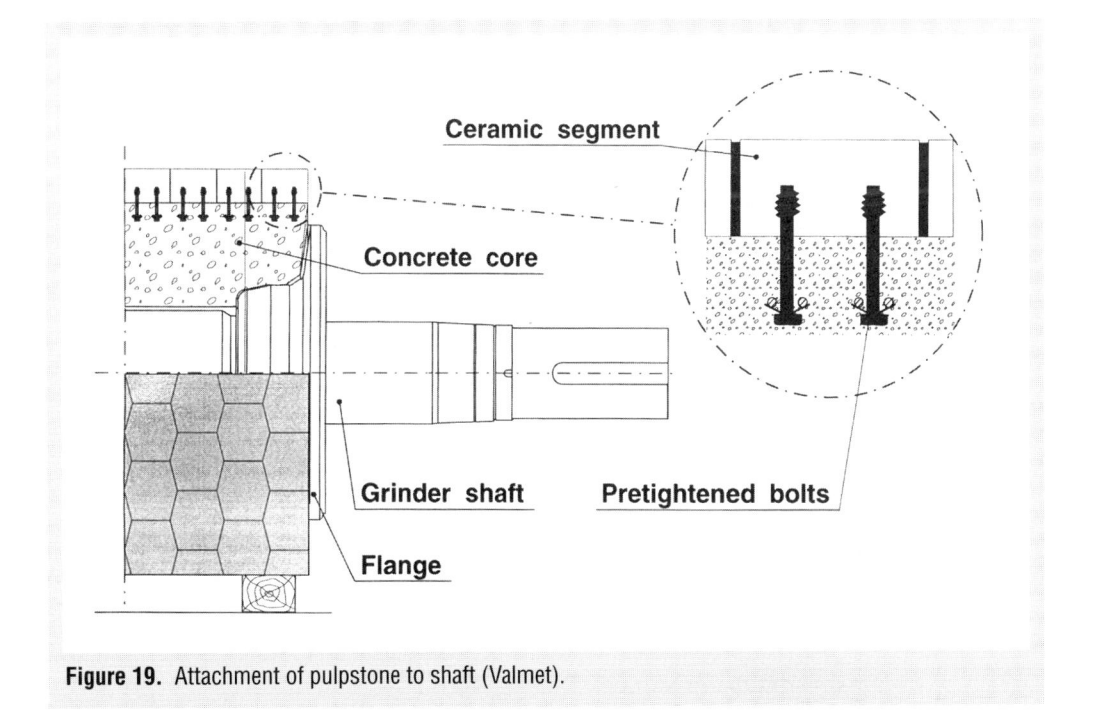

Figure 19. Attachment of pulpstone to shaft (Valmet).

4.2 Abrasive specifications and their influence on pulpstone behavior

Pulpstone abrasives are specified using the Standard Marking System of the Grinding Wheel Institute. The specification consists of six fields as in the following example for the A601-N7VG stone from Norton[26]:

Field no.	1	2	3	4	5	6
Example	A	601	N	7	V	G

Field 1: Abrasive type
Field 2: Grit size
Field 3: Grade
Field 4: Structure
Field 5: Bond type
Field 6: Manufacturer's symbol (of specific bond or characteristic of ceramics)

The following information discusses the range and properties as well as influence on pulpstone behavior for each field of the marking system.

Abrasive type

There are two main types of abrasives used in pulpstones:

A = aluminum oxide (ALUNDUM)

C = silicon carbide (CRYSTOLON)

This code letter can be preceded by a more detailed specification of abrasive. The detailed abrasive specification is manufacturer-specific. For instance, Norton uses numbers and Carborundum letters.

Aluminum oxide is mostly used in pulpstones. Silicon carbide abrasive is sharper, harder, and more brittle than aluminum oxide. Kärnä found that a silicon carbide pulpstone in pressure grinding produced more long fibers and somewhat stronger pulp, but a lower scattering coefficient. Also the specific energy consumption to reach the same pulp freeness was clearly increased when using a silicon carbide pulpstone as opposed to an aluminum oxide pulpstone[27, 28].

Grit size

The grit size of the abrasive is classified by a figure corresponding to the number of openings per lineal inch in the device that has been used to screen the grits. For instance, grits of size 60 will pass through a screen to which there are 60 openings/inch and not through a screen in which there are 70 openings per inch. Generally grit sizes from 36 to 80 are used for pulpstones. Table 3 gives nominal grit diameters for some of most common grit sizes.

Table 3. Nominal particle sizes for some grit classes according to FEPA standards.

Grit class	Particle size mm
46	0.297–0.420
54	0.250–0.354
60	0.210–0.297
70	0.177–0.250
80	0.149–0.210

The grit size is usually coded with three numbers. The first two indicate the size class of the grits. The pulpstone segments are often made with a combination of grit sizes. The third number specifies the grit blend by the manufacturer. For instance, Norton 702 pulpstone contains 50% of grit size 70 and 50% of grit size 80. The average grit size is thus 75.

The grit size for the pulpstone is chosen mainly on quality requirements set by the paper grade and the used wood species. The finer the required pulp (lower freeness) is, the finer pulpstone is normally utilized. Pulpstones used for paper grades generally have grit sizes from 54 to 80. On the other hand, pulpstones for board grades have grit sizes ranging from 36 to 54.

Grit size has an influence on vibration frequency, on heat distribution in the wood, and on friction forces that release the fibers from the wood matrix. With larger grits, the pressure pulses go deeper into the wood and the heated wood layer is deeper. Kärnä compared pulpstones having grit sizes 60 and 70 mesh. His findings indicate that size 70 produces smaller long fiber fraction, higher light-scattering coefficient, higher tensile index, and lower specific energy consumption when compared at the same freeness level[28].

Sandås and Lönnberg mixed different shares of 36-mesh grits into a 60-mesh abrasive. When the share of 36-mesh grits was increased, specific energy consumption at certain freeness was increased and the pulp strength properties were improved[29].

Grade

The grade or hardness of the ceramic segment is the strength with which the bond holds the grits. It does not relate to the hardness of either abrasive or the bond material, but only the resistance the bond offers against letting the abrasive be torn out of the ceramic. The grade is described with alphabets in sequence – J indicating soft and S indicating hard. Generally pulpstone grades range from K to P. The harder the grade is, the lower the wear of stone surface is and the more difficult it is to sharpen the pulpstone. If the pulpstone is too soft, it has a tendency to be self-sharpening. That means it wears out so fast that new grits come to the surface frequently and the pulpstone surface never reaches the level that would require sharpening. A pulpstone behavior of this kind is difficult, if not impossible, to control[26].

Structure

The structure is a measure of the porosity of the segments. The numerals indicate the relative spacing of grits ranging from a minimum of 3 to a maximum of 12. Experience has proven that there is a standard structure which best complements the grade hardness and the grit size of the segment. The structure seen most frequently in the pulpstones is 7[26].

The porosity of the segment partly determines how well the pulpstone is transporting water to the grinding zone. However, even if the pulpstone is very porous, the surface of the segments becomes plugged rather quickly. This is caused by fine material, lignin, resins, and extractives that are separated from the wood matrix.

Type of bond

All pulpstones have vitrified bonds. The letter indicating this bond type is V[26].

Manufacturer's symbol

The last symbol is the manufacturer's record mark and indicates specific modification of bond or other characteristic of the segment employed[27].

4.3 Pulpstone treatments

The purpose of the pulpstone grinding surface treatment is to control the production of the grinder and/or to make changes in pulp quality. During grinding, the surface of the pulpstone wears (although very slowly). The tops of the grits in the surface become rounded and can loosen from the ceramic bed. Additionally pores and cavities at the surface will be covered by a thickening layer of resins, lignin, and fines. These phenomena cause change in the quality of the produced pulp and in the operating point of the grinder or, in other words, the production level and the power consumed by the grinder change. At some point, a situation is reached where the pulpstone needs to be resharpened because of too low freeness and/or too low production. In most cases, the time between two sharpenings – the sharpening cycle – is for burr sharpening 1–3 weeks. With the newer waterjet pulpstone conditioning technique, the typical interval between sharpenings is from 1 day to 1 week. Pulpstones are treated using sharpening equipment fastened to the grinder body.

Sometimes the pulpstone also needs to be dulled. This happens, for instance, when the sharpening is made too vigorously or when the freeness target of the produced pulp is decreased in a paper grade change.

A third type of pulpstone treatment is truing (dressing) of the stone surface. The purpose of the truing is to make the pulpstone perfectly round and concentric to the grinder axis. Thus truing is always performed on a new pulpstone. Also periodic truing is necessary. Because of wear in normal use, the pulpstone circumference becomes smaller, providing space for fewer sharpening grooves and lands. When the grinding time since the last truing gets long enough, it is certain that some pattern disturbance and partial breakdown will occur. The wear of the pulpstone surface is also less at stone edges, which are outside the grinding area. Thus periodic truing is required to keep pulpstone surface "straight." Typically truing is done 1–3 times per year.

4.4 Burr treatments for pulpstones

Sharpening of the pulpstone traditionally has been made with metallic burrs. For this purpose, there is a sharpening lathe installed on top of the grinder on pocket grinders or on the side of the grinder on chain grinders. The burr itself is installed into a burr holder, which is fastened to the sharpening lathe.

Grinding and pressure grinding

The Valmet sharpening lathe (Fig. 20) has a horizontal sledge running on protected linear slides. The sledge is driven by a double-speed brake motor with gear belt and ball screw transmissions. A vertical piston is fastened to the sledge. The vertical movement is hand driven. Under the hand wheel, there is a micrometer scale to indicate the vertical movement of the piston and burr for the operator to be able to control sharpening depth.

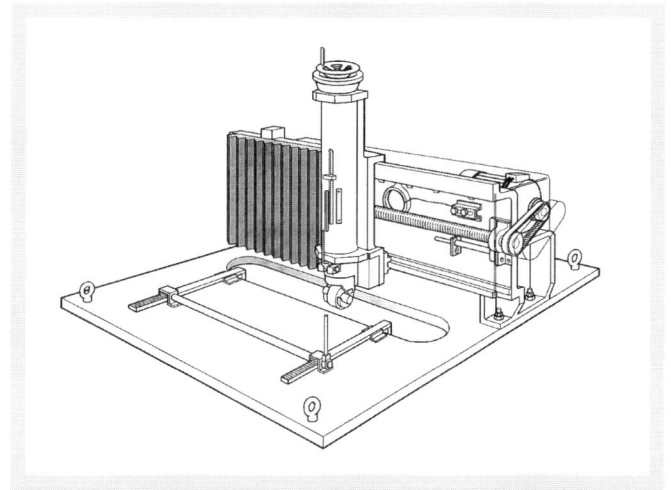

Figure 20. Sharpening lathe for (pressure) grinder (Valmet).

4.4.1 Pulpstone truing

Truing of the pulpstone is done with low lathe speed using diamond burrs. In truing, the depth of treatment is increased in small steps until the contact of the burr to the stone is even over the whole pulpstone width. After that, the surface can be treated with a finer, worn diamond burr to get the micro roughness of the stone surface to the suitable level.

4.4.2 Pulpstone sharpening

The burr sharpening produces a spiral pattern having grooves and land areas, in turn, in the pulpstone surface (Fig. 21). The burr sharpening deepens the grooves and reduces the land area, the area that contacts the wood. It also exposes fresh abrasive grits and removes impurities from the stone pores and surface. Smaller land area provides higher unit grinding pressure, resulting in a coarser pulp having higher freeness. The deeper grooves having higher void area at the stone surface bring more water to the grinding zone and are able to carry more pulp out of the grinding zone.

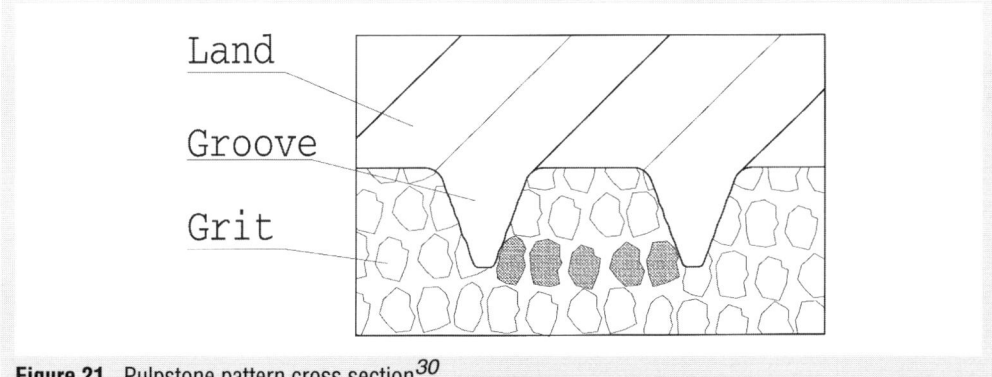

Figure 21. Pulpstone pattern cross section[30].

CHAPTER 6

The most important parameters that can be varied in the sharpening pattern are tooth frequency of the burr or groove frequency of the pattern, tooth angle or pitch of the burr, and the sharpening depth. The burr specification includes the two former ones. The burr number gives tooth frequency as teeth per inch and tooth angle of lead, which is specified in degrees.

Table 4. Recommended finest burr numbers for some pulpstone grit sizes[27].

Grit size	Burr number
46	10
54	10
60	10
70	12
80	16

For each pulpstone type, there is a minimum width for the base of the land to be strong enough to support the grinding load (Table 4)[27]. If finer burrs are used, it is very probable that the lands break and pulp quality is impaired.

Spiral burrs are manufactured with several tooth angles. The burr having an angle of 28° is generally used. The larger angle of 45° gives the fibers a longer treatment time on the land before the relaxing groove. This can result in some more long fibers and higher pulp strength at equal freeness.

The depth of the sharpening pattern is chosen according to the pulp freeness target, used burr, and experiences by each mill. Generally depths of 0.2–1.0 mm are being used for paper grade pulps.

The pattern depth can also be controlled by using a so-called controlled depth burr. The teeth of the controlled depth burr are much shorter than those of standard burr. The teeth are intended to be of the same depth as the desired pulpstone pattern. The burr is designed to be bottomed out on the pulpstone. This means that the land area between the burr teeth will touch the land area on the stone surface. Therefore, this type of burr also treats the lands of pulpstone pattern. The risk with the controlled depth burr is that the bottoming of the burr breaks the land area of the sharpening pattern, which will lead to a lower than expected pulp quality[30].

The traverse speed of the sharpening lathe is calculated with Eq. 2:

$$BT = k \times \frac{60 \times Ws}{Ss \times Wb} \qquad (2)$$

where BT is time of burr traverse, s
  Ws width of pulpstone, mm
  Ss pulpstone rotating speed, 1/min
  Wb width of burr, mm
  k coefficient for overlap.

In Valmet grinders, the constant sharpening speed of the lathe is designed for 40% overlap.

In order to get successful sharpening results, the sharpening equipment should be kept in good condition and the burr sharpening should be performed by experienced and skillful persons. By doing so, it is possible to get the grinder to operate close to the target operational point and to get pulp of specified quality shortly after the sharpening. The repeatability of the sharpening performed "manually", on the other hand, has often been relatively poor. Either the stone gets oversharp or it stays unnecessarily dull after the treatment.

Burr sharpening operations have been automated by Tampella and Roberts. Both systems measure the vertical position of the sharpening lathe fork and the burr diameter and keep in memory the previous sharpening depth. The target vertical positioning of the burr is calculated by adding the desired or programmed incremental depth. The Tampella device makes the vertical positioning of the burr during sharpening automatically driven by a motor. The Roberts device gives an operator target reading for the vertical burr position in the local panel. These equipment are able to reduce the variation caused by contact point detecting problems or by variable operator actions. The methods are sensitive to temperature changes and the mechanical condition of the sharpening equipment. The use of automated sharpening makes it possible to improve the sharpening to a level comparable to that of a skillful sharpener and to conduct light sharpenings fairly consistently[51,52].

4.4.3 Dulling of pulpstone

Dulling can be made with a brick fastened into a wooden arm or by using a dull diamond burr. Dulling should be avoided, if possible, because there is no controllable way to do it and dulling can decrease pulp strength.

4.4.4 Pulpstone grooving

The pulpstone transports a certain amount of shower water to the grinding zone. Part of the water is transported in the grooves of the sharpening pattern, and part forms a thin water film at land surfaces. In the grinding zone, fiber material is mixed into the water and, because of this, the thickness of the water film tends to increase. When the film gets thick enough in the end part of the grinding zone, it becomes more difficult to get the required wood to stone surface contact. The grinding action of the pulpstone is reduced. This can be counteracted by increasing the pressing force, but only to some extent. This phenomenon has been compared to hydroplaning. The pulpstone is not able to make full grinding action. The situation gets worse when a dullish pulpstone has low depth of sharpening pattern and when grinder production is increased.

According to calculations by Sollinger, the amount of water going to the grinding zone is about 100 ml/m^2. When producing low freeness pulp, a very shallow sharpening pattern is normally used. The groove volume of the pattern might be only half of the amount of water coming into the grinding zone. This means that there can easily exist a 0.1-mm thick water film on the stone surface. To reduce the risk of hydroplaning, the water pressure between pulpstone and wood has to be lowered. By increasing the groove volume at the stone surface, it becomes easier for the water and fiber material to move away from between the stone and wood[31].

The easiest way to increase the groove volume is to make the pattern deeper, which leads either to a higher freeness level or to reduced production. The best way to increase water transportation capacity is to make large enough voids in the stone surface. Both cut or molded grooves and holes have been used for this purpose with good results[32, 33].

4.5 Pulpstone surface control with ultrahigh pressure water

Conventional burr sharpening of pulpstone surface is difficult to control, and repeatability is even more difficult to achieve. As a consequence, quite wide freeness and quality deviations from the target level can often occur during a sharpening cycle. Variations in the average pulp quality will be reflected in variations in paper quality. Also some production is often lost. At the beginning of the sharpening cycle, the grinder power has to be limited to avoid producing too coarse pulp. Production is low at the end of the sharpening cycle because the pulpstone is dull.

These problems in quality and production ability of the grinder can both be avoided if the stone surface condition can be kept virtually constant over long periods. To achieve this, the pulpstone surface conditioning method has to be gentle, accurate, and easily controllable. The Valmet Water Jet technique is based on ultrahigh-pressure water, which is sprayed on the stone surface. The efficiency of the treatment is controlled by adjusting the water pressure.

This treatment is based on the idea that the dulling of the pulpstone, in reality, does not wear the pattern. Instead, it is thought to be caused by the fact that the grooves and the porous surface of the land will be filled with fines and resins. These materials fill the spaces between the grits on the land between grooves, decreasing the micro roughness of the grinding surface. Also the grooves between the lands get at least partly filled which, then, decreases the water and fiber transporting capacity of the pulpstone.

With the waterjet system, the pulpstone surface is cleaned by removing this "extra" material in such a way that the specific sharpness and condition of the surface of the stone will remain. Supposedly the waterjet also removes some bond material from between the grits, exposing them more or exposing totally new grits, and loosens some of the rounded worn grits (Fig. 22).

The target is to perform the waterjet conditioning so often that the change in the sharpness of the pulpstone is at a minimum. That is possible because the pressure of the waterjet is directly related to the conditioning force and it can be controlled accurately. This means that the production of the grinder can be kept at a high level without changing the freeness and the quality of the pulp considerably. Additionally there are no production losses caused by the sharpening itself because the conditioning is performed under normal grinding[34].

Grinding and pressure grinding

BEFORE SHARPENING

Resin + other impurities

BEFORE CONDITIONING

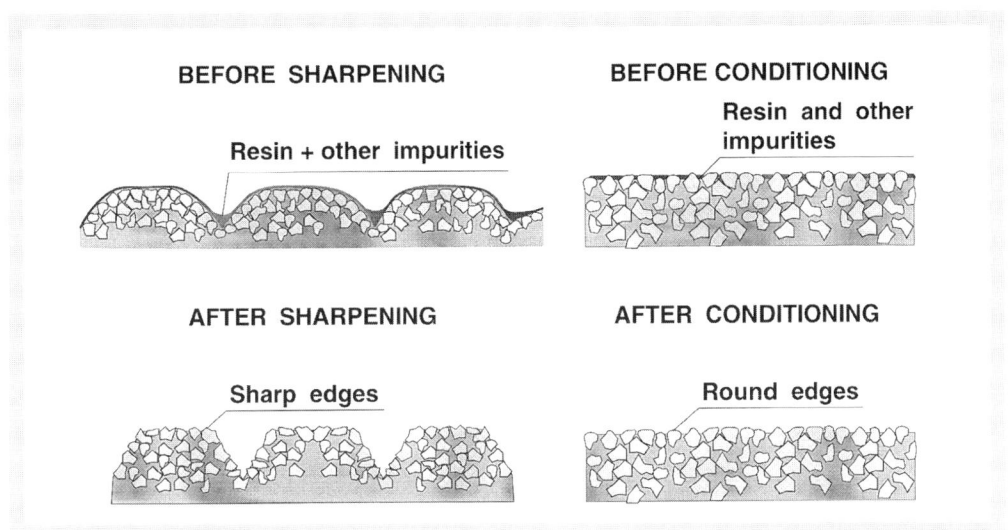

Resin and other impurities

AFTER SHARPENING

Sharp edges

AFTER CONDITIONING

Round edges

Figure 22. Sharpening pattern of the pulpstone surface before and after burr sharpening (left) and waterjet conditioning (right)[34].

The plugging of the stone surface will decrease the heat transfer ability of the stone surface because the shower water cannot penetrate through the plugged stone surface and transfer the heat out from the grinding zone anymore. With the waterjet conditioning, it is believed to be possible to keep the stone surface more open than with the conventional burr sharpening. In this case, the smaller temperature difference in the radial direction of the pulpstone segment will help to prevent heat-shock damages.

In the conventional sharpening, the inelastic sharpening lathe and burr combination causes a big mechanical stress against the segments of the pulpstone. The mass of the jet is so small that it does not cause any stress to the fastening construction with which the segment is fixed to the body of the pulpstone.

4.5.1 The waterjet equipment

The ultrahigh-pressure unit produces a water pressure of up to 2400 bars. The desired pressure can be set from 500

WATERJET - SYSTEM

2 jet devices

User display remote control

Control valves

Ultrahigh pressure unit

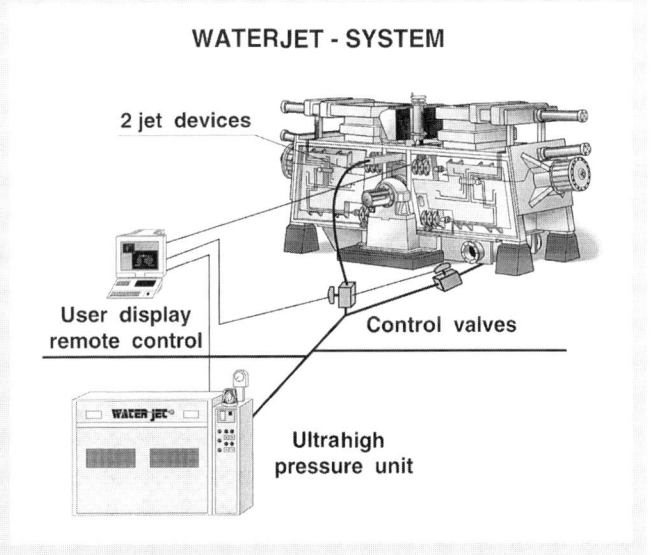

Figure 23. Equipment for waterjet conditioning system (Valmet).

to 2400 bars, and the internal logic controls the pressure accurately. The special tubing leads the ultrahigh-pressure water to the grinders. There are two jet devices that are mounted on the side walls of the grinder, one on each side. The jet device has an oscillating nozzle arm, which is connected to the tubing with special valve and hose construction. The operation sequence is totally remote-controlled by the mill DCS system (Fig. 23).

4.5.2 Pulpstone treatment with waterjet

The pulpstones that are waterjet conditioned, are also in most cases grooved to give the stone surface without any spiral burr pattern enough volume for removal of the defiberized material from the grinding surface. The new pulpstone is trued, and a shallow diamond burr pattern is normally cut into the surface to bring the stone surface into target grinding condition. The coarseness of the diamond burr has to be selected case-by-case depending on the used pulpstone type, wood species, target freeness, etc. The waterjet treatment, interval, jet pressure, and number of treatment cycles have to be chosen for each stone to meet the targeted operation point.

4.5.3 Impacts of waterjet conditioning on grinding process variables and pulp properties

Research has shown that waterjet conditioning stabilizes the grinding process significantly. Variations in stone sharpness have been reported to decrease by 50%–60%, in motor power by 50%, in production rate by

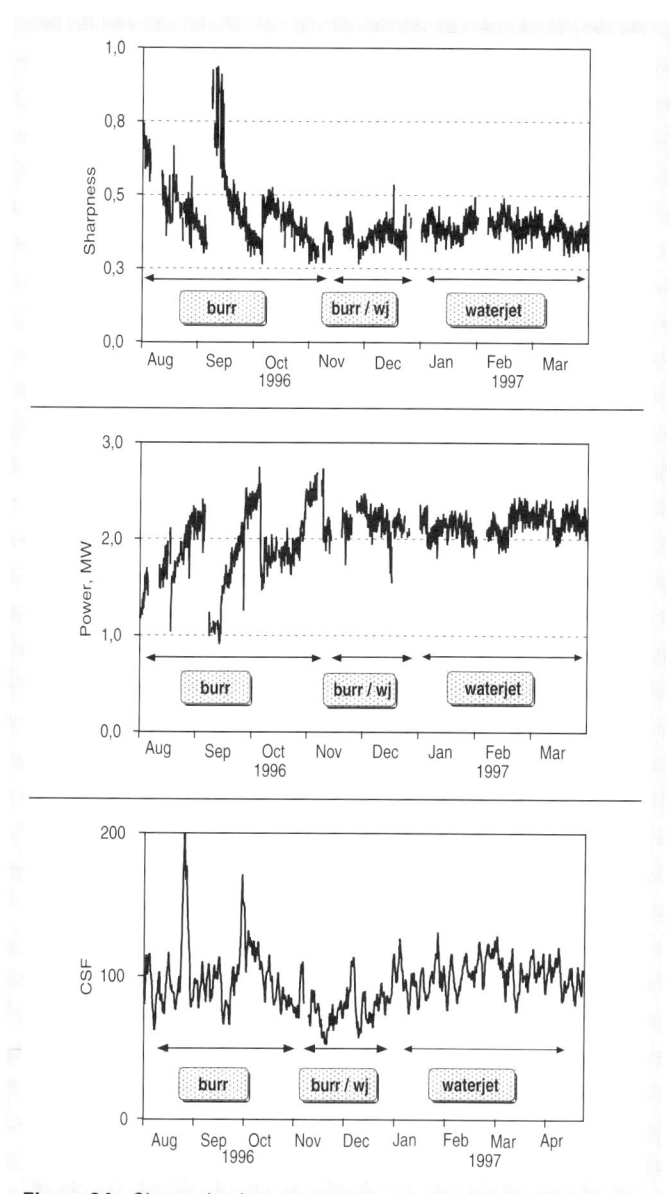

Figure 24. Change in sharpness, motor power, and pit freeness of the grinder during the burr sharpening and waterjet conditioning periods[36].

45–60%, and in specific energy consumption by 27%–80%. This reduction in process variations means that the operating point of the grinder can constantly be kept quite close to the target, and it is easier to optimize other grinding variables, like pulpstone showering and pulpstone type[35, 36].

A more stable grinding process promotes higher production as well as a more stable and better pulp quality (Fig.24). Production has been reported to increase substantially. Puurunen estimates a 10%–20% increase in production potential of the grinders. With waterjet sharpening, freeness variations have been reduced by 24%–32%. The variations in shive content and fiber fractions decreased as well, except for the +14 fraction[35,36].

Manner et al. found that the fine fraction under waterjet conditioning contains a higher proportion of fibril particles than pulp manufactured using burr sharpening. The fine material seemed to work well in increasing bonding between chemical pulp fibers. Laboratory sheets having 10% kraft pulp and 90% PGW pulp had better tensile index when the PGW pulp was made under waterjet conditioning. Consequently, the installation of the waterjet conditioning was followed by a stepwise increment in paper tensile strength[34]. Puurunen found that the character of the long fiber fraction of atmospheric groundwood pulp was improved considerably with waterjet sharpening (Table 5)[35, 36].

Table 5. The effect of the waterjet method on the character of long fibers (CSF 50–60 ml)[35].

The effect of waterjet conditioning	
Tensile index	+ 15%
Tear index	+ 15%–20%
Apparent density	–30%

5 Grinder feeding systems

The pulp production of the grinders requires a uniform and continuous wood supply. The wood processing system is also supposed to provide an uninterrupted flow of clean logs with correct size to the grinder room to meet the requirements for pulp and paper quality. Logs have to be clean because already minor quantities of bark cause black spots in paper. Log dimensions have to be within the allowable limits: a too short log causes poor piling on the conveyor line and a too long log disables line filling and grinder charging. Also too large log diameter and too poor log shape can cause problems with the system function. For this purpose, the logs need to be sorted after the debarking drum in the woodroom.

Today there are log feeding systems of two main types. One is for continuous operating grinders, and the other one is for the pocket grinders. The feeding systems for continuous operating grinders pile the logs on a feeding line with successive belt conveyors and the logs are supplied to the grinders through the gaps between the conveyors. The batch-type charging systems make log batches of suitable size with the help of the conveyors and batch forming units. Then the system supplies the batches to the pocket grinders.

The operation of a modern log feeding system is highly automated. This is a necessity to ensure a continuous wood supply to the grinders. The systems take care of the whole wood feed to the grinders, that is, line filling, batch forming, and wood supply onto the grinders. The operator is required only in case there are disturbances. The

CHAPTER 6

whole charging system is remotely supervised from the grinder control room by means of a DCS control system and a closed-circuit TV system which monitors the most critical points of the system.

5.1 Feeding systems for the chain grinders

The feeding system for continuously operating grinders (Fig. 25) offers an automatic flow of logs, which ensures steady feeding of the grinders. This allows even some accidental fluctuations on wood intake because the entire conveyor line serves for a log storage.

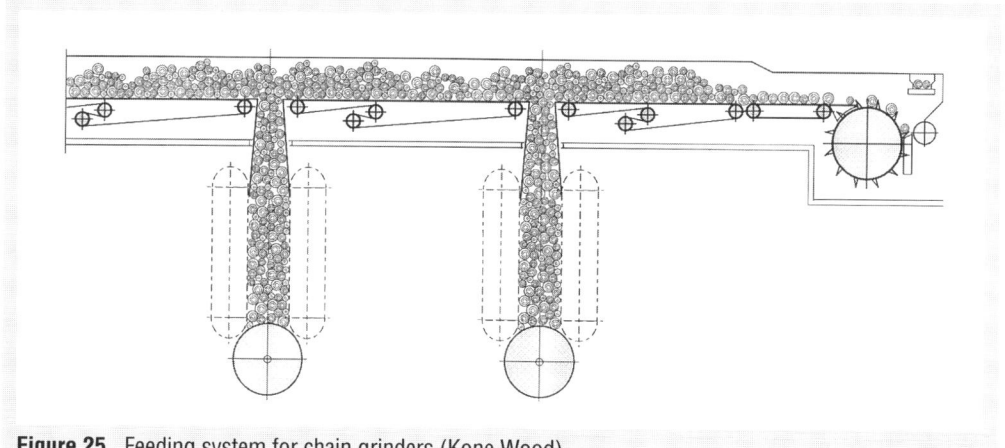

Figure 25. Feeding system for chain grinders (Kone Wood).

The feeding conveyor line is installed above the grinders. The line consists of a series of belt conveyors designed so that there is an opening above each grinder between two successive belts. Speedup rollers and a log aligner bring the aligned logs onto the first belt conveyor. The first two belt conveyors pile the logs to a height of approximately 1.2 m, and the logs are stored at the same height along the entire conveyor line. To enable this, the running time of the conveyors is the longest at the head of the system, and is gradually reduced toward the end of the line. Running times can be adjusted within wide tolerances from the control panel[37].

The log stream moves slowly forward on the conveyor line and downward into the feeding chutes of the grinders. The entire row of fixed belt conveyors also serves for wood storage. Conveyors start and stop automatically by pre-selected time sequences. Each of them can be separately driven from the control panels in either direction, if so required[37].

The operator's main function is to supervise and control the feeding and piling of logs on the conveyor line. All other operations are automatic. One person can supervise the operation of two adjacent feeding lines.

5.2 Charging system for the pocket grinders

The automated grinder charging system for pocket grinders consists of a speedup roller conveyor, log aligner, belt conveyors, and batch forming units[38] (Fig.26).

Grinding and pressure grinding

Figure 26. The log feeding system for the pressure grinders at Metsä-Serla Kirkniemi Paper Mill.

The speedup roller conveyor straightens and separates the incoming logs and feeds them correctly aligned into the log aligner pocket. In addition, the conveyor in most cases incorporates a washing station and a short-end removal trap. Washing is advantageous for several reasons. It removes sand coming with the logs taken from the stockpile for debarked wood. It makes the surface of logs more slippery, thus making it is easier to transport them further in the feed system. Washing also reduces the amount of debris going into the grinder.

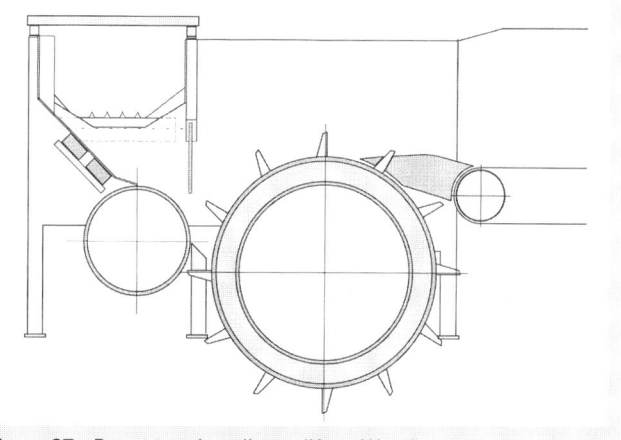

Figure 27. Drum-type log aligner (Kone Wood).

The log aligner aligns the logs and lifts them onto the belt conveyor line. The drum-type log aligner consists of two rotating drums: a smaller feeding drum and a larger flight drum (Fig. 27). When rotating, the smooth feeding drum presses the log against the flight drum which, in turn, lifts the log to the first belt conveyor. Compared to the older jackladder type log aligner, the drum-type log aligner is more silent and offers a more reliable and safer operation with less maintenance[38].

CHAPTER 6

The conveyor line includes several belt conveyors that are located before and between the batch forming units (Fig. 28). The first conveyor in the line is a double belt conveyor. Its first part runs only forward, and the second part forward and backward according to the pre-selected time sequences. In this way, the logs pile up to a stack approximately 1.2 m high. The rest of the conveyors move the logs forward in the line, keeping the line full and providing extra storage for the system[38].

The level control system keeps the logs at a sufficiently high level on the conveyor line all the time. The through-beam photo cells placed on the conveyor skirtboards detect the height of the log pile. According to the information from the cells, the automatic Level Control System runs the conveyors, wood intake, piling, and line filling so there are always enough logs above each operating grinder[38].

The batch forming units take care of automatic forming of the log batches. The units are located by each grinder pocket in the grinder charging line. The batch forming unit consists of hydraulically driven grippers and sinking boards, pushing devices, and feed channels[38].

To form the batch (Fig. 28), the sinking board rises to the upper position and the grippers open. Then the sinking board lowers to the batching position and the grippers close when the batch is full. When closing, the grippers remove possible extra logs from the sinking board, ensuring a correct batch size. In addition, the grippers prevent the logs from falling down when there is no batch forming going on[38].

After the grippers have closed, the sinking board goes down to the lower position. Then the batch is pushed through the feed channel onto the grinder feed gate or into a reserve position located beside the gate. All the sequences from forming the batch to transferring the logs onto the grinder gate are automatic[38].

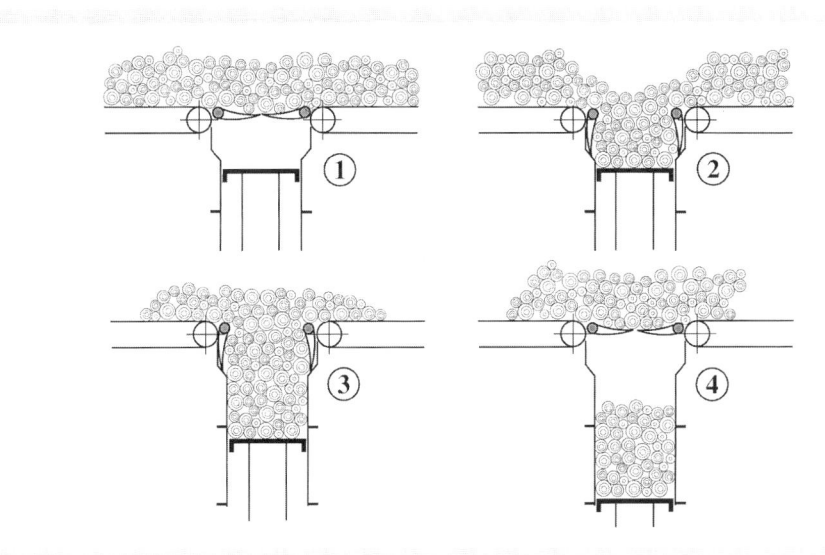

Figure 28. Stages of automatic batch forming for a pocket grinder: (1) sinking board into upper position, (2) opening of grippers, (3) sinking board into batch forming position, and (4) closing of grippers and sinking board into lower position.

6 Process control

6.1 Introduction

The process control of the grinding process can be divided into two sections based on their target and action times. For pocket grinders, the load controls (GLC) are taking care of the short-term variations related to a single grinding stroke with typical action times of 2–10 seconds.

Usually there is more than one grinder that delivers pulp for a certain paper or board machine. The grinder group control (GGC) takes care of long-term variations and controls the grinders so that the pulp production and quality demands will be satisfied with the lowest possible quality variations. Typical action times for this control are 2–10 minutes.

This chapter discusses sources of variations, the classical grinding model, and realizations of grinder load controls as well as grinder group controls.

6.2 Sources of variation

The variations in the groundwood process can be classified according to their lifetime to short- and long-term variations. The sources of short-term variation are:

- Changes in wood diameter
- Different operation of front and back pocket (pocket grinder)
- Changing batch size from batch to batch
- Batch density changes as function of grinding stroke face (pocket grinder)
- Tilted patch (chain grinder)
- Different grinding speeds in the patch sides and center (chain grinder).

The list of the things that could be used as the control variables in short-term operation are numerous:

- Pressure inside the grinder (PGW)
- Shower water temperature
- Shower water flow
- Stone immersion in the pit
- Grinding shoe pressure
- Speed of chains.

Because most of the variables not only change the pulp quality but also have effects on process stability, energy balance, and stone life, only the control of grinding shoe pressure in the pocket grinders and the chain speed in the chain grinders can be regarded as applicable.

CHAPTER 6

The sources of long-term variation are:

- Pulpstone type
- Stone sharpening pattern (burr type and depth of the pattern)
- Wood quality and moisture
- Pulpstone sharpness.

The pulpstone type is chosen based on the end product as described earlier in this chapter. A stone change is also a very laborious and capital intensive task, and therefore cannot be regarded as a control factor.

Using the wood quality and moisture for quality control is not so easy either. These things are not easy to measure or even to characterize, nor are they controllable after they have changed from the initial stage. This does not mean that one can neglect this issue, but merely that the variations in raw material ought to be taken care of before they enter the grinder room. The sharpening pattern is chosen by the stone grit size and is then usually kept constant because the stone should be dressed before a new pattern could be introduced to it.

Pulpstone sharpness changes with grinding time. When the stone gets too dull to keep up the desired pulp production and quality, the stone will be resharpened. Traditionally, this is done by a burr. Because this act is not so easy to handle, it usually changes the pulp quality quite drastically. The new stone condition technique with high-pressure water, waterjet, drastically reduces the variation caused from this act[34-36].

6.3 The classic grinding model

To be able to understand the principles of grinder controls, one has to know how the process works. This can be done by building a model that links the process variables and pulp quality together. The principles described below apply both for two-pocket and chain grinders.

The most commonly used model, which determines the state of grinding stone in the process, is called the stone sharpness model (Eq. 3). This term is somewhat misleading because it does not depend only on the grinder stone surface, but merely determines the interactions between the stone and wood.

$$m = A \cdot S \cdot \left[\frac{P}{u}\right]^\alpha \cdot u^\beta \qquad (3)$$

where m is Pulp production rate
S Stone sharpness
A Constant
P Grinding power
u Stone peripheral speed
α Exponent, usually 1.6
β Constant

For mill scale operation, Eq. 3 can be reduced to Eq. 4. This model was first stated by Bergstrom[39] for a laboratory grinder. Later it was verified in mill scale by Paulapuro[40].

$$m = S \cdot p^{\alpha} \tag{4}$$

where m is   Pulp production rate
        S Stone sharpness
        P Grinding power
        α Exponent, usually 1.6

Kärnä has presented Eq. 4 in a somewhat different form[41]. He found that the alfa factor is not constant, but tends to decline when the grinding energy increases, i.e., the stone becomes duller. Equation 5 describes this relation.

$$m = S \cdot P^{a - b \cdot S} \tag{5}$$

where m is   Pulp production rate
        S Stone sharpness
        P Grinding power
        a Constant
        b Constant

Equations 3–5 do not have a time dimension. It was found by Paulapuro that the stone sharpness decrease rate is increasing with increasing grinding energy. Based on his studies, a model for stone sharpness decrease was generated (Eq. 6)[40].

$$\frac{\partial S(t)}{\partial t} = -c \cdot P(t) \cdot (S(t) - a) \tag{6}$$

Integration of Eq. 6 results in Eq. 7, which gives a stone sharpness at a certain time, when initial sharpness, final sharpness, and grinding energy since resharpening are known.

$$S = a + be^{-c \cdot E} \tag{7}$$

where S is   Stone sharpness
        a Limit sharpness for the stone, value depends on grit size etc.
        $a+b$ Initial sharpness of the stone after sharpening ($E=0$)
        c Constant for sharpness decreasing rate
        E Energy used for grinding since sharpening

CHAPTER 6

In his studies, Bergstom *et al.*[39] found that most pulp quality values depend on the specific energy which is applied to the pulp during production. Based on this observation, he formulated an experimental formula (Eq. 8).

$$Q = K \cdot S^a \cdot \left[\frac{P}{u}\right]^b \cdot u^c \tag{8}$$

where Q is Pulp quality parameter, i.e., CSF
K, a, b, c Constant
S Stone sharpness
P Grinding power
u Stone peripheral speed

For mill scale grinders, the stone peripheral speed is constant and Eq. 8 can be reduced to Eq. 9. In this formula, Q can be any pulp quality parameter. This model has been found to be most applicable for freeness, tear index, and burst index[42].

$$Q = K \cdot S^a \cdot P^b \tag{9}$$

6.4 Load controls of a two-pocket grinder

The implementations of grinder load control are numerous, but the basic principle in all the load controls is the same – to control the opening degree of the high-pressure valve (Fig. 29).

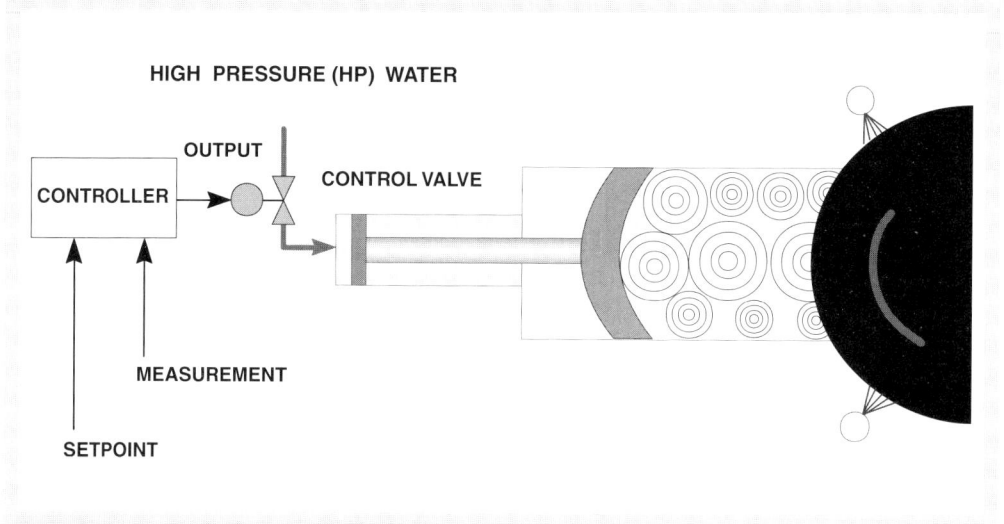

Figure 29. Principle of two-pocket grinder load control.

Grinding and pressure grinding

It is characteristic for this control that the actuator has only one effective direction. It can add pressure in the hydraulic cylinder, but the pressure decreases only when the grinding takes place. In early implementations, there was only one hydraulic valve, which gave equal pressure for both pockets or even for two grinders. This arrangement, in many cases, did lead to a significant imbalance between pockets because the grinding circumstances are not equal for both pockets or separate grinders. This phenomenon is the main reason why the load control with pressure feedback yielded to other load controls. It would be ideal to control the specific energy consumption during the grinding stroke because it has been found to have a good relation to pulp characteristics[41]. In any case, this kind of control is not very applicable because there is no practical way to divide grinding power between individual pockets.

Currently the most common way to implement the load controls for modern two-pocket grinders is to use shoe speed as the objective. Figure 30 shows the principle of constant shoe speed control. In this mode, the shoe speed is measured with either a pulse sensor or an absolute sensor. Using this signal as the feedback input for the controller, the hydraulic water flow to the cylinder is controlled in order to maintain the set value given by operator or computer. Because of limited grinding power, this control mode also has a watchdog system that changes the controller to power mode, when the power limit is reached. The mode is then changed back to speed mode when the power yields or when the shoe speed goes over the setpoint.

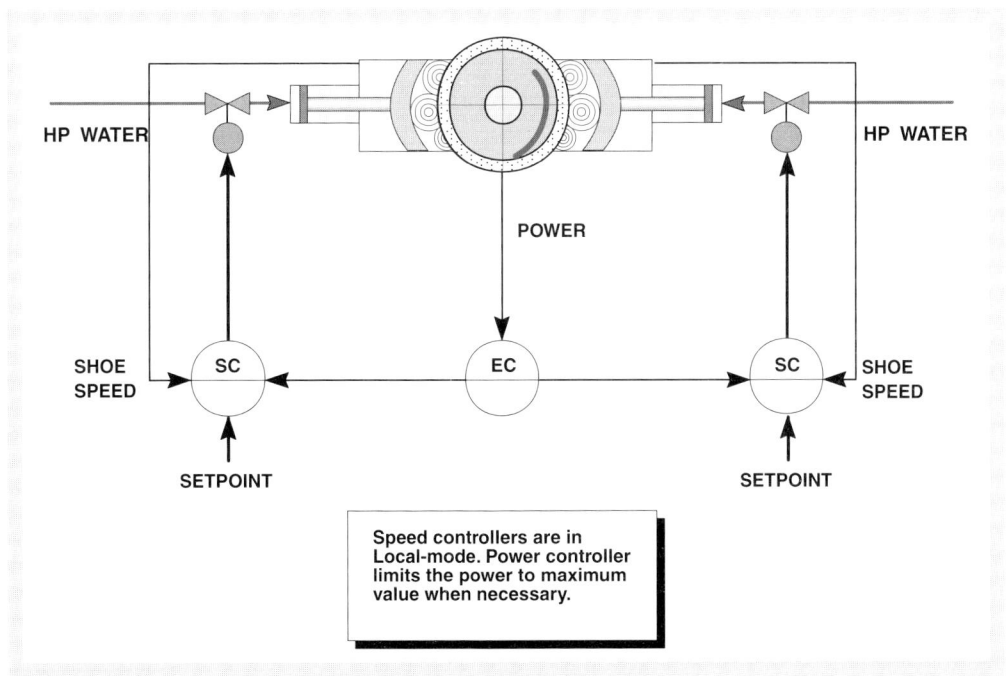

Figure 30. Principle of two-pocket grinder shoe speed control.

CHAPTER 6

Figure 31. Principle of two-pocket grinder power control.

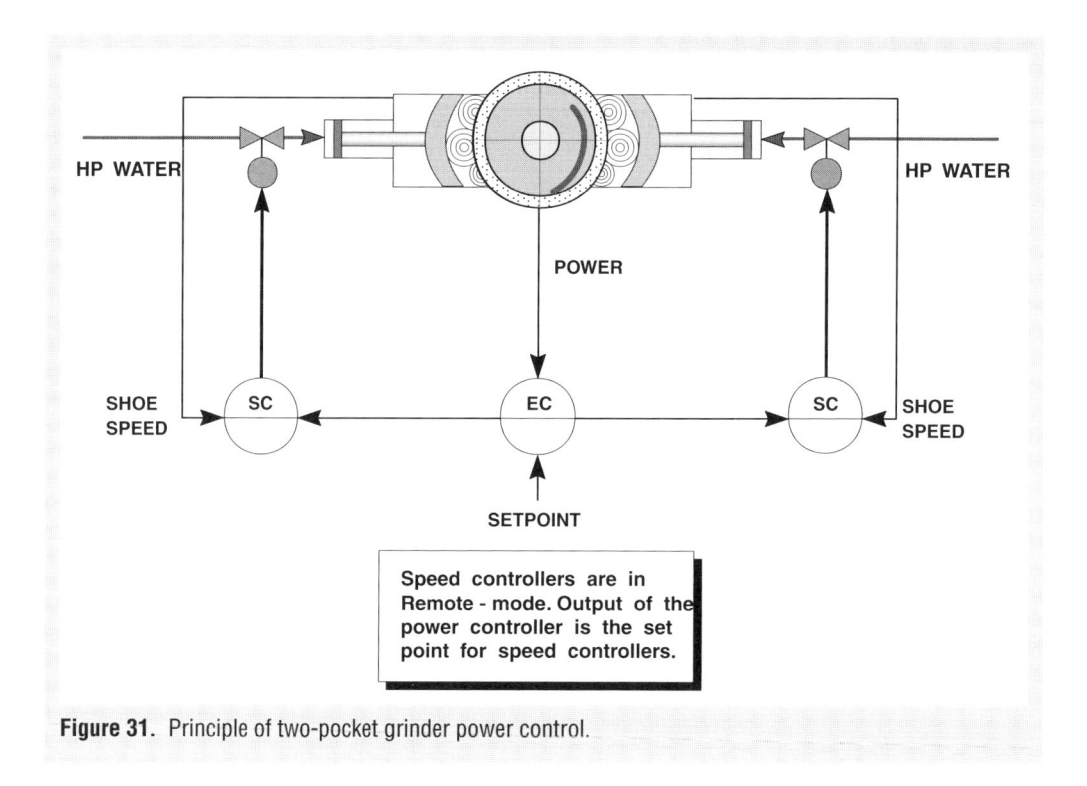

Another commonly used control mode is grinding power control, where grinder power is used as the primary setpoint. Figure 31 shows the modern implementation. The shoe speed controllers act as inner loops in this cascade control, ensuring balance between individual grinding pockets.

The constant shoe speed mode has one major drawback. While the shoe speed is kept constant, the density in the log batch charge will increase as can be seen in Fig. 32[43]. This phenomenon has been taken into account in so-called programmed shoe speed control. In this control mode, instead of using constant shoe speed, the setpoint is varied during the grinding stroke, as illustrated in Fig. 33. In the beginning of the stroke, the shoe speed is increased by 30% and then slowly decreased toward the end of the stroke until it is 30% lower that the setpoint.

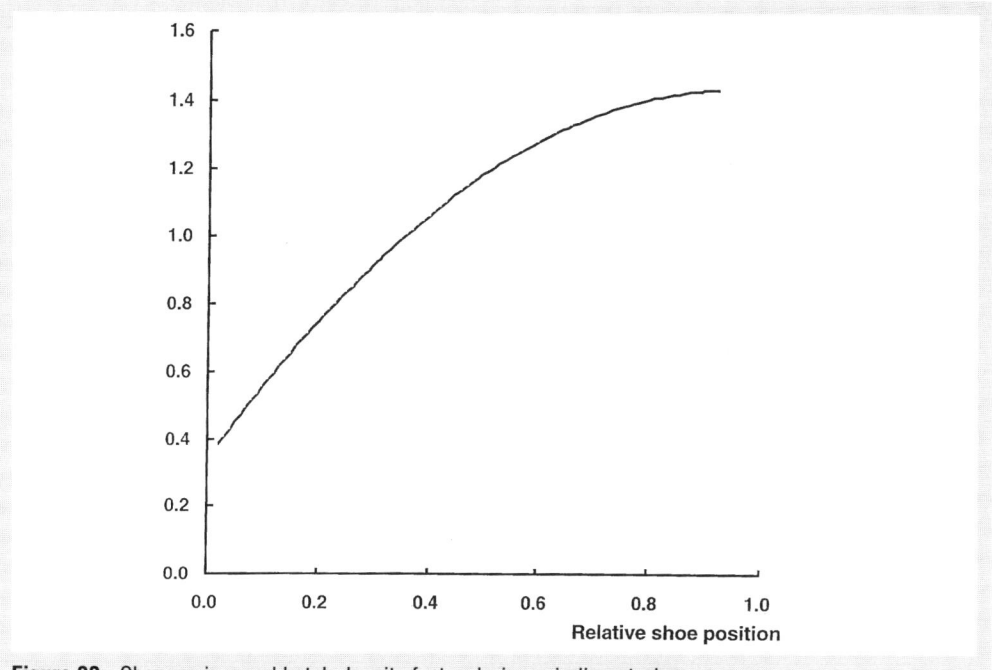

Figure 32. Changes in wood batch density factor during grinding stroke.

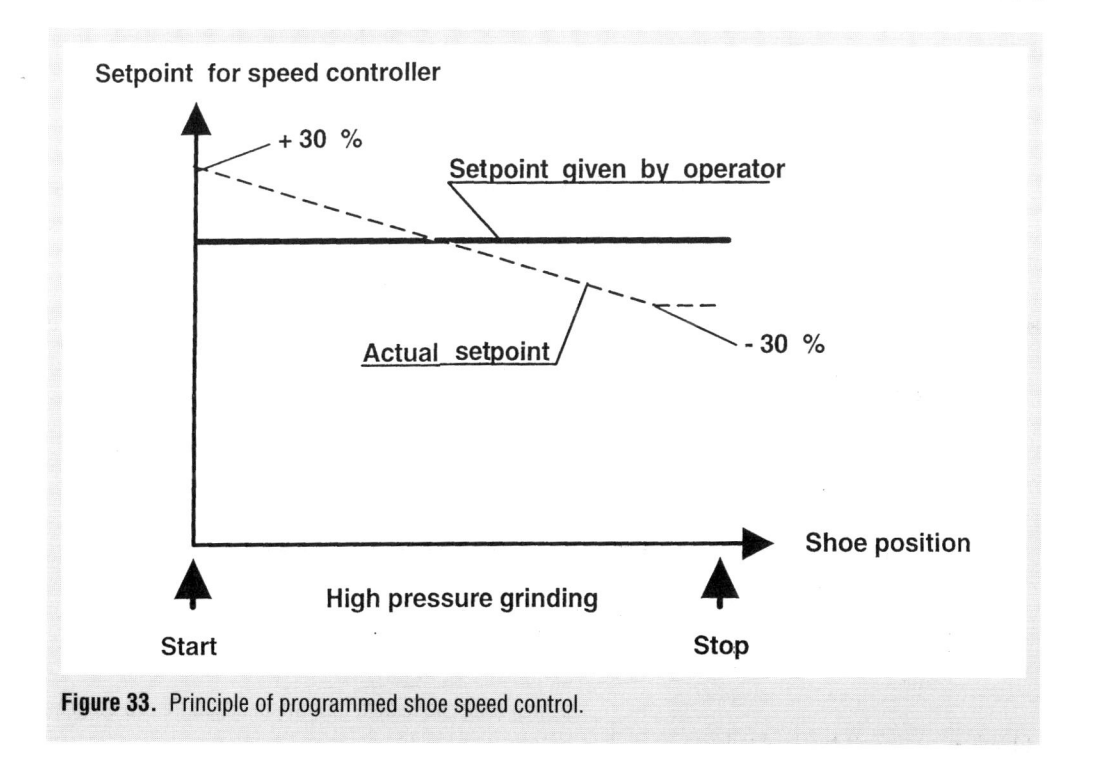

Figure 33. Principle of programmed shoe speed control.

CHAPTER 6

6.5 Pros and cons of different control modes

Figures 34 and 35 show the behavior of grinding power and pulp production during one grinding stroke in different load control modes.

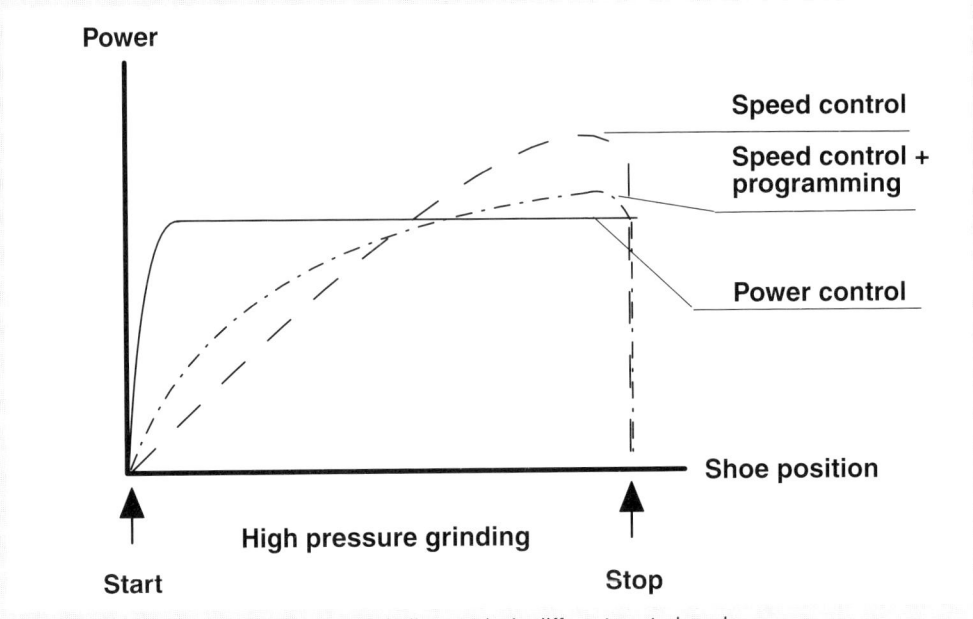

Figure 34. Grinding power during one grinding stroke in different control modes.

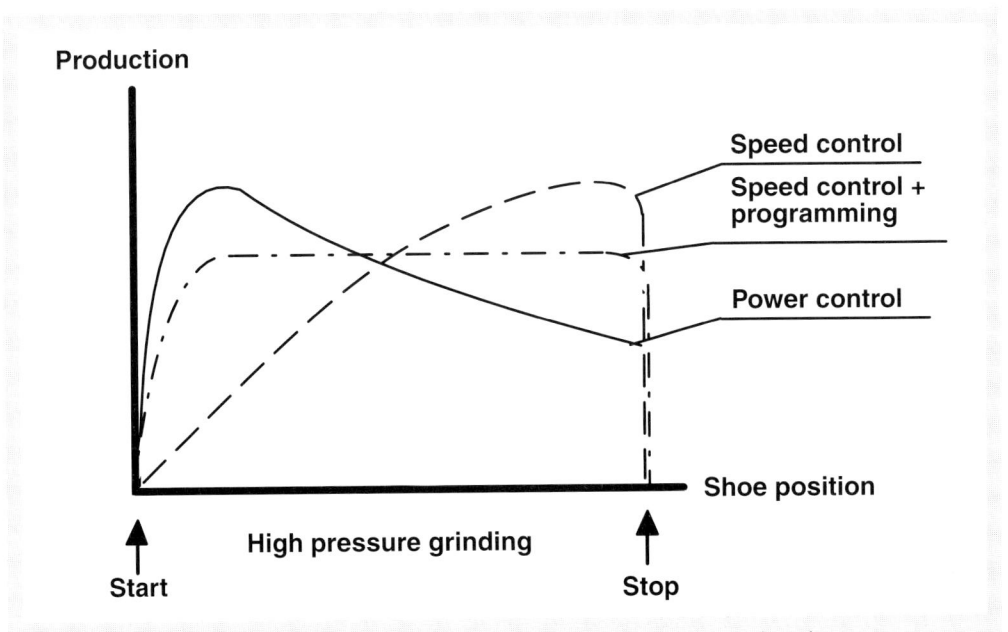

Figure 35. Grinder production rate during one grinding stroke in different control modes.

When the grinder is run in the power mode, shoe speeds are controlled to achieve the desired grinding power. In the beginning of the stroke, wood logs are loosely packed and the actual grinding area is low. Using constant grinding power, high freeness will be produced because the pressure shoe has to be pressed strongly to achieve desired grinding power. In the end of the stroke, the wood logs are tightly packed and grinding area is big. Now grinding with constant power will give low pulp freeness. Despite the quality drawback, this load control mode generally yields the biggest pulp production.

For the constant shoe speed mode, the beginning of the grinding stroke will yield low freeness pulp because the grinding area is small. At the end of the stroke, the logs are tightly packed; and maintaining the same shoe speed results in high pulp production, high power, and high freeness.

The programmed shoe speed control mode is considered to be the best in terms of short-term pulp quality variations. This mode takes the batch packing factor during the grinding stroke into account by changing the shoe speed around the setpoint as illustrated in Fig. 33. This mode yields almost constant production rate and uniform pulp quality during the grinding stroke. In this control mode, the grinding power also tends to increase toward the end of the grinding stroke. To supply the best possibilities for the programmed shoe speed control, some of the grinder capacity has to be sacrificed in order to minimize the pulp quality variations.

6.6 Load controls of a chain grinder

As explained earlier in this chapter, the massive chains between the wood patch and grinder body generate the wood feeding force in the chain grinder. DC electric motors normally drive these chains. Because the grinding power is usually the restricting factor in a chain grinder, the power control gives the biggest production and, therefore, is most commonly used. In this mode, the output of feed motor controllers are adjusted based on grinder power feedback[45].

Even though there are no pocket recharges in a chain grinder, the wood density varies substantially in the wood patch. The feed chains only have a grip of the logs against the side walls; therefore, feed rate tends to be lower inside the wood batch.

Through the years, more sophisticated control modes for chain grinders also have been introduced[45]. Many of them have the inadequacy of the chain speed giving only an arbitrary estimation of grinder production. Therefore they have not been much more successful than standard power control.

6.7 Grinder group controls

Usually there is more than one grinder that delivers pulp for a certain paper or board machine. The grinder group control (GGC) is bound to take care of long-term variations and control the grinders so that the pulp production and quality demands will be satisfied with the lowest possible quality variations. The same principles apply both for two-pocket grinders and chain grinders.

In Fig. 36, there is a graphical interpretation of the grinder model described earlier. The graph consists of constant value curves for grinder varying stone sharpness

CHAPTER 6

and pulp freeness. The principle of grinder group controls can be explained by using this model. The state of a single grinder is defined by the constant sharpness curve and grinding power. At a certain time, the state of a grinder can only be changed along this constant sharpness curve (e.g., from A to B).

The sharpness constant value curves cross the constant value freeness curve. When the grinding power is increased, the operating point of the grinder moves along the sharpness constant value curve, crossing higher and higher freeness curves. In other words, the pit pulp freeness can be increased by increasing the grinding power. In the same manner, the pulp freeness can be decreased by lowering the grinding power. This means that the grinder production and pulp quality are tied together with stone sharpness; it is not possible to change pulp quality without changing the production. To keep up the desired production and quality, the grinder group controls compensate long-term variations by changing the production targets of grinders in the group based on the grinding model and/or direct process measurements.

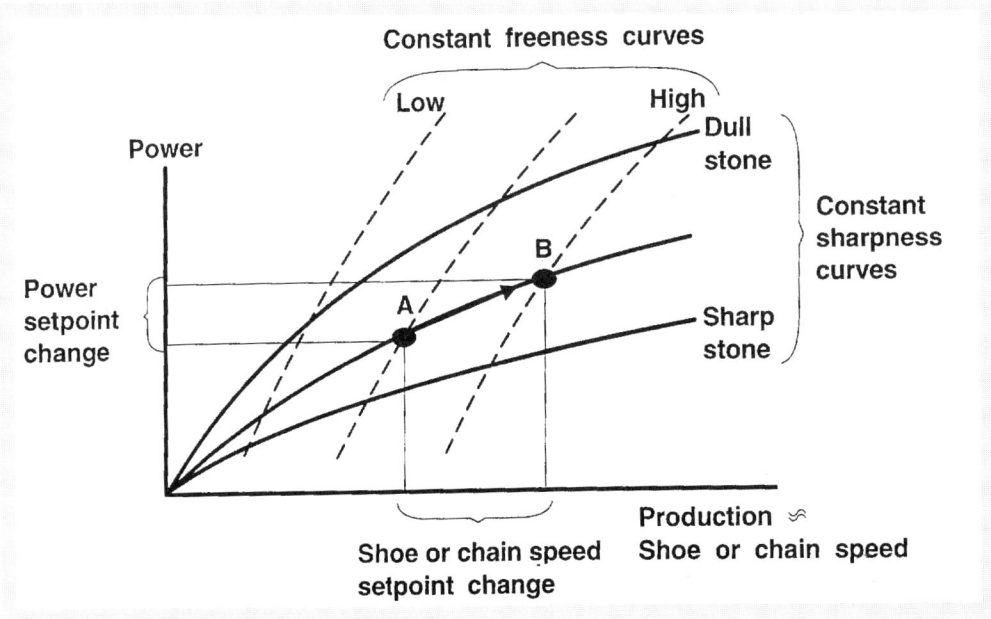

Figure 36. Constant value curves for sharpness and pulp freeness in grinding model.

Figure 37 illustrates a simplified grinding process flow sheet. In many commercial implementations, pulp quality (freeness) and production targets are set for the final pulp. This control loop changes the decker pulp freeness target into the combined pit pulp target, assuming the screening stage as a black box with constant gain. This target is then changed to setpoints for individual grinders. This procedure, in most cases, is based on grinder models.

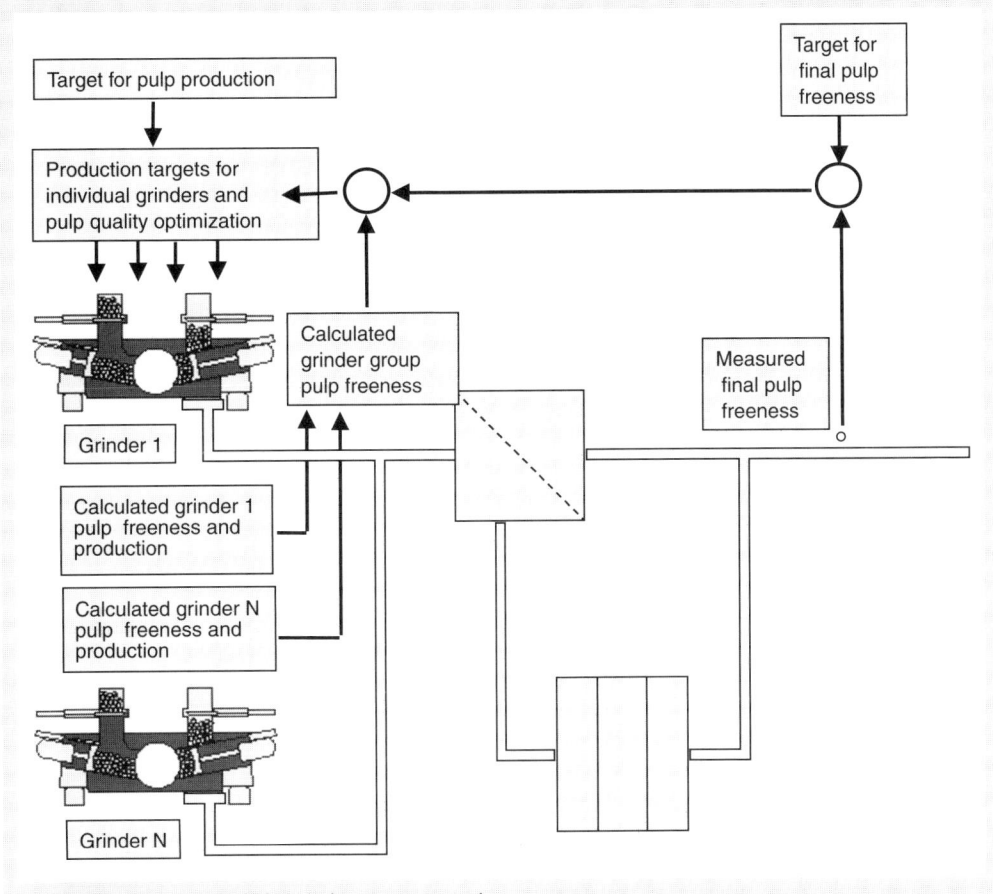

Figure 37. Base principle in grinder group control.

6.8 Commercial products for grinder group controls

Through the years, there have been numerous approaches for grinder group control. Most of them were based on classic grinding models. AGMO was one of the novel systems to integrated grinder group control in the 1970s[44]. It was based on classic grinding models and had the ability to simulate and compare various control strategies beforehand. This system never attained broad commercial interest.

The KAJAANI 4000 (later KAJAANI 64G) system was developed in the early 1980s, and it was one of the first commercial grinding control systems. Characteristic for this system is the target area control, which intended to ensure that the pulp from the grinders in the group do not differ too much from the target. The idea of target area is revealed in Fig. 38[46]. The shaded acceptable pulp quality area is bordered by minimum and maximum grinding power lines, minimum and maximum freeness lines, and minimum and maximum sharpness curves. Inside this target area, the control system performs its strategies in order to satisfy the setpoint for pulp production and freeness. It also optimizes the pulp quality at a certain freeness level by loading the dullest stones the most

CHAPTER 6

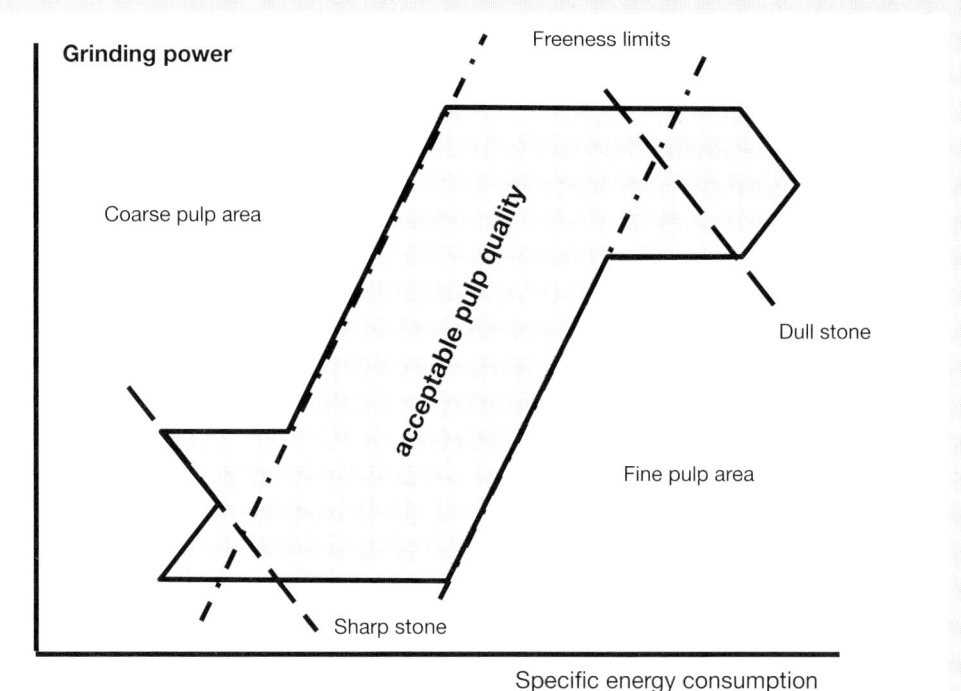

Figure 38. Target area in KAJAANI 4000 grinder group control system.

and keeping the sharper ones in lower production. This optimization can also be done based on laboratory measured pit pulp qualities, if these measures are available. The HONEYWELL GRINDERMAN is in many ways similar to the KAJAANI system.

7 The PGW pulp family and various paper grades

As described earlier in this chapter, pressure grinding can be performed under different grinding conditions. The conditions have a great effect on the pulp properties, and pressure grinding provides a range of pulps (i.e., PGW pulp family) based on grinding conditions used. Which process alternative is chosen depends on the paper grade to be produced as well as on the wood species. In this chapter, all the given data come from grinding Norway Spruce *(Picea abies)*. All other wood species are dealt with in the chapter "Raw materials."

The paper and board grades can be classified in three different groups in regard to the use of pressure groundwood processes:

- Newsprint type papers, such as newsprint and improved newsprint
- Magazine papers or value added papers, such as SC and LWC
- Folding boxboard (FBB).

Groundwood pulps, to some extent, are also used for WF (fine) papers, but here these are handled in the value added paper group. PGW mills have nowadays the following process alternatives: PGW70, PGW95, and PGW-S95/120. In the laboratory, even higher temperatures and pressures have been studied, but grinders for these conditions have not yet been manufactured[47].

7.1 Pulp properties

References[10, 12, 48, 49, 50] present properties of pulps ground at various process conditions. Table 6 provides typical differences between various pit pulps from the references. The basis of the comparison is the PGW95 pulp, which corresponds to the value 100% in all properties. Values for GW (atmospheric groundwood) pulp are also given.

Table 6. Relative pit pulp properties at CSF 80 ml compared to PGW95 pulp = 100.

Pulp type	GW	PGW70	PGW95	PGW-S120
Long fibers (R14 &R28)	62	86	100	121
Fines (P200)	118	100	100	93
Density	103	103	100	99
Wet strength	77	n.a.	100	121
Tensile strength	82	97	100	107
Tear strength	74	94	100	113
Internal bond strength	100	103	100	116
Light scattering ability	102	101	100	98
Brightness	102	105	100	97

When the grinder pressure and the shower water temperature in pressure grinding are decreased, the long fiber content and to some degree also the pulp strength will be reduced. Simultaneously, the optical properties improve. An increase in the temperature and pressure has the opposite influence on the pulp properties. No clear change in the fines content or in pulp density takes place, but PGW-S120 has less fines than other pressure groundwood pulps. The GW pulp has more fines and clearly lower long fiber content and pulp strength than the pressure groundwood pulps, but its optical properties are equal to or only slightly better than those of PGW70.

The changes in the pressure groundwood pit pulp properties are also seen in the screened pulps. The pit and decker pulps from Ref.[48] are presented in Table 7.

The PGW70 pulp has the highest light scattering and the lowest shive and coarse fiber content. The PGW-S120 pulp has the best strength, but its light-scattering power and brightness are reduced. The properties of the PGW-S95 (as well as PGW95) pulp are between those two extremes.

8 Requirements of various paper grades

The requirements set on mechanical pulp for different paper grades can be summarized as follows:

- News pulps should have high pulp strength (high amount of long fibers), low linting tendency, and moderate optical properties.

Table 7. Pit and decker pulp properties of PGW pulp family[48].

PIT PULPS	PGW70	PGW95	PGW-S95	PGW-S120
CSF, ml	90	90	90	90
R 14 & R 28 fraction, %	29	33	34	37
P200 fraction, %	28	28	26	27
Air resistance (G-H), s	55	75	82	92
Density, kg/m^3	415	420	425	420
Tensile index, Nm/g	39	41	42	44
Tear index, mNm^2/g	5.7	6.0	6.2	6.9
T.E.A., J/m^2	35	37	39	41
Light scattering, kg/m^2	64	63	63	61
Brightness, %	65	63	63	60
DECKER PULPS	**PGW70**	**PGW-S95**	**PGW-S120**	
CSF, ml	40	35	33	
Shives (Pulmac #0.1), %	0.08	0.15	0.17	
R14 fraction, %	1.3	3.3	3.9	
P14/R28 fraction, %	20	22	22	
P200 fraction, %	30	30	32	
Air resistance (G-H), s	70	150	200	
Density, kg/m^3	470	485	515	
Tensile index, Nm/g	45	48	53	
Tear index, mNm^2/g	4.9	5.3	5.4	
Burst index, kPam^2/g	2.2	2.6	2.6	
Light scattering, kg/m^2	68	66	63	
Brightness, % (dit.)	68	70	67	

- Magazine paper pulps should have good optical properties, a low amount of shives and coarse fibers, low porosity, and low fiber roughening tendency in offset printing.
- Folding boxboard pulps should have a high bulk-bond ratio and good creasing ability.

With modern gap formers, paper grades also need good internal bond strength. Additionally, all grades require good formation ability.

8.1 Suitability for various grades

By combining the data in Tables 6 and 7 and the set pulp requirements, Table 8 is constructed. This table presents the suitability of different pressure groundwood pulps from spruce/fir for various paper grades regarding to the pulp characteristics.

Table 8. Suitability of pressure groundwood pulps for various paper grades (N = news, M = magazine, B= FBB; 0 = suitable, + = very suitable).

Paper grade	PGW70 N	PGW70 M	PGW70 B	PGW95 N	PGW95 M	PGW95 B	PGW-S95 N	PGW-S95 M	PGW-S95 B	PGW-S120 N	PGW-S120 M	PGW-S120 B
Shives	+	+	+	+	+	+	+	0	+	+	0	+
Coarse fibers	+	+	+	+	+	+	+	0	+	+	0	0
Long fibers	0	+	+	0	+	+	+	+	+	+	+	+
Fines	+	+	+	+	+	+	+	+	+	+	0	+
Air resistance	0	0	+	+	+	+	+	+	+	+	+	+
Density	+	+	+	+	+	+	+	+	+	+	+	+
Tensile strength	+	+	+	+	+	+	+	+	+	+	+	+
Tear strength	0	+	+	0	+	+	0	+	+	+	+	+
Internal bond	+	+	+	+	+	+	+	+	+	+	+	+
Brightness	0	+	+	+	+	+	+	0	+	+	0	0
Light scattering	+	+	+	+	+	+	+	+	+	+	0	+

The PGW70 and PGW95 pulps fit best for magazine grades and boxboard. The lower air resistance of the PGW70 pulp has to be compensated by lowering the freeness. The PGW-S95 and PGW-S120 pulps are most suitable for newsprint. The final decision on the pulp type depends, however, on many other factors, which are handled as follows.

8.2 Other influencing factors

In addition to the paper grade and the wood species, at least the following factors should be considered, when choosing the pulp type:

- Share and price of chemical (kraft) pulp in paper furnish
- Initial and final brightness
- Bleaching method and costs
- Washing requirements and effluent load
- Costs for effluent handling
- Use of heat from possible heat recovery system
- Investment costs.

These factors can have different importance at each mill and, thus, no general rule can be applied. However, the conclusion presented in Ref.[49] is probably suitable in most cases: "The PGW70 process yields excellent optical and good printability properties as well as good strength and low energy consumption for low freeness pulps. The optical properties of PGW-S pulps are somewhat lower than those of PGW70, but the strength of the pulp and long fibers is maximized, while maintaining low energy consumption. This combination is very favorable for newsprint type papers"[49].

CHAPTER 6

References

1. Carpenter, C. H., "The history of mechanical pulping," TAPPI Mechanical Pulping Committee report, 1989.

2. Klemm, K. H., Modern Methods of Mechanical Pulp Manufacture, Lockwood Trade Journal Co., New York, 1958.

3. Leask, R., Pulp and Paper Manufacture, Vol.2: Mechanical Pulping, TAPPI/CPPA Textbook Committee, Atlanta, 1987.

4. Salakari, H., Paperi ja Puu 59(5):401 (1977).

5. Lindahl, A. and Haikkala, P., Pressurized Groundwood Manufacture Saves Energy and Improves Quality, SPCI, Stockholm, 1978.

6. Aario, M., Haikkala, P., Lindahl, A., "Pressure grinding – A new method to produce mechanical pulps," CPPA 1979 Annual Meeting Notes, CPPA, Montreal, p.A33.

7. Aario, M., Haikkala, P., Lindahl, A., Tappi J. 63(2):139 (1980).

8. Kärnä, A., Paperi ja Puu 62(1):27 (1980).

9. Pietarila, V., Mitchell, G, Haikkala, P., Tuominen, R., "PGW at high temperatures and pressures," 1987 International Mechanical Pulping Conference (Vancouver) Preprints, CPPA, Montreal, p.19.

10. Pasanen, K., Peltonen, E., Haikkala, P., Liimatainen, H., Tappi J. 74(12):63 (1991).

11. Kärnä, A., Pap. Technol. Ind. 27(1):23 (1986).

12. Murtola, C. and Peltonen, E., "Industrielle Druckschlifferzeugung mit niedrigen Spritzwassertemperaturen in einer Papierfabriken, " PTS-TUD-Symposium Papierzellstoff und Holzstofftechnik ´95, PTS Verlag, München, 1995.

13. Haikkala, P., Liimatainen, H., Tuominen, R., "Impact of superheated shower waters on pressure groundwood pulp properties," Appita 1988 Conference Preprints, Appita, Charlton, p. A24.1.

14. Burkett, K. and Tapio, M., Tappi J. 73(6):117 (1990).

15. Honkanen, T. and Yrjövuori, R., "The first years of the first PGW-S mill," 1993 International Mechanical Pulping Conference Preprints, PTF, Oslo, p. 44.

16. Henrich, H., Wochenbl. Papierfabr.122(1):14 (1994).

17. Haikkala, P., Liimatainen, H., Tuominen, R., "The impact of process on mechanical pulp properties, " TAPPI 1988 Pulping Conference Proceedings, TAPPI PRESS, Atlanta, p. 631.

18. Haikkala, P., Liimatainen, H., Manner, H., Tuominen, R., Paperi ja Puu 72(4):393 (1990).

19. Kärnä, A., Das Papier 38(10A): 147(1984).

20. Lucander, M., "Competitive pressure groundwood," Kestävä Paperi 14, KCL, Espoo, Finland, 1997, 24 pp.

21. Meinecke, A., Süttinger, R., Schmidt, K., Wochenbl. Papierfabr. 9(5):307 (1984).

22. Meinecke, A., "Thermogrinding," 1985 International Mechanical Pulping Conference Proceedings, STFI, Stockholm, p. 136.

23. Schweizer, G. and Böck, A., Das Papier 42(10A):V134 (1988).

24. Henrich, H., Das Papier 42(11):603 (1988).

25. Pearson, J., Pulp Pap. Int. 30(6):61 (1988).

26. Bechamp, R., "Fundamentals of abrasives and pulpstones," CPPA 1987 Mechanical Pulping Course Notes, CPPA, Montreal, p. 1.

27. Norton International Inc., "Norton Pulpstones," technical brochure, 1985.

28. Kärnä, A., "Studies of pressurised grinding," dissertation, Helsinki University of Technology, Espoo, 1984.

29. Sandås, E. and Lönnberg, B., Paperi ja Puu 72(8):765 (1990).

30. F. W. Roberts Manufacturing Company, "Pulpstone Sharpening," technical brochure.

31. Sollinger, H. -P., "Die Temperatur und der Druck als Einflussgrössen auf die Holzschliffqualität beim Holzschleifverfahren," dissertation, Fakultät für Maschinenwesen, Rheinisch-Westfälische Technische Hochschule, 1982.

32. Lucander, M., Lönnberg, B., Haikkala, P., "The effect of stone surface modifications on groundwood properties," TAPPI 1983 International Mechanical Pulping Conference Proceedings, TAPPI PRESS, Atlanta, p. 79.

33. Widdifield, D., Frazier, W., Stationwala, M., Jardine, C., Broderick, L., "Grooved pulpstones: a statistically based mill evaluation," TAPPI 1995 International Mechanical Pulping Conference Proceedings, TAPPI PRESS, Atlanta, p. 49.

34. Manner, H., Reponen, P., Vahteri, R., Kärpänen, T., "Effect of water jet pulpstone conditioning on mechanical pulp and paper quality at Anjala paper mill," TAPPI 1995 International Mechanical Pulping Conference Proceedings, TAPPI PRESS, Atlanta, p. 35.

35. Puurunen, A., "Effect of water jet conditioning on low freeness pulp at the Rauma paper mill," CPPA 1997 Annual Meeting Notes, Technical Section CPPA, Montreal, p. B 157.

36. Puurunen, A., Blomqvist, S., Ora, M., "First mill scale experiencies of the Water Jet pulpstone conditioning system at UPM-Kymmene Voikkaa and Rauma Paper Mill," TAPPI 1997 Pulping Conference Proceedings, TAPPI PRESS, Atlanta, p. 317.

CHAPTER 6

37. Kone Wood Oy, Automated Grinder Feeding Systems for the Chain and High Magazine Grinders, Woodprints, newsletter from KONE WOOD, Winter 1996.

38. Kone Wood Oy, Automated Grinder Charging Systems for the Pocket Grinders, Woodprints, newsletter from KONE WOOD, Winter 1996.

39. Bergstöm, J., Hellstöm, H., Steenberg, B., Svensk Papperstid. 60(11):409 (1957).

40. Paulapuro, H., Paperi ja Puu 58(1):5 (1976).

41. Kärnä, A. and Liimatainen, H., Pulp Paper Can. 86(12):T377 (1985).

42. Luhde, F., Pulp Paper Mag. Can. 50(10):659 (1959).

43. Kärnä, A. and Liimatainen, H., Finnish pat. 813842 (1.12.1981).

44. Ryti, N., Paulapuro, H., Manner, H., Paperi ja Puu 55(11):811 (1973).

45. Droscher, H., WochenBl. fur Papier. 108(11/12):388 (1990).

46. Kajaani 4000, product information sheet, courtesy of Kajaani Automation.

47. Lucander, M., "Recent investigations of pressure grinding at elevated grinder pressure," PTS-Symposium Holzstofftechnik, PTS, Dresden, 1993.

48. Valmet, Groundwood Systems, Pressure Groundwood, Tampere 1996, 14 p.

49. Tuovinen, O. and Liimatainen, H., "Fibers, fibrils and fractions – an analysis of various mechanical pulps," 1993 International Mechanical Pulping Conference Proceedings, PTF, Oslo, p. 324.

50. Asunmaa, P., "The selection, process flowcharts, optimization and results obtained from the new PGW-S Mill at the Kaukas Oy Voikkaa Paper Mill in Finland," CPPA1993 Annual Meeting Notes, CPPA, Montreal, p. B71.

51. F. W. Roberts Manufacturing Company, Brochure "Sharpening Control & Management System."

52. Roberts, C., Chapman, R., "Precise control of the sharpening burr" CPPA 1995 International Mechanical Pulping Conference Proceedings, CPPA, Montreal, p. 45.

CHAPTER 7

Thermomechanical pulping

1	**Introduction to thermomechanical pulping**	**159**
1.1	The principles of refiner mechanical pulping	159
1.2	Main refiner types	160
1.3	The thermomechanical pulping process	160
1.4	Energy consumption and yield	161
2	**Handling and pretreatments of chips**	**162**
2.1	The chip washer	163
2.2	Dewatering of chips	164
2.3	Chip wash water system	164
2.4	The preheating	165
2.5	Chip impregnation systems	166
3	**The refiner**	**168**
3.1	Refiner concepts of different manufacturers	168
3.2	The single-disc (SD) refiners	169
	3.2.1 Andritz SB 150 and 170 single-disc refiners	169
	3.2.2 Sunds Defibrator RGP 200 Series single-disc refiners	171
	3.2.3 Sunds Defibrator SD 65 single-disc refiner	172
	3.2.4 Other single-disc refiners	173
3.3	Large-capacity TMP refiners	174
	3.3.1 The Sunds Defibrator RGP CD refiner	175
	3.3.2 Andritz Twin refiner	177
3.4	Double disc (DD) refiners	179
3.5	Processing pulp and steam in the refiner line	181
3.6	The effects of refining conditions and equipment design parameters	183
4	**Refiner segments – the "heart" of the refining process**	**185**
4.1	Refiner segments design parameters	187
4.2	Materials of refiner segments	189
5	**Main TMP process types**	**190**
5.1	The "standard" TMP line	190
5.2	Single-stage refining	193
5.3	TMP with no separate reject refiner	193
5.4	Two-stage refining in the same refiner	194
5.5	TMP using high-speed refining	194
5.6	Thermopulp	195

6	**How refiner type and process conditions affect pulp properties**	**196**
6.1	Intensity of refining	196
6.2	Influence of refiner type	197
6.3	Refiner speed	199
6.4	Consistency	200
6.5	Production rate	201
6.6	Preheating and steaming of chips	201
6.7	Temperature in preheating and refining	202
6.8	Special combinations of process conditions	203
	6.8.1 The multistage process	204
	6.8.2 The RTS process	204
	6.8.3 Thermopulp	205
7	**Process control**	**206**
7.1	Main principles	207
7.2	Control of operating variables	208
	7.2.1 Plate gap clearance	209
	7.2.2 Production rate	209
	7.2.3 Refining consistency	209
	7.2.4 Steam pressure	210
8	**Heat recovery**	**211**
8.1	Amount and composition of TMP steam	212
8.2	Reboiler for condensing TMP steam and generating clean steam	213
8.3	TMP steam vent/feed water exchanger	214
8.4	TMP condensate/feed water exchanger	214
8.5	TMP condensate/white water exchanger	215
8.6	Startup scrubber for TMP steam	215
8.7	Surface condenser for low-pressure TMP steam	215
8.8	Turpentine recovery	218
	References	219

CHAPTER 7

Taisto Tienvieri, Erkki Huusari, Jan Sundholm, Petteri Vuorio, Juha Kortelainen, Harri Nystedt, Arvi Artamo

Thermomechanical pulping

1 Introduction to thermomechanical pulping

1.1 The principles of refiner mechanical pulping

Today the thermomechanical pulping (TMP) process is the dominating refiner-based mechanical pulping process. It originated on one hand from the refiner mechanical pulping (RMP) process, which used open discharge refiners, and on the other hand from the Asplund Defibrator fiberboard process, which is the original "thermomechanical" process. Chapter 3 "History of mechanical pulping" further describes the historical development.

In refiner mechanical pulping, the raw material – normally in the form of wood chips – is ground or said to be refined in the narrow gap between two metal discs with a grooved pattern. At least one of the discs is rotating with high speed, normally 1500 or 1800 rpm. Figure 1 illustrates the principle involved. When entering into the refiner through the opening in the right disc in the picture, the chips are accelerated by the large bars near the center of the fast-moving opposite disc. The chips start to rotate with the revolving disc and are simultaneusly broken up to smaller fragments and fibers due to the heavy mixing and to the abrasive action of the

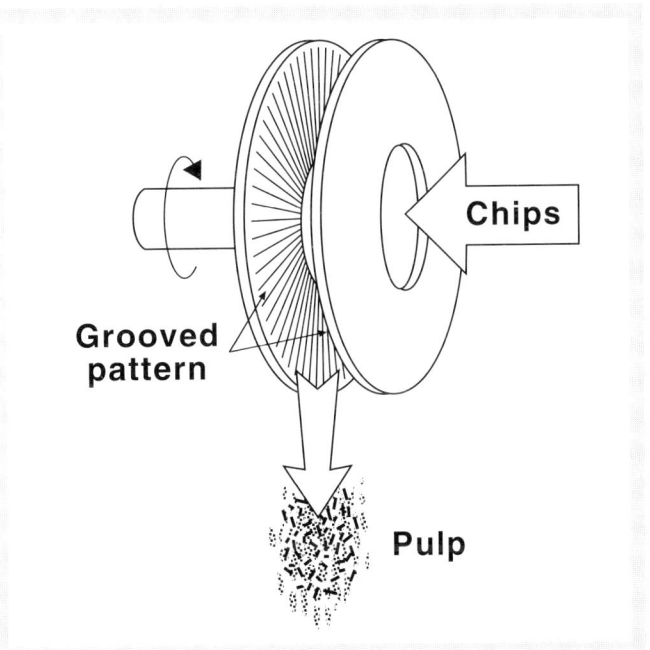

Figure 1. Principle of refiner mechanical pulping. The refiner has two opposite discs, of which in this case one is revolving (a single-disc refiner, SD refiner). Wood chips are fed into the refining zone between the refiner discs through an opening in the center of the stationary disc.

bars of the opposite disc. When these wood particles start to rotate with high speed, a considerable centrifugal force starts to act on them. Because of this centrifugal force, the wood raw material is pressed outward through the narrowing gap between the discs and is refined between the finer and finer bars to a pulp of desired fineness (normally measured as Canadian Standard Freeness, see chapter 15). The residence time in the refining zone is less than a second. Chapter 4 "Fundamentals of mechanical pulping" discusses the refining mechanisms of the TMP process in more detail.

1.2 Main refiner types

Today almost 100% of TMP is produced using disc refiners or disc refiners with a conical section. Currently there are only two full conical high-consistency refiners in operation. There are two main types of disc refiners: single-disc (SD) and double-disc (DD) refiners. Of the single-disc refiners, there are refiners with one or two plate gaps. Those with only one gap are most common. Of the single-disc refiners with two plate gaps, there is one special refiner type with two-stage refining in the same refiner. The CD refiner is a common single-disc refiner which incorporates both flat disc and conical disc refining zones in the same refiner. The various refiner types are discussed in detail later in this chapter.

1.3 The thermomechanical pulping process

Originally thermomechanical pulping meant a process with pressurized preheating of chips followed by either pressurized or open discharge refining. It was believed that the preheating at elevated temperatures, which softens the wood, was the key to the exceptional strength properties of TMP. But at the end of the 1970s, it was determined to be essential that the refining – not the preheating – should be performed at elevated temperature. Since the early 1980s, this idea has led toward the development of the TMP process toward lower temperatures in preheating and higher temperatures in the refiner, also known as the PRMP process (pressurized refiner mechanical pulping)[24]. However, the name has remained TMP.

    Figure 2 shows a simplified flowsheet of a typical TMP plant for producing pulp for mechanical printing papers. The wood logs are debarked and cut down to chips of a certain size. The chips are steamed in an atmospheric steaming vessel at 100°C, after which the chips are washed in hot circulating water. The warm and wet chips are fed by a plug screw feeder into a pressurized preheater with relatively low pressure and temperature. After the preheater, the chips pass a second plug screw feeder on their way into the first refiner, which operates at relatively high pressure and temperature (usually 300–500 kPa over pressure and 143°C–158°C temperature). After this first-stage refining, the coarse pulp is blown to a steam separator from which it is fed to the second-stage refiner. From the second-stage refiner, which operates at about the same pressure and temperature as the first-stage refiner, the pulp is blown to a second steam separator. After passing the plug screw feeder at the bottom of the second steam separator, the pulp falls down into a pulper for removal of latency. Then follows screening and reject refining, and usually also bleaching before storage and transport to the paper mill.

Thermomechanical pulping

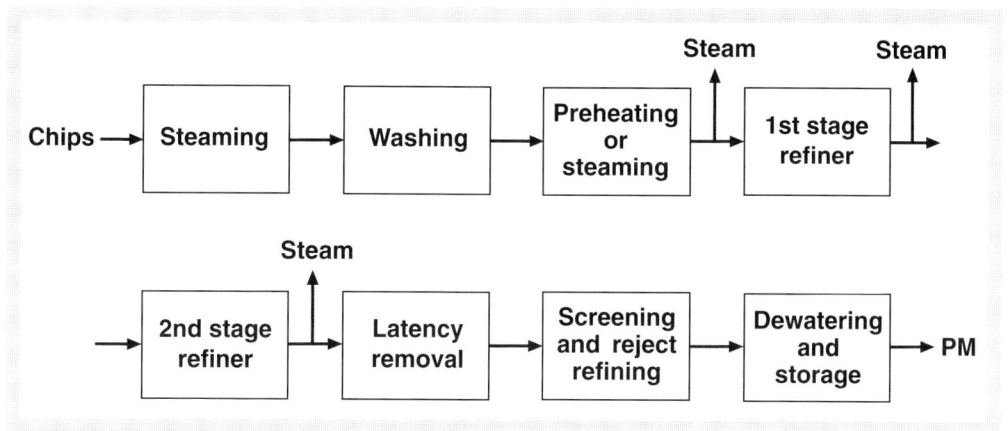

Figure 2. Simplified flowsheet of a typical TMP plant.

1.4 Energy consumption and yield

Figure 3 shows the total electric energy usage per air dry metric ton in producing TMP, PGW, and SGW for different end uses. The raw material is Norway spruce. Of the total energy consumption, normally only 200–300 kWh/t is used for pumping, screening, etc., which means that around 90% of the energy is used for defiberizing and refining (reject refining included). The energy consumption is high for all mechanical pulping methods, but especially so for TMP. The electric energy need for producing magazine grade TMP is in many cases well over 3 MWh/t. The energy costs, however, are substantially lowered in all modern TMP mills by the use of effective heat recovery systems (described later in this chapter). The

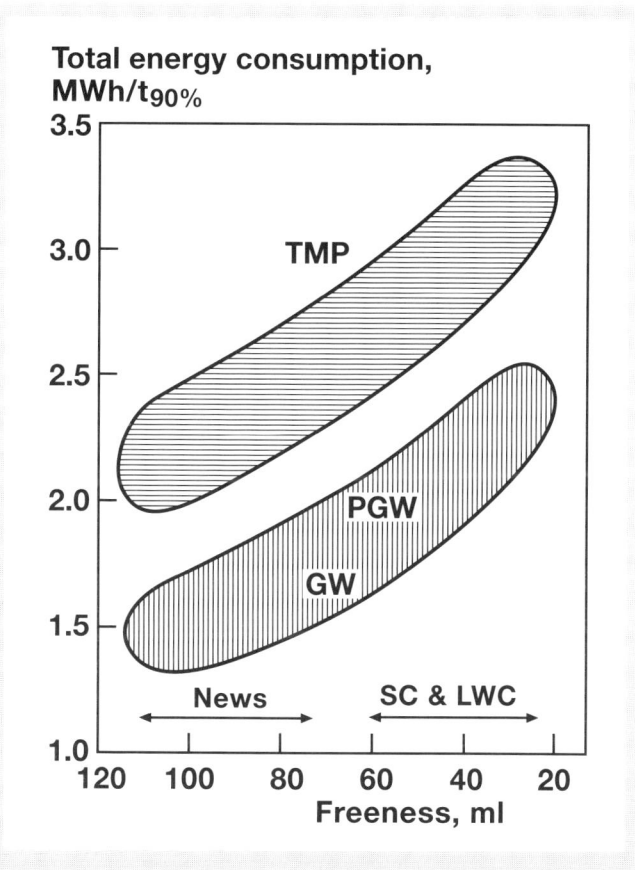

Figure 3. The total electric energy consumption for GW, PGW, and TMP as a function of freeness and end use. Norway spruce, Finnish mills.

heat recovery is possible because in practice almost all electric energy, which is consumed, is transformed to thermal energy in the form of hot saturated steam.

The chapter "Fundamentals of mechanical pulping" discuss the reasons for the high energy consumption, and the chapter "Future outlook" discusses the implications on the development of the industry.

The yield based on fresh debarked spruce wood is between 96%, when manufacturing TMP in laboratory or pilot conditions using only fresh water, and over 98%, when manufacturing TMP in mill conditions using very small amounts of water (< 10 m^3/t).

2 Handling and pretreatments of chips

The TMP process before refining can be divided into the following subprocesses:

- Chip handling and conveying to chip silo or chip pile
- Atmospheric presteaming (not necessary)
- Chip washing
- Chip preheating (atmospheric or pressurized).

Chips are stored either in a chip pile or more often in a chip tower to decrease the detrimental effect of storing on TMP quality. Covered storages (silos) are primarily used when there is need for well controlled storage, and also severe climate conditions call for covered storage. The volume of a chip silo is normally 5 000–8 000 m^3 and a two tower system is often used. This also makes it possible to use a classification system for the wood raw material. The chips are discharged from the bottom of the silo onto a belt conveyor which feeds the chips to the chip recieving bin. Some mills uses steam, especially in winter time, to warm up the chips before washing. This prevents icy chips entering the process and also, to some extent, facilitates the water uptake and temperature rise in the chip washing system. In mill trials, good presteaming increased the moisture content of washed chips from 59% to 62%, compared to the poor presteaming. The steam used can be blow-back steam from the refiners or steam from the main heat recovery. The use of steam in the presteaming bin is the most effective way to increase the temperature and the moisture content of the chips.

The chips from the chip receiving bin are discharged to the chip washer where chips are soaked in the hot water and then pumped to the dewatering screws (Fig. 4). The wash water is recycled from dewatering screws via bow screens (which remove sawdust and bark) back to the chip washing tank and then to the chip washer. A purge stream from the chip wash water tank decreases excessive buildup of colloidal material in the washing system. However, the chip washing water is heavily contaminated. In a mill trial, measured values from chip wash water was 160 mg/l lipophilic extractives, 120 mg/l lignans and 250 mg/l lignin, and TOC 1050 mg/l. The temperature in chip washing water has been kept at 70°C–85°C by adding contaminated condensate from main heat recovery or from the surface condenser (first-stage heat recovery). The temperature of washing water affects the steam consumption and condensation in the preheater before refiners. A high and stable temperature decreases refining consistency

Thermomechanical pulping

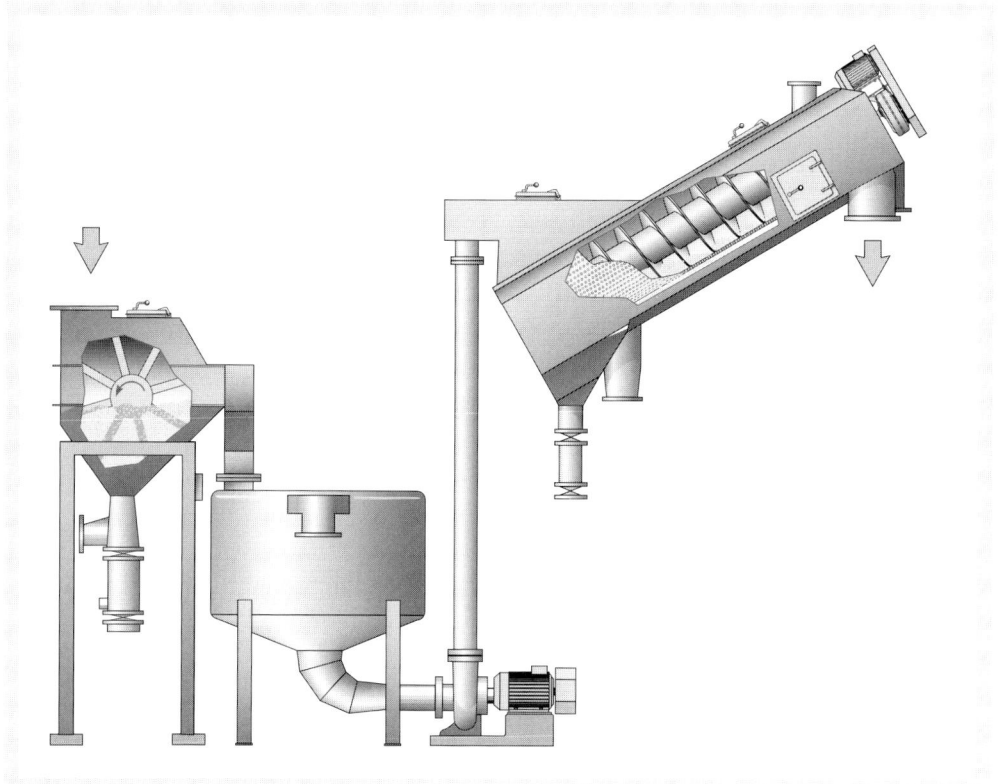

Figure 4. Chip washing system (Sunds Defibrator).

variation and improves the heat recovery efficiency of the refiners. A temperature in the range of 60°C–90°C does not affect the moisture content of the washed chips.

The targets of chip washing system are to:

- Remove heavy contaminants from the chip flow, e.g., stones, sand, and metal-scrap
- Remove sawdust and bark from the chips
- Somewhat add the moisture content of the chips and to make the moisture level uniform
- Increase the temperature of the chips.

2.1 The chip washer

The chip washer in Fig. 4 consists of a rotating paddle wheel in a vat. The chips fed to the chip washer are immersed by the rotating paddle wheel, and the heavy contaminants (like stones and metal particles as well as knots) are dropped to the conical bottom of the vat. The rotation of the wheel and the water flow lead the chips to the feed chute of the chip pump. The dumping system of the washer is automatically sequence

CHAPTER 7

controlled. The heavy contaminants are purged to a container for transportation to the dump. An old type of chip washer uses a horizontal screw with low rotation speed to separate heavy contaminants.

2.2 Dewatering of chips

The warmed and washed chips are pumped as a chip/water mixture at a consistency of 3%–4% to the screw drainer (as shown in Fig. 4) either directly or via a cyclone. To ensure good operation of the chip pump, the chip "slurry" should be well mixed and the "consistency" well controlled in the feed chute before pumping. A hydrocyclone can be used to separate remaining foreign particles and sand. The screw drainer also removes sand and fines; therefore, some mills have found the use of cyclones unnecessary. Also some mills use two screw drainers in a series to improve dewatering and fines removal. Vibrating screens are no longer used for dewatering because of high maintenance costs; nearly all TMP plants have replaced the vibrating screens by screw drainers.

2.3 Chip wash water system

The wash water from the chip draining screw is fed to the chip washing water tank (Figs. 4 and 5). Because it contains fines and sand, the water is purified with sidehill screens. Some mills use a separate wash water cleaning circulation system where the water from the chip washing water tank is pumped to the hydrocyclone or to the sidehill

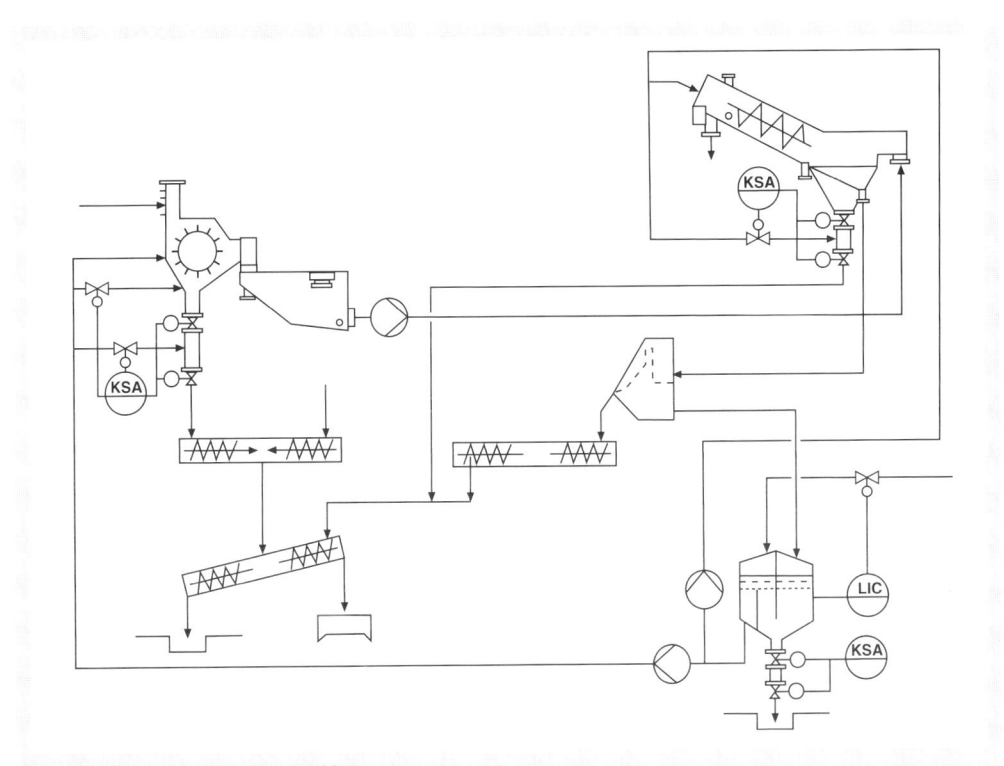

Figure 5. Flow sheet of a chip washing system.

screens and then back to the tank. The difference between these two systems is that, in the former case, all chip washing water is purified and, in the latter case, only part of the water is purified. The chip washing water tank often has a conical bottom section designated as a sedimentation tank and the same purging system for the sand as the chip washer has. The fines and bark and sand removed in sidehill screens and in cyclones are fed to the container or to the sump tank. The fines content of the chips decreases about 30% in the chip washing system (e.g., from 1.4% to 1.0%). The water level of the chip washing water tank is controlled by the clear or cloudy filtrate from the TMP disc filter. Normally excess water is not needed because the condensate flows from heat recovery are enough to cover the purging streams from chip washers, sidehill screens, and cyclones and from the chip washing tank. Normally there is a small overflow from the wash water tank to decrease an excessive buildup of dissolved and colloidal material in the system.

2.4 The preheating

The chips from the dewatering screw are dropped to a screw conveyor and fed to the preheater of the refiner or to an impregnator system. Some screw conveyor systems have an overflow of washed chips back to the chip receiving bin. The amount is controlled manually. This system ensures an even feeding of chips to all refining lines. Some systems use a chip level controller at the preheater to adjust the chip flow to the washing system and, in this case, a backflow of chips is not necessary. The feeding system to the preheater or presteaming tube depends on the pressure of the presteaming. When atmospheric pressure (temperature 80°C–95°C) is used, no special feeding devices are needed. The chips are normally fed to the pressurized preheater by a plug screw feeder. Also pocket or rotating valve feeding systems are used. The plug feeder keeps a typical overpressure of 50kPa to 110 kPa

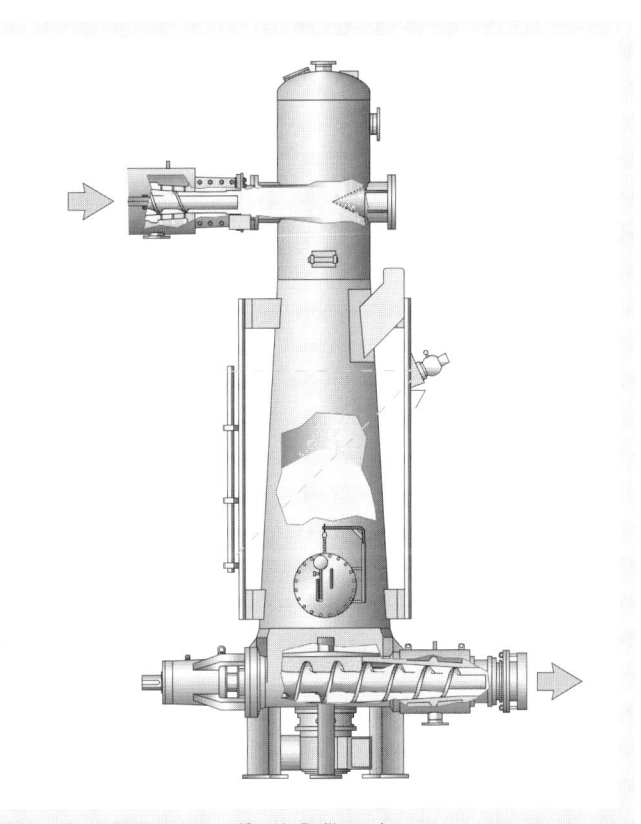

Figure 6. A TMP preheater (Sunds Defibrator).

CHAPTER 7

(temperature 105°C–122°C) in the preheater (Fig. 6). High temperatures are normally not needed in the preheater to reach optimal TMP properties. The target of the chip preheating is to warm the chips and equalize the moisture content of the chips before refining. It is also possible to fine tune the pulp properties with the temperature. A higher temperature will give better strength properties and a lower shive content, and a lower temperature gives better optical properties. The chips are preheated either by blow-back steam from the first-stage refiner or by steam from the heat recovery.

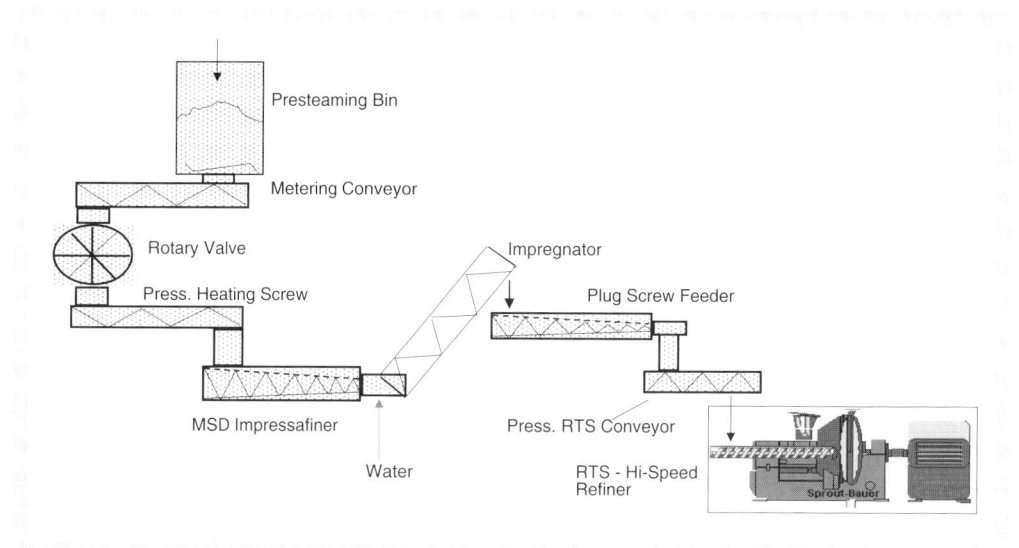

Figure 7. The high temperature and high chip compression RT pretreatment process (Andritz).

The retention time in the preheater is normally 1–3 minutes. When, like in the RTS process, a shorter retention time is needed, the preheating is done in a screw conveyor tube ("Press. RTS conveyor") as shown in Fig. 7. This also makes it possible to adjust the preheating time by controlling the rotation speed of the screw.

The chips from the preheater are discharged using a plug screw or an open pressurized screw conveyor to the refiner feeding system. When an open screw conveyor is used, the feed pressure in the feeding zone of the refiner is the same as the pressure in the preheater. The plug screw feeder enables the use of different pressures in the preheater and in the feeding of the refiner. The discharging screw has a variable-speed drive, and the production rate of the refiner is controlled by the rotation speed of the screw.

2.5 Chip impregnation systems

Separate chip impregnation systems are used in CTMP and APMP processes. An impregnator is also sometimes used to equalize the moisture content of the chips or to reduce the extractives content of the chips.

The PREX-impregnator system is widely used in CTMP processes (Fig. 8). Chips from the chip washing system are preheated at atmospheric pressure before the PREX plug screw feeding to the impregnator. The water squeezed out from the chips has a

Thermomechanical pulping

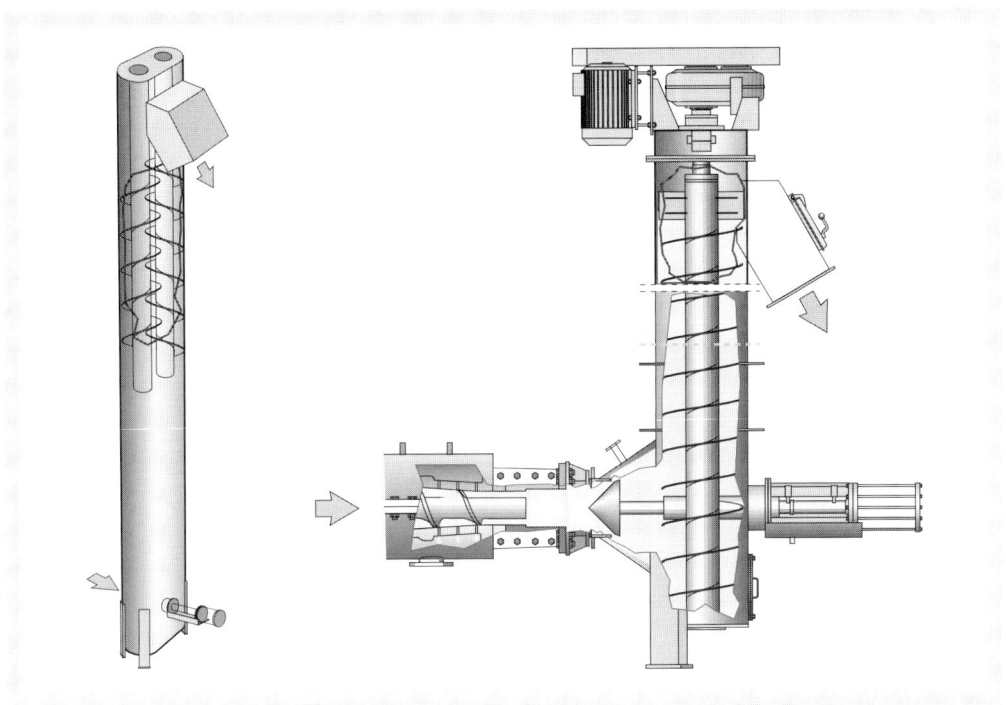

Figure 8. The PREX impregnator (Sunds Defibrator).

high content of dissolved and colloidal substances and therefore is led to the effluent. The pressed chips are fed to a vertical screw impregnator where the chips are impregnated with water or sodium sulfite solution. The level controller adjusts the flow of the liquor to the impregnator. The amount of water impregnated is about 0.5–1.0 m^3 per ton of pulp. The moisture content of the chips increases about 6%–7% units. The screws are available with two different compression ratios, 2:1 or 3:1, respectively. Fig. 9 shows another conical screw impregnator system.

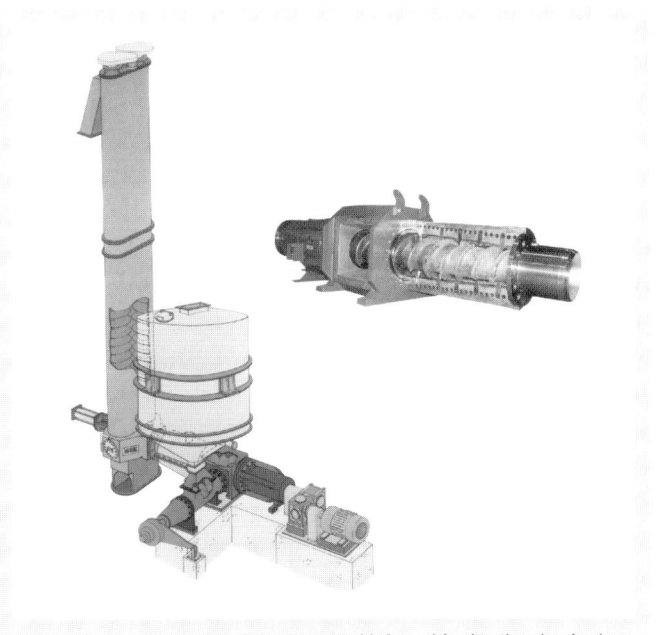

Figure 9. A conical screw impregnator with forced feed to the plug feeder (Hymac).

CHAPTER 7

The Impressafiner (Fig. 10) is used in the APMP process for the penetration of chemicals into the chips. The APMP system often consists of multiple stages of impregnation. Washed chips are at first presteamed atmospherically to soften the wood and to prevent fiber damage when chips are compressed (Fig. 7). The Impressafiner squeezes out the air and free water, then passes chips into a pool of liquor containing chemicals. The compression ratio in the compaction zone of the screw press is 4:1 or higher.

The Impressafiner is also used in the RTS process for prefiberizing chips to enhance water impregnation and heating of the wood raw material before the high temperature high speed refining. The impressafiner can be operated at atmospheric pressure or in a pressurized mode using a rotary valve as in Fig. 7. The atmospheric Impressafiner is used in some conventional TMP plants for prefiberizing of chips before the preheating in order to homogenize the chip size distribution, to enhance the heating of the chips, and to reduce the energy consumption in refining.

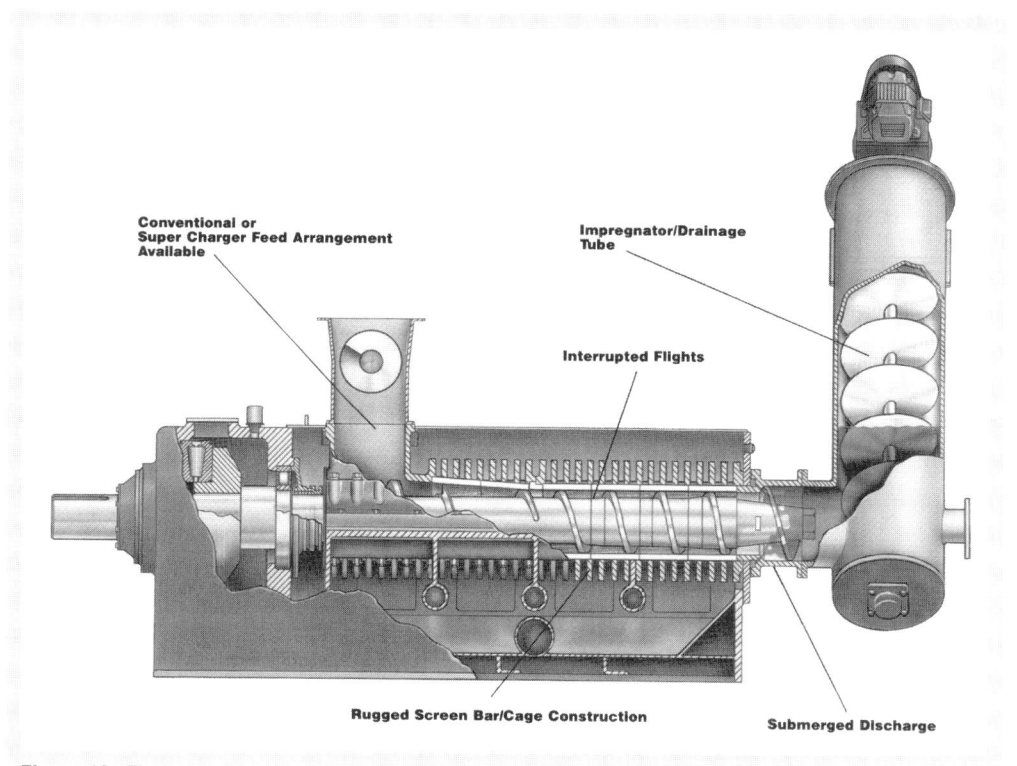

Figure 10. The Impressafiner (Andritz).

3 The refiner

3.1 Refiner concepts of different manufacturers

Today the two main manufacturers of TMP refiners are Sunds Defibrator and Andritz. Prior to the formation of the Sunds Defibrator Company, Sunds and Defibrator operated independently. Sunds marketed double-disc refiners for markets outside North America and Defibrator concentrated on systems utilizing single-disc machines. Another manufacturer of single-disc refiners, Jylhävaara Engineering Works, joined with Sunds Defibrator in 1987. An earlier Jylhävaara licensed manufacturer – Hymac Ltd. in Canada – marketed its single-disc-type refiners from a Jylhävaara-originated design. In 1997, after Hymac merged with Kvaerner, Andritz bought from Kvaerner the rights for the Hymac refiners. Earlier Andritz had bought Sprout-Bauer, a company which was formed in 1986 by merging Sprout – a company manufacturing single-disc refiners (including Sprout's Twin concept) – and C-E Bauer, a company manufacturing double-disc refiners mainly for North American markets.

The following text only discusses those TMP refiners which are marketed by the two remaining manufacturers of TMP refiners: Sunds Defibrator and Andritz. Refiners are presented in three groups: the single-disc refiners, the large-capacity TMP refiners (CD concept of Sunds Defibrator and Twin concept of Andritz), and the double-disc refiners.

3.2 The single-disc (SD) refiners

The operation principle of different single-disc refiners used in TMP refining is the same: A ribbon-type screw feeds the material into the eye of the rotor disc, which feeds the material into the gap between rotating and stationary discs. The motor load is controlled by adjusting the refiner gap clearance. Steam generated in high-consistency refining transports the pulp forward in the refining process. SD refiners are used in TMP lines with lower production capacity (a maximum of 250 tons/day). Usually, in one refiner line, there are two mainline refiners in a series and screen rejects are refined in a separate reject refining system. Another application of SD refiners is in reject refining (see chapter 10).

3.2.1 Andritz SB 150 and 170 single-disc refiners

Andritz SB 150 and 170 single-disc refiners (Fig. 11) are designed for high-consistency mechanical pulping and reject refining applications. Material to be refined is fed into the refiner by means of a side entry screw. The rotating assembly incorporates a large-diameter shaft-mounted feeder and rotating disc mounted between two roller bearings and a fluid film thrust bearing. This design means that the feeding screw and the rotor disc have the same rotational speed. A servo-based hydraulic loading system allows the refiner operation in constant loading pressure or in fixed disc position mode. A short, compact refiner thrust path is reached by pulling the rotating disc against the stator disc. A plate gap monitor provides rapid feedback of process variations and is engineered to prevent plate contacts. The following chart provides nominal disc diameter and rated power of SB 150 and 170 refiners at 1500 rpm:

	Rated power MW	Disc diameter mm	Disc diameter inch
SB 150	10	1520	60
SB 170	14	1680	66

CHAPTER 7

For high-speed operation in its RTS system, Andritz developed a new version of the SB 170 refiner for a speed of up to 2300 rpm and max. 18 MW, as well as the new single-disc refiner MRSD (Fig. 12). Disc diameter of the new refiner is up to 62 in., operation speed is 1500–2300 rpm, max. design pressure is 14 bars, and connected power is up to 9 MW.

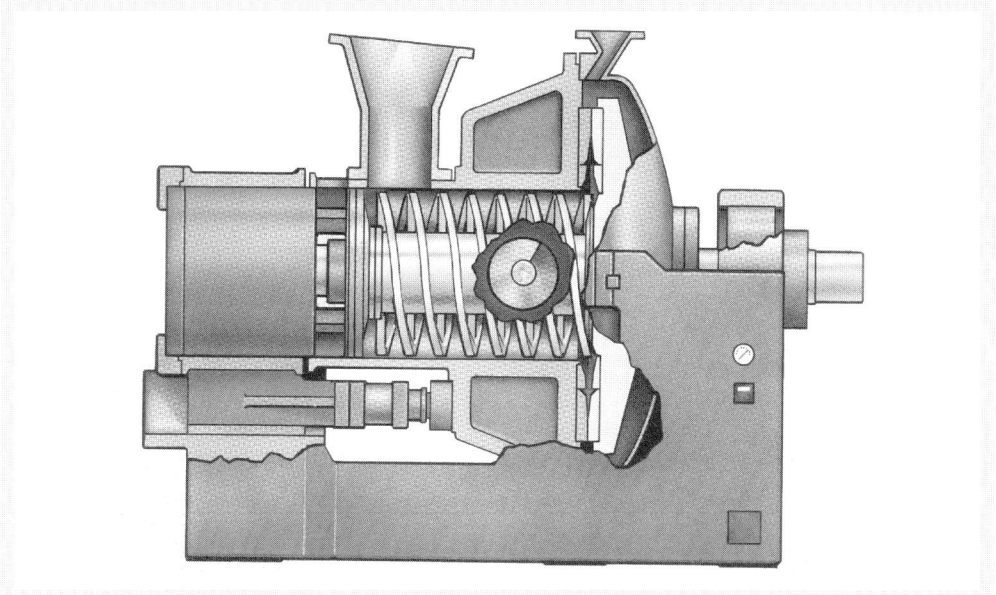

Figure 11. The Andritz SB 150/170 refiner.

Figure 12. The Andritz MRSD refiner.

Thermomechanical pulping

3.2.2 Sunds Defibrator RGP 200 Series single-disc refiners

Figure 13 presents the Sunds Defibrator RGP 200 series refiners. A more detailed picture of the largest Sunds Defibrator single-disc refiner RGP 268 is in Fig. 14.

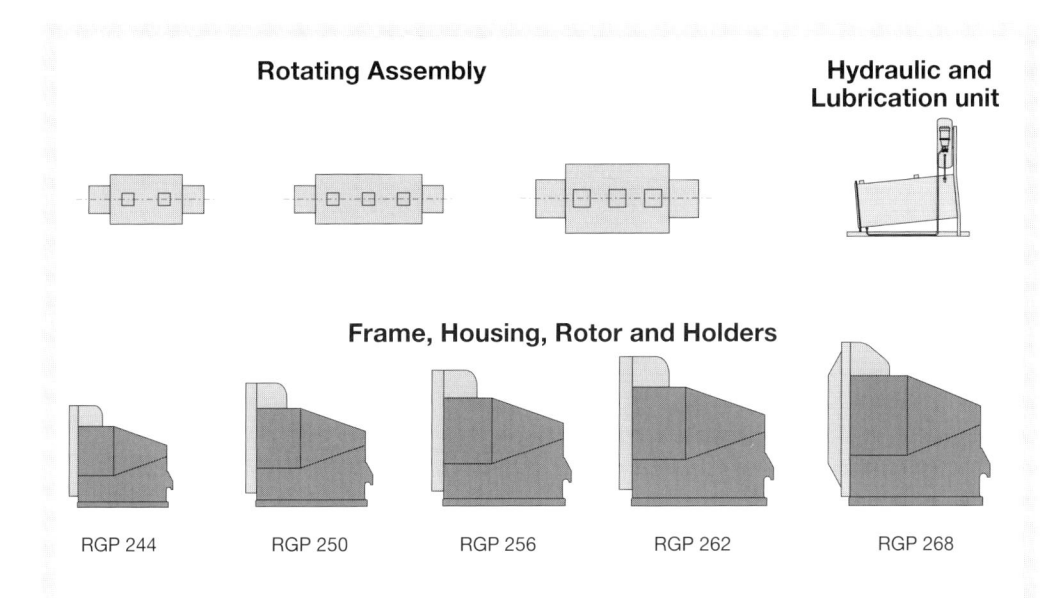

Figure 13. The RGP 200 series refiners (Sunds Defibrator).

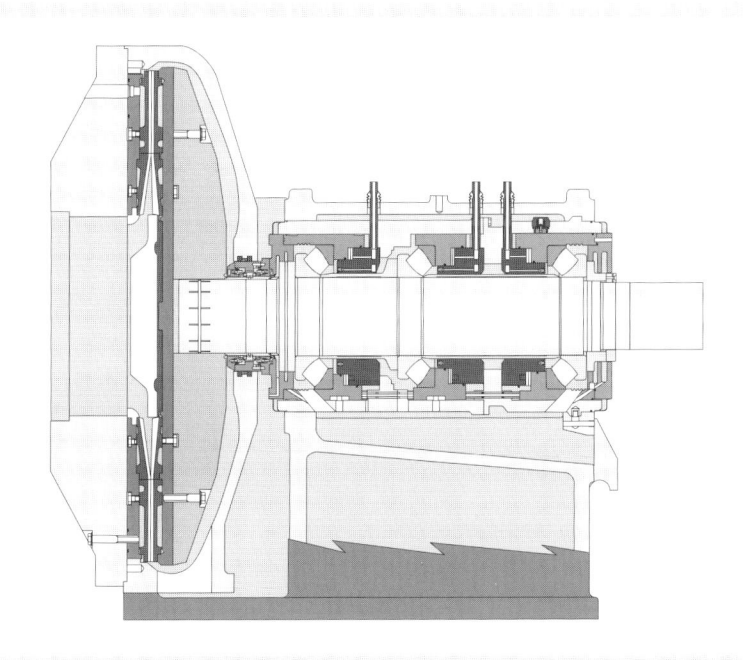

Figure 14. The RGP 268 refiner (Sunds Defibrator).

CHAPTER 7

The nominal disc diameter and rated power (at 1500 rpm) of the Sunds Defibrator RGP 200 single-disc refiners is as follows:

Type of refiner	Motor size MW	Disc diameter inch	Disc diameter mm
RGP 244	2,5	44	1180
RGP 250	5	50	1270
RGP 256	7	56	1422
RGP 262	10	62	1575
RGP 268	15	68	1728

All RGP 200 series refiners, in principal, have very similar mechanical construction. When maintenance is needed, the entire package of bearing and shaft assembly and hydraulics is simply lifted out and replaced with a spare unit. A complete segment change can be accomplished in less than four hours. After removing the upper housing cover, the segment holder is lifted out and replaced with another holder already equipped with a new set of segments. High rigidity of the machine is maintained by attaching the main stator plate, the refiner housing, and the bearing assembly to a massive cast-iron frame. The refiner housing is of cast stainless steel and the front cover is of solid stainless steel plate. Design of the frame and housing are symmetrical so that thermal expansion and pressure forces do not affect parallelism of the stator and rotor discs.

The rotating assembly is a rugged ready-to-mount cartridge unit, complete with rotor, shaft seal, bearings, and hydraulic rotor positioning system. The large-diameter shaft is supported by spherical roller bearings, which eliminate the need for a separate thrust bearing.

The system ensures very short overhang of the rotating disc. Lubrication of bearings is ensured by an oil circulation systems incorporating two parallel pumps, one driven by an electric motor and the other driven by the refiner shaft. In case of a power failure, the axially driven pump provides adequate lubrication until the refiner runs down. The material to be refined is fed by means of a side feeding arrangement into the double flighted infeed screw driven by a separate motor. The design of the infeed screw permits backflow of steam without disturbing the refiner feed.

The refiner gap is controlled by moving the rotor assembly by means of a hydraulic system, which makes it possible to control the refiner gap with high precision. A true disc clearance (TDC) sensor is used to monitor the gap. In case of feed disturbance, a feed guard activates the hydraulic disc-positioning system for rapid opening of the gap.

3.2.3 Sunds Defibrator SD 65 single-disc refiner

The original design of the Jylhävaara single-disc refiner SD 65 is still used in some of the latest TMP installations. The design of the SD 65 refiner (Fig. 15) maintains good parallelism of the refining gap. The rotational symmetric rigid stainless steel front cover holding the stationary segments is supported to the base frame by means of a rotational symmetric conical frame.

The large-diameter feeding unit is separated from the front cover and is rigidly mounted on the same foundation as the refiner. The rotor assembly with the rigid stainless steel rotor disc has mechanical seal, large-diameter cylindrical roller bearings and a 900 kN hydrodynamic thrust bearing. The gap between the rotor and the stator segment is controlled by moving the rotor assembly along the refiner frame sideways by means of an electromechanical loading device. In the case of a refiner segment clashing alarm, the loading device is activated to open the gap.

The refiner is opened for segment change by means of a screw mechanism which moves the front cover apart from the refiner. The maximum diameter of the SD 65 refiner is 68 inches, and the maximum motor size is 15 MW at 1500 rpm.

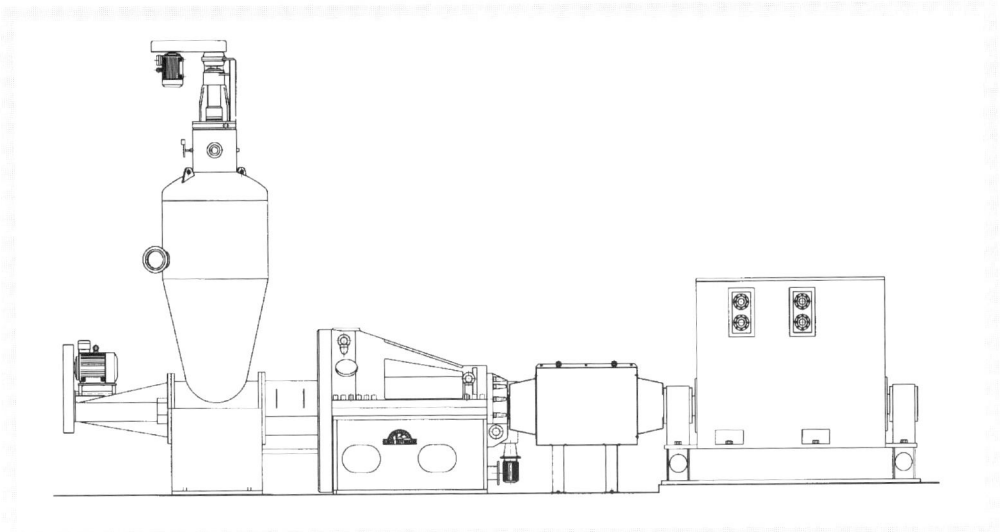

Figure 15. The Jylhävaara type Sunds Defibrator SD 65 refiner.

3.2.4 Other single-disc refiners

A special kind of modern single-disc refiner is Hymac's HXD 64 refiner (Fig. 16). In this refiner, two single-disc refining stages in a series are placed inside one refiner frame. The chips are fed into the first SD-refining zone and then blown to the steam separation cyclone, after which pulp is fed back into the refiner and another single-disc refining is made on the other side of the rotor disc. This special refiner concept originates from Hymac's earlier licensee Jylhävaara, and the first HXD 64 refiner is now in operation at Modo's Hallstavik mill in Sweden.

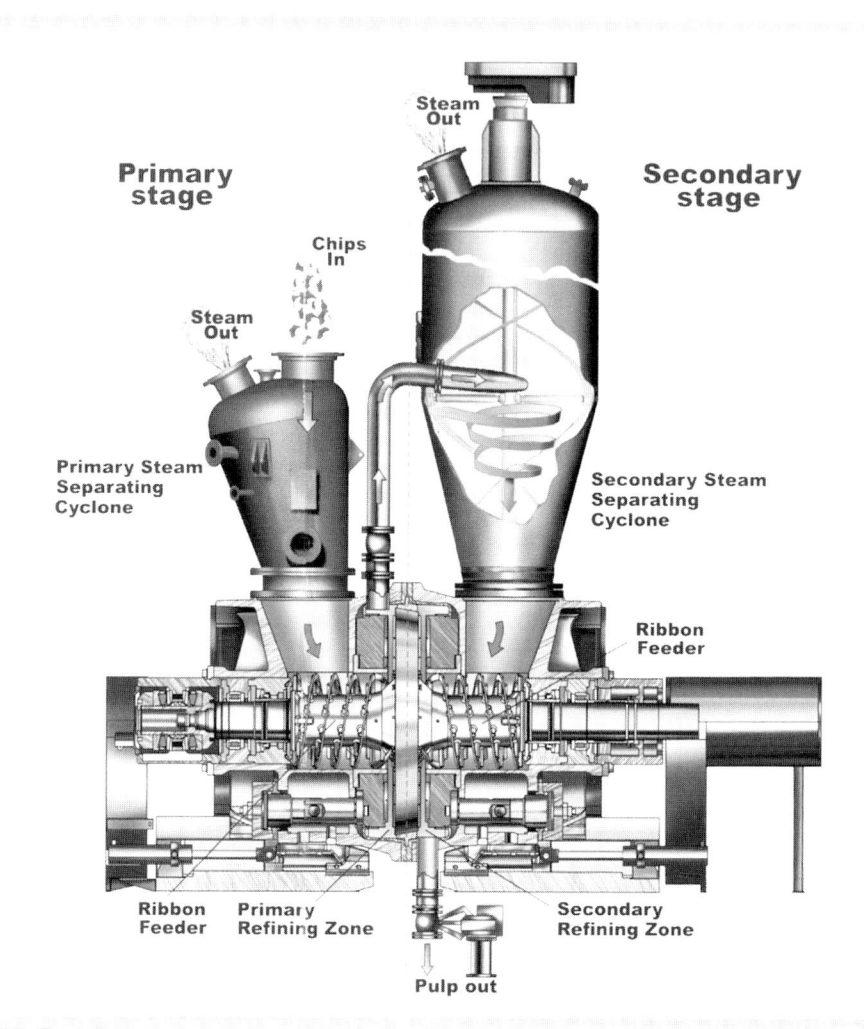

Figure 16. Two refining stages in the same HXD 64 refiner (Hymac Andritz).

3.3 Large-capacity TMP refiners

The explanation behind the rapid breakthrough of the large-capacity refiners is the investment cost: the larger the refiner is, the lower are the capital expenditures required per produced ton of pulp. First Sprout introduced its Twin concept and, soon after, Sunds Defibrator developed its RGP CD concept. The Twin refiner has two parallel single-disc refiner gaps, and the CD concept has a single-disc refining gap followed by a conical disc refining gap inside one refiner frame. Both the Twin and CD refiners have a large production capacity. One refiner line with two refiners in series has a production capacity of 450 tons/day at 1500 rpm and 550 tons/day at 1800 rpm of TMP for newsprint.

3.3.1 The Sunds Defibrator RGP CD refiner

The first CD refiner was built in 1980 for the TMP system at Hallsta newsprint mill in Sweden. Designated the RLP 70 CD, it was equipped with a straight-through shaft. Later it was designated RGP CD and the refiner concept was changed to a cantilever-type design. The first machine of this type was called the RGP 70 CD and was started up at Norske Skog in 1984. The high capacity of the CD refiner is explained by a conical refining surface positioned at the periphery of the flat refining zone. With the addition of the conical outer zone, it is possible to significantly increase the refining surface without making any substantial increase in the rotor diameter (Fig. 17). The latest and largest model of the RGP CD refiners is the RGP 82 CD. Technical data of the three models of RGP CD refiners are:

		CD 70	CD 76	CD 82
Motor at 1500 rpm	MW	15	19	22
Motor at 1800 rpm	MW	17	23	26
Refining surface	m^2	2,4	2,7	3,2
Weight	tons	22	25	27

Its mechanical design is based on modification of the RGP 60 flat discs refiner to accommodate both the flat and the conical discs. The design integrates the refiner frame, the refiner housing, the stator disc, and the rotating assembly to provide a stable machine construction. The rigidity built into the machine minimizes deflections under heavy refiner loads. Shaft rigidity is achieved with a compact, large-diameter shaft supported by three well dimensioned bearings. The bearing closest to the rotor disc is mounted only 300 mm from the center of the gravity of the rotor disc. This arrangement improves the operation stability. The RGP CD refiner is designed for rapid removal and replacement of the entire shaft assembly. For maintenance, the shaft – complete with bearings and hydraulic components – is simply lifted out and replaced with a new shaft assembly. Refiner segments can be changed in less than four hours by lifting out and replacing the segment holders as a complete unit.

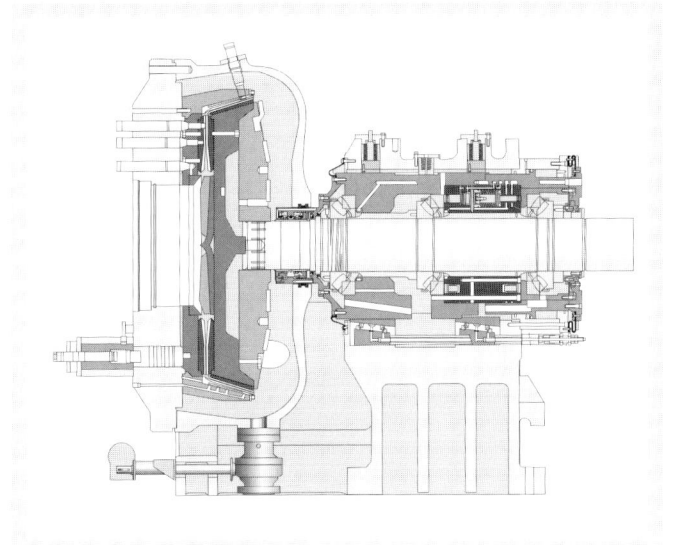

Figure 17. The RGP 76 CD refiner (Sunds Defibrator)

CHAPTER 7

The water and oil hydraulic units can be mounted in a separate room outside the refiner. The lubricating system incorporates two parallel oil pumps operating independently: One is driven by an electric motor, and the other is driven by the refiner main shaft. In the event of a power failure, the axially driven pump provides adequate lubrication until the refiner stops running. Refiner gap stability is reached through mounting the refiner and the motor on a common heavily reinforced concrete base resting on rubber mats above the foundation. Since the RGP CD is based on the RGP flat disc design, most parts of these machines are interchangeable, e.g., in case of the need to use the RGP CD design as a mainline refiner and the RGP flat disc design as a reject refiner. The refiner is normally started and stopped by means of the built-in control system, and it takes less than 2 minutes to attain the full refiner load. The key of the refiner control system is the True Disc Clearance sensor (TDC), which inductively measures the distance between the refiner plates.

The side feed screw feeds preheated wood chips into the double flighted infeed screw of the refiner. The design of the infeed screw permits backflow of steam without disturbing the refiner feed. The rotor disc feeds wood chips into the refiner gap, where chips are defibrated and refined to single fibers and fiber fragments. In the CD refiner, there are two refiner gaps in series: The first gap is between the flat rotor and stator discs, and the second gap is between the conical rotor and stator discs. The clearance of both gaps can be adjusted individually by means of separate gap adjustment systems. Transroller screws mounted on the stator adjust the clearance between the flat discs. An electric motor operates the transroller screws, and a common gear synchronizes the operation of the gap adjustment screws. Changing the position of the rotor via a hydraulic gap-setting mechanism allows control of the clearance between the conical discs. TDC sensors monitor clearance of both gaps. The gap-setting mechanism is provided by the feed guard system which, in case of a feeding disturbance, moves the refiner discs apart. The two refining zones of the CD refiner provide a great opportunity to influence the course of events in refining. Pulp quality can be determined by selection of the most suitable segment pattern and also by an optimal power split between the two zones. In the ordinary disc refiners, pulp is transported through the refiner gap by means of forces (centrifugal force, steam velocity) which increase in relation to increasing radius and rotational speed. In the conical refiner, the effects of centrifugal force and steam velocity are different. In the conical zone, centrifugal force transports the pulp against the stator discs, thereby allowing free passage of steam through the grooves of the rotor segment. Therefore segment patterns of the stator and rotor discs are different and are designed to meet the specific operational requirements of the stator and rotor segments.

In a typical CD refiner system, a pressurized or open preheater feeds chips by means of a plug screw to the feeder of the primary CD refiner. From the primary refiner, pulp is blown to the steam separation cyclone at the feed of the secondary refiner. After secondary refining, pulp is blown to another steam separation cyclone from where pulp is fed to latency removal. Adding dilution water allows control of refining consistency. The number of dilution points is dependent on the material being refined and the pro-

Thermomechanical pulping

cessing conditions. A separate reject treatment system handles screenroom rejects. TMP steam from the refiners is blown to a heat recovery system for generation of fresh steam for paper drying. In the modern Thermopulp® systems, the refining pressure of the secondary CD refiner is increased from the conventional 3 bar-level to approximately 7 bars. High pressure reduces the volume of steam in the refiner gap. This reduces the gap clearance which, in turn, reduces the need of specific refining energy.

3.3.2 Andritz Twin refiner

Sprout-Waldron originally designed the Twin refiner, which currently is used in TMP systems sold by Andritz. The Twin refiner (Fig. 18) is designed for a large production capacity and incorporates two flat disc refining zones in a single refiner frame. Power is transmitted through the refiner shaft and distributed to two refining zones. The Twin refiner's double-sided disc is mounted between two stationary discs. In this way, axial thrust is balanced on both sides of the rotating disc. Therefore this type of refiner requires no high-capacity thrust bearings. Balanced axial forces, symmetrical design, and sturdy construction help maintain a good parallelism of the refining zones. The rotating assembly consists of the shaft, the disc, two ribbon feeders, and the bearing seal cartridge. The bearings and the mechanical seals are easily accessible from the end of the machine.

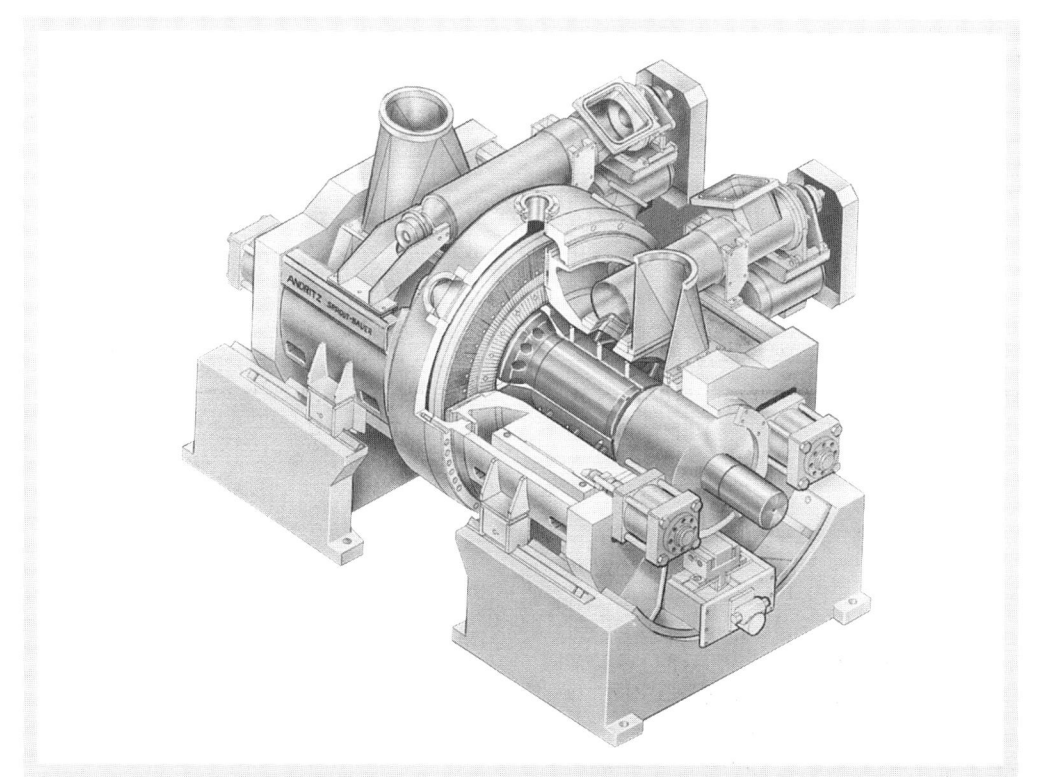

Figure 18. The Twin refiner (Andritz).

CHAPTER 7

Two load-sensing conveyors transfer the raw material from the feeding device into two refiner inlets. The refiner feed screws on both sides of the rotating disc are of the ribbon-type design. The ribbon is attached to the main shaft. The ribbon is welded to the spokes which, in turn, are welded to a stainless steel sleeve on the shaft. The ribbon feeder rotates at the same speed as the refiner shaft. Due to the high speed, material is transported at the periphery of the feeder barrel, thus allowing steam to escape in the middle along the shaft. In this way, blow-back steam does not disturb the feed of material into the refiner. In the eye of the refiner, material is fed into two parallel refining zones, one on each side of the rotor disc.

The double-sided rotor disc is mounted between two stationary plate holders which maintain balanced thrust on both sides of the rotating disc. The loading of the refiner is accomplished by four hydraulic cylinders, two of which are installed on each end of the refiner and connected to the movable plate holder assembly. Hydraulic pressure is provided by the hydraulic system, which is designed in such a way that the same hydraulic loading is applied on each of the four cylinders. This assures equal loading on both sides of the refiner. Later designs allow for a certain difference in hydraulic pressure respectively loading the two refining zones up to the capacity of the axial thrust bearing in order to compensate for variations in operating parameters such as feed rate and refiner plate condition. The plate protection system measures the acceleration of the vibration created when the refiner plate bars pass each other. This provides a signal which can be used to indicate the gap clearance between the plates. Through the use of a programmable controller, the plate protection system can anticipate the plate contact and consequently unload the refiner. In addition, plate gap sensors are installed in both refining zones, which establishes a minimum plate gap setting and provides refiner plate protection. The Twin refiner can be opened in the center to provide space for changing the refiner plates with access from both sides of the rotor disc. The rotor disc diameter of Twin refiners varies from 45 to 66 inches and a motor size from up to 30 MW at 1 800 rpm. Nominal disc diameter and rated power for the Twin 60 and 66 refiners at 1 800 rpm are as follows:

	Rated power MW	Disc diameter mm	Disc diameter inch
Twin 60	24	1520	60
Twin 66	30	1680	66
TC 66 (high speed refiner)	30	1680	66

In a typical Andritz two-stage Twin refiner system, chips are presteamed in an open atmospheric chip bin. After passing the chip bin, chips are fed by means of a plug screw feeder to the feed system of the refiner. Between refining stages and after secondary refining, steam and pulp are separated in "swept orifice"-cyclones. Steam from the cyclones and also blow back steam from the feeding screw is taken to the heat recovery system, which utilizes heat of TMP steam for generation fresh steam, e.g., for paper drying. It is often typical of the Andritz TMP systems that screen rejects (after dewatering) are refined in the secondary stage of the mainline refiners. Co-refining of rejects together with mainline pulp reduces the required number of refiners. The two-stage refining system is typical in news-grade TMP; but, in higher grades such as SC and LWC, a typical Andritz system has three refining stages in mainline, which usually consists of two Twin refining stages and third-stage single-disc refining.

In modern RTS systems, the speed and pressure of the primary refiner is increased from conventionally used 1500/1800 rpm and 3 bars to 2300 rpm and 5–6 bars. Increased refiner speed reduces the specific energy consumption in refining. The pressure and temperature are increased to eliminate excessive shortening of fibers in high-speed refining.

3.4 Double disc (DD) refiners

The double-disc refiner has two counter rotating discs mounted on cantilevered shafts, each driver by a separate motor. Chips are fed into the refiner gap through openings in one of the rotating discs. The rotors of a double-disc refiner rotate at differential speeds of 2400 rpm (2 x 1200 rpm) at 60 cycles electrical supply and 3000 rpm (2 x 1500 rpm) at 50 cycles current. The high differential speed of the DD refiner discs makes it possible to use single-stage refining in mainline refining instead of traditional two-stage single-disc refining when manufacturing, e.g., news grade TMP. When refined to the same freeness, DD refining uses at 1500 rpm approximately 15% less energy, when compared to the standard single-disc 1500 rpm refining. At the same freeness, DD pulp has somewhat shorter fiber length, approximately the same bonding properties (tensile) and higher light scattering. Rated motor power of DD refiners at different speeds and disc diameters are:

Speed	rpm	1500	1500	1800	1800
Segment diameter	mm	1650	1800	1650	1800
Motor	MW	15	18	20	25

The RGP 65/68 DD (Fig. 19) is the double-disc version of the Sunds Defibrator family of RGP-refiners. The RGP 65/68 DD refiner is based on the use of two facing units of the RGP single-disc refiner. The refiner is powered by two separate external motors coupled to independent shaft assemblies. For servicing the shaft, complete with bearings and hydraulic components, it is lifted out and replaced with a new shaft assembly. For a rapid change, segments are mounted on the holders which are lifted out and replaced with the holder of new set of segments. The RGP 65/68 DD is equipped with the disc clearance sensor (TDC) as a standard feature. The TDC sensor is a key component of the refiner control system, permitting faster startup and loading of the refiner and protection against plate to plate contact. The Sunds Defibrator double-disc refiners are operating at 1500 rpm in Europe using segment diameters 1650 mm (65 in.) and 1800 mm (68 in.).

CHAPTER 7

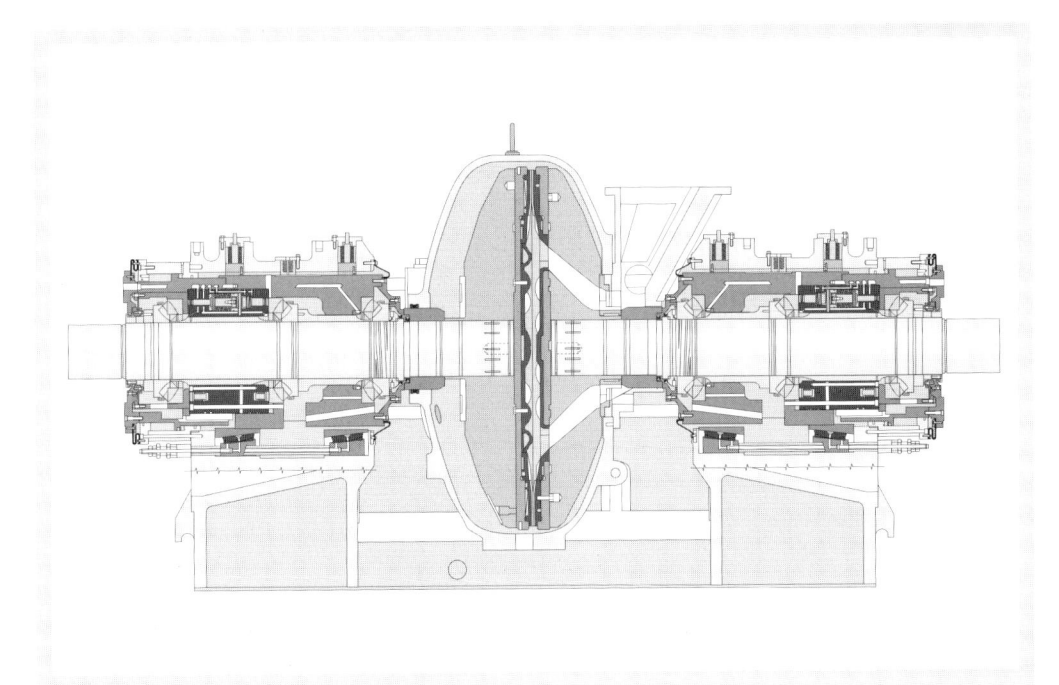

Figure 19. The RGP 68 DD refiner (Sunds Defibrator).

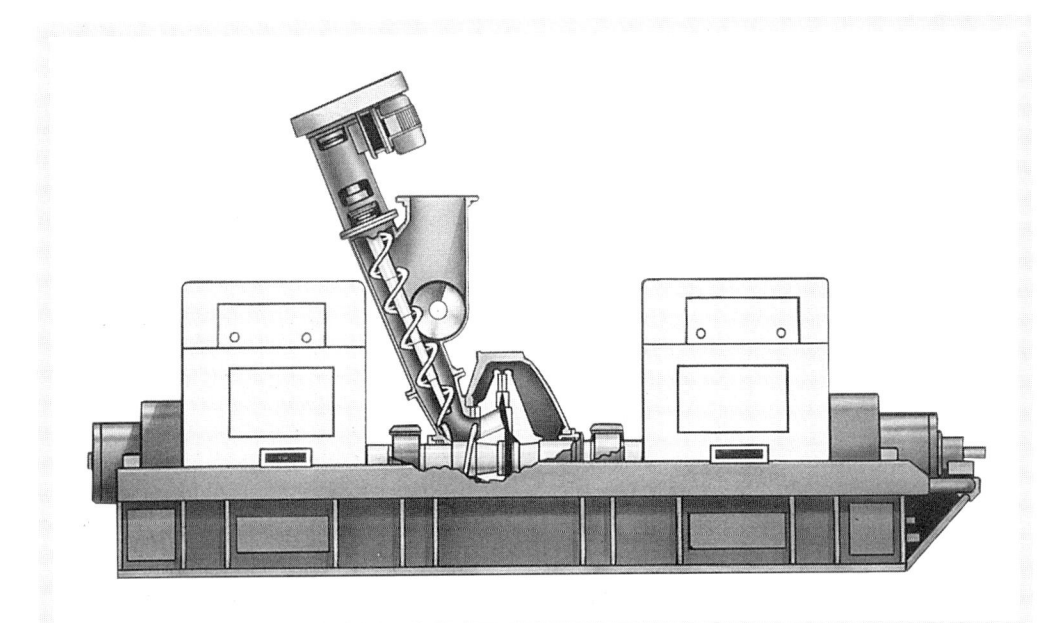

Figure 20. The SB 160/190 double-disc refiner with disc diameter 1600–1850 mm and a maximum rated power 20–26 MW (Andritz).

The Andritz (earlier Sprout-Bauer) double-disc refiner (Fig. 20) has motors integrated with the refiner: The refiner and the motor on both sides are installed on a common shaft and bearings. The Andritz DD refiners are so far operating only at 1200 rpm in North America, except some trials have been made with DD refiners operating at 1800 rpm at feed side when the other side is operating at 1200 rpm. This mode of operation is claimed to give an approximate 20% energy saving compared to standard 2 x 1200 rpm operation.

3.5 Processing pulp and steam in the refiner line

Large volumes of steam are generated in the high-consistency TMP refining. Efficient separation of pulp and steam is important for the consistent, trouble-free operation of TMP refining. The feeding equipment of the refiner and the steam separation cyclones are designed to handle large volumetric flows of steam and pulp. In pressurized systems, the volume of steam is reduced – the higher the pressure is, the smaller is the volume of steam compared to atmospheric refining. Degrading pulp quality, in particular lowering brightness, can be avoided by minimizing the residence time of pulp in the elevated pressure and temperature and by presteaming chips at lower pressure – even atmospheric preheating is used, while refining is accomplished at higher pressure. In the Sunds Defibrator Thermopulp®-process, the pressure of secondary refining is very high, up to 7 bars. The use of different pressures in different parts of TMP refining necessitates the use of the pressure-sealing devices. Plug screw feeders are used between the preheater and the primary refiner and before the secondary Thermopulp refiner.

The discharge consistency of refined pulp has a major effect on the amount of steam generated in refining – the higher the consistency is, the more steam is generated, e.g., increasing discharge consistency from 35% to 50% adds the amount of TMP steam by approximately 25%. In addition to the amount of steam, the refining consistency has an influence on the refiner gap clearance, which in turn has a great impact on pulp quality. It is therefore very important to keep refining consistency at its target value. This is accomplished by adding dilution water into the refiners. Addition of water makes fibers heavier which helps operation of the steam separation cyclones. In some cases, water sprays are used at the outlet of the cyclones to prevent fiber carryover with steam into the heat recovery system. The pulp blowlines between the refining stages as well as steam lines to heat recovery must be dimensioned to match the amount and volume (pressure) of generated steam. Too small pipelines will result in excessive pressure losses; on the other hand, fiber deposits on the pipe walls will cause clogging of too large blow lines. In separation of steam and pulp, different type of cyclones are used.

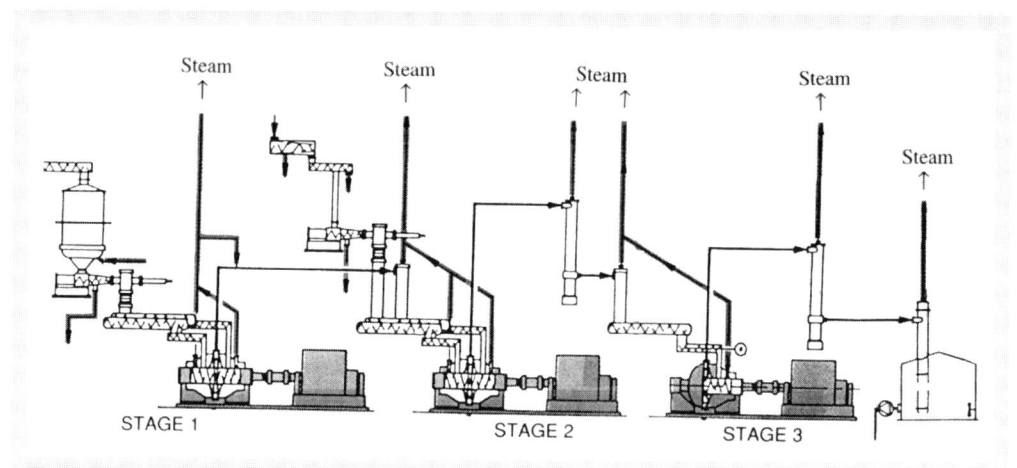

Figure 21. Andritz TMP system with small-diameter cyclones.

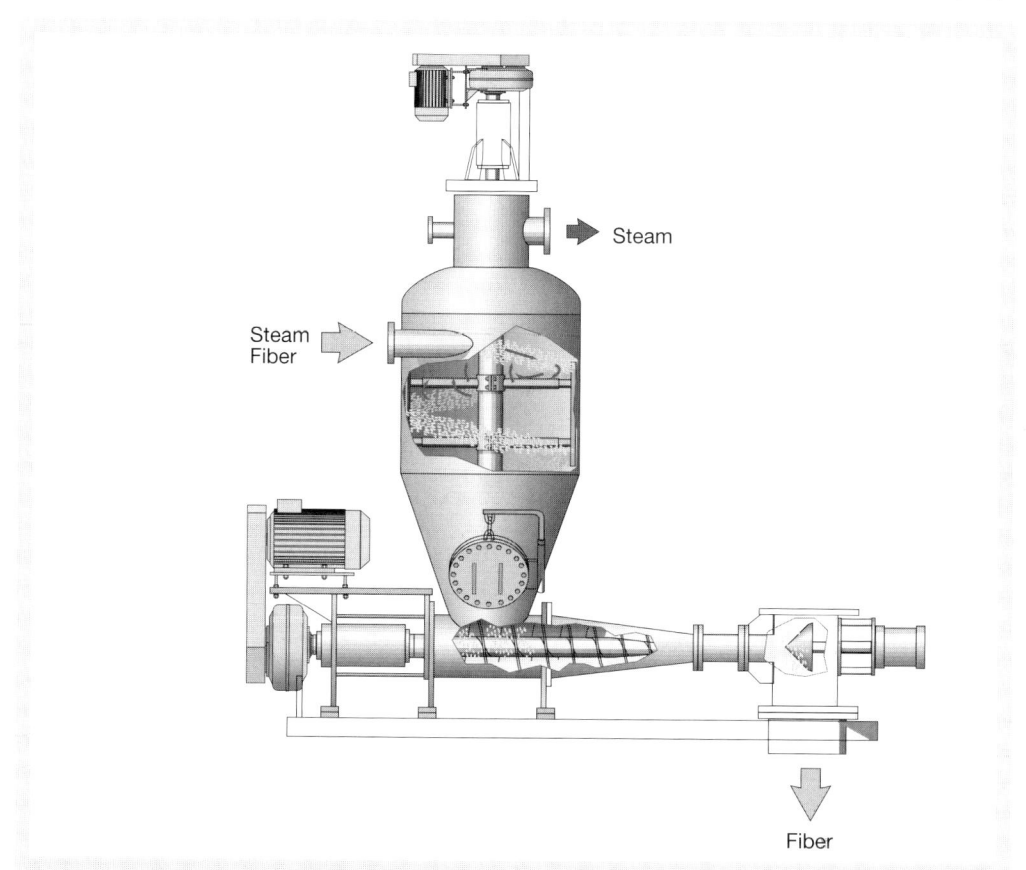

Figure 22. Large-diameter cyclone (Sunds Defibrator).

In Andritz systems (Fig. 21), a small-diameter cyclone is used where higher rotational speed separates fibers and steam. Steam is separated at the upper part of the cyclone and pulp is discharged from the bottom by means of a "Swept Orifice Discharger." An impeller rotates within the discharger to prevent fiber buildup or plugging at the orifice. A small amount of process steam conveys the fibers through the orifice into the latency chest. Sunds Defibrator uses larger-diameter cyclones to separate pulp and steam (Fig. 22). Here the combined effect of centrifugal and gravity forces separates pulp from steam. To prevent accumulation of pulp, a scraper screw is used to clean the inside walls of the cyclone. Pulp is discharged by means of the plug screw discharger. Density of the plug is controlled by adjusting the pressure of the counter cone, which disintegrates the pulp and discharges it into the latency chest.

After refining, TMP fibers have curl setting which is called latency. Curly fibers change the properties of pulp: Freeness increases and strength properties are deteriorated. Screening efficiency of curly fibers is poor. Therefore latency has to be removed before pulp is taken into screening. Latency is removed by disintegrating pulp in low consistency (2%–4%) and in relatively high temperature (70°C–80°C). Often latency removal is made in two stages: At first, pulp is diluted and agitated vigorously for a few minutes in a small sampling tank placed under the steam separation cyclone. Separate pumping chests after each refiner line enable latency removal and pulp sampling needed in individual pulp quality control of the refiner lines. The major part of the latency is removed in the sampling chest, and the remaining latency is then removed in the common latency chest placed before screening. In order to reduce investment costs, some mills have installed a high-consistency storing of pulp between mainline refining and screening. This kind of arrangement is not recommended because too high of a storing temperature reduces pulp brightness and makes removal of latency from pulp more difficult.

3.6 The effects of refining conditions and equipment design parameters

The key parameters affecting pulp quality and SEC of refining are the specific energy consumption and the intensity of refining. Figure 23 schematically presents interaction of the key parameters of TMP refining. The specific energy consumption (SEC) is the major factor influencing pulp quality. SEC is determined by the motor load and by the production rate of fiber material through the refiner. In addition to SEC, the intensity of refining is another key parameter of TMP refining. There is no generally accepted definition for the refining intensity, but the refining intensity is increased when the energy input of single impacts of refining is increased. This can be reached by reducing the refiner gap clearance. The refiner gap is reduced by speeding up the pulp flow through the refiner gap. The process conditions increasing the speed of pulp flow through the gap are, e.g., increased motor load (higher SEC), reduced consistency, increased pressure (and temperature) of refining, reduced differential pressure between the casing and feed of the refiner, increased production rate at constant SEC of refining, and increased refiner speed.

CHAPTER 7

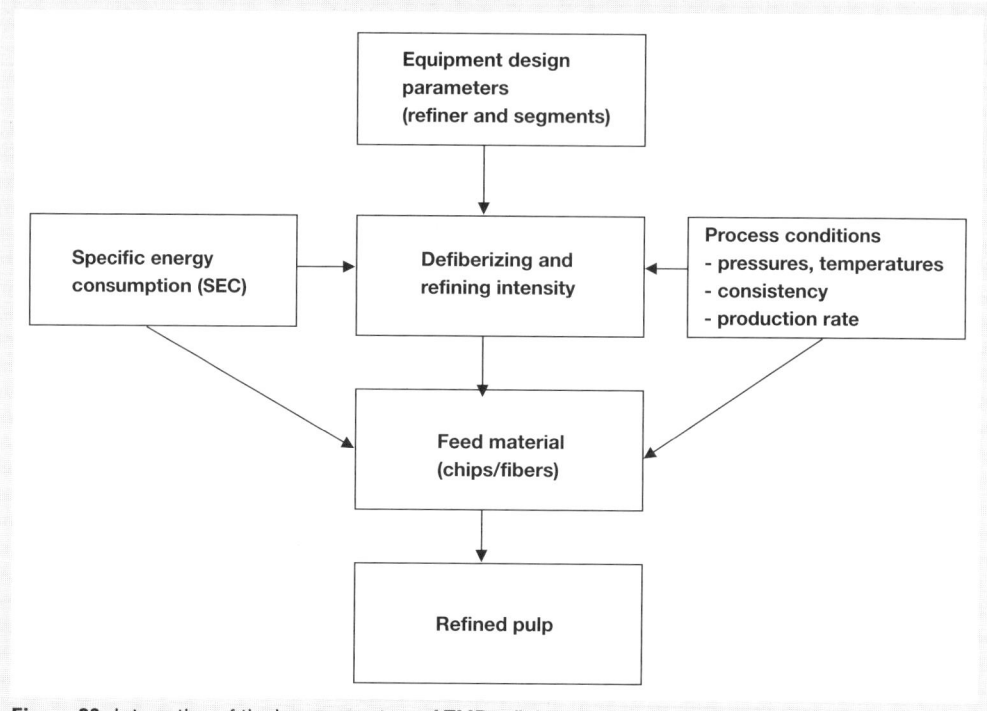

Figure 23. Interaction of the key parameters of TMP refining.

The performance of all refiners is influenced by different disturbances of the refining process. These disturbances arise due to variations, e.g., in chip quality (species mix, chip size distribution, and moisture content), in production rate (bulk density variations), in refining consistency (chip moisture content and dilution water flow), and in the condition of the refiner plates. These disturbances affect pulp quality by changing specific energy and intensity of refining.

The production rate is a difficult parameter to control. The chip feeding equipment is based on the volumetric feed of material into refining. An accurate measurement of the mass flow can only be achieved by combining the volumetric feed rate with an accurate measurement of the bulk density and the moisture content. The problem is that reliable devices for the bulk density and the moisture measurements of chips are not available. Exact measurement of the production rate based on the pulp flow and consistency after the mainline refiner line is just as problematic because the consistency measurement devices are sensitive to different disturbances such as pulp quality, temperature, and air content of pulp. Poor performance of the mass flow measurement is a major problem of consistent SEC control of TMP refining.

In addition to the refining conditions, the design of the refiner, refiner process, and refining segment have an impact on the intensity of refining. Of the single-disc refiners, the conical refiner (RGP CD) operates usually at a larger refiner gap clearance than the other single-disc refiners because centrifugal force transports the fibers at the conical gap against the stator segment and not toward the gap outlet as is the case in flat disc refining. For this reason, the conical disc refiner usually produces pulp with longer fiber length than the flat disc refiner.

In DD refining, the direction of pulp rotation is changed when pulp flows inside the gap from the feeder side disc to the control side disc. This change of rotation reduces the effect of centrifugal force which in turn increases the retention time of fibers and the disc gap clearance when compared to ordinary flat disc single-disc refining. In spite of the larger gap clearance, DD disc refiner produces shorter fibers than single-disc refining. The exact reason for the different behavior is not known; the higher relative disc speed and different orientation of fibers across the segment bars could be the reason for the reduced length of fibers. Compared to single-disc refining, DD pulp has at constant freeness shorter fiber length, a higher light-scattering coefficient, and 10–15% lower SEC.

The refiner speed has a strong effect on the refining intensity and on the development of the pulp properties. Tests have proved that the effect of rotational speed is stronger in the primary stage refining. Increasing the speed of the primary refiner from 1500 to 2400 rpm reduces SEC by 10%–20%. Simultaneously, fiber length and fiber length-related pulp properties are deteriorated; on the other hand, light scattering is improved.

In the modern high-intensity processes such as the Sunds Defibrator Thermopulp® and the Andritz RTS processes described later on in this chapter, an excessive shortening of fibers is avoided by increasing the processing temperature. This is reached by increasing the refining temperature. Higher temperature reduces the rigidity of fibers with the result of a reduction in the shortening of fibers in high-intensity refining. In order to avoid darkening of pulp, the residence time of fibers in elevated temperature should be as short as possible.

The refiner segment design has a major impact on the intensity of refining. Different refiner concepts and process conditions require specific segment design. In general, the refining intensity can be increased by using forward feed-type segment bars and grooves, which feed pulp and steam forward in the gap. Thus residence time of pulp and refiner gap clearance are reduced, resulting in increased intensity of refining. In segment design, one always has to make a compromise between reducing the refining energy and maintaining the fiber length.

4 Refiner segments – the "heart" of the refining process

The invisible but very important part of every refining process is the refiner segment. The type of refiner segment chosen ensures reaching the pulp quality and the production level requirements set for the TMP mill by the paper machine. These requirements depend on the type of produced paper and on the control strategy of the paper machine and are therefore different in different mills, which accordingly require different types of segments. The varying refiner positions (first-stage, second-stage, and reject) in the same mills also need basically different designs of refiner segments. To make this even more complicated, there are different refiner types, refiner processes,

CHAPTER 7

and operating conditions – the combinations of which need unique types of refiner segments.

Figure 24 shows refiner segments used in the Sunds Defibrator RGP 268 single-disc refiner. The diameter of this refiner is 68 inches, and it is designed to be able to use an axial load of 90 tons and a refiner power of 15 MW. In this type of refiner, the refiner segments are mounted on segment holders and do not have bolt holes on the front side, which reduces the active refining surface of the refiner. Refiner segments form the stator and rotor rings as shown in Fig. 24. In both the stator and rotor, there are separate periphery segments (p-segments) and center segments (c-segments). On the rotor side, there is also a center plate. In this case, there is a hole in the stator p-segment for the plate gap measurement device (TDC sensor).

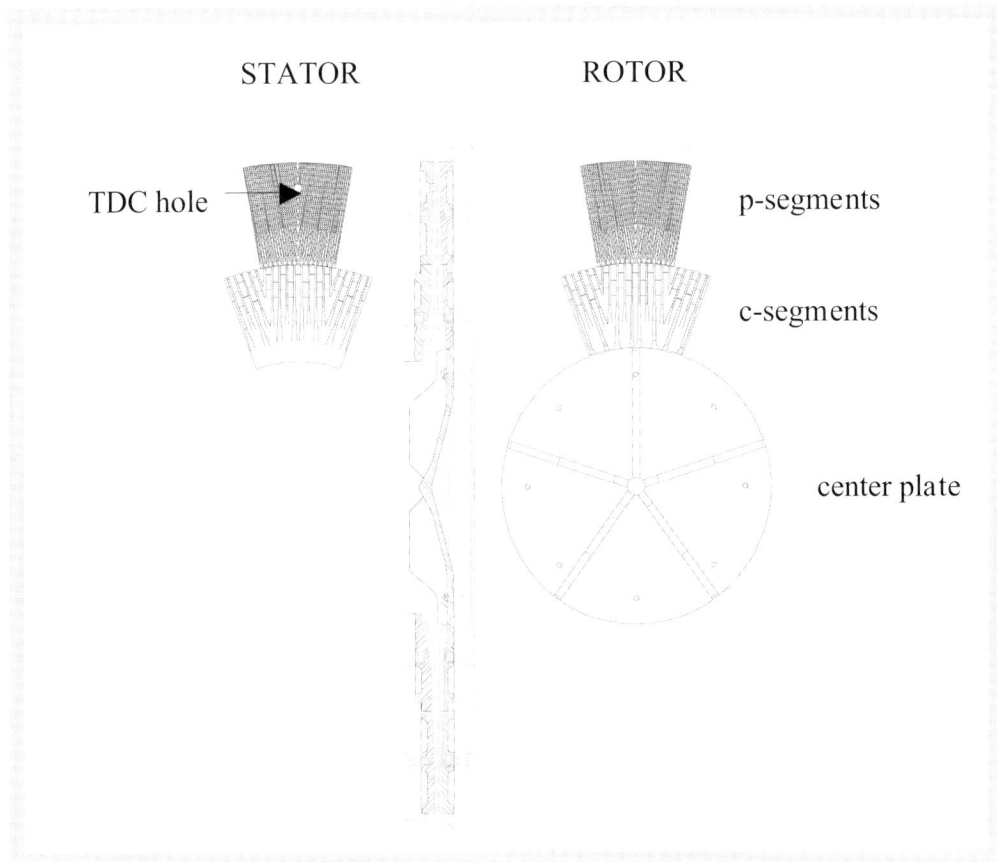

Figure 24. Refiner segments in the Sunds Defibrator RGP 268 single-disc refiner.

There are always some exceptions from these segment types in different types of refiners. For example, in the double-disc refiner there is no stator, and the terms "infeed" and "control side" are used to separate the disc which has the openings for the feeding from the disc which is used to control the plate gap. The term "breaker bar" is commonly used for the c-segment in the double-disc refiner. In the CD refiners, there is also a conical zone with cd-segments (see Fig. 17). The conical refining zone makes it possible to have a reasonable low axial thrust and a good control of the plate gap at high production and refiner power levels.

In some pulp mills, there are also refiners for low-consistency refining (reject refining or post refining). These refiners threat pulp in a consistency of 3%–5%. The most common theories used to control low-consistency refiners are the specific edge load and the specific surface load theories[1]. These theories also take into consideration the effect of basic segment design parameters. Together with these basic design parameters, there are many details in the segment geometry that have to be considered when optimizing segments for different applications. For example, it has been possible to increase the capacity of the refiners with a new type design of the segments at the refiner inlet zone (Sunds Defibrator Maxiflo segment). The chapter on thickening, storage, and post refining describe low-consistency refining in more detail.

4.1 Refiner segments design parameters

Refiner segment design parameters are different for different types of refiners and for the special selection of the refiner operating conditions. Some basic rules, however, can be presented. During operation, the conditions between the refiner segments are very special. The steam is flowing toward the inlet of the refiner or toward the refiner housing, depending on the location in the refining zone. The pulp is fed into the refiner against the steam flow, transported through the plate gap, and refined to improved quality. Successfully designed refiner segments not only produce pulp with the desired quality, but also have stable runnability over the total operating range of the refiner. To reach these goals, the steam removal from the plate gap must be effective, the pulp must be fed evenly through the refiner, and the fibers must be refined to target properties.

Refiner segments have traditionally been developed through experimenting, but theories to support this work have also been published. Theories about the defibration of the chips[2], the pulp flow in the refiner[3], and the concept of the refining intensity[4,5] have been used to explain the phenomena in the plate gap. However, these theories describe the refiner in a simplified way[6,7] and cannot be used to design refiner segments in detail. An improved theory, taking into account the difference between the rotor and the stator as well as the refiner inlet, has recently been presented[8]. Improved theories, together with modern measurement devices[9–11] and a basic knowledge about the disturbances in the process[12–14], will widen our knowledge about the pulp and steam flow in the plate gap. In today's theoretical models, there are still some parameters that need to be developed before models can be used to design refiner segments without mill-scale trials.

CHAPTER 7

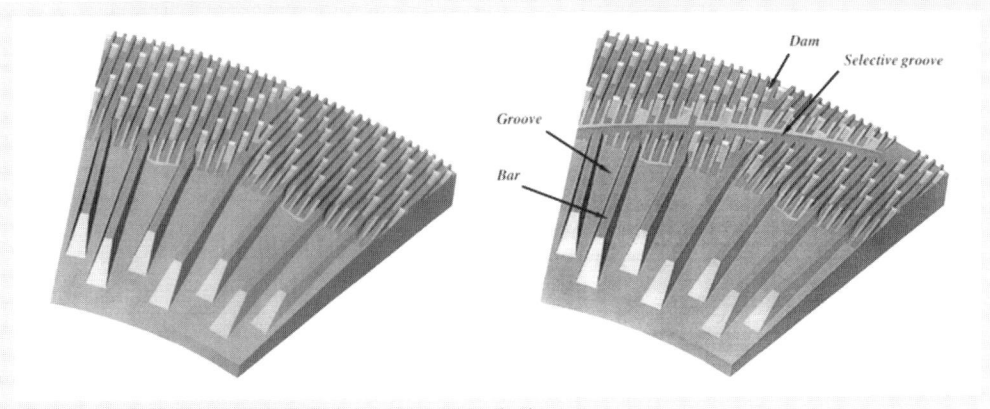

Figure 25. Typical refiner segment with and without the selective groove.

The basic design parameters of the refiner segments are: the width of the bars and the grooves; the height of the bars; number, placement, and design of the dams; selective grooves; taper; and angle of the bars. For the infeed section of the refiner (the c-plate), design parameters are different from those of the refining zone.

The width of the bars and the grooves has traditionally been the main design parameter for the refiner segments. Wide grooves and narrow bars often reduce specific energy consumption in the refining. High open volume in the grooves makes the plate gap narrower, and reduction in the pulp quality might occur. Wide bars and low groove volume increase specific energy consumption and improve the quality of the produced pulp. However, when the volume in the grooves is reduced, the removal of steam is more difficult and the axial load of the refiner is higher, the infeed of the fibers is more difficult, and the refiner load might be unstable.

The height of the bars has a very similar effect on steam removal as the groove volume does. The higher the bars are, the more open volume there is in the grooves and the more effective is the steam removal. Low bar height forces fibers to the plate gap, and the pulp quality is improved. Bar height in the rotor side also has an important effect on the transportation of the fibers. High bars in the rotor side force fibers to a rotating motion, and the centrifugal force affecting them becomes higher.

Number, placement, and design of the dams have an important effect on pulp quality. Dams force pulp from the grooves into the plate gap, and the residence time of the fibers in the refiner increases. A large number of dams in the segments improves fiber quality but also makes steam removal more difficult.

Grooves in the rotor side are used to remove steam from the plate gap to the refiner housing. These selective segments need less axial thrust. Together with selective grooves, the segment can be designed with wider bars, narrower grooves, lower bars, more dams, and smaller taper. These, together with improved steam removal, improve pulp quality without runnability problems.

The taper is ground into the segments so that the plate gap is most narrow at the outlet of the refiner and wider at the inlet. There are often two radial zones in a segment. The outer zone is the refining zone with a small taper and the inner zone with a steeper taper. The taper is used to compensate the possible bending of the rotating disc at the refining zone and to improve the feeding of the chips into the refining zone. Too short of a refining zone might cause a reduction in pulp quality, and too long of a zone might increase the axial load.

The angle of the bars from the radial direction causes force on the pulp and steam flow. When the rotor bars form a pumping angle, the fibers are forced to a faster and less turbulent flow and the residence time of the fibers in the refiner decreases. This leads to reduced energy consumption in the refining. Using this theory, Sunds Defibrator has designed SC LE segments for refining with low specific energy consumption. This theory has also been used to design segments that produce pulp with different fiber length distribution[15]. The angle of the bars is not only important on the rotor side, but the intersecting angle between opposite bars also has an important effect on refining results.

4.2 Materials of refiner segments

Refiner segments are used in exceptionally difficult conditions, and this puts great demands on the material used in the segments. Usually the lifetime of the segments in a TMP mill varies from 1000 to 3000 operating hours. Wear normally shows as rounded edges of the bars at the refining zone of the segments. Worn segments produce pulp of poor quality. The better the control of the refiner and the cleaner the process, the longer the segment lifetime will be. Plate clashing (contact between the opposite plates) might also be a reason for a segment change. Also, to prevent clashing, a good control of the refiner is of significant importance.

Refiner segments are manufactured by casting normally from martensitic stainless steels. The combination of material characteristics is adjusted by composition and heat treatment. Refining process (stability, impurities in pulp), pattern design, and place in the segment ring are the factors which determine what combination of material characteristics is needed. Different material demands mean that the compositions used vary on quite a wide scale. The basic factors affecting material characteristics are carbon and chromium contents. The composition ranges used are approximately 0.2%–3% C and 15%–30% Cr.

The most important material characteristics affecting plate life are abrasive wear resistance, impact strength, resistance against cavitation, and corrosion resistance. When selecting a suitable material for each application, the following factors are considered.

Abrasive wear resistance, like all material characteristics, is based on the microstructure of the material. The structure of martensitic stainless steels is formed of hard chromium carbides and martensitic matrix. The amount of hard carbides, and thus the wear resistance, increases when the carbon content is increased. High carbon qualities are used when there are abrasive impurities in the pulp.

Impact strength is needed to ensure that the probability of bar breakage is minimized to an acceptable level. Increasing the carbon content and, thus, wear resistance will have an opposite effect on impact strength. A high amount of hard and brittle carbides in the structure lowers the impact strength. When estimating the requirement for impact strength, the probability of having trash in the pulp or plate clashing should be considered.

Cavitation can occur mainly at the inlet section of the refiner. The most important parameters affecting the occurrence of cavitation are the absence of free water and the design of the inlet geometry. The damage rate, however, also depends on material selection. The cavitation resistance increases when hardness increases, but only to a certain level. The optimum material composition range is found around the carbon content of 0.5%–1%.

Corrosion is normally not a problem in TMP refiners. However, because of the complexity of the wear phenomenon in refiners, corrosion resistance cannot be neglected totally. When combined with other wear mechanisms, it can increase the wear rate of the segments. In normal refining processes, the corrosion resistance of martensitic stainless steels is on an adequate level.

5 Main TMP process types

As a rule, a TMP line is always custom made for the mill and paper grade in question. Because of this, there are no "normal" TMP processes. Still, there are some main developmental lines, which should be discussed.

5.1 The "standard" TMP line

At the end of the 1980s, a sort of standard layout for TMP for news and magazine papers had developed. As stated earlier, the pressurized preheating is to a large extent replaced by atmospheric steaming combined with a short low-pressure preheating, and the refining is performed in two pressurized stages at relatively high temperatures. The refiners are of the single-disc type.

An example of the result of this evolution is given in the simplified Fig. 26. The layout is quite simple. The chips are first steamed, then washed, and then steamed again before the two-stage pressurized refining in single-disc refiners. The refiners in this example (Fig. 26) are of the type with two gaps (Andritz Twin 60). After latency removal, the pulp is screened and cleaned before going to the dewatering filter and storage. The reject is pressed to high consistency and refined at pressurized conditions. The refined reject is screened and cleaned and routed directly to the accept line. The pressurized steam from the main line refining, as well as the steam developed in the reject refining, is collected and reused in the heat recovery system.

Figure 27 shows a typical modern newsprint TMP line. Washed chips are pre-steamed in the preheater. The chips are then fed to the two-stage mainline refining with RGP CD refiners. From the cyclones located before and after the secondary refiner, TMP steam is taken to the heat recovery. After mainline refining, screening is in this case performed at elevated consistency (3%), which reduces pumping volumes and simplifies pulp thickening. Reject is refined in a separate reject refining system.

Thermomechanical pulping

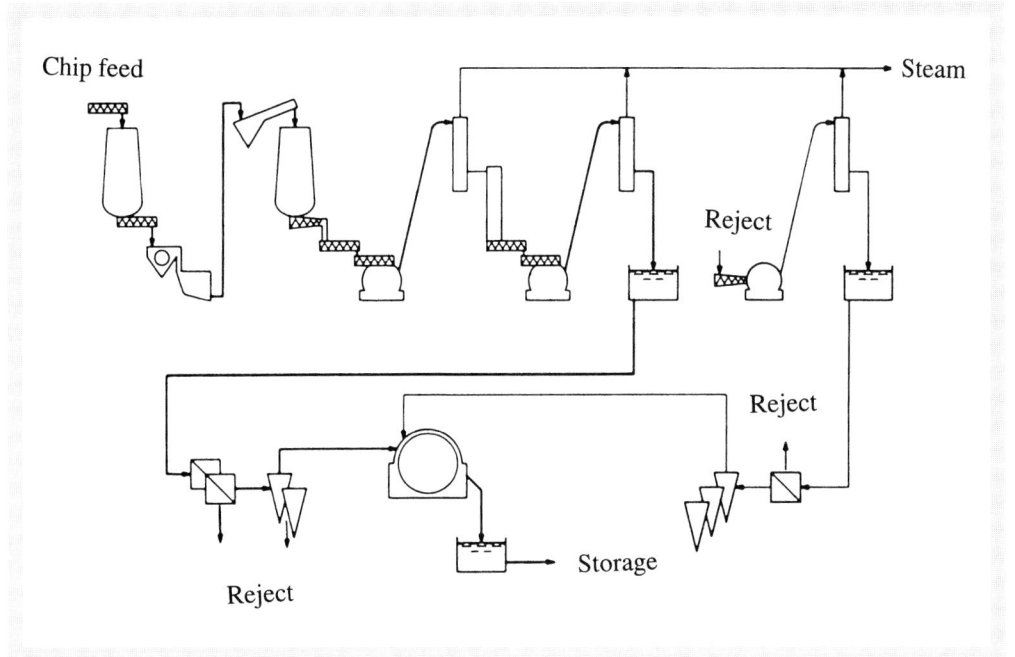

Figure 26. A two-stage TMP line for newsprint at MoDo Braviken, Sweden (Andritz).

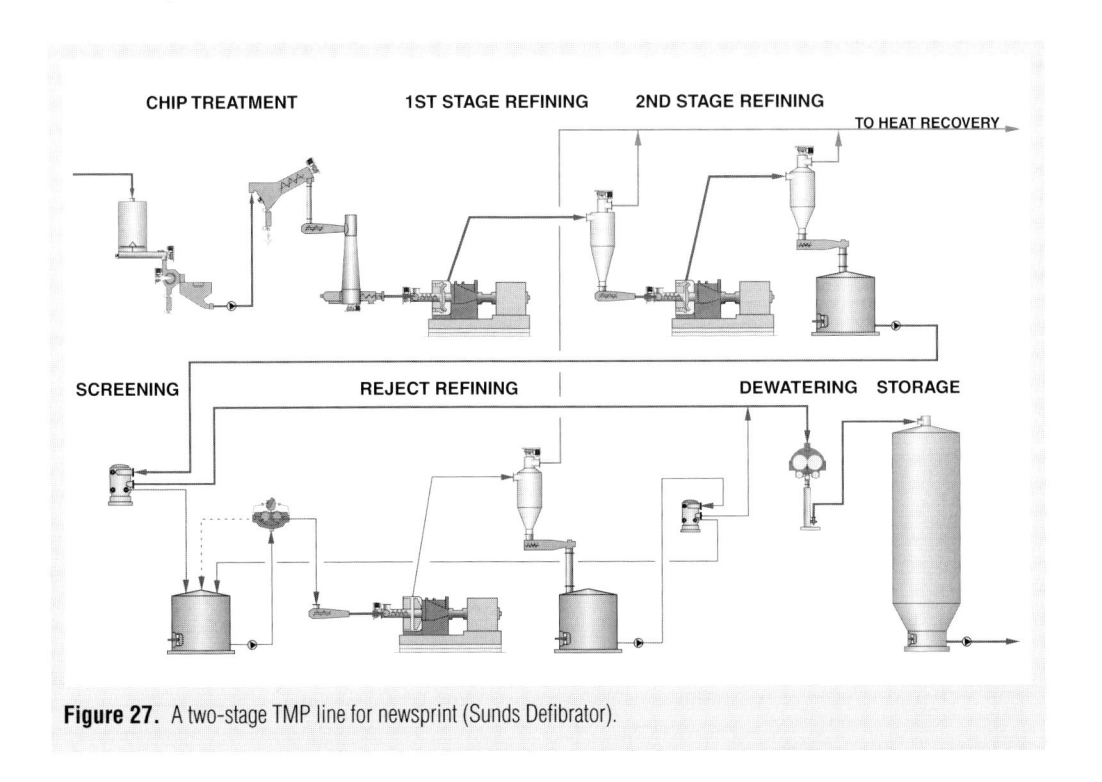

Figure 27. A two-stage TMP line for newsprint (Sunds Defibrator).

CHAPTER 7

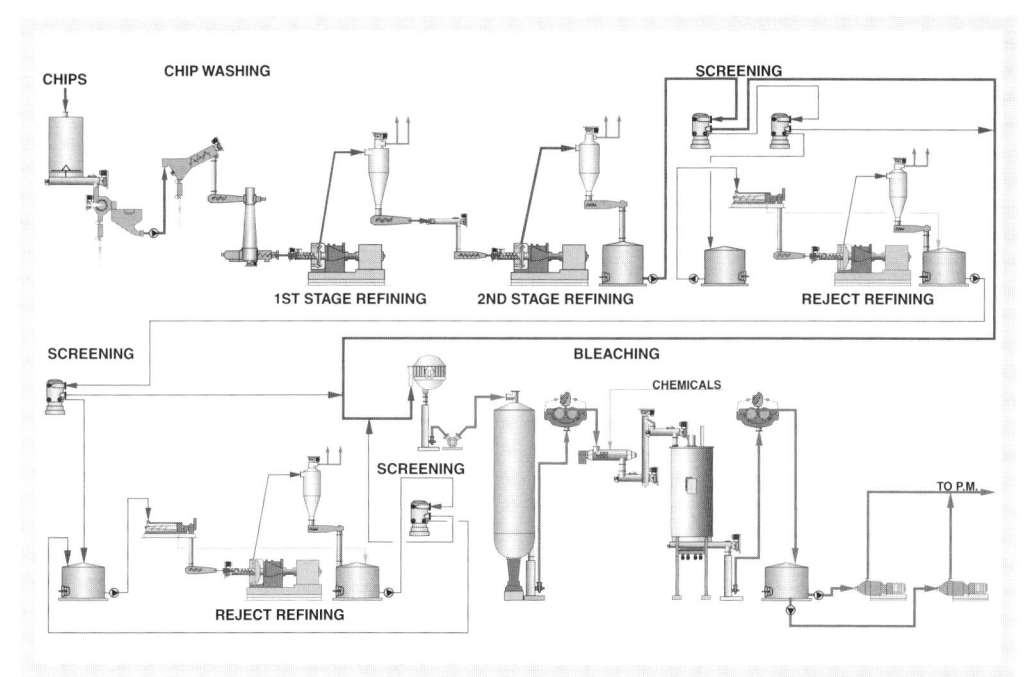

Figure 28. A two-stage TMP line for LWC paper (Sunds Defibrator).

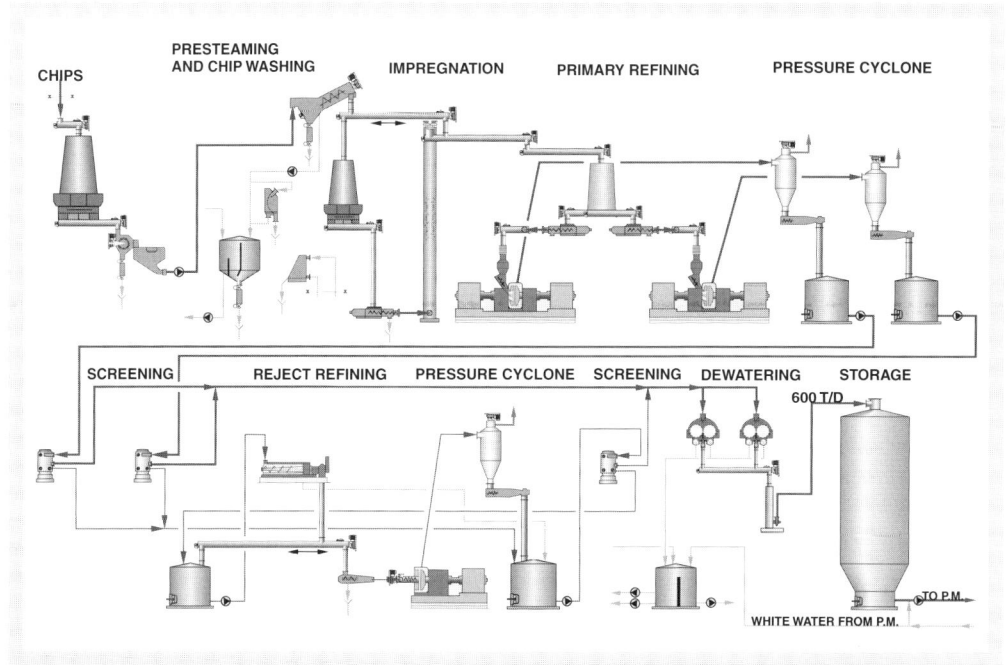

Figure 29. Single-stage TMP at Union Bruk, Norway, using double-disc refiners (Sunds Defibrator RGP 68 DD).

Thermomechanical pulping

New TMP mills making pulp for higher paper grades (SC, LWC) are still in many cases built like this, with most developments being made mainly in the screening and reject refining (Fig. 28). The load split is today normally about 50%–60% in the first stage and 40–50% in the second stage. The refining consistency (outlet consistency) is normally 40%–50%. The screens of today with very narrow slots have in many cases made it possible to completely leave out the cleaning stages. The higher emphasis on the reject and coarse fiber refining means that today the reject refining is often performed in two stages.

5.2 Single-stage refining

Since the beginning of TMP development, there have been systems with only single-stage refining. This is possible using double-disc refiners, which can be loaded to reach 150 ml CSF after one single stage of refining. Figure 29 shows a modern double-disc-based, single-stage system. In this example, the pressurized preheating is omitted, but both main line and the reject refining is done using pressurized refiners.

5.3 TMP with no separate reject refiner

In the 1980s and the early 1990s, several TMP lines with no separate reject refining were built. Fig. 30 shows an example of such a process configuration. The layout is otherwise typical of those for a modern standard TMP line, but the reject pulp from the screening room is fed to the second-stage refiners and there is also a third refining

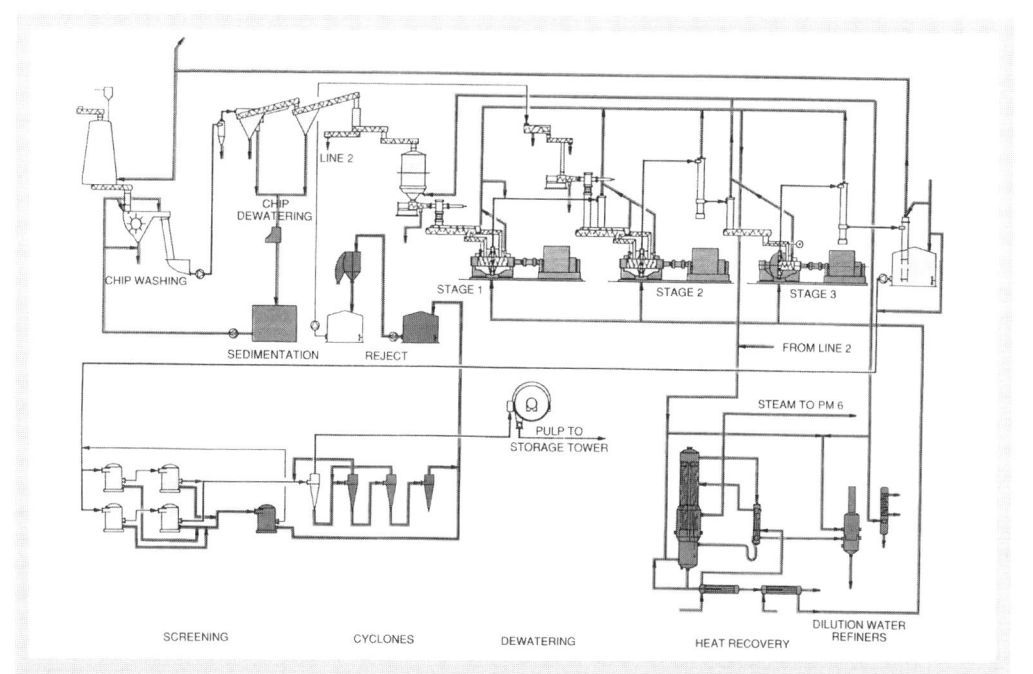

Figure 30. Three-stage TMP with no separate reject refining at Norske Skog Halden, Norway (Andritz Twin 66 and SB 170 refiners).

CHAPTER 7

stage. The thinking of today suggests, however, that the addition of the screen reject to the second stage refiners is not the optimal arrangement from a quality point of view.

5.4 Two-stage refining in the same refiner

The special Hymac refiner for two-stage refining inside the same refiner was presented earlier (Fig. 16). The load split for the one refiner operating at Hallsta in Sweden is normally 60%–65% in the first stage and 35%–40% in the second stage. The refining zone consistency is 45%–48%.

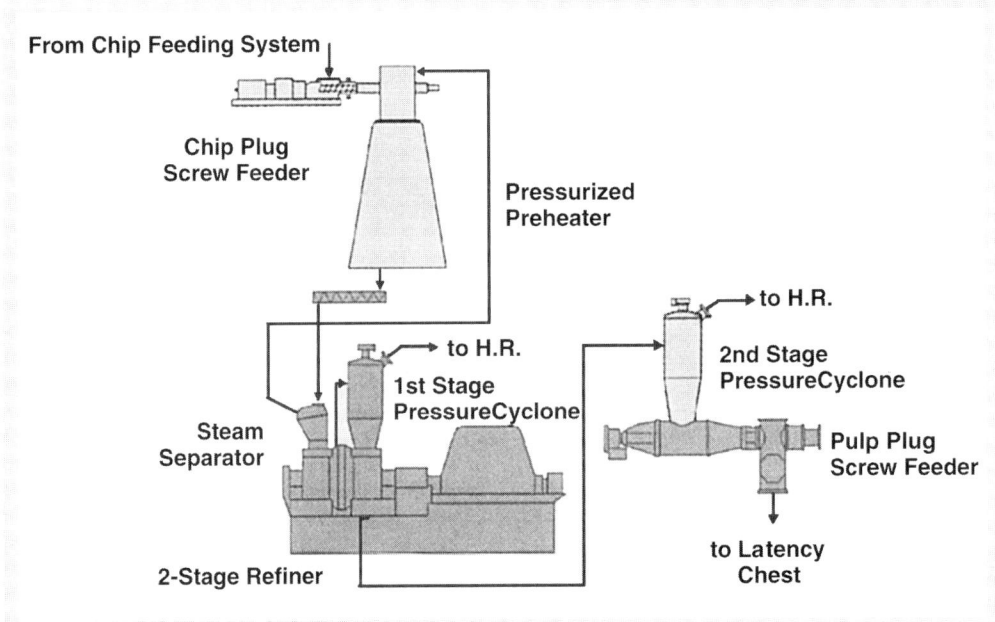

Figure 31. Special two-stage refining with only one refiner at MoDo Hallsta, Sweden (Andritz Hymac HXD 64 refiner).

5.5 TMP using high-speed refining

Pilot scale research in the 1980s showed that it would be possible to substantially reduce the energy consumption in chip refining by increasing the disc speed[16] and, in the early 1990s, some high-speed TMP systems were installed[17]. The high-speed refining has some drawbacks, though, the most important being the reduced fiber length and tear strength. But by restricting the high-speed operation to only the first stage and at the same time raising the temperature during this first stage, it was possible to preserve the fiber length[18]. Andritz has developed the "RTS"-process, a TMP process according to these lines. In RTS, the R stands for retention time, the T for temperature, and the S for speed. Figure 32 shows the RTS TMP concept. The first RTS refining stage has a refiner, which usually operates at 2300 rpm and 550 kPa overpressure. The first RTS installation was at Perlen Papier AG, Switzerland, in 1996, and the operating experience suggests that an energy reduction in the order of 15% is possible using this system.

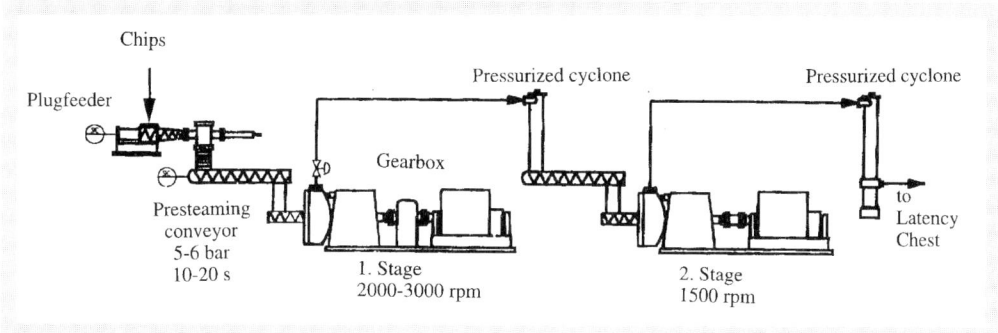

Figure 32. The RTS TMP process (Andritz).

5.6 Thermopulp

In the mid-1990s, Sunds Defibrator developed the Thermopulp process, of which several lines went into operation both in Europe and North America. In this two-stage process, the first refining stage is performed at a relatively low temperature (according to the theory to avoid fiber separation in the middle lamella). The pressure and temperature are raised before the second stage refining, which is performed at very high pressure and temperature (up to 700 kPa and 170°C). Energy savings of 10%–20% have been reported[36]. The energy saving can be explained by a higher fiber wall breakdown rate at high temperature and by the reduced plate gap due to the compressed steam and the softer fibers.

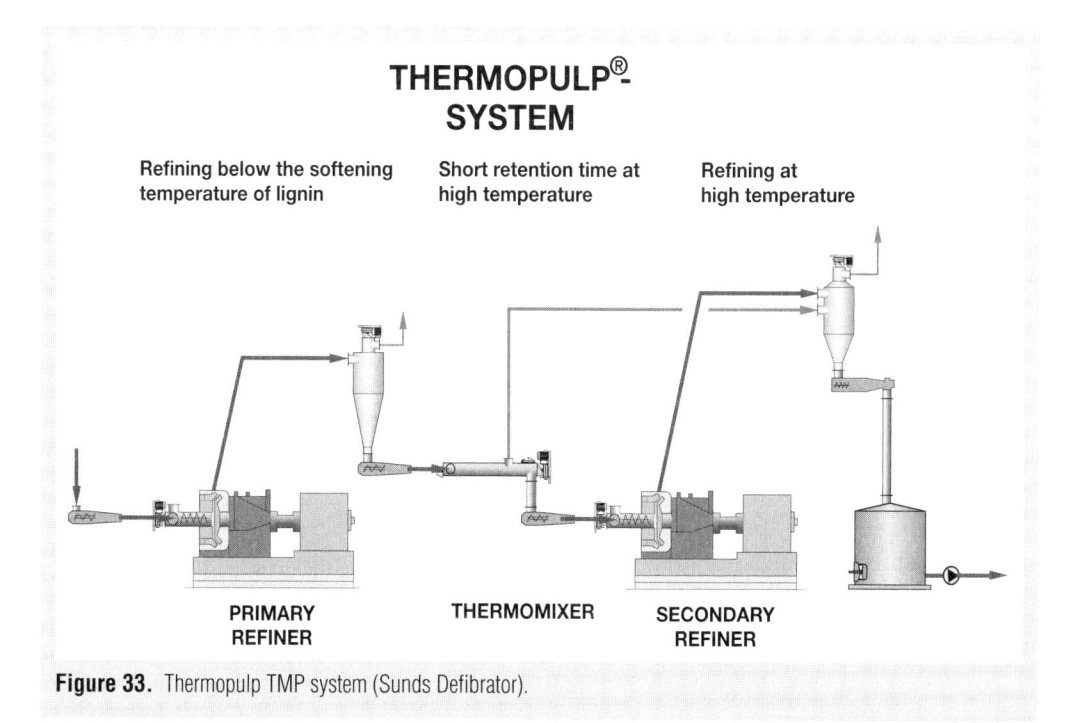

Figure 33. Thermopulp TMP system (Sunds Defibrator).

6 How refiner type and process conditions affect pulp properties

There are two parts in the refining process, the fiber and the refiner, and thus there are also two principal ways in which process conditions influence pulp properties. First, we can affect the way in which the refiner applies the energy on the fiber, and secondly we can affect the fiber itself, how it responds to the forces applied by the process (illustrated in Fig. 34).

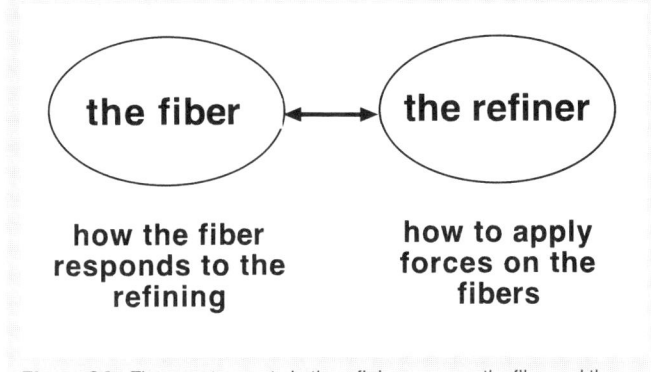

Figure 34. There are two parts in the refining process, the fiber and the refiner.

By varying the process temperature, the moisture content, or by adding chemicals, we can affect the softness and toughness of the wood raw material and thus how the wood fibers and the different constituents of the fiber (different parts of the fiber wall) react to the refining. There are numerous different operating and design variables that affect the way the refiner applies forces on the wood raw material, such as SEC of refining, refining consistency, steam pressures, plate pattern, and disc speed. These variables determine the refiner gap clearance and the refining intensity. The two independent variables of TMP refining are:

- The amount of energy applied
- The intensity of defiberizing and refining.

Together with the process parameters that affect the way the wood raw material responds to the refining, the amount of energy applied and the intensity of the refining largely determine the properties of the produced pulp.

6.1 Intensity of refining

In the low-consistency refining ("beating") of chemical and mechanical pulp the "edge load" theory originally introduced by Wultsch and Flucher in 1958[19] has been most important in the development of low-consistency refiners and for the designing of low-consistency refining lines. The edge load theory is a tool to calculate how intense or harsh the refining is. The edge load theory, however, cannot be used for high-consistency refining because the refining zone in an LC refiner is filled with water in which the pulp is dispersed. The flow of pulp, though, is determined by the flow of water. This means that the residence time in an LC refiner is largely determined by the speed at which the water is pumped through the refiner (even though there is a certain amount of backward flow in the LC refiner). In HC refining, the steam present cannot act as such a flowing medium, and the steam and pulp travel at different speeds and even in opposite directions.

Thermomechanical pulping

It was not before 1989 that Miles and May in their now well known "refining intensity" theory developed a way to calculate the refining intensity[3]. The theory states that the radial velocity of pulp through the plate gap in the refiner depends only on the centrifugal force, the radial friction between pulp and discs, and the drag force of steam developed during refining. From the velocity, the residence time in the refiner can be calculated. And, finally, from the residence time and the power input, the specific energy per impact, i.e., the intensity can be calculated.

In most cases, the TMP process can be contolled in a satisfactory way by just measuring the SEC, the freeness, and the long fiber content. Figure 35 shows how the specific energy consumption and the refining intensity affects freeness and long fiber content in a TMP mill[20]. With increasing specific energy, the freeness naturally drops and the amount of long fibers decreases. To increase the long fiber fraction at a certain freeness, the energy must be applied in a less intense fashion.

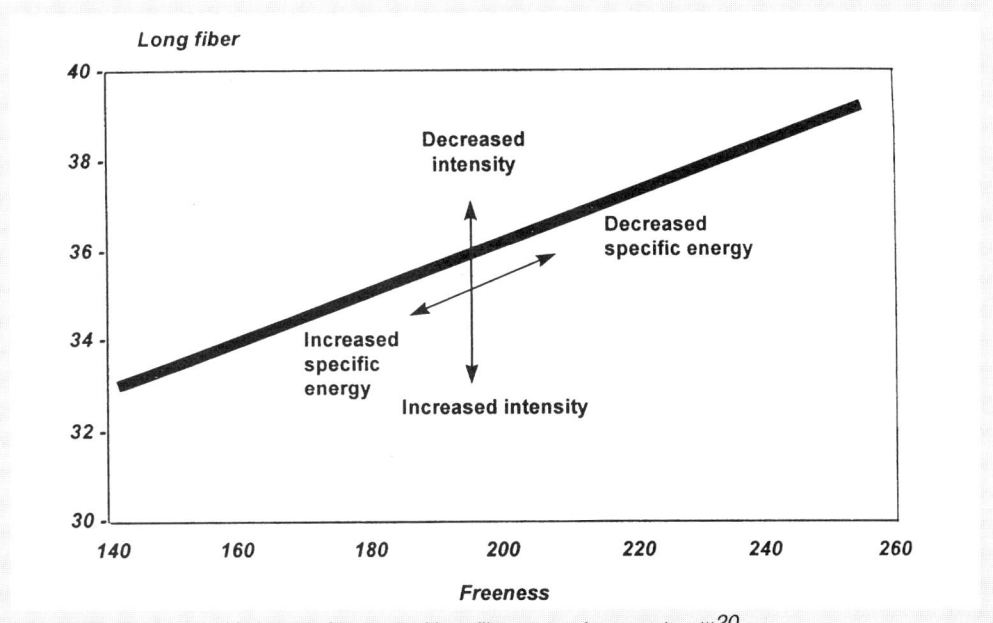

Figure 35. The relationship between freeness and long fiber content for a certain mill[20].

6.2 Influence of refiner type

The three major refiner types – the single-disc (SD), the double-disc (DD), and the conical refiners (CD, SC) – can all be used for the production of all grades of TMP, but the refiner type does have a clear influence on the pulp properties. These differences, to a large extent, can be explained by the fact that the refiner types apply the refining forces in different ways and thus with different intensity. Table 1 gives the pulp quality for three commercial Sunds Defibrator refiners, and Table 2 provides the pulp quality for SD and DD TMP produced on the same pilot refiner at KCL.

CHAPTER 7

It is obvious that the DD refiner has the highest and the CD refiner has the lowest refining intensity. In practice, these differences are not always easily distinguished because, during its development path, each refiner type has been optimized in its own way. Thus differing disc sizes, refining areas, breaker bar sections, plate patterns, etc., make comparisons difficult.

Table 1. Pulp quality of single-stage TMP from commercial CD, SD, and DD refiners producing newsprint type pulp and using the same kind of wood raw material[21].

Refiner Explanation	CD 70 single disc with outer conical section	RLP 58S single disc	RGP 65 DD double disc
Disc speed, rpm	1500	1500	1500
Freeness, ml	200	145	105
Energy consumption, MWh/t	1.68	1.65	1.68
Long fiber fraction (+30), %	53	40	41
Fines fraction (−200), %	20	24	31
Sheet density, kg/m³	320	340	355
Tensile index, Nm/g	36	36	36
Tear index, mNm²/g	10.2	8.2	8.0
Light scattering, m²/kg	52	53	57

Table 2. A comparison of single- and double-disc TMP produced on the same RGP 42 SD/DD pilot refiner[22]. HT-DD stands for high-temperature, double-disc refining.

Process	SD	DD	HT-DD
Disc speed, rpm	1500	1500	1500
Temperature in first stage refining, °C	143	143	170
Freeness, ml	100	100	100
Energy consumption, MWh/bdmt	2,4	2,1	2,0
Minishives, %	3,5	3,7	0,7
Fiber length, mm	1,7	1,4	1,5
Sheet density, kg/m³	420	420	450
Tensile index, Nm/g	39	35	40
Tear index, mNm²/g	7,8	6,4	7,4
Light scattering, m²/kg	56	60	58
Brightness, %	59	60	57

6.3 Refiner speed

Using normal electrical drives and direct coupling, the refiner speed is the same as that of the motor. The rotational speed of the motor depends on the frequency of the electric supply, which in Europe normally is 50 Hz and in North America 60 Hz. The frequency of 50 Hz enables motor speeds of 3000, 1500, 1000, etc., and the frequency of 60 Hz enables speeds of 3600, 1800, and 1200 respectively. The refiner processes were developed in North America originally with double-disc refiners at 1200 rpm and later with single-disc refiners at 1800 (because of feeding problems with SD refiners at 1200 rpm). In Europe, 1500 rpm is used both for single- and double-disc refiners.

It was not before 1987 that it generally became clear that the speed of the refiner disc has a large influence on energy consumption as well as pulp properties. Reliable research results could be produced because KCL in 1984 started up a 24-in., 500-kW Jylhävaara pilot refiner designed for disc speed trials[16]. The trials with this refiner showed that an increase in the refiner speed means lower energy consumption, lower fiber length, lower tear strength, and higher light scattering to a certain freeness level. These results have since been verified both with larger pilot refiners and in mill trials.

The first mill installation of a high-speed single-disc refiner started up in 1994 in Perlen, Switzerland (before that some North American DD-refiners' speed had been raised to that of the single-disc refiners, or up to 1800 rpm). The results from Perlen are compared to results from KCL pilot trials using the institute's second-generation high-speed refiner in Table 3. It can be seen that there is agreement between the pilot and mill results, even though the pilot results show smaller energy savings with high-speed refining. It is also clear that the refiner speed, because it affects the refining intensity, has a large influence on fiber length (Fig. 35) and strength properties. With higher speed, a pulp with lower fiber length and better light-scattering power is produced.

Table 3. The effect of refiner speed on TMP energy consumption and pulp properties[17, 22].

Process	KCL-SD pilot	KCL-SD pilot	KCL-SD pilot	Twin 60 mill	TC 66 mill (Perlen)
1 stage rotational speed	1500	2400	2400	1500	2300
2 stage rotational speed	1500	1500	2400	1500	1500
Freeness, ml	96	97	114	90	96
SEC, MWh/bdmt	2,4	2,1	1,9	2,2	1,7
Fiber length, mm	1,7	1,6	1,5		
Tensile index, Nm/g	44	43	41	42	40
Tear index, mNm^2/g	7,5	7,1	7,0	8,2	7,4
Light scattering, m^2/kg	56	57	57	53	55

6.4 Consistency

The consistency affects the pulp quality in two quite opposite ways. First, the moisture affects the viscoelastic properties of the wood and thus how the fibers respond to refining. A higher moisture content makes the fibers less brittle, which in practice means higher fiber length.

The second effect of moisture is on the wet mass of the fiber. With higher moisture content, the wet mass of the fiber grows as does the centrifugal force acting on the fiber. Assuming that the friction forces and steam forces are more or less unaffected by the consistency, this means that the residence time in the refiner gets shorter and the refining intensity higher, which means shorter fiber length. Consistency also has an effect on the pressure in the gap, e.g., additional dilution water reduces the gap clearance by reducing the steam volume in the gap.

The concept of refining consistency is somewhat unclear. The energy used in refining is converted to heat, which is consumed in evaporating water. Thus the consistency in the refining zone rises from the inlet to the outlet of the refiner, as can be seen in Fig. 36. Normally, "refining consistency" stands for the outlet consistency, mainly because the outlet consistency is easy to measure. This, as shown in Fig. 36, can lead to misconceptions. The three cases in the figure have all the same outlet or "refining consistency," but their actual average refining consistencies differ quite a bit.

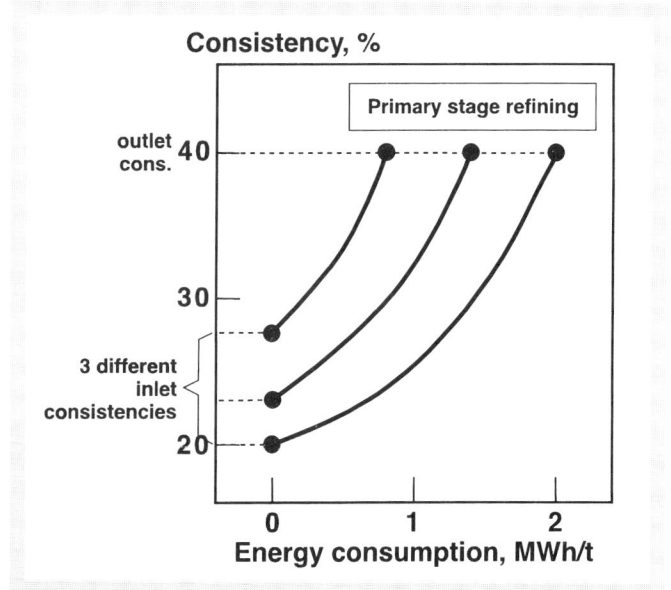

Figure 36. Calculated refining consistencies as a function of energy applied.

Mill trials from the Daishowa's Quebec City mill have indicated that the energy consumption of TMP can be reduced by lowering the refining consistency. In these mill trials, the energy reduction was 7% when the refining consistency was reduced from 50% to 38%. Other properties were not adversely affected[33]. These effects are in line with the Miles and May theory. Pilot trials at KCL in the early 1980s using a 40-in. Bauer double-disc refiner showed that the optimal outlet consistency for this pilot refiner was 24%–34% for single-stage newsprint TMP[23]. Also in these pilot trials, the energy consumption to a certain freeness was reduced with lower consistency over the whole range investigated, but fiber cutting occurred at consistencies under or above the optimal range. At the highest consistencies, when the centrifugal force got too small to transport the pulp through the refining zone, the pulp actually burned. The conclusion is that all refiners have their own optimal consistency range, which gives the optimal centrifugal force for the refiner in question. Each refiner should then be run at the lower limit of this optimum consistency range in order to achieve the lowest possible energy consumption at acceptable strength properties.

6.5 Production rate

The production rate of a TMP refiner is shown to have a considerable effect on both energy consumption and pulp properties, as shown in Table 4. Increasing production rate will reduce the energy consumption to a certain freeness but will also somewhat reduce fiber length and strength properties. The probable explanation is that the plate gap will decrease when throughput is increased. Härkönen also reports that the temperature and the volume fraction of fiber in the refining zone increases with higher production rates[9]. Similar effects as shown in Table 4 for SD refiners have also been reported for DD refiners[6].

Table 4. The effect of the production rate on TMP energy consumption and pulp properties. Mill trials with Scandinavian spruce[6, 9]. The first column gives values from a single-stage TMP plant and the second column the average results from two separate mills, each with a two-stage TMP process.

Production rate change, t/h	3.2 → 4.2	6.5 → 9.5
Production rate change, %	+31	+46
Refiner type	SD 60 in.	SD 65 in.
Freeness, ml	175	120–130
Change in energy consumption, %	–10	–10
Change in tear strength, %	–12	–6
Change in tensile strength, %	–3	–5

6.6 Preheating and steaming of chips

Originally thermomechanical pulping meant a process with pressurized preheating of chips followed by pressurized or open discharge refining. It was believed that the preheating at elevated temperatures, which softens the wood, was the key to the exceptional strength properties of TMP. But at the end of the 1970s, it was found out that it was the refining – not the preheating – that should be performed at elevated temperature. KCL pilot trials initiated by Mannström showed that completely omitting the preheating stage did have very small effects on pulp quality while, to some extent, reducing the energy consumption[24]. These results suggested that the preheating step in TMP was unnecessary, and KCL suggested that a process without preheating should be called PRMP (pressurized refiner mechanical pulping) to distinguish it from TMP. Since then the TMP process has developed toward shorter preheating times and lower temperatures in preheating combined with higher temperatures in the refiner, or toward the PRMP process.

Even if the pressurized preheating is of relatively small importance in the TMP process, the chips must be warmed up to about 100°C before entering the refiner in order to avoid fiber breakage and shive production during the initial chip breakdown to match stick-like fragments in the feeder and the breaker bar section. This warming up of the chips is normally performed in an atmospheric steaming bin and also in the chip washer.

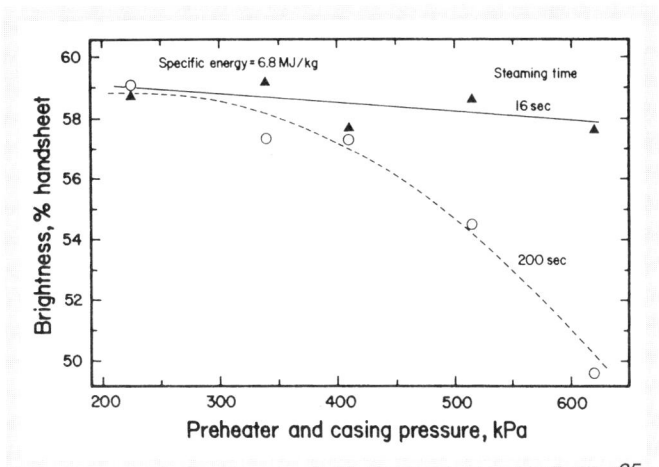

Figure 37. TMP brightness at two chip preheating times (16 and 200 s)[25].

At high temperatures, the pressurized preheating of chips has a negative influence on pulp brightness. In Figure 37, we can see that a short preheating time of under 20 s has little effect on brightness even at temperatures around 160°C, while a 3-min longer preheating time caused a severe drop in brightness.

6.7 Temperature in preheating and refining

The refiner housing is filled with saturated steam with a low air content. Thus the steam pressure and the steam temperature are connected to each other as shown in Table 5.

Table 5. The connection between the pressure and temperature of saturated steam. The pressure gauges normally indicate atmospheric pressure as zero and thus give overpressure. If not otherwise stated in the text, the pressure is given as overpressure (p_e).

Absolute pressure	Overpressure	Overpressure	Temperature of saturated steam at this pressure
bar	bar	kPa	°C
1	0	0	99
2	1	100	120
3	2	200	133
4	3	300	143
5	4	400	151
6	5	500	158
7	6	600	164
8	7	700	170
9	8	800	175
10	9	900	179
11	10	1000	183

As already stated, the preheating is of relatively small importance, and there are results that indicate that atmospheric preheating is optimal when using high pressure and high temperature in the refiner, as can be seen from Fig. 38. Table 6 shows how refiner housing steam pressure affects the strength and optical properties of TMP. The higher the temperature of the steam in the refiner housing is, the better strength properties are attained for the TMP. Table 2 earlier in this chapter shows recent results for DD refining at 3 bar and 7 bar steam pressure, respectively. By raising the temperature, it was in this case possible to make a DD-TMP with strength and optical properties not very far from those of SD-TMP, but still at a lower level of energy consumption.

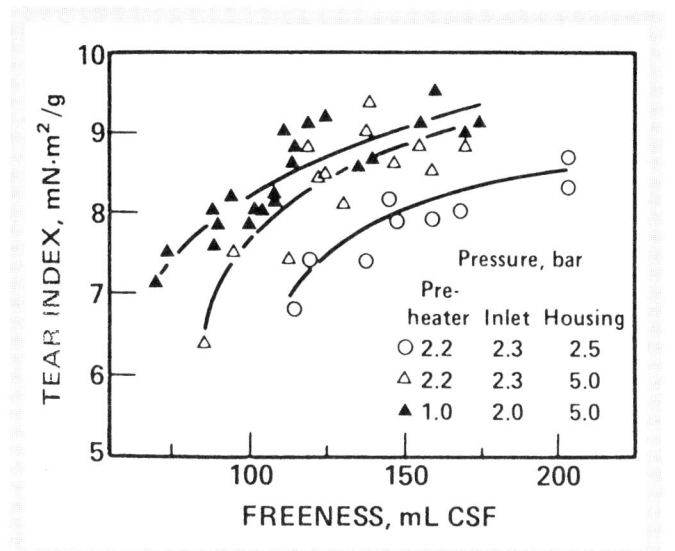

Figure 38. Tear strength of TMP in mill trials with varying pressure in preheater and refiner housing[26]. The refiner was a 7 MW Sunds Defibrator RLP-58S. The pressure values are given as absolute pressure.

Table 6. How refiner housing steam pressure and temperature affect the strength and optical properties of TMP at freeness 150 ml[23]. Pilot trials with a 40-in. Bauer 411 DD refiner in the early 1980s at KCL. Strength properties tested from 100 g/m² handsheets.

Steam pressure, bar	0	1	2	3
Temperature, °C	99	120	133	143
Energy consumption, MWh/bdmt	1.5	1.7	1.9	2.0
Tear index, mNm²/g	6.5	7.5	8.0	8.5
Tensile index, Nm/g	39	41	43	45
Light-scattering coefficient, m²/kg	53	53	53	52

6.8 Special combinations of process conditions

As was stated earlier, there are two main "parts" in the TMP process: the fiber and the refiner. The process conditions thus both affect how the refining forces are applied on the wood and fiber, and how the wood and fiber react to these forces. Another point that should be made is that the process "from wood to pulp" can be divided into several steps, such as shredding of chips, liberation of fibers, delamination of fibers, and fibrillation of fibers, which all probably require different process conditions. During the last 10 years, efforts have been made to optimize the TMP process according to these lines and, in the following text, three such attempts are discussed: the KCL multistage process, the Andritz RTS process, and the Sunds Defibrator Thermopulp process. The latter two have already been shortly described earlier in this chapter.

6.8.1 The multistage process

In the KCL multistage process, the chips are defibered at high intensity (high rate of rotation or double disc refining)[18]. To avoid fiber cutting, the fibers must be kept as soft and flexible as possible by keeping the temperature high. At the subsequent step of delamination, the fiber wall must be harder, and therefore the temperature is lower. Refining is also gentler at normal rotational speed and preferably conical plates. After this, the pulp is fractionated into a coarse and a finer fraction. These fractions are then disc refined at atmospheric pressure and normal intensity. Table 7 gives the results from pilot trials with KCL's RGP 42 pilot refiner. In these trials, the multistage process was simplified with only normal single-disc refining and no fractionation[27]. As can be seen, the multistage process resulted in 10%–15% lower energy demand, a much lower shive content, and higher tensile strength. But the optical properties were not as good as for normal TMP. It was concluded, however, that it should be possible to improve the optical properties by minimizing the time the fibers spend at high temperature[27]. In these trials, the total dwell time at 165°C was in the range of 25 seconds.

Table 7. Comparison of KCL multistage refining process with normal TMP. Normal TMP was in this case considered to be two-stage SD refining at 3 bar overpressure. The process conditions of the multistage process were as follows. First stage: SD, 2000 rpm, 165°C; second stage: SD, 1500 rpm, 140°C; third stage: SD, 1500 rpm, 100°C.

Multistage process in comparison with normal TMP	
Energy consumption	−10 to 15%
Minishives	−60 to 70%
Tensile strength	+ 10%
Tear strength	+/−0%
Light scattering coefficient	−2 to 3%
Brightness	−3%-units

6.8.2 The RTS process

The Andritz RTS process – in which R stands for retention, T for elevated temperature, and S for high speed – is similar to the multistage process in that it combines high disc speed with elevated temperature. The process incorporates atmospheric presteaming of chips followed by a quite low retention time (10 –20 s) at high temperature (steam pressure 5.5–6.0 bar) in the feeding system of the high-speed refiner (2000 –2500 rpm) operating at the elevated temperature (Fig. 32). The short retention time at the elevated temperature minimizes the loss of brightness always caused by high temperatures. Another reason for the good brightness values reported for RTS pulps is the short retention time at the temperature peak in the refining zone. This short retention is caused by the high centrifugal force in high-speed operations. Table 8 gives some pulp properties from the first commercial RTS operation. The operating experience suggests that an energy reduction in the order of 15% is possible using this system.

Table 8. RTS TMP properties compared to high-speed TMP. First refining stage using a 2300 rpm Andritz Twin TC 66 refiner and second stage using a 1500 rpm Andritz SB 150 refiner[28].

Process	High-speed TMP	RTS TMP
Energy consumption, MWh/t	1.84	1.85
Freeness, ml	96	94
Breaking length, km	3.52	3.74
Tear strength, cNm/m	129	144
Brightness, %	55.7	55.5
Light scattering coeff., m^2/kg	60.9	58.8

6.8.3 Thermopulp

The Sunds Defibrator Thermopulp process (Fig. 33) has two refining stages. In the primary stage, refining is carried out below the softening temperature of lignin but, in the secondary refining, temperature and pressure are raised to 160°C–170°C and 600–700 kPa. The high pressure increases the refining intensity by compressing the steam in the gap which reduces the refiner gap clearance. In order to make fibers less brittle to withstand the high refining intensity, temperature of fibers is raised to 160°C–170°C before entering the Thermopulp refining. Dynamic testing of wood samples indicates that the breakdown of wood fibers is more efficient at higher temperatures[29].

In the Thermopulp system, the distribution of refining energy is focused on the secondary refining stage, where the higher refining intensity enables savings of overall refining energy. In mill scale Thermopulp refining trials, 10%–20% energy savings compared to conventional TMP have been reached[36]. Table 9 shows a comparison of TMP and Thermopulp properties in a North American mill.

Table 9. Mill comparison of Thermopulp and standard two-stage TMP. The pulp samples are collected after the mainline RGP 76 CD refiners.

Process	TMP	Thermopulp
Freeness, ml	180	160
SEC, MWh/t	2.15	1.75
Tensile index, Nm/g	37.5	37.1
Tear index, mNm^2/g	9.8	9.6
Light scattering, m^2/kg	54.3	54.8
Brightness, %	55.1	54.5

The retention time of fibers at the elevated temperature is kept short (a few seconds only) in order to avoid brightness loss of pulp. In addition to reduced brightness, too high a temperature also increases the amount of dissolved substances. This reduces the strength properties of the pulp. Therefore the temperature of Thermopulp refining should not exceed 170°C.

7 Process control

This chapter presents the principles of automatic on-line controls used in TMP refining. There is no common control strategy for all the TMP plants due to the differences in the raw material, in the process equipment, and in the operating philosophy. The target of the controls is to keep the process at the desired operating point. If this works on the long term, optimization of the operating point becomes possible. Usually the controls do not work well enough due to variation in the raw material, changes in the refiner plate condition, and uncertainties in the process measurements.

Some common parameters are needed to define the operating point of the refiner. The control variables of the refiners are the production rate, the dilution water flow rates, the steam pressures, and the plate gap clearances. The most important controlled variables are the specific energy consumption (SEC) and the refining consistency. Refining intensity is defined as the average energy of the impacts directed to fibers[3]. With a constant SEC, a reduced residence time leads to increased refining intensity. Using on-line controls, the refining intensity at constant SEC can be varied with production rate, power ratio between refiner stages, steam pressures, and refining consistency. One must notice that there are only a few empirical tests made concerning the residence time, and the refining intensity is still a theoretical quantity[31].

Figure 39 presents the control of the main line refiners. First, it is important to minimize the variation in the chip properties as much as possible. This enables keeping the volumetric efficiency of the chip screw stable and reducing the variation in the production rate. The pressure controls are kept at the constant set values, but problems can occur due to a deficient heat recovery system. For the control of the most important measurements of the refiners, the dilution water flow rate and the plate gap clearance are used. These basic refiner controls can be cascaded with the upper level pulp quality controls. Variation in the dilution water temperature affects the properties of the refined pulp and can disturb the on-line measurements.

In many cases, the variations of the pulp quality have been successfully decreased with the basic refiner controls stabilizing the refining consistency. Occasionally, this has improved the manual control of the pulp quality so much that the uncertainty of the on-line quality analyzers becomes significant and the quality measurements should be averaged.

The most advanced control solutions use pulp quality measurements as feedback. These solutions are presented in the process control book of this same textbook series. The most common on-line quality measurements are the freeness and the average fiber length. Also the fiber length distribution and the shive content are used. The measurement cycles range from some minutes to about one hour because the analyzers usually measure several different positions in the TMP plant. The faster variation in the pulp quality must be compensated with the basic controls.

Only few automatic controls of pulp quality are in active use, although trials have been made in most TMP plants. The most common problem has been the unreliability of

Thermomechanical pulping

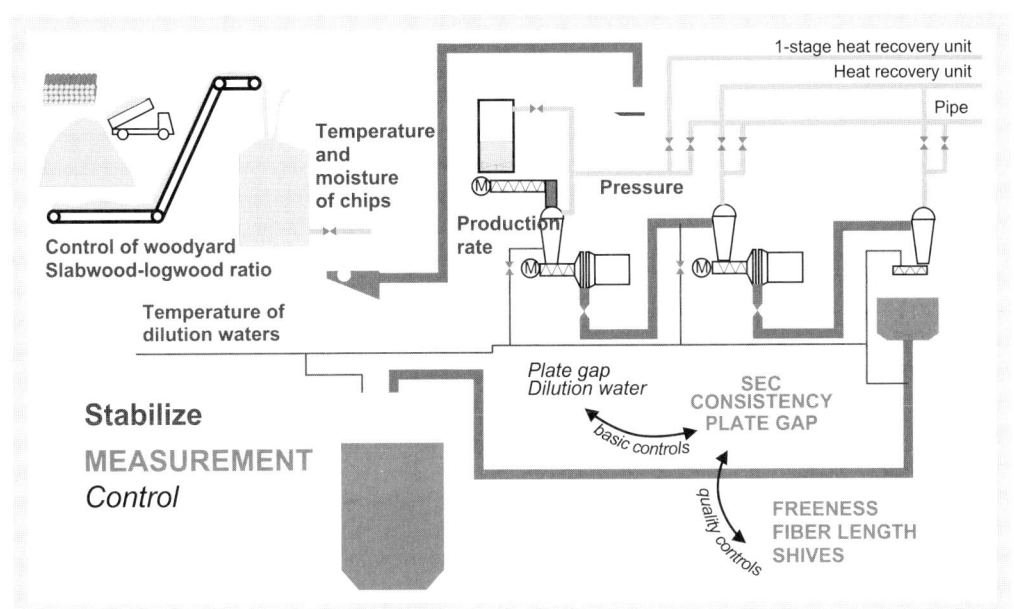

Figure 39. Control of main line TMP refining. The process variables that should be stabilized are indicated (in bold) in the figure: pressure, production rate, etc. The most important measured variables for the control of the main line refining are indicated with capitals. The main control variables are indicated with italics.

the on-line analyzers. Another problem is that only the freeness was controlled: For example, if the disturbances are coming from the chip moisture variation and the compensation of the freeness variation is controlled by the plate gap, the variation in the fiber length can increase.

7.1 Main principles

The aim of the on-line controls is to keep the selected measurements at their set values. This can be done basically in two ways: First by directly compensating the process variable that was varying and, second, by compensating the effect of the variation using some other process variables. The first case is always preferable.

Table 10. Typical relationships for the control of the TMP refiner.

Control Variable	Refiner Load	Refining Consistency
Increase in plate gap clearance	Decreases	Decreases
Increase production rate	Increases	Increases
Increase dilution water flow rate	Decreases	Decreases

Table 10 shows the effect on refiner load and refining consistency. An increase in the plate gap naturally decreases the refiner load, which causes the refining consistency to decrease because the amount of produced steam decreases. An increase in the production rate (with no control operations on plate gap and dilution water) increases the refiner load and the refining consistency. An increase in the dilution water decreases the inlet consistency, which decreases the refiner load and refining consistency.

All of these relationships are empirically proven, and many refining theories are based on these. It must be noted that there is a lower limit for the plate gap clearance. Below this, strong cutting of fibers occurs and it has been reported that tightening of the plate gap will reduce refiner load[32]. There is also an upper limit for the refining consistency and above it fiber cutting occurs[13]. Above this consistency limit, a reduction of the dilution water decreases the refiner load[34]. Fiber cutting is always undesired because of low quality of pulp, and the relationships in the table above are valid in the normal operation range of refining. When fiber cutting is due to too small plate gap, it can be directly indicated from the plate gap clearance or the plate vibration measurement. Too high refining consistency can be directly indicated only from the blowline consistency measurement. The relationship between freeness and fiber length naturally also indicates the fiber cutting.

One must notice, that in many refiner models there are, e.g., several dilution waters. Some of them are fed directly into the plate gap, and the effect of them can be different. In some refiner types, there are also two plate gaps – parallel or cascaded.

The refiner naturally either compensates or strengthens some of the feed disturbances. Because of the typical relationships presented in the table above, the effect of the production rate disturbances on the specific energy consumption (SEC) is partly compensated[35]. On the contrary, the effect of the feed disturbances on the refining consistency is amplified. If the input consistency is increased, the refiner load is also increased and the output consistency increases more than the input consistency.

7.2 Control of operating variables

The target of the basic refiner controls is to compensate the effects of the feed disturbances. This does not always work, and in some cases use of the control can even make things worse. The following list shows the basic refiner control strategies presented in the literature:

1. Keeping the refiner load in setpoint by controlling the production rate
2. Decreasing the variation in the refiner load by controlling the dilution water flow rate
3. Keeping the refiner load in setpoint by controlling the plate gap clearance
4. Keeping the measured refining consistency at the setpoint by controlling dilution water flow rate
5. Keeping the measured plate gap at the setpoint.

Usually a combination of several controls is needed[14]. The drawbacks are the increased expenses in the installation and the maintenance for the measurements needed by the controls. The next section discusses the controls in more detail.

7.2.1 Plate gap clearance

This includes the automatic plate gap setting based on some of the following process measurements: plate gap clearance, refiner load, or specific energy consumption. The principle of setting the refiner load constant by automatic plate gap control is not recommended when the production rate is variating. The SEC control has uncertainties regarding the measurement of the production rate.

When the axial thrust of the refiner is driven hydraulically, it is essential that the control of the plate gap is based on direct measurements. Measurements of the plate gap are difficult to calibrate and the absolute value cannot be continuously measured as the refiner plates are aging. In all cases, automatic plate gap opening sequences are needed to avoid plate clashing. The indication is usually made from plate vibration measurements. Plate gap controls have also been used in the automatic startup of the refiner.

7.2.2 Production rate

For the main line refiners, the production rate is controlled by the speed of the chip screw, which usually is the preheater discharge screw. The measurement of production rate has significant uncertainty, especially if the quality of chip or pulp varies a lot. Thus the production rate usually cannot be kept exactly at its set value. However the short-time variation of production rate might be compensated for by having an indication of the motor load of the chip screw[37] or by the refiner load[38, 39]. In the latter case, the control of the refining consistency would be needed also.

The actual measurement of the production rate is made downstream of the refining line at the flow of the refined pulp. The product of the consistency and the flow rate is calculated. For the control of the refiner load, a relative uncertainty of at most 2% in the measurement of production rate would be beneficial. This currently is not possible due to the uncertainty in the pulp consistency on-line measurement devices. The different consistency meters are disturbed by the variation in the pulp quality, fines content, flow rate, and temperature. The flow rate of the pulp is usually measured with magnetic flowmeters which also perform the needed accuracy better.

7.2.3 Refining consistency

The consistency increases in the refining and is not constant in the plate gap[31, 40]. Currently, the refining consistency cannot be measured during normal operation. It must be assumed that the refining consistency can be controlled by the refiner discharge consistency in the blowline. The optimal level for discharge consistency can be found by empirical tests. In some refiners, part of the dilution water can also be fed directly inside the plate gap or between the sequential plate gaps. In these cases, the ratio of division for the different dilution waters is also essential, although the optimal value might be difficult to find.

Existing refining consistency controls are based either on measured or calculated values of the refiner discharge consistency. The measurement principle of the most commonly used refining consistency on-line analyzer is based on the measurement of the absorption of the infrared light[41]. The calibration of discharge consistency is a problem because the pulp samples taken from the blowline are not always representative.

The simplest consistency estimation is based on the refiner load and the production rate only[42]. Improvements have been made by using the energy and the mass balance equations with several process measurements. Also almost all measurements, including refined pulp properties, have been used[43]. In this case, the estimation uses statistical modeling of factor network analysis. Using the balance equations, the outlet consistency of the refiner is quite accurately estimated if the inlet consistency is known. This is the case for the secondary stage refiner, when there is a measurement of the discharge consistency of the first stage refiner[44, 45]. However, in addition to the unknown moisture of the chips, the ratio of the steam wash water flow into the heat recovery system and back into the refiner feed is also unknown.

Dilution water control performance is limited for the compensation of the fast disturbances in the refining consistency. In some cases, the main reason for the fast consistency variation is the false operation of the dilution water controls due to agressive tuning. The solution is an additional dilution water with a smaller sized and faster operating control valve. To achieve optimal refining consistency control, an adaptive control algorithm is needed because the process gain in the consistency control loop is alternated by the operating point of the refiners.

7.2.4 Steam pressure

Many important properties of refining are set by the steam pressure levels. The amount of chip heating is a compromise of several dependencies on the properties of the refined pulp. The temperature during the refining zone is controlled by the outlet pressure of the refiner. The pressure difference over the refiner has an effect on the residence time of the refining and so on the refining intensity. The pressure difference affects the steam flow inside the refiner significantly.

In the normal operation, the pressure controls usually operate quite well. However, in some mills, the refining pressures cannot be controlled due to the small capacity of the heat recovery system. If the necessary pressure levels cannot be achieved at all, the constructional modifications are needed on chip heating, steam line constructions, plug screws, plate pattern, etc.

Preheating

In most cases, the blow-back steam of the primary refiner is used for the pressurizing of the preheater. For the plug screw type of the preheater discharge screw, the refiner feed and the preheater can be set to different pressure levels. In one type of solution, all the blow-back steam is directed through the preheater and the refiner feed pressure is controlled by a pressure valve between the preheater and the heat recovery system. In most cases, part of the blow-back steam is directed to the preheater and the rest is directed to the heat recovery system or air. If the chips are icy when entering the preheater, all the blow-back steam might be condensed and the pressure of the refiner feed and the preheater decreases.

In some cases, the preheater steam is taken from the main steam flow after the primary refiner. The blow-back steam might be forced to condense on the chips or to flow to the blowline using a pressure balancing steam line between the feed and the outlet of the primary refiner.

Refining pressures

Occasional disturbances in the refiner pressures are caused during the variations in the heat recovery system. The plate gap and the refiner load are significantly disturbed even by a 20 kPa change in the pressure difference over the refiner. Variations are seen also in the pulp quality.

The tuning of pressure controls is a compromise between several problems, and moderate values for the PI-controller parameters have to be selected. Due to the large range in the steam flow, big valves are needed and control of valve position is inaccurate. Fast operation would be needed in the case of large disturbances, but usually a conservative tuning is preferable to prevent unnecessary control operations. Due to the large range in the amount of produced steam and in the pressure of the heat recovery, the operating range of control valves changes a lot. The nonlinearity of the valve characteristic means that the optimum values for the control parameters are dependent on the valve position. The improvements in the PI controller could be found from the linearization of the control output, using more exact valve positioners or by adding a parallel steam line with a small size control valve.

The closed loop dynamics of steam pressure control circuit can also be dependent on the operation of the other refiner lines due to the shared heat recovery system. Improvements might be found from the multivariate and the nonlinear control principles based on both traditional and fuzzy logic.

8 Heat recovery

A heat recovery unit plays an essential role in the operating economy of a TMP mill. Normally around two-thirds of the refining energy can be recovered in the form of clean steam. The TMP steam generated in the refiners is separated from the fibers in the cyclones and then condensed in the reboiler against vaporizing clean steam. Figure 40 shows the main components of a typical heat recovery unit. The purpose of the additional heat exchangers is to increase the clean steam yield. The feed water is preheated first against TMP condensate and then against TMP steam vented from the reboiler. The TMP condensate is further cooled against white water. The purpose of the spray scrubber is to wash the TMP steam free of fibers when the TMP steam is blown to the atmosphere. In addition, the scrubber works as noise silencer.

Often the heat recovery unit is provided with a turpentine recovery unit to recover the turpentine present in the TMP steam vent from the heat recovery. If required, the clean steam can be boosted to a higher pressure level with a thermocompressor (steam ejector) or with a centrifugal compressor.

CHAPTER 7

HEAT RECOVERY SYSTEM FOR TMP-STEAM

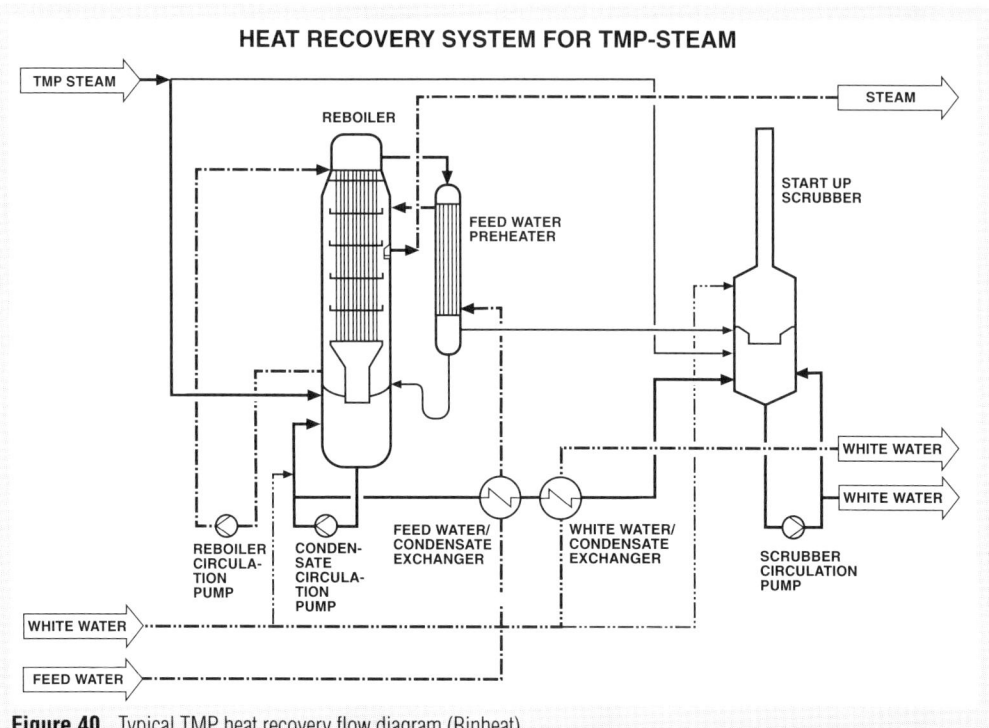

Figure 40. Typical TMP heat recovery flow diagram (Rinheat).

8.1 Amount and composition of TMP steam

The specific refining energy is in the range of 2–3.5 MWh/t. In an electric boiler, 1 MWh gives 1.55 t of clean steam at 3 bar$_{(e)}$ with a feed water temperature of 100°C. In a TMP process, 1 MWh gives roughly 1 t of TMP steam, while the rest of the energy is used to heat the pulp and the dilution water in the refiner.

In the refiner, the mechanical refining energy is converted into heat, which generates steam from the dilution water and from the moisture contained in the wood. At the same time, the main part of the volatile organic compounds (VOC) from the wood chips are evaporated. Thus TMP steam contains turpentine, i.e., volatile organic oils and other VOCs like methanol, acetic acid, propionic acid, and formic acid. In addition, TMP steam contains some air, which enters the refiner with the chips and is dissolved in the dilution water. The separation of fibers from TMP steam in the cyclone is not complete; therefore, TMP steam always contains some fibers. Due to organic acids, the pH of the TMP condensate is in the range of 2 to 4. The impurities present in the TMP steam make the heat recovery unit necessary because the direct use of TMP steam is difficult.

The turpentine is usually quite easily evaporated from the chips. That is why the blow-back steam tends to have the highest turpentine concentration. On the other hand, the TMP steam from the reject refiners is virtually free of turpentine. The amount of turpentine in the TMP steam follows the turpentine content of the wood chips, but the pretreatment has a remarkable effect on the turpentine concentration of the TMP steam.

Thermomechanical pulping

8.2 Reboiler for condensing TMP steam and generating clean steam

The purpose of the reboiler is to condense the TMP steam against vaporizing clean steam at the lowest possible pressure loss. The Rinheat 3R Reboiler shown in Fig. 41 is the most common reboiler with more than 60 operating units throughout the world. The pressure difference between the TMP steam and clean steam is normally 0.5 bar. The clean steam is normally used at the paper machine. For a 3 bar(e) clean steam

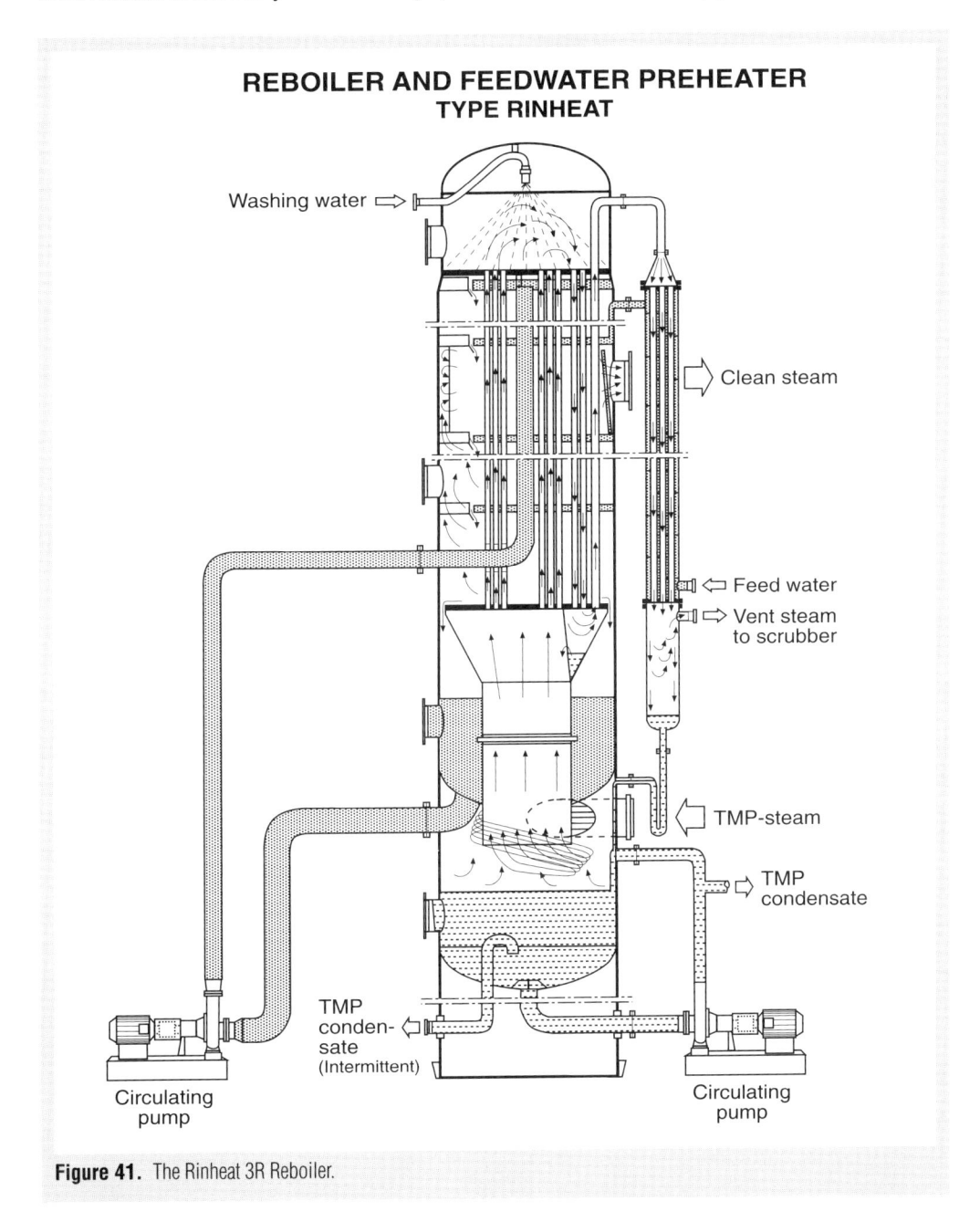

Figure 41. The Rinheat 3R Reboiler.

pressure, the TMP steam pressure must thus be 3,5 bar(e). The capacities of single reboilers vary from 7 t/h to more than 100 t/h.

A reboiler for TMP steam must be capable of handling steam which contains fibers, fouling organic compounds, and inert gases. The fibers should be separated before TMP steam enters the tubes and the fibers taken away by TMP condensate. Inert gases require adequate venting and flow velocities. For fouling steam, washing with mild caustic solution is the best cure.

The 3R reboiler is a vertical tubular heat exchanger where the TMP steam is condensed in the tubes and the clean steam is generated on the shell side. The TMP steam inlet in the lower part of the reboiler is designed like a cyclone with a tangential inlet and the central pipe as a vortex finder. In order to separate fibers, the TMP steam flows upward in the tubes. The downflowing condensate keeps the tube walls wet and prevents fouling of the heat transfer surface with organic compounds. In the second pass, the TMP steam flows downward and the TMP condensate from the second pass is taken into the main TMP condensate stream through a water seal. The second pass is small, less than 10% of the tubes, in order to increase the velocity of the TMP steam as the concentration of turpentine and inert gases are growing when more steam is condensed. The turpentine behaves in the reboiler like an inert gas.

The volume of the TMP-condensate in the bottom of the reboiler is fairly large in order to provide buffer capacity for the eventual fiber bursts. In order to prevent the settling of fibers in the reboiler, it is provided with a TMP condensate recirculation to achieve good agitation.

The clean steam generation takes place in the falling film of the circulating water on the outside of the tubes. In the 3R reboiler, the tubes are supported and the circulation water is redistributed at about two-meter interwalls. This feature ensures the wetting of all tubes and prevents vibration. The 3R design also features a built-in droplet separator and a large volume of circulating water.

8.3 TMP steam vent/feed water exchanger

The purpose of the TMP steam vent/feed water exchanger is to improve the clean steam yield by preheating the incoming feed water with TMP steam vent from the reboiler. The condensation of all TMP steam in the reboiler is not possible because of the inert gases (VOC+air). But by cooling the vent steam against the feed water, more TMP steam is condensed and thus the clean steam yield is normally increased by 2%–3%.

The TMP steam vent/feed water exchanger is a vertical tubular exchanger, where the TMP steam flows downward in the tubes and feed water upward on the shell side guided by baffles. The inert gases are separated from the condensate in the lower end and then led to the spray scrubber or to the turpentine recovery. The exchanger is usually mounted on the reboiler.

8.4 TMP condensate/feed water exchanger

The purpose of the TMP condensate/feed water exchanger is to improve the clean steam yield by preheating the feed water against TMP condensate. A normal improvement in the yield of the clean steam is 3%–5%. A horizontal tubular heat exchanger with fixed tubesheets is the usual selection. The TMP condensate flows in the tubes and the clean feed water in the shell side. Both ends are usually flanged in order to make the mechanical cleaning of the tubes possible.

8.5 TMP condensate/white water exchanger

The purpose of the TMP condensate/white water exchanger is to cool the TMP condensate below boiling point and to utilize the heat available to heat white water. This heat exchanger does not boost the clean steam yield, but it helps to save steam by recovering heat. Because both streams can contain fibers, both sides of the heat exchanger must lend itself for mechanical cleaning. A tubular heat exchanger with floating head or a spiral plate heat exchanger meets this requirement. Clear filtrate or the dilution water for refiners can also be heated against TMP condensate instead of white water.

8.6 Startup scrubber for TMP steam

The purpose of the startup scrubber is to wash TMP steam free of fibers before blowout to atmosphere. In addition, an insulated spray scrubber works as an efficient noise silencer for the blowout TMP steam. It also works as a flash tank for the hot TMP condensate in case the TMP condensate coolers are bypassed.

The spray scrubber is an atmospheric vertical vessel provided with a stack to bring the blowout steam high enough so that the steam does not harm surrounding buildings or equipment. Figure 42 shows the operating principle. The bottom part is designed like a cyclone and the upper part like a spray scrubber. The lower end works as pump tank for the spray water and TMP condensate. In order to prevent settling of fibers, the scrubber is either provided with condensate recirculation or the condensate is entrained by on-off control.

8.7 Surface condenser for low-pressure TMP steam

The surface condenser is used to condense TMP steam against hot water. It has mainly been used for low-pressure TMP steam such as: blow-back steam from the first-stage refiner or TMP steam from the reject refiner. The surface condenser is a vertical tubular heat exchanger with TMP steam condensing on the tube side and water being heated on the shell side (Fig. 43). The tube side construction is very much like the reboiler with cyclone-type TMP steam inlet and upward flow of TMP steam in the tubes. Inert gases from the TMP side are vented to the scrubber or to the turpentine recovery.

The TMP steam side of the surface condenser can be run pressurized or atmospheric. The TMP condensate is normally removed with a pump having a circulation to prevent settling of fibers. The shell side of the surface condenser is like an ordinary tubular heat exchanger with baffle plates. The water on the shell side can be once through or a closed circulation. The heat recovery can also be accomplished with a simple spray condenser, but then the hot water and TMP condensate are mixed. This system is often used for TMP reject refiner steam.

CHAPTER 7

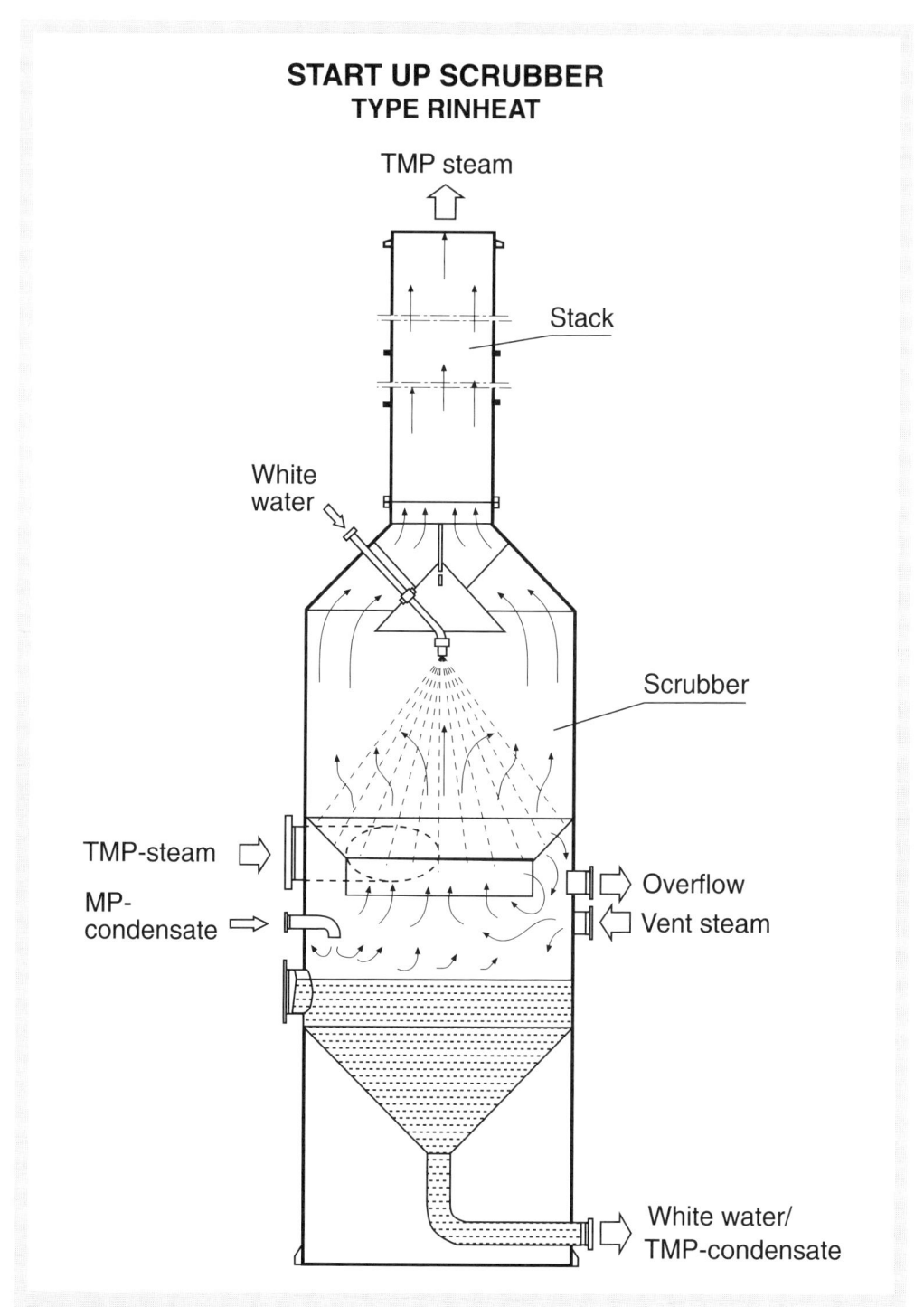

Figure 42. Spray scrubber.

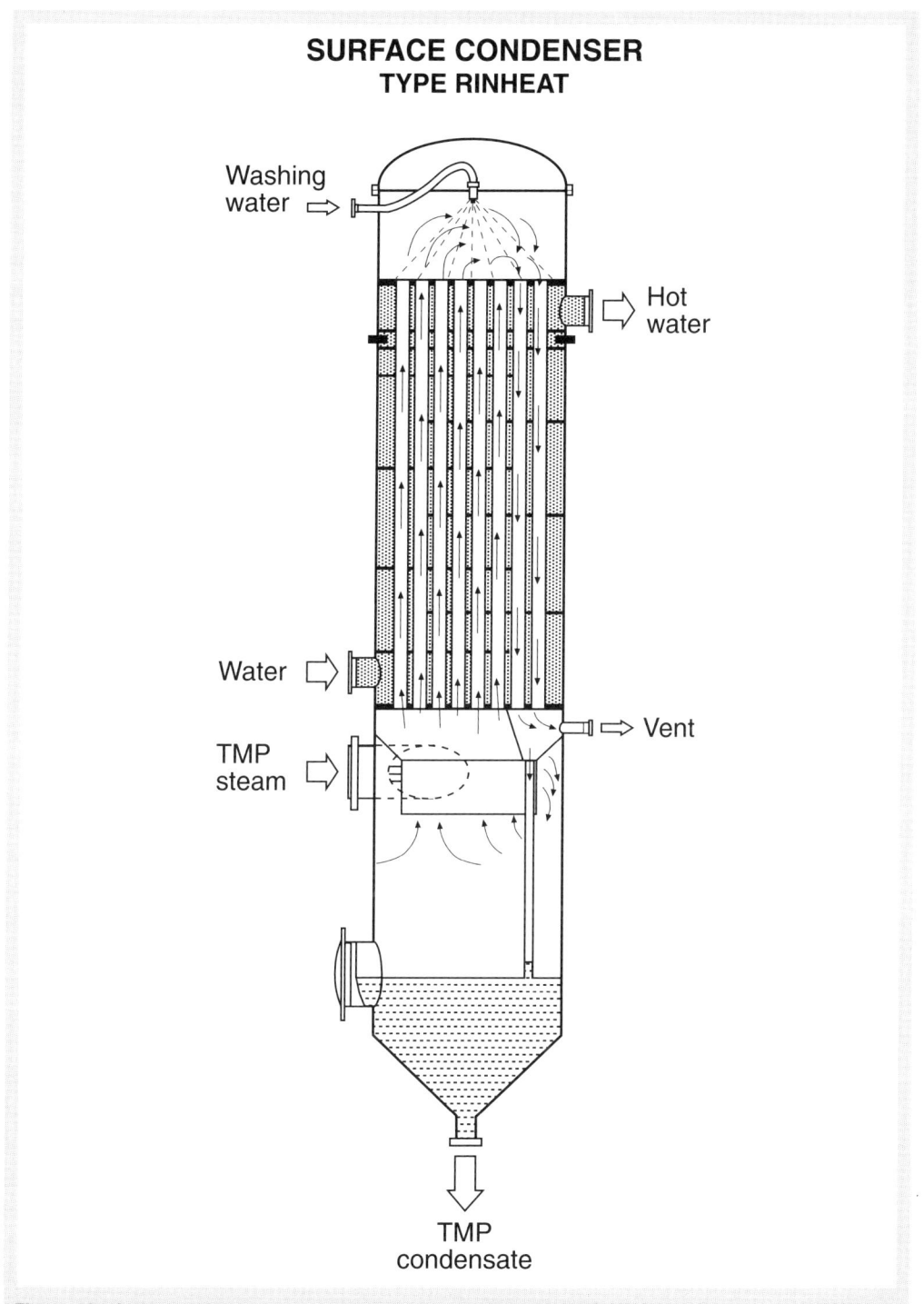

Figure 43. Surface condenser.

8.8 Turpentine recovery

The turpentine recovery is based on the fact that turpentine is virtually insoluble in water. In the turpentine recovery process, the turpentine rich vent steams are condensed and the turpentine is separated from the condensate by decanting. The turpentine vapor present in the TMP steam is not condensed in the main heat recovery nor in the surface condenser due to the presence of inert gases. The boiling point of pure turpentine is around 130°C at atmospheric pressure. When a gas mixture consisting of steam turpentine and air is cooled, then first steam starts to condense. The concentration of the turpentine is increasing until the turpentine dew point is reached and also the turpentine starts to condense. Figure 44 shows a typical flow diagram of a turpentine recovery. The idea of the precondenser is to condense the main part of the steam without condensing virtually any turpentine, thus the condensate can be led to the sewer. In the turpentine condenser, both the remaining steam and turpentine are condensed and the mixture is taken to the decanter for turpentine separation. In order to eliminate odors, the inert gas vent from the decanter is often washed with cold water in a packed bed column.

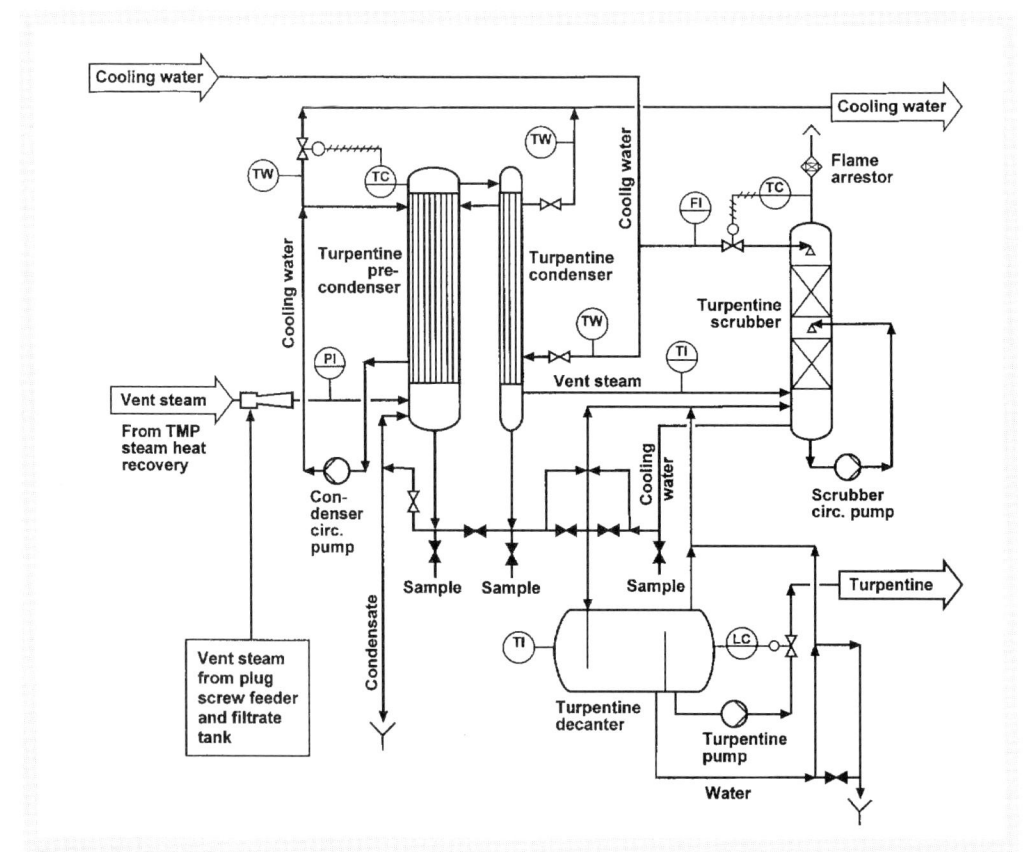

Figure 44. Typical flow diagram for turpentine recovery.

CHAPTER 7

References

1. Lumiainen, J., "Post refining of mechanical pulps," PIRA 1997 International Refining Conference Proceedeings, Pira International, Leatherhead, p. 125.

2. May, W. D., Pulp Paper Mag. Can. 74(1):70 (1973).

3. Miles, K. B and May, W. D., "The flow of pulp in chip refiners," CPPA 1989 Annual Meeting Notes, Technical Section CPPA, Montreal, p.177.

4. Miles, K. B., Paperi ja Puu 72(5):508 (1990).

5. Miles, K. B., Paperi ja Puu 73(9):852 (1991).

6. Strand, B. C., Mokvist, A., Falk, B., Jackson, M., "The effect of production rate on specific energy consumption in high consistency chip refining," 1993 International Mechanical Pulping Conference Proceedings, PFI, Oslo, p. 143.

7. Allison, B. J., Isaksson, A. J., Karlström, A., "Grey-box identification of a TMP refiner," 1995 International Mechanical Pulping Conference (Ottawa) Proceedings, CPPA, Toronto, p. 119.

8. Härkönen, E., Ruottu, S., Ruottu, A., Johansson, O., "A theoretical model for a TMP refiner," 1997 International Mechanical Pulping Conference Proceedings, SPCI, Stockholm, p. 95.

9. Härkönen, E. and Tienvieri,T., "The influence of production rate on refining in a specific energy," 1995 International Mechanical Pulping Conference (Ottawa) Proceedings, CPPA, Toronto, p. 95.

10. Gradin, P., Johansson, O., Berg, J., Nyström, S., "Measurement of the power distribution in a single disc refiner," 1997 International Mechanical Pulping Conference Proceedings, SPCI, Stockholm, p. 83.

11. Oullet, D., Bennington,C. P. J., Senger, J. J., Borisoff, J. F., Martiskainen, J. M., "Measurement of pulp residence time in a high consistency refiner," 1995 International Mechanical Pulping Conference (Ottawa) Proceedings, CPPA, Toronto, p. 171.

12. Kortelainen, J. and Nystedt, H., "Disturbance analysis of the TMP process," 1995 International Mechanical Pulping Conference (Ottawa) Proceedings, CPPA, Toronto, p. 143.

13. Strand, B. C., "The effect of refiner variation on pulp quality," 1995 International Mechanical Pulping Conference (Ottawa) Proceedings, CPPA, Toronto, p. 125.

CHAPTER 7

14. Kortelainen, J., Nystedt, H., Parta, J., "Optimizing the refiner conditions with on-line controls," 1997 International Mechanical Pulping Conference Proceedings, SPCI, Stockholm, p. 103.

15. Hoydahl, H-E., Solbakken, M,. Dahlquist, G., "TMP for SC-grades – A challenge in fiber modelling," 1995 International Mechanical Pulping Conference (Ottawa) Proceedings, CPPA, Toronto, p. 233.

16. Sundholm, J., Heikkurinen, A., Mannström, B., "The role of rate of rotation and frequency in refiner mechanical pulping," 1987 International Mechanical Pulping Conference (Vancouver) Proceedings, CPPA, Toronto, p. 45.

17. Münster, H. and Dahlqvist, G., "Operational experience with the first commercial high speed refiner at Perlen Papier AG, Switzerland," 1995 International Mechanical Pulping Conference (Ottawa) Proceedings, CPPA, Toronto, p. 197.

18. Sundholm, J., Mannström, B., Heikkurinen, A., and Särkilahti, A., Finnish pat. no. FI-89610 (Oct. 25, 1993).

19. Wultsch, F. and Flucher, W., Das Papier 12 (13/14):334 (1958).

20. Strand, B., "Quality control of high consistency refiners," 1997 International Mechanical Pulping Conference Proceedings, SPCI, Stockholm, p. 127.

21. Jackson, M., Falck, B., Ferrari, B., Danielsson, O., Design Features and Pulp Quality Aspects of Single and Double Disc Refiners, Sunds Defibrator pamphlet, Sundsvall, 1987.

22. Eriksson, A. and Sundholm, J., internal KCL report, 1995.

23. Sundholm, J., internal KCL report, 1982.

24. Sundholm, J. and Mannström, B., Paperi ja Puu 64(1):8 (1982).

25. Lunan, W. E., Miles, K. B., May, W. D., Harris, G., Franzen, R., "High pressure refining and brightening in thermomechanical pulping," TAPPI 1983 Pulping Conference Proceedings, TAPPI PRESS, Atlanta, p. 239.

26. Jackson, M. and Akerlund, G., Tappi J. 67(1):54 (1984).

27. Sundholm, J., "Can we reduce energy consumption in mechanical pulping?" 1993 International Mechanical Pulping Conference Proceedings, PFI, Oslo, p. 133.

28. Aregger, H. J., The TMP system at Perlen using the RTS process," 1997 International Mechanical Pulping Conference Proceedings, CPPA, Stockholm, p. 251.

29. Salmén, L., Tigerstrom, A., Fellers, C., J. Pulp Paper Sci. 11(3):J68 (1985).

30. Jackson, M., Danielsson, O., Falk, B., "Thermopulp^{TM} – A new Energy efficient mechanical pulping process," International conference on new available techniques. World pulp and paper week. Proceedings, SPCI, Stockholm p. 229.

31. Härkönen, E., Routtu, S., Routtu, A., Johansson O., "A theoretical model for a TMP-refiner," 1997 International Mechanical Pulping Conference Proceedings, SPCI, Stockholm, p. 95.

32. Allison, B. J., Ciarnello, J. E., Dumont, G. A., Tessier, P. J., "Automatic refiner load control," U.S. pat. no. 5,500,088 (March 19, 1996).

33. Alami, R., Boileau, I., Harris, G., Lachaume, J., Karnis, A., Miles, K. B., Roche, A., "Evaluation of the impact of refining intensity on energy reduction in commercial size refiners; The effect of primary stage consistency," 1995 International Mechanical Pulping Conference (Ottawa) Proceedings, CPPA, Montreal, p. 203.

34. Hill, J., "Process understanding profits from sensor and control developments," 1993 International Mechanical Pulping Conference Proceedings, PTF, Oslo, p. 201.

35. MacDonald, J. E. and Guthrie, J. J., "Chip mass flow meter," 1987 International Mechanical Pulping Conference (Vancouver) Proceedings, CPPA, Montreal, p. 187.

36. Höglund, H., Bäck, R., Falk, B., Jackson, M., "Thermopulp™ (A new energy efficient mechanical pulping process," 1995 International Mechanical Pulping Conference (Ottawa) Proceedings, CPPA, Montreal, p. 213.

37. Kortelainen, J. and Nystedt, H., "Disturbance analysis of the TMP-process," 1995 International Mechanical Pulping Conference (Ottawa) Proceedings, CPPA, Montreal, p. 143.

38. Cluett, W. R., Guan, J., Duever, T. A., Pulp Paper Can. 96(5):31 (1995).

39. Fournier, M., Ma, H., Shallhorn, P. M., Roche A. A., "Control of chip refiner operation," TAPPI 1991 International Mechanical Pulping Conference (Minneapolis) Proceedings, TAPPI PRESS, Atlanta, p. 91.

40. Atack, D., Karnis, A., Stationwala M. I., "What happens in refining (part II)," 1985 International Mechanical Pulping Conference Proceedings, SPCI, Stockholm, p. 35.

41. Pietinen, P. and Tiikkaja E., "Developments in on-line analysers for refiner control," 1993 International Mechanical Pulping Conference Proceedings, PTF, Oslo, p. 44..

42. Perkola, M., "TMP-jauhatuksen teoreettinen ja kokeellinen tarkastelu," Licentiate Thesis, Tampereen teknillinen korkeakolu, Tampere, 1985.

43. Strand, B. C., Tappi J. 79(10):140 (1996).

44. Evans, R., Sutinen, R., Saarinen, K., Pulp Paper Can. 96(5):36 (1995).

45. Evans, R., Saarinen, K., Sutinen, R., Pulp Paper Can. 96(6):76(1995).

CHAPTER 8

Chemimechanical pulping

1	**What is chemimechanical pulping?**	**223**
2	**Process alternatives**	**224**
2.1	Position of the chemical treatment stage in the process	224
2.2	Type of chemical treatment	225
3	**The chemical stage**	**226**
3.1	Sulfonation chemistry	226
3.2	Effect of process variables on the sulfonate content	228
4	**The mechanical stage**	**229**
4.1	Dynamic mechanical properties of chemically treated wood and fibers	231
4.2	Kind of rupture at defibration	231
5	**Process design and operation**	**233**
5.1	General schemes	233
5.2	Chemical pretreatment	235
5.3	Refining	237
6	**Fiber and pulp properties**	**238**
6.1	Fiber properties	238
6.2	Pulp properties	239
7	**End uses**	**243**
8	**Special processes**	**244**
8.1	Interstage sulfonation	245
8.2	Reject sulfonation	245
8.3	Alkaline peroxide mechanical pulping (APMP)	245
	References	248

CHAPTER 8

Carl-Anders Lindholm, Joseph A. Kurdin

Chemimechanical pulping

1 What is chemimechanical pulping?

Chemimechanical pulping involves a gentle chemical treatment stage combined with mechanical defibration, such as disc refining, in order to defiber wood and develop the necessary paper or board properties of the resulting pulp. Chemimechanical pulps are generally defined as that category of pulps with yields in the range of 80 to 95 percent. Their properties are intermediate between those of high-yield chemical pulps and mechanical pulps. The chemimechanical pulps can be further divided into two subgroups:

- Chemithermomechanical pulp (CTMP), which is produced with pressurized refining. Relatively low chemical doses are applied and the yield is typically above 90%.

- Chemimechanical pulp (CMP), which can be produced with refining at atmospheric pressure. The chemical treatment stage is more severe than in the CTMP process, and the yield is typically below 90%. It is also the general name for all chemimechanically produced pulps.

The first pulping lines which could be called chemimechanical pulping began operation in the 1950s and 1960s. The processes aimed at producing printing paper from hardwoods that could not be used in mechanical pulping. The production for this type of pulp, however, remained low. The breakthrough in chemimechanical pulping took place during the 1970s. This rapid development was a result of the improved thermomechanical pulping technology. The key subprocess in chemimechanical pulping is refining, and thus all developments of the thermomechanical process could also be utilized in chemimechanical pulping. This resulted in a rapid growth of the production of softwood CTMP pulps during the late 1970s and 1980s.

In Canada, there was considerable activity in the 1970s to develop chemimechanical processes that could replace sulfite pulp as reinforcement pulp in newsprint. This developmental work led to processes using higher chemical doses and longer pretreatment times than the CTMP process. The process, known by the general name of CMP (chemimechanical pulping) applies to both softwood and hardwood.

In summary, there are various chemimechanical pulping processes that have been developed for different purposes:

CHAPTER 8

- CMP and CTMP processes for hardwood were developed to make it possible to produce very high-yield pulp grades from this type of raw material.
- The CTMP process for softwood was developed in order to produce very high-yield pulps with improved properties, especially cleanliness, absorbency, and strength properties, compared to mechanical pulp.
- The CMP processes for softwood were developed to produce high-yield reinforcement pulp, saving wood raw material and cooking chemicals. In some cases, e.g., when converting sulfite mills without chemical recovery into CMP, the effluent load was also reduced.

2 Process alternatives

2.1 Position of the chemical treatment stage in the process

A chemical treatment stage can be inserted at various positions in a mechanical pulping process[1]. A two-stage refining process with separate reject refining especially offers several possibilities (Fig. 1).

Pre-treatment has the greatest potential in modifying the pulp properties because this type of treatment has an impact on the initial defibering stage. This is by far the most common chemimechanical pulping method. Normally, unless otherwise stated, chemimechanical pulping refers to this process alternative. Processes of this type are in general referred to as:

- *CTMP – ChemiThermoMechanical Pulp (low chemical doses and pressurized refining)*
- *BCTMP – Bleached CTMP*
- *CMP – ChemiMechanical Pulp (high chemical doses)*
- *BCMP – Bleached CMP*
- *CRMP – Chemi Refiner Mechanical Pulp (low chemical doses and atmospheric refining).*

Refiner treatment. Addition of sulfite liquor to the first refining stage can result in a pulp that combines some of the properties of TMP and CTMP pulp[2]. This process has been named:

- *DWS – Dilution Water Sulfonation process.*

Interstage treatment. In a two-stage refiner mechanical process, the coarse pulp can be chemically modified between the two mechanical stages. This process concept is referred to as interstage treatment. Proposed processes based on this concept are:

- *OPCO – A process developed by Ontario Paper Company*
- *TMCP – ThermoMechanicalChemical Pulp.*

Post-treatment can be combined with all types of mechanical pulping, i.e., applied in both the grinding and refining processes. Such a treatment, however, has no effect on the release of fibers from the wood matrix in a defibering stage; therefore, the possibilities of modifying the fiber and pulp properties are limited. Peroxide bleaching can enhance the bonding properties of otherwise low-bonding mechanical pulps such as hardwood PGW or TMP.

Reject or long fiber treatment is in principle an interstage treatment, but it is subjected only to the screening reject prior to reject refining. This process modification can be applied both in the grinding and the refining processes. Abbreviations used for processes of this type are:

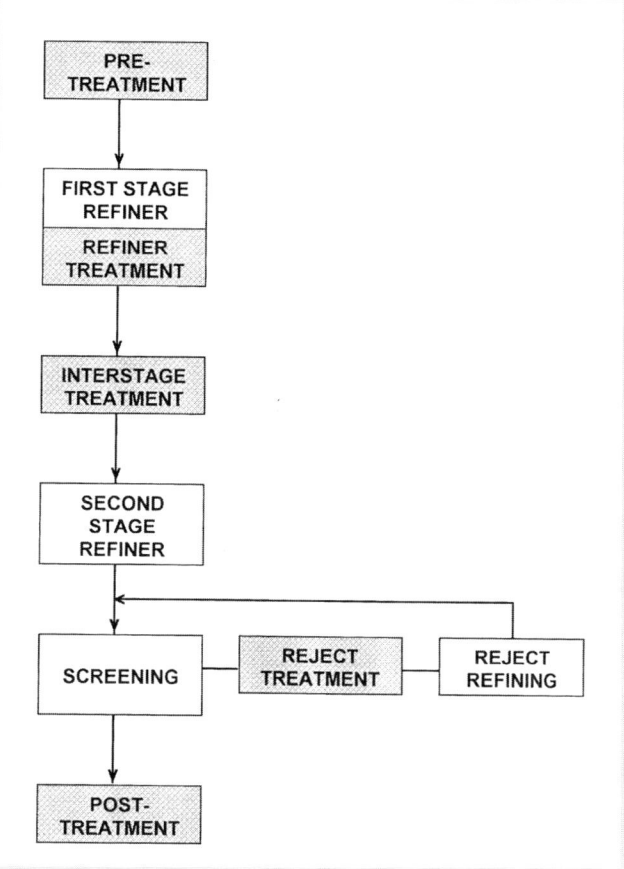

Figure 1. Various possible positions of chemical treatment in a two-stage refiner mechanical process.

- LFCMP – Long Fiber CMP
- CTLF – Chemically Treated Long Fibers
- SLF – Sulfonated Long Fibers
- G-CTMP – Groundwood CTMP
- CTMP$_R$ – Reject CTMP.

2.2 Type of chemical treatment

Chemimechanical pulps, in principle, can be produced by various combinations of chemical treatments and mechanical defibration. A wide variety of chemicals has been proposed for the chemical modification stage[3]. In practical operation, however, sodium sulfite is the dominating chemical in softwood pulping, and sodium hydroxide and/or

CHAPTER 8

sodium sulfite are the common chemicals in hardwood pulping. Alkaline hydrogen peroxide is used in some special processes:

- APMP – Alkaline Peroxide Mechanical Pulp
- APTMP^{TM} – Alkaline Peroxide ThermoMechanical Pulp
- APP/BCTMP – Alkaline Peroxide Bleached ChemiThermoMechanical Pulp.

The chemical dose applied in sulfite-based chemimechanical processes can vary considerably from very low charges of 1–3 percent Na_2SO_3 on b.d. wood to relatively high doses in the range of 10 to 20 percent Na_2SO_3 on b.d. wood. The CTMP process is normally based on low sulfite doses. A common name for processes based on high sulfite doses is *CMP – ChemiMechanical Pulp*. Other abbreviations used for specific pulps within the CMP category are:

- SCMP – Sulfite or Sulfonated ChemiMechanical Pulp
- UHY, UHYS, UHYSP – Ultra High-Yield Sulfite Pulp
- VHY, VHYS, VHYSP – Very High-Yield Sulfite Pulp.

3 The chemical stage

3.1 Sulfonation chemistry

The wood matrix can be chemically modified in various ways that affect the behavior in refining and change the fiber properties. Lignin swelling can be improved by addition of hydrophilic groups like sulfonate and carboxylic groups. Also the carbohydrates can be chemically modified through deacetylation, hydrolysis, partial dissolution, etc. Much research has been carried out over several decades to understand conventional low-yield sulfite pulping. However, the factors which affect the initial, relatively rapid lignin sulfonation taking place in chemimechanical pulping have been less thoroughly investigated.

Figure 2 depicts the major overall reaction introducing sulfonate groups into lignin. The reactive groups in lignin can be divided into various categories: B, X, and Z (Fig. 3). The groups of the B type are sulfonated in the pH range of 1 to 2, whereas the groups of the X and Z type can be sulfonated in the pH range of 4 to 9[5].

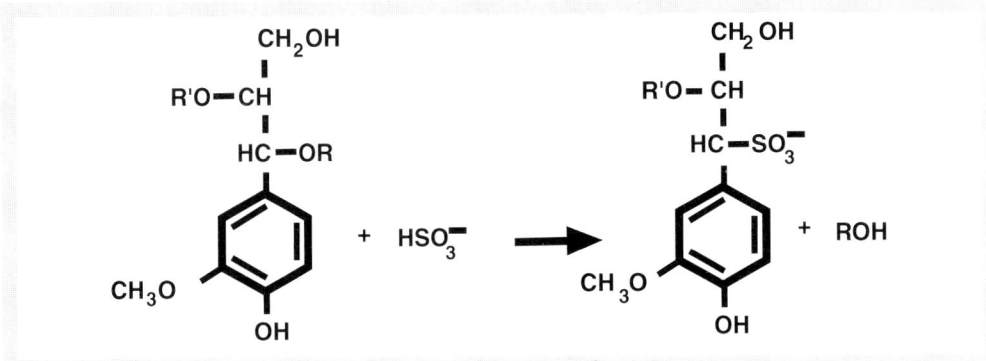

Figure 2. The basic reaction introducing sulfonate groups into lignin[4].

Figure 3. Different basic groups in lignin that can be sulfonated under suitable conditions[5].

The pH in sulfite treatment in the chemimechanical processes is normally clearly higher than 1–2, which means that mainly groups of the X and Z type are sulfonated. About 15% of the phenyl propane units in lignin are of the X type and 15% are of the Z type. Based on this, it can be calculated that the theoretical maximum sulfonate content in softwoods with lignin contents of 25%–30% would be 2.5%–3% on b.d. wood. However, because some lignin starts to dissolve at a high level of sulfonation, the maximum attainable sulfonate content in practical operation is somewhat lower, i.e., 2.1%–2.2% on wood.

Kinetic studies have further indicated that the groups within category X are reacting relatively fast, whereas groups of the Z type are reacting more slowly. Based on this, it can be assumed that mainly groups of the X type are sulfonated in the relatively gentle conditions used in the CTMP processes, whereas groups of both the X and Z types are sulfonated in the more severe conditions used in the CMP processes.

Hardwoods contain less lignin than softwoods, which limits the attainable sulfonate content level. In addition, the structure of hardwood lignin is different from that of softwood lignin, which further restricts sulfonation reactions. Figure 4 illustrates this general difference between softwood and hardwood species and shows the difference in the sulfonation rate for spruce and birch.

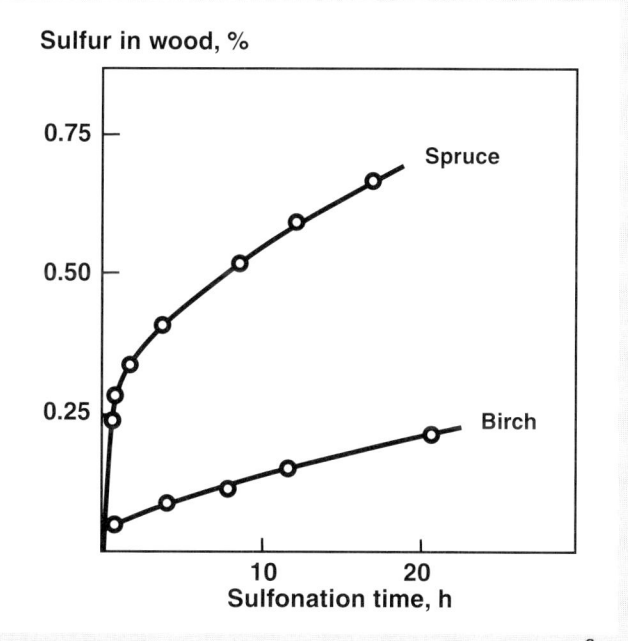

Figure 4. Comparison of the sulfonation rate of spruce and birch wood[6].

In producing chemimechanical pulp from hardwoods, therefore, it is important not to modify lignin only, but also the carbohydrates, if the wood is going to be treated to give as good a pulp as possible[7]. By a relatively mild alkaline treatment, some hardwood hemicelluloses are stabilized against degradation, partly because acetyl groups are split off. Such a moderate alkaline treatment can also improve the strength properties, possibly because the cell walls can be fibrilized without simultaneous rupture of the fibers. However, a sulfonation of the lignin is also needed in order to make it more hydrophilic and thus easier to soften. The sulfonation reaction proceeds fastest at low pH values, but then there is also a risk for too much dissolution of lignin and carbohydrates. Too high pH values, on the other hand, should also be avoided because of slow sulfonation, darkening of the lignin, and simultaneous dissolution of carbohydrates.

For optimum results, hardwood should therefore be treated in two separate stages. The hydroxide stage should be the first one, since the sulfite from the second stage can be partly left in the wood during defibration as a protection against oxidation and alkaline degradation[7]. For practical reasons, however, the two stages are normally combined in commercial installations.

The effect of sulfite treatment can be followed by determining the sulfonate content of the pulp. This can be accomplished by determining the content of sulfur in the pulp gravimetrically[8]. A more convenient method is to determine the amount of sulfonate groups directly by conductometric titration[9]. Under the assumption that all sulfur in the sample is in the form of sulfonate groups, it can be calculated that:

The concentration of sulfonate groups = 2.5 x the concentration of sulfur.

3.2 Effect of process variables on the sulfonate content

The conditions in the pretreatment stage have to be chosen to ensure a suitable degree of sulfonation of the wood chips when they enter the mechanical stage. Thus the optimum conditions in the CTMP type process, operating with sulfonate concentrations of 0.25%–0.75%, to some extent differ from those suitable for CMP processes operating with sulfonate contents of 1%–2%.

Figure 5 shows examples of the effect of the pretreatment conditions on the degree of sulfonation. The *sulfite dose,* or more correctly the *sulfite concentration,* of course affects the treatment result. Increasing the sulfite concentration from 10 to 20 g/l, corresponding with increasing the sulfite dose from about 2 to about 4% in CTMP process conditions, increases the sulfonate content by 40%.

The CTMP process normally operates at slightly alkaline conditions. A pH level of 8 results in slightly higher sulfonate contents than higher pH levels. In practical operation, pure sodium sulfite, Na_2SO_3, is often used, which results in an initial pH of about 10. The pH decreases to about 8, i.e., close to the optimum, during the pretreatment stage.

Increasing the temperature speeds up the reactions, and the degree of sulfonation increases twofold when the sulfonation is performed at 130°C, as compared with 70°C. Increasing the temperature further will still increase the degree of sulfonation, but too high temperatures reduce the yield. The chemical treatment in CTMP pulping is thus performed in the temperature range of 120°C–135°C.

Figure 5 shows that there is a very steep increase in the sulfonate content in the initial stage. From a process point of view, it is interesting to observe that about 75% of the sulfonation achieved in 15 minutes is already achieved during the first minute. This rapid initial stage might be explained by sulfonation of lignin groups of the X type (cf. Fig. 3). Even after this rapid initial stage, the degree of sulfonation increases if the treatment time is prolonged. For example, at pH 9.5 and 130°C, extending the treatment time from 3 to 100 minutes will increase the degree of sulfonation from about 0.4% to about 0.8%. This slow sulfonation phase is normally not fully utilized in the CTMP process, in which relatively short treatment times of 2–10 min are applied.

4 The mechanical stage

The changes in the wood substance brought about by the reactions between sulfite liquors and wood influence wood softening, defibration, and refining in a rather complicated manner. The scheme in Fig. 6 shows an attempt to describe the phenomena involved. The reactions between sulfite and lignin lead to softening

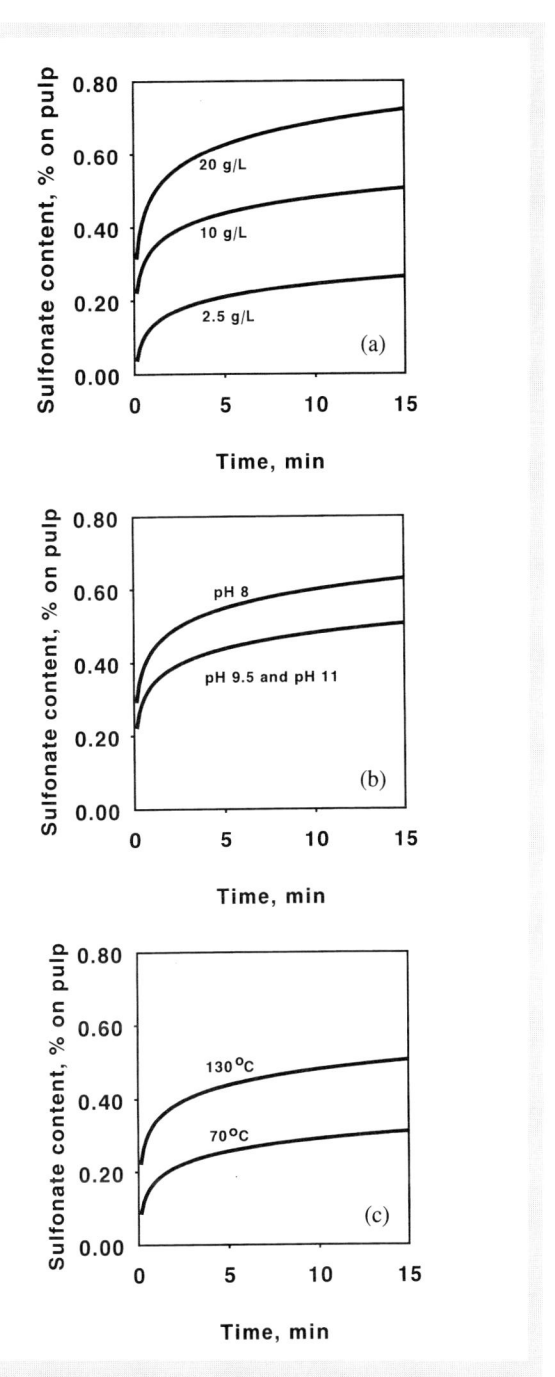

Figure 5. Degree of sulfonation in wood meal from Norway spruce as a function of time at various conditions [10]: (a) 130°C and pH 9.5. Variable: Na_2SO_3 concentration; (b) 130°C and 10 g/l Na_2SO_3. Variable: pH; (c) 10 g/l Na_2SO_3 and pH 9.5. Variable: temperature.

CHAPTER 8

of lignin that changes the dynamic mechanical properties of wood and fibers. This affects the energy consumption in the mechanical stage.

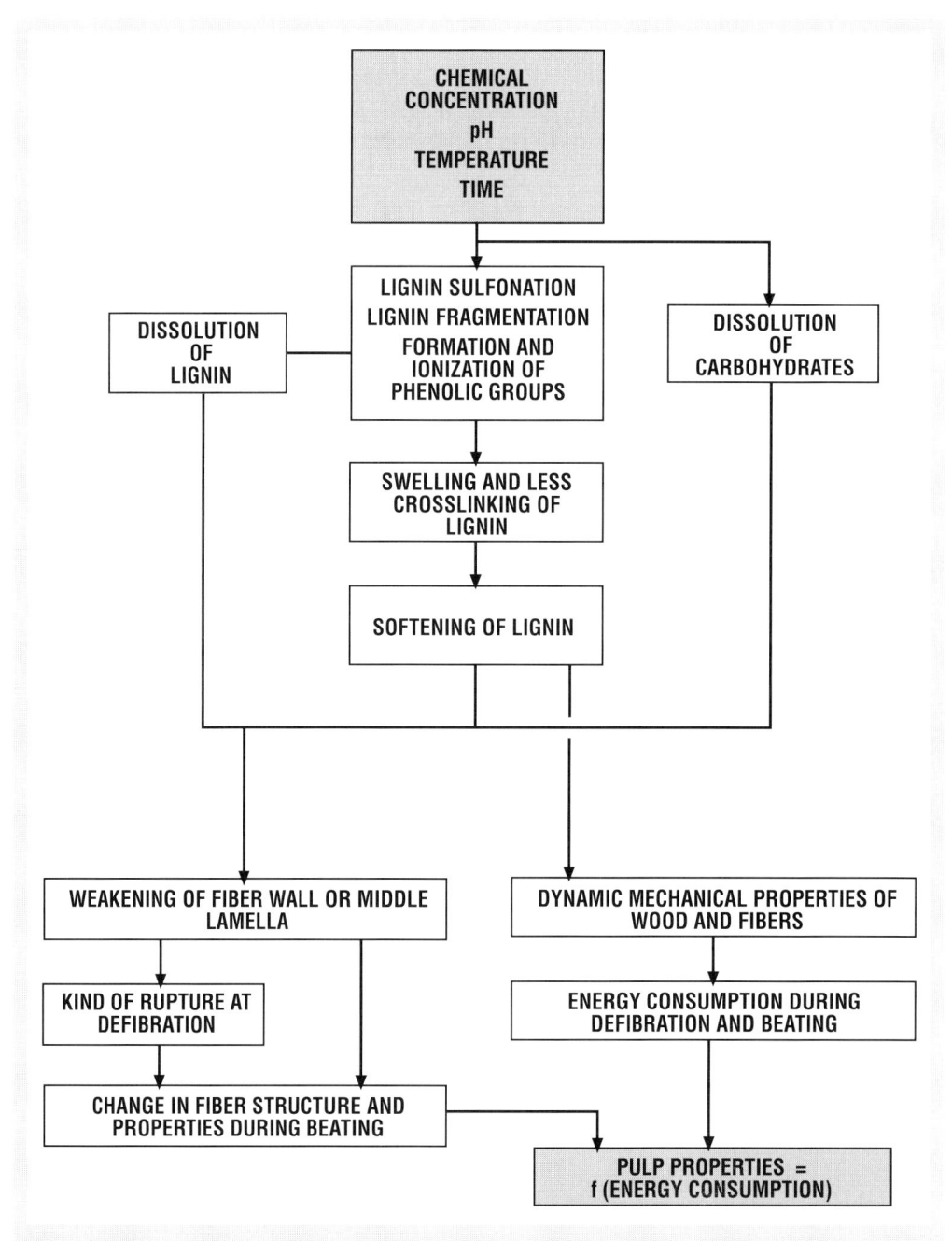

Figure 6. Influence mechanism of pretreatment with sulfite on wood softening and on defibration and refining results [11].

Softening of lignin together with dissolution of some lignin and carbohydrates also weakens the middle lamellae and makes the fiber walls tougher and less brittle. This can change the kind of rupture at defibration and affects the development of the fiber properties during further refining in the mechanical stage.

4.1 Dynamic mechanical properties of chemically treated wood and fibers

Chemical treatment changes the properties of the wood matrix in a manner that can change the optimum conditions in the mechanical stage. Some researchers argue that the temperature in the refiner should preferably be chosen so that the internal friction is near its maximum because then a maximum amount of the energy applied can be absorbed by the the wood substance. This is evident from the following relationship[12]:

$$E_a/E_e = \lambda/(1+\lambda/2) \tag{1}$$

where E_a is the energy absorbed by the fibers
E_e the energy consumption of the refiner
λ the internal friction of the wood and fibers

Sulfonation reduces the temperature at which the internal friction reaches its maximum (the "softening temperature"). Atack et al.[13] showed that the extent of reduction of the softening temperature appears to be a unique function of the degree of sulfonation regardless of the type of sulfite liquor used (Fig. 7). Highly sulfonated wood can thus be refined at clearly lower temperatures than untreated wood[14,15].

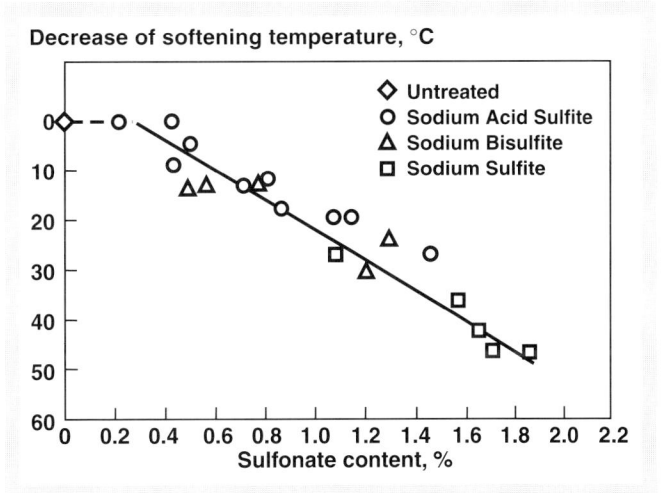

Figure 7. Softening temperature reduction of residual lignin in black spruce as a function of the degree of sulfonation[13].

4.2 Kind of rupture at defibration

Weakening of the fiber wall and middle lamella changes the kind of rupture in defibration. Much more of the rupture takes place in the primary layer and in the middle lamella (Fig. 8). This results in high portions of long undamaged fibers. Due to the softening effect of the sulfite treatment on the lignin, less wood material is broken down into fines as compared with purely mechanical pulping. In purely mechanical pulping, cracks across the cell wall as indicated for RMP in Fig. 8 often result in liberation of bundles of several fibers, i.e., shives. To reduce the amount and size of these shives, purely mechanical pulps have to be refined to a relatively low freeness level (Fig. 9). In chemimechanical pulping, the different behavior of the fibers in defibration results in a low amount of shives already at high freeness levels. These pulps can therefore be used for certain products at high freeness levels. On the other hand, chemimechanical pulp fibers can be considerably modified through further refining, thus changing the properties of the pulp and making it suitable for various end products.

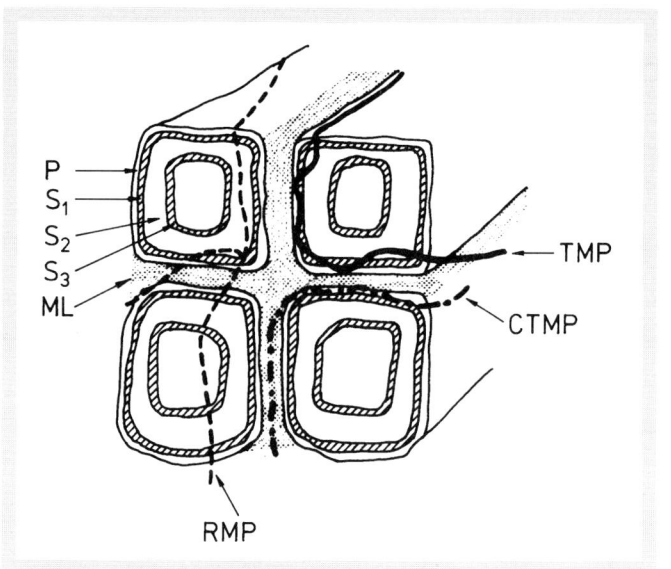

Figure 8. Typical kinds of fiber rupture in different high-yield pulping processes. P is the primary wall; S_1, S_2, S_3 are the secondary wall; and ML is the middle lamella[16].

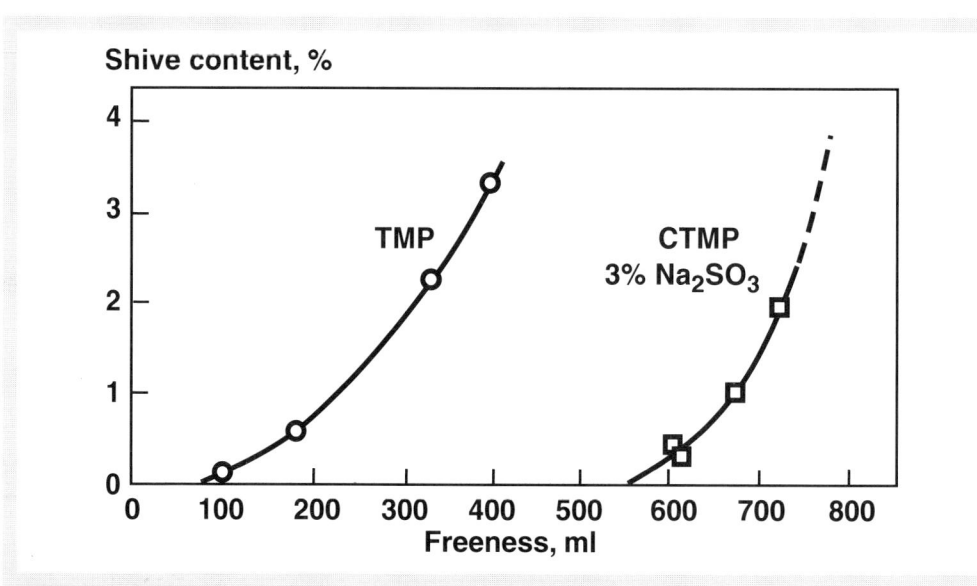

Figure 9. Development of the shive content as a function of pulp freeness in mechanical (TMP) and chemimechanical (CTMP) pulping[17].

5 Process design and operation

5.1 General schemes

Figure 10 shows a general block diagram of chemimechanical pulping. The process closely resembles that of thermomechanical pulping, but it includes subprocesses necessary for chemical impregnation and chemical treatment at elevated temperatures.

The chips used as raw material in the process are either purchased or produced at the mill site. In the latter case, additional equipment for debarking, chipping, and chip screening is needed. An efficient impregnation stage is important, especially in the CTMP process.

The design of the equipment for the preheating stage can vary significantly depending on the process alternative. In the CTMP process, the retention time in the pretreatment stage is short and a relatively small pretreatment vessel is needed. In production of CMP pulp, longer retention times and higher temperatures are applied. In this case, reactors that can be considered as small digesters are used. The digester or reactor design can vary considerably between different process alternatives.

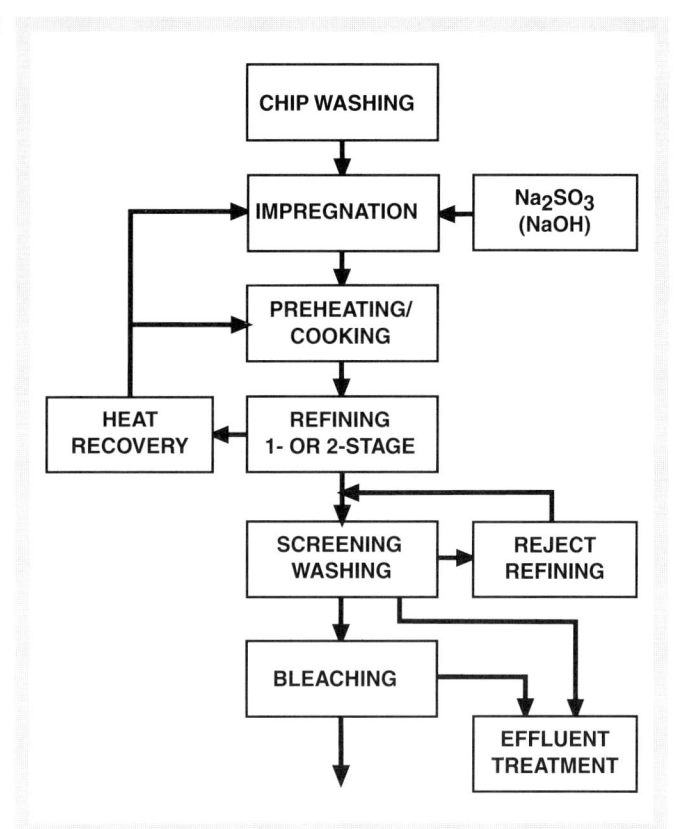

Figure 10. General block diagram of chemimechanical pulping.

The refining operation in chemimechanical pulping can be based either on one or two refining stages. The refining energy that can be applied in one stage can be sufficient in processes producing high-freeness pulps, but production of lower-freeness pulp requires refining in two stages. CMP processes using high chemical doses and possibly recovery of spent chemicals can have a washing stage between the two refining stages.

Screening of chemimechanical pulps does not essentially differ from that of thermomechanical pulps. In many cases, the screening operation can be less exacting because of the comparatively low shives content of chemimechanical pulps.

After screening, the pulp is washed – normally in two or three stages with disc filters or screw or roll presses. The washing efficiency can vary, depending on the requirements. If the pulp is going to be bleached, it is necessary to have an efficient washing stage in order to minimize the carryover of pretreatment chemicals and dissolved material to the bleachery[18].

CHAPTER 8

Figures 11–13 show examples of modern CTMP and CMP lines.

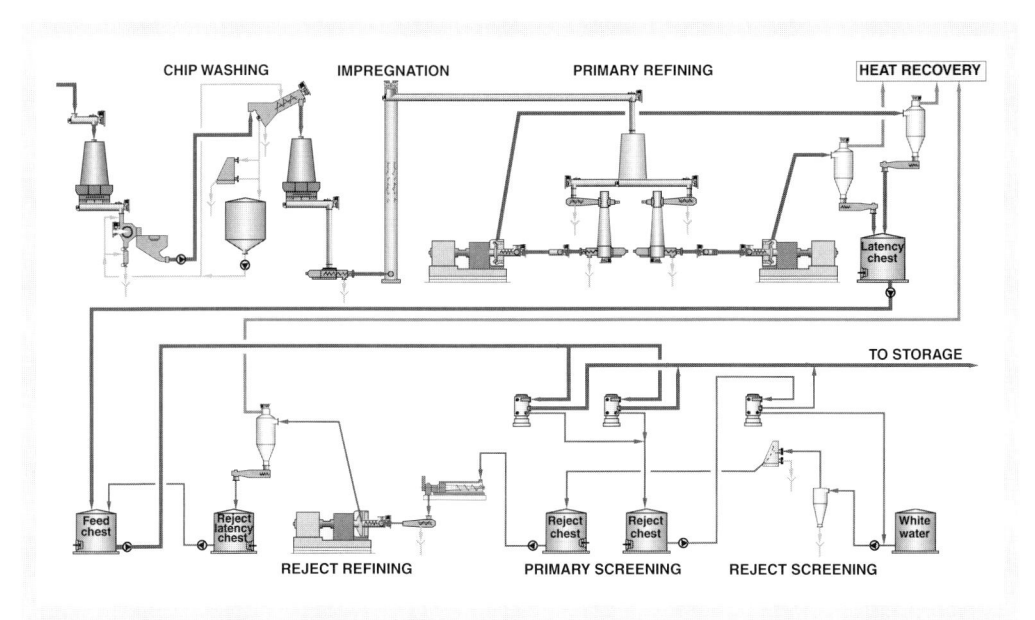

Figure 11. A modern line for softwood board grade CTMP (Sunds Defibrator).

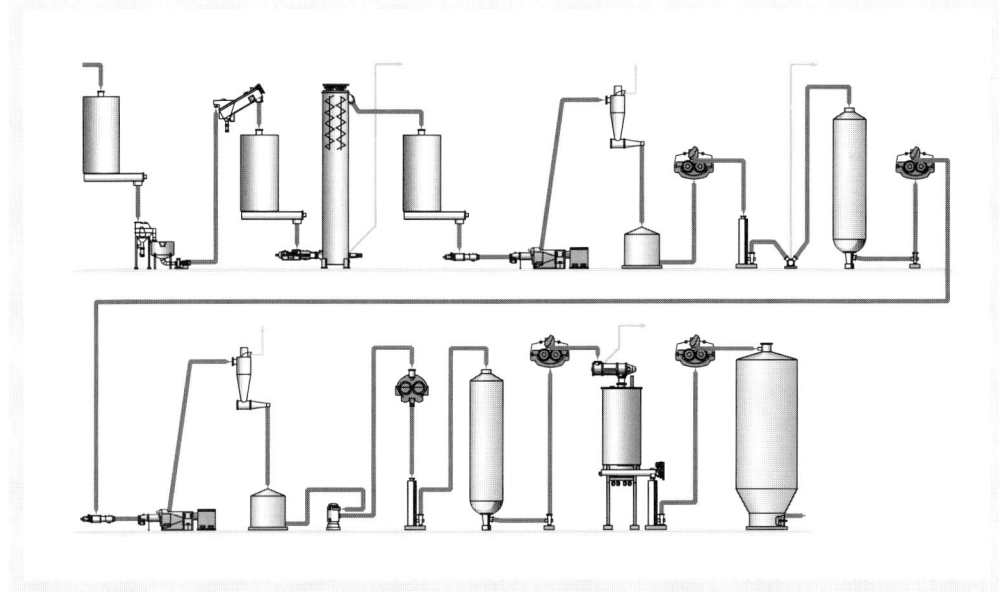

Figure 12. A modern BCTMP line (Sunds Defibrator).

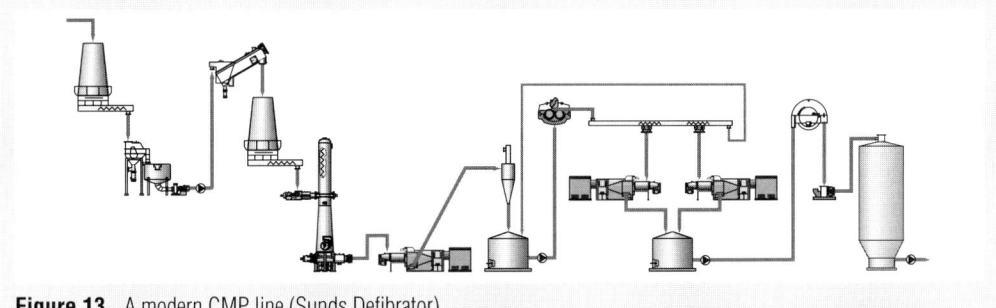

Figure 13. A modern CMP line (Sunds Defibrator).

5.2 Chemical pretreatment

An efficient impregnation stage is important in the chemimechanical process, especially in the CTMP process because the pretreatment is performed in steam phase and the short treatment time does not allow much diffusion within the chips. Incomplete impregnation can result in clearly increased shive contents, even if the average sulfonate content is reasonable high[19].

Various methods can be used for impregnation of chips with chemicals[20]:

- Spraying of chemicals onto the chips
- Steaming the chips and then soaking them in a cold sulfite solution
- Compressing the chips mechanically, followed by expansion in a sulfite solution
- Addition of chemicals to the refiner.

The first three methods have been found to be efficient. The compression method gives the highest liquor-uptake[20]. With this method, some of the water in the chips will be replaced by sulfite solution. Another advantage is that differences in chip moisture is equalized, which will give a more even impregnation.

Different equipment concepts based on the compression/expansion technique are commercially available. Figures 14 and 25 show some examples.

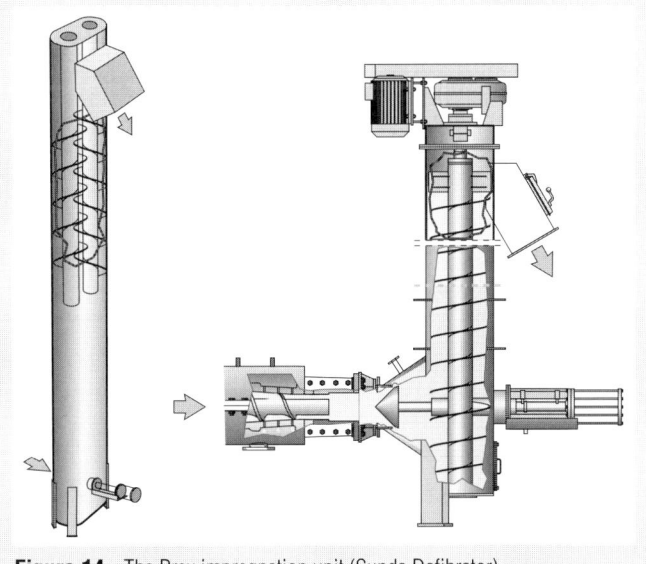

Figure 14. The Prex impregnation unit (Sunds Defibrator).

CHAPTER 8

The type and amounts of chemicals applied varies depending on the wood species (softwood/hardwood) and the type of process (CTMP/CMP). In production of softwood CTMP, the conditions are rather standardized[17, 21]:

- Na_2SO_3 charge 2%–4% on b.d. wood
- pH 9–10
- Temperature 120°C–140°C
- Retention time 2–15 min.

The conditions used in CMP processes are more severe than those used in the CTMP process. Table 1 shows some recommended conditions as well as some conditions reported for commercial installations. Various pH levels are applied, and some of the processes operate at acid conditions. The reaction temperatures are clearly higher and the pretreatment times longer than in the CTMP process.

Table 1. Reaction conditions in some commercial and recommended softwood CMP processes[21–23].

Process	CMP	CMP	SCMP	BCMP	UHY
Sodium sulfite/bisulfite, % on wood	12–20	10–15	13–16	10.5–15	6
pH		7–10	7.5–9.5	5.9	4.5
Reaction temperature, °C	140–175	140–160	150–160	168	147
Reaction time, min	10–60	30–60	30–45	50–60	240

The chemical and morphologial composition of different hardwood species varies much more than those of softwood species. Only a few hardwood species, such as aspen or poplar, can be used as raw material for purely mechanical pulp. Most species like eucalyptus, gmelina, and birch need some kind of chemical treatment. Modification of the carbohydrates in a highly alkaline stage, combined with modification of lignin in a sulfite stage, can be useful in pretreatments of hardwoods. Despite that, processes based on either alkali or sulfite chemicals only can also be successfully used for some species. Table 2 shows pretreatment conditions recommended for production of CTMP and CMP pulps from hardwood.

Table 2. Recommended conditions for production of CTMP and CMP pulps from hardwoods[21,23].

Process	CTMP	CMP	
Chemical dose, % on wood	0–4% Na_2SO_3 + 1%–7% NaOH	10%–15% Na_2SO_3	4-8% NaOH
pH	12–13	9–10	12–14
Reaction temperature, °C	60–120	130–160	50–100
Reaction time, min	0–30	10–60	30–60

5.3 Refining

The most important feature of the chemimechanical processes is connected to the changed behavior of the wood substance in the mechanical refining stage, as outlined in Fig. 6. The intensity of the chemical pretreatment stage also affects the softening temperature. Strong treatments resulting in sulfonate contents over 1% on wood can reduce the softening temperature below 100°C. This means that atmospheric refiners can be used instead of pressurized ones.

The most important variable in the refining stage is the specific energy consumption (SEC). The freeness of the pulp and the specific energy consumption are dependent on each other, but chemical pretreatment can have a significant effect on the development of freeness as a function of the specific energy consumption. In softwood pulping, as a rule, chemical pretreatments with yields above 85% increase the energy consumption needed for refining to a certain freeness level (Fig. 15). Conversely, a strong interstage chemical treatment can decrease the energy consumption in the second refining stage. Normally, chemical

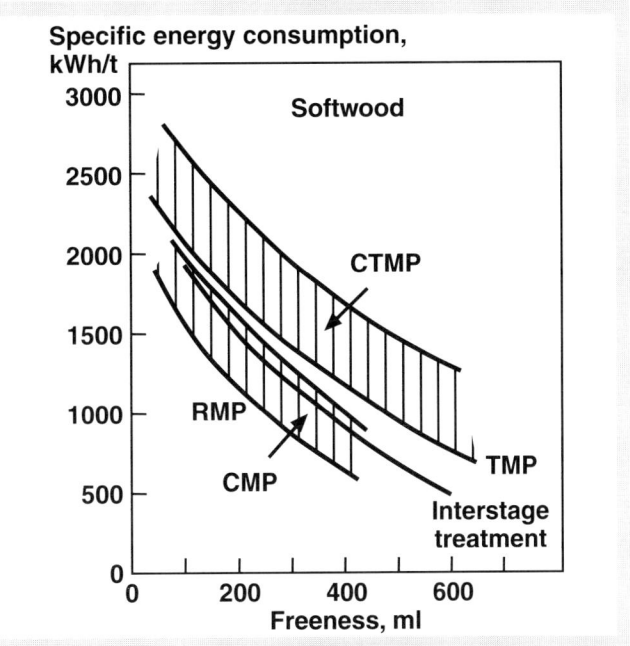

Figure 15. The specific energy consumption in various softwood high-yield pulping processes[21].

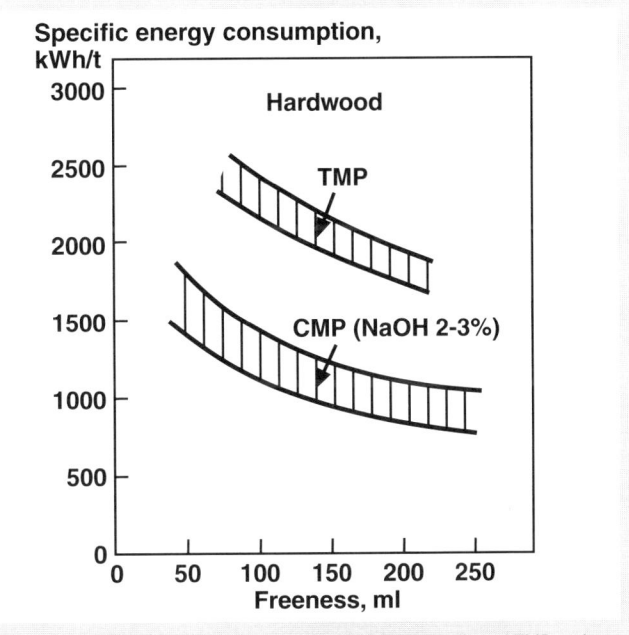

Figure 16. The specific energy consumption in production of TMP and CMP from hardwood[21].

CHAPTER 8

pretreatment of hardwood pulps clearly reduces the energy consumption (Fig. 16).

The energy consumption in the mechanical refining stage or stages is adjusted to result in a suitable pulp quality, mostly by controlling the freeness level of the pulp. Depending on the energy consumption needed for refining to the desired freeness levels, one or two stage systems can be used.

6 Fiber and pulp properties

6.1 Fiber properties

A chemical pretreatment modifies the pulp properties both through changing the fiber size distribution and by changing the specific properties of the individual fibers.

The effect of the chemical pretreatment on the behavior of wood and fibers in the refiner affects the "pulp geometry" in the following way (compared with mechanical pulping):

- The amount of shives decreases.
- The amount of long fibers increases.
- The amount of fines decreases.

Thus the fiber size distribution of chemimechanical pulps is between that of mechanical and chemical pulps (Fig.17). It should also be noted that the fiber size distribution can vary significantly between pulps refined to different freeness levels (Table 3). The increased proportion of long fibers and the reduced fines content, of course, increases the average fiber length of chemimechanical pulps. CTMP pulps of high freeness can have average fiber lengths approaching that of kraft pulp (Table 4).

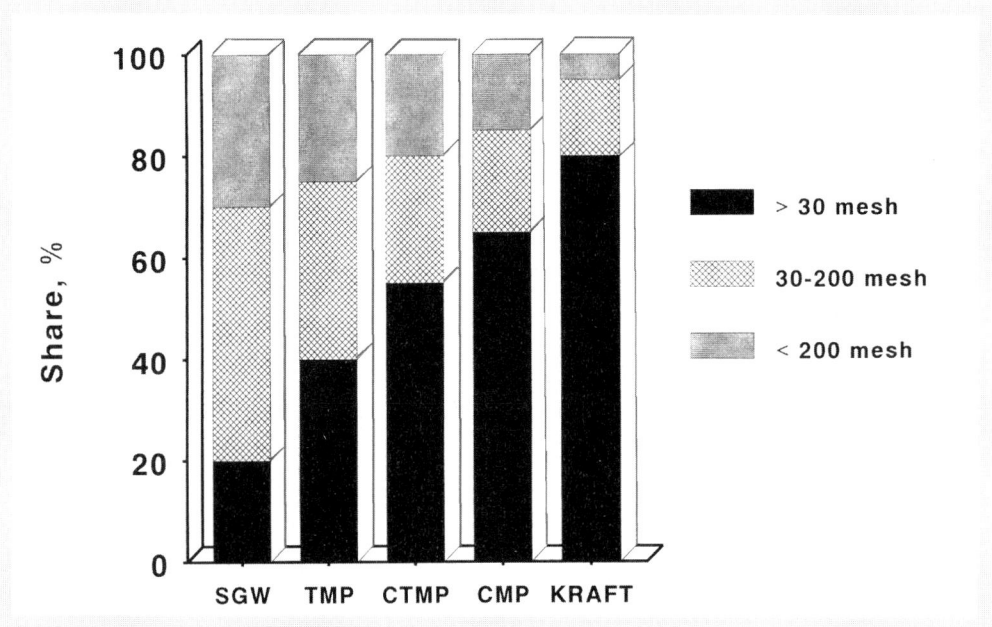

Figure 17. Fiber size distribution (BauerMcNett) of various pulp types.

Chemimechanical pulping

Table 3. Examples of fiber size distributions of various softwood CTMP pulps.

	Fluff Grade	Board and Tissue Grade	Newsprint Grade	LWC Grade
Freeness, ml	650–700	250–500	80–100	40–50
Shive content, %	1.5	0.2	0.1	0.05
BauerMcNett:				
> 30 mesh, %	65	60	45	30
< 200 mesh, %	10	15	25	30

Chemical pretreatment affects the specific properties of the individual fibers through various mechanisms by changing the properties of the fiber surface and by sulfonating the lignin in the cell wall, thus increasing the flexibility and collapsibility of the fibers. Both mechanisms can have a significant impact on the pulp properties. The mechanisms involved in changing the properties of the fiber surface are:

- The kind of rupture that can change the fiber wall layer constituting the surface of the fiber particle

- Sulfonation of lignin on the fiber surface which increases the hydrophilicity and improves bonding ability and local plasticity of the surface.

Table 4. Average fiber length of various softwood pulps produced from the same type of raw material (fiber length in wood c. 3.0 mm)[24, 36].

Pulp type	Average fiber length, mm
TMP, freeness 100 ml	1.3–1.7
CTMP, freeness 100 ml	1.3–1.8
CTMP, freeness 200 ml	1.4–1.9
CTMP, freeness 300 ml	1.5–2.0
Softwood kraft, unbeaten	2.2–2.4

6.2 Pulp properties

The changes in fiber properties brought about by chemical pretreatment are also reflected on the sheet properties. In comparing the properties of chemimechanical pulps with those of corresponding thermomechanical pulps, it must be considered that the chemical pretreatment changes the freeness-SEC-relationship (cf. Figs.15 and 16). The comparison situation is thus not unambiguous, but the manner of com

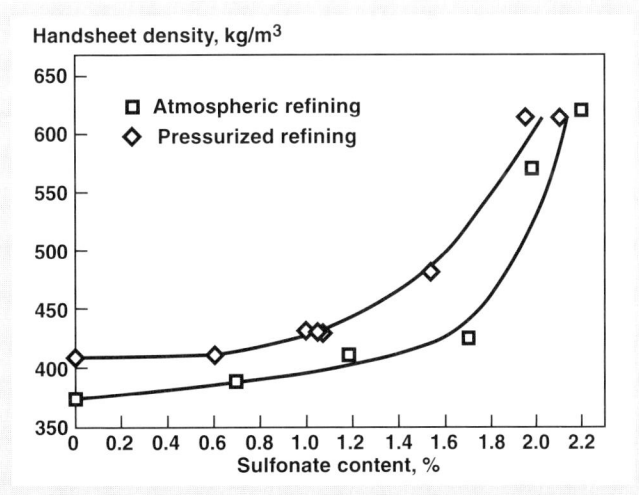

Figure 18. Handsheet density of chemimechanical pulps from black spruce at a freeness level of 100 ml as a function of the sulfonate content[26].

parison must be stated. Pulps can be compared, for example, at the same freeness level after refining with the same specific energy consumption or, preferably, as used in the same kind of products.

Atack, Heitner, and co-workers[25-27] made extensive pilot plant studies on the effects of the conditions on the properties of softwood CTMP and CMP pulps. Although the atmospheric and pressurized refinings were not done in the same refiner, which might have had some impact on the result, some general trends can be seen from the results.

Pretreatment to sulfonate contents below 1% does not significantly affect the sheet density (Fig. 18). The probable explanation of this is that the increased flexibility of the fibers in this range is counteracted by the reduced quantity of fines.

Compared at a constant input of refining energy, sulfite treatments to sulfonate contents below 1% do not change the tensile strength significantly (Fig.19). Thus high-freeness CTMP pulps do not necessarily have higher tensile strength than TMP pulps. Compared at constant freeness, sulfonate levels below 1% improve the properties of CMP/CTMP pulps, as compared with pure mechanical pulps (Figs. 20 and 23).

Pressurizing the refiner, as in the TMP process, increases the maximum attainable tear index more than sulfonation to a sulfonate content of about 1% (Fig. 21). The effects

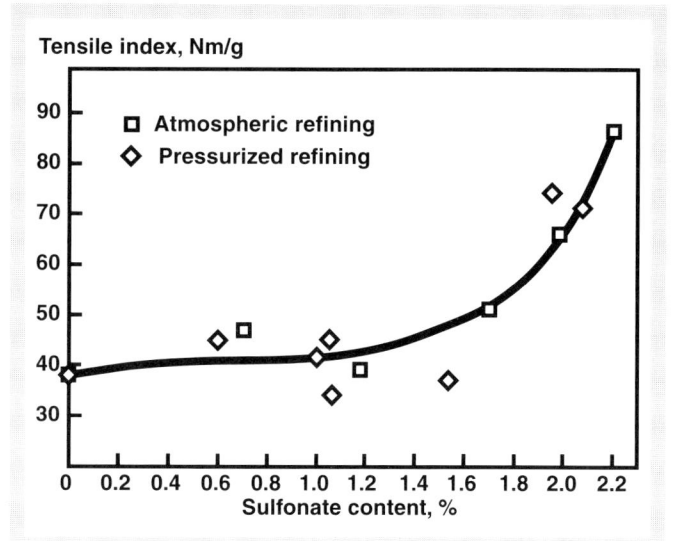

Figure 19. Tensile index of chemimechanical pulps from black spruce refined at a specific energy of 1.94 MWh/metric ton as a function of the sulfonate content[26].

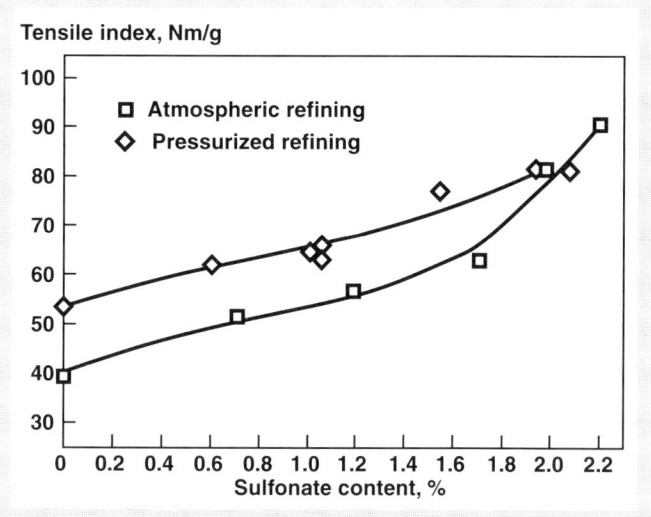

Figure 20. Tensile index of chemimechanical pulps from black spruce at a freeness level of 100 ml as a function of the sulfonate content[26].

of pressurizing the refiner and chemical treatment, on the other hand, are additive, so the far highest tearing resistance can be attained with the CTMP process. The CTMP and CMP pulps compared in Fig. 21 have about the same sulfonate content. Normally CMP pulps are pretreated to clearly higher sulfonate contents than CTMP pulps. This will reduce the difference between the tearing resistance between the pulp types because the maximum attainable tear index increases with increasing sulfonate content.

The light-scattering coefficient of chemimechanical pulps with a relatively low sulfonate content behaves like that of mechanical pulps. Increasing the specific energy increases both the light scattering coefficient and the tensile strength (Fig. 22). A high degree of sulfonation reduces the light-scattering coefficient to a level close to that of chemical pulps. The development of the light-scattering coefficient and tensile strength also starts to resemble that of chemical pulps: An increase in refining energy increases the tensile strength but decreases the light-scattering coefficient.

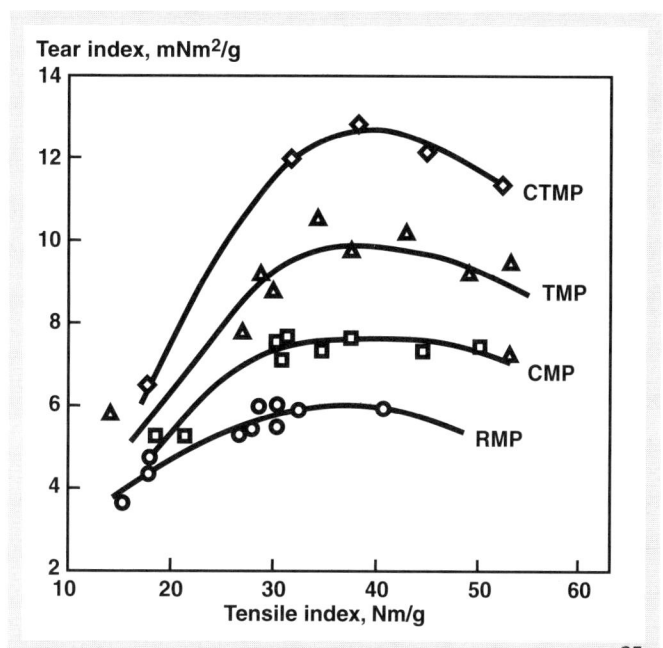

Figure 21. Tear index versus tensile index for various high yield pulp types[25].

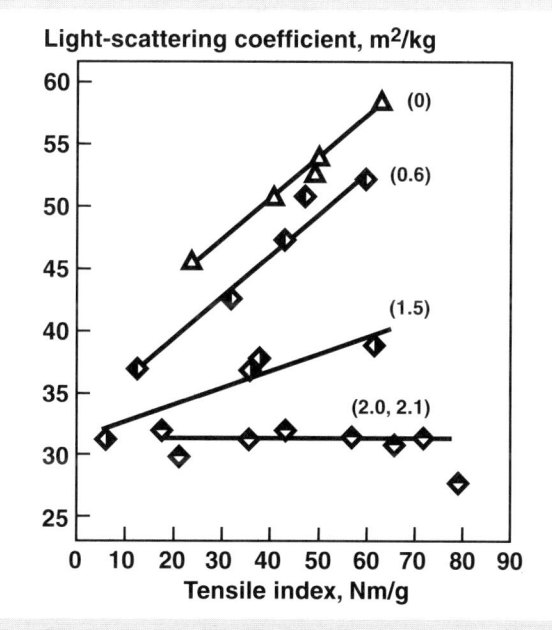

Figure 22. Development of the light-scattering coefficient and tensile index in refining of pulps pretreated to various sulfonate content levels (denoted in parentheses)[26].

CHAPTER 8

Figures 23 and 24 show typical strength properties of CTMP pulps produced from Norway spruce (*Picea abies*) compared with those of groundwood and thermomechanical pulp. The strength level, especially the tensile strength of CTMP pulps, is highly dependent of the freeness level. Various types of pulps can therefore be produced with this process.

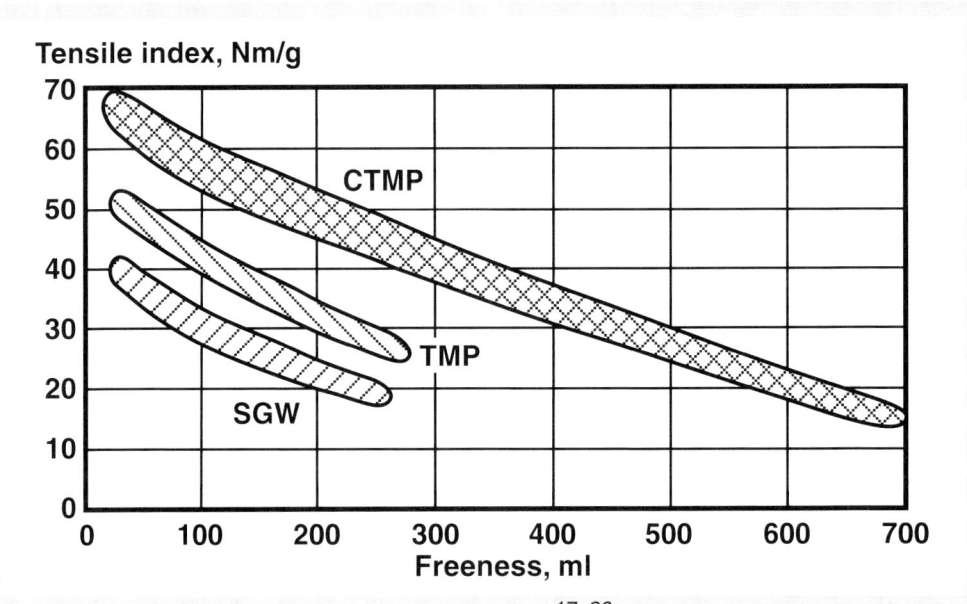

Figure 23. Typical tensile index levels for various high-yield pulps [17, 36].

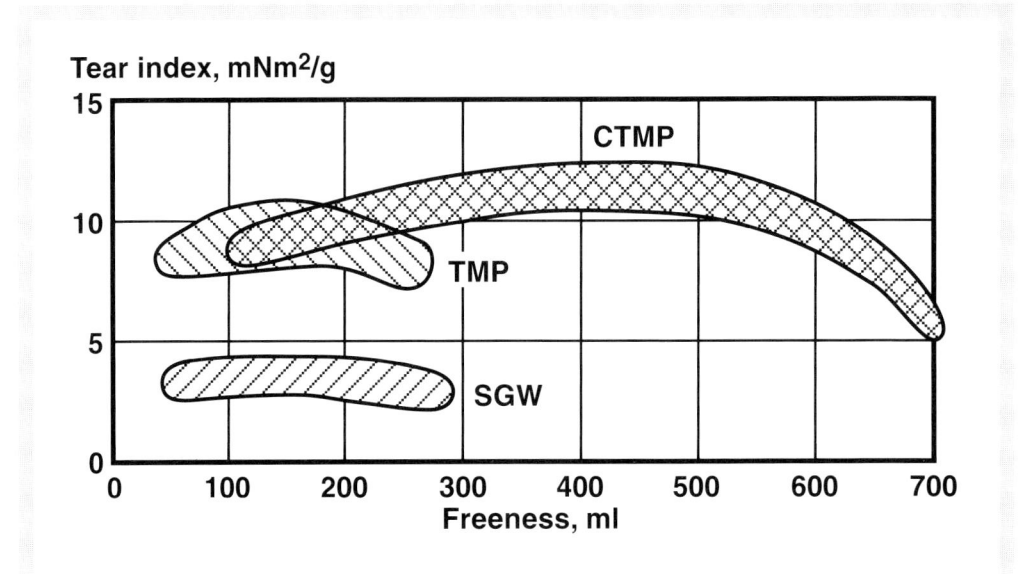

Figure 24. Typical tear index levels for various high-yield pulps [17, 36].

The brightness of CTMP pulps depends on the wood raw material. Treatment with sulfite has a gentle bleaching effect; thus, unbleached CTMP pulps are often a few units brighter than corresponding thermomechanical pulps.

The following properties can be considered typical for softwood CMP pulps, although they can vary somewhat depending on the process conditions[28,29]:

- Yield 80%–90%
- Freeness 300–400 ml
- Tensile index 50–60 Nm/g
- Tear index 8–12 mNm^2/g
- Light-scattering coefficient 30–40 m^2/kg.

The strength properties are somewhat higher than for CTMP pulps, but the light-scattering coefficient is clearly lower. The low light-scattering coeffecient lowers pulp brightness. For example, a reduction of the light-scattering coefficient from 70 to 35 m^2/kg can reduce pulp brightness by more than 10% ISO at the same content of colored material in the pulp.

The papermaking properties of hardwood chemimechanical pulps vary depending on the wood species and the process conditions[29]. The alkali charge has a strong effect on the strength properties and the light-scattering ability. The strength properties are normally lower than those of softwood chemimechanical pulps and are about on the same level as those of softwood groundwood pulps.

7 End uses

Chemimechanical pulps can be characterized as intermediate pulps between pure mechanical and chemical pulps. They can be used to replace mechanical or chemical pulps or different combinations of them in various paper grades.

Softwood CMP pulps are normally refined to freeness levels of 300–400 ml and used as reinforcement pulps in newsprint to replace either sulfite or kraft pulp.

Hardwood CTMP and CMP pulps are used for replacing mechanical pulp in printing papers in countries where softwood is not available for production of pure mechanical pulps.

The softwood CTMP process can produce pulps of widely varying properties by varying the specific energy consumption[29, 30]. Thus grades suitable for different products can be produced (Table 5).

Fluff grade CTMP pulps are refined with a specific energy input of about 1.0 MWh/metric ton resulting in a freeness level of 650–700 ml. Due to the sulfite treatment, the shive content is reasonably low at this high of a freeness level and the pulp has a high content of long fibers and low content of fines. This gives the pulp good absorption properties, which is essential for this type of pulp.

In the production of *tissue grade CTMP,* specific energies of about 1.4 MWh/metric ton are applied, resulting in freeness levels of 350–500 ml. The density and tensile strength are clearly higher, as compared with fluff grade pulps. Still the content of long

CHAPTER 8

fibers is high and the content of fines low, resulting in good absorption properties. The amount of CTMP in tissue furnishes is typically 20%–30%.

Board grade CTMP pulps are frequently used as the middle layer in multilayer board and for the production of liquid packaging board. Advantages of CTMP in these products are high stiffness, good dimensional stability, and a low content of extractives. Board CTMP pulps are refined to freeness levels of 250–500 ml with a specific energy of about 1.4 MWh/metric ton.

By applying high specific energies of 2–3 MWh/metric ton, low-freeness pulps suitable for various *printing paper grades* can be produced with the CTMP process. Most of the chemimechanical pulps used in printing papers, however, are not CTMP pulps but CMP pulps used as reinforcement pulp in newsprint.

Table 5. Typical properties of softwood CTMP pulps for various end uses.

	Fluff Grade	Tissue Grade	Board Grade	Newsprint Grade	LWC Grade
Freeness, ml	650–700	350–500	250–500	80–100	40–50
Shive content, %	1.5	0.2	0.15	0.1	0.05
Density, kg/m^3	240–260	330–370	330–370	400–480	435
Tensile index, Nm/g	15–20	35–45	30–35	45–60	50
Tear index, mNm^2/g	5–8	8–12	7–10	8–9	6
DCM extractives, %	0.25	0.15	0.10		

8 Special processes

Chemimechanical processes based on pretreatment with sulfite chemicals are by far the most common ones. In addition to these processes, those based on other concepts have also been developed. Sulfite treatment can be applied in other stages of the process (cf. Fig. 1):

- Between the two mechanical stages in a two-stage refiner process (interstage treatment)
- Directed to the screening reject fraction only.

Sulfite treatment after the mechanical stages – post-treatment – has also been investigated, but this type of treatment seems to have the lowest potential.

In addition, a variety of different chemical treatments have been investigated. Of those, pretreatment with alkaline peroxide is the most promising.

A novel process, based on a short high-temperature vapor phase pretreatment, called "explosion pulping" has also been proposed for producing chemimechanical pulp[35].

8.1 Interstage sulfonation

In interstage sulfonation, the sulfite treatment is performed in a reactor between the two refining stages (cf. Fig. 1). No special equipment is needed for impregnation in this process because the wood material is already defibered. The possibilities of affecting the pulp properties in interstage sulfonation are more limited than in pretreatment because the main part of the fiber liberation work has already taken place in the first refiner stage[27]. Relatively severe conditions compared with pretreatment are therefore recommended for interstage treatment:

- Chemical charge 10% Na_2SO_3 on pulp
- Temperature 160°C
- Retention time 60 min.

A special feature of interstage treatment is that the wet web work to rupture is clearly increased as compared with pretreatment. Interstage sulfonation also clearly reduces the specific energy consumption needed for refining to a certain freeness level (cf. Fig. 15).

8.2 Reject sulfonation

In reject sulfonation, the chemical treatment is inserted ahead of the reject refiner (cf. Fig. 1). Because most of the long fibers are enriched in the reject fraction, this process makes it possible to direct the chemical treatment to the long fibers only, without affecting the finer fractions. Contrary to chemical pre- and interstage treatment, the reject sulfonation can be applied also in groundwood processes.

Reject sulfonation can clearly improve shive reduction in reject refining and can reduce the specific energy needed[31]. It can also clearly improve the strength potential of the long fibers but, because only part of the total pulp stream is treated, the effect on the properties of the final pulp is more limited.

8.3 Alkaline peroxide mechanical pulping (APMP)

An alternative to sulfonation of lignin for increased paper strength is carboxylation of the lignin with oxidants, such as alkaline peroxide. Similar to sulfonation, carboxylation results in lignin with carboxylate groups which are capable of participating in hydrogen bonding, which increase fiber bonding in papermaking. Alkaline peroxide treatment of wood chips softens the lignin, resulting in easier fiber separation and less fines generation during refining. The typical discoloration of the pulp by the alkaline treatment, as found in the "cold soda" process, is avoided by adding the peroxide to the caustic solution. An added benefit of the process is that alkaline peroxide in the refiners can potentially eliminate the need for separate postbleaching. The low efficiency of the peroxide can be improved by impregnating the wood chips with chelating agents as a first step, to remove or deactivate metal contaminants from the wood.

CHAPTER 8

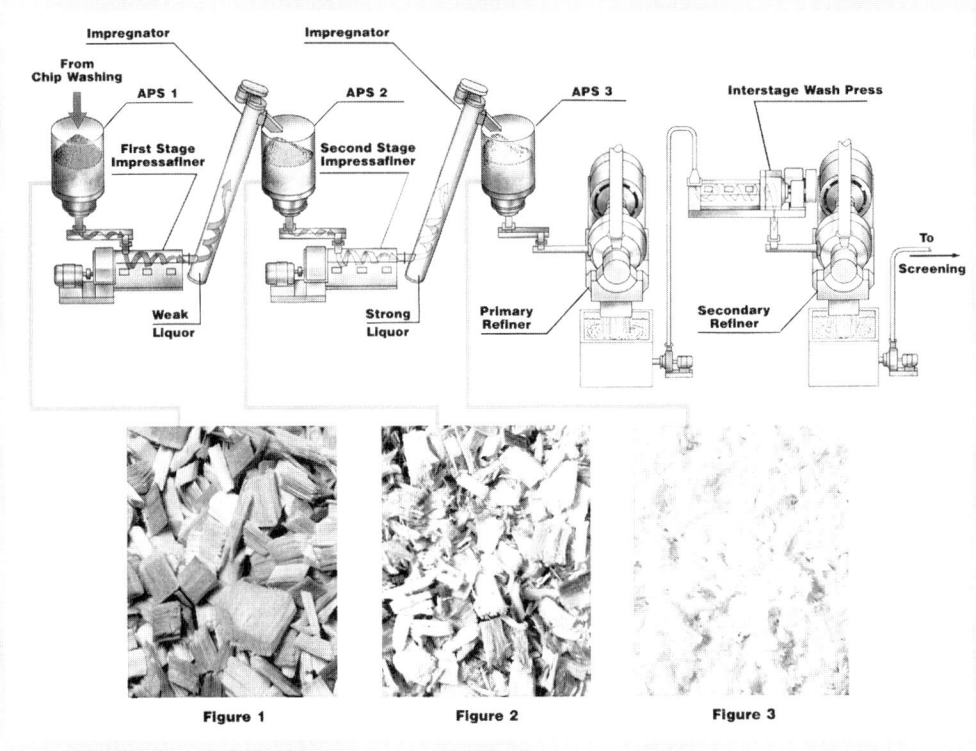

Figure 25. APMP Impregnation and refining system (Andritz).

Alkaline peroxide mechanical pulping (Fig. 25) has become a fast-growing alternative to sulfite chemicals in production of bleached chemimechanical pulp[32-34], and is well suited for softwoods, but will more significantly improve strength properties of low density hardwoods like aspen, poplar, cottonwood, and willow and improve its brightness. Table 6 compares aspen APMP to aspen BCTMP.

Chemical impregnation of the wood chips in the APMP

Table 6. A comparison of the two major processes for producing chemimechanical aspen pulp.

	BCTMP	APMP
Chemicals applied, %		
Sulfite	1.4	
Caustic	1.8 – 4.3	5.8
Peroxide	4.0	4.0
Energy consumption, MWh/t	1.72	1.22
Freeness, ml	77	77
Density, kg/m^3	555	558
Tear index, mNm^2/g	6.3	6.3
Tensile index, Nm/g	58	60
Brightness, %	82.8	83.5
Opacity, %	80	81.8
Light scattering coefficient, m^2/kg	39	43

Chemimechanical pulping

process is carried out by compression in a "Prex," an "Impressafiner," or similar equipment. The impregnation is carried out in two or three steps. The first stage liquor consists of a mixture of chelating agents and residual caustic and peroxide. The residual chemicals come from the inter stage washing of the pulp after the primary refiner. The impregnated chips are steamed for 10 minutes in an atmospheric bin. From here the chips are sent to the second impregnation stage, where the caustic and peroxide are applied. The caustic assures the strength development, while the peroxide improves the brightness of the pulp. The second stage impregnation press squeezes out some of the wood constituents that have been identified as inhibitors to bleaching reactions. The effluent from this stage contains 25 to 30% of the BOD and COD generated by the process. The impregnated chips is allowed to steep for up to 60 minutes in an atmospheric bin, at 60 to 80°C. This allows the pulping and bleaching reactions to proceed. The chemical doses applied are typically 1%–5% hydrogen peroxide and 3%–6% sodium hydroxide. In addition some DTPA, sodium silicate, and epsom salt are added for stabilizing peroxide.

Increasing the caustic charge will result in substantial improvements in bonding strength and density (Fig. 26). With aspen this can be similar to hardwood kraft. As the chemical charge increases, the generation of surface area diminishes, resulting in loss of scattering coefficient (Fig. 27). At the same time, refining energy is reduced for the same freeness.

The brightness of the pulp as it discharges from the secondary refiner will be dependent on the amount of peroxide applied in the impregnation stage. The efficiency of brightening reactions in the APMP process is similar to the response expected from a modern high consistency bleaching system. Brightness of 84 to 85% with aspen is economically achievable with 3.5 to 4.5% peroxide applied.

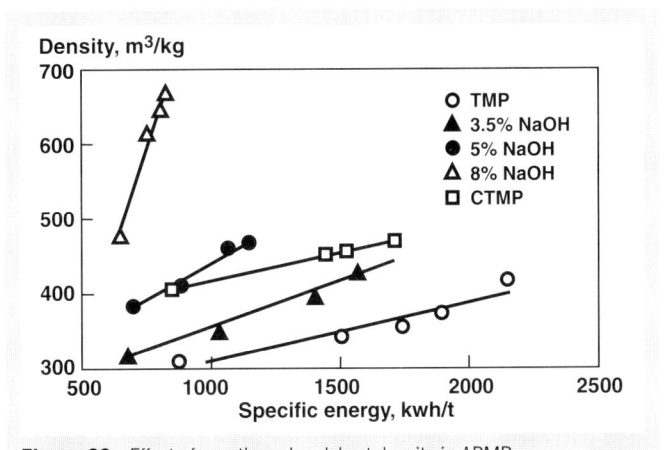

Figure 26. Effect of caustic on handsheet density in APMP.

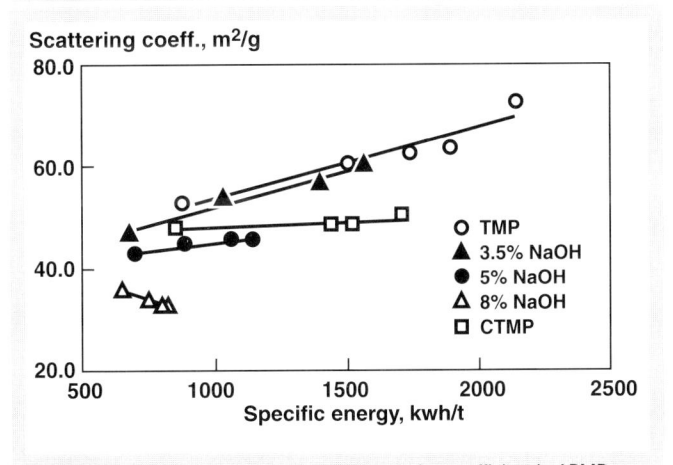

Figure 27. Effect of caustic on the light scattering coefficient in APMP.

247

CHAPTER 8

References

1. Mackie, D. M. and Taylor, J. S., Pulp Paper Can. 89(2):T64 (1988).

2. Richardson, J. D., Appita 49(1):27 (1996).

3. Lindholm, C. -A., Paperi ja Puu 61(3):141 (1979).

4. Beatson, R., Heitner, C., Atack, D., J. Pulp Paper Sci. 10(1):J12 (1984).

5. Adler, E., Lindgren, B. O., Saedén, U., Svensk Papperstid. 55(7):245 (1952).

6. Wennerås, S., "On the sulfonation of undissolved lignin in sprucewood and birchwood," thesis, NTH, Trondheim, 1962.

7. Janson, J. and Mannström, B., Pulp Paper Can. 82(4):T111 (1981).

8. Standard Method G 28, Total sulfur in pulp, paper and paperboard, Canadian Pulp and Paper Association, Technical Section (1970).

9. Katz, S., Beatson, R. P., Scallan A. M., Svensk Papperstid. 87(6):R48 (1984).

10. Engstrand, P., Hammar, L. -Å., Htun, M., "The kinetics of sulfonation reactions on Norwegian spruce," 1985 International Symposium on Wood and Pulping Chemistry Notes, CPPA, Montreal, p. 275.

11. Vikström, B. and Hammar, L. -Å., "Softening of spruce wood during sulfite pulping and its relevance for the character of high yield pulps," 1981 International Symposium on Wood and Pulping Chemistry Notes, SPCI, Stockholm, p. V:112.

12. Höglund, H., Sohlin, U., Tistad, G., Tappi 59(6):144 (1976).

13. Atack, D. and Heitner, C., Trans. of Technical Section CPPA 5(4):TR99 (1979).

14. Giertz, H. W., "Basic wood raw material properties and their significance in mechanical pulping," 1977 International Mechanical Pulping Conference Proceedings, The Finnish Paper Engineers´Association, Helsinki, p. 1:1.

15. Higgins, H. G., Puri, V., Garland, C., Appita 32(3):187 (1978).

16. Htun, M. and Salmén, L., Wochenblatt für Papierfabrikation 124(6):232 (1996).

17. Åkerlund, G. and Jackson, M., "CTMP – The pulp of the future," SPCI 1984 World Pulp and Paper Week Proceedings, SPCI, Stockholm, p. 42.

18. Hägglund, T. -Å. and Lindström, L.-Å., Tappi J. 68(10):82 (1985).

19. Kurra, S., Lindholm, C. -A., Virkola, N., "Effect of uneven impregnation in chemimechanical pulping," 1985 International Mechanical Pulping Conference Proceedings, SPCI, Stockholm, p. 80.

20. Ferritius, O. and Moldenius, S., "The effect of impregnation method on CTMP properties," 1985 International Mechanical Pulping Conference Proceedings, SPCI, Stockholm, p. 91.

21. Rahkila, P., "CMP/CTMP-nykytekniikka ja kemihierreprosessien ominaispiirteet," Insinöörijärjestöjen koulutuskeskus, Publication 75-86, p. III 1.

22. Collicutt, S. A., Frazier, W. C., Holmes, G. W., Joyce, P., Mackie, D. M., Torza, S., Tappi 64(6):57 (1981).

23. Jackson, M., Paper Tech. Ind. 26(6):258 (1985).

24. Levlin, J. -E., Sundholm, J., Paulapuro, H., "Kemihierteen paperitekninen potentiaali," Insinöörijärjestöjen koulutuskeskus, Publication 75-86, p. V1.

25. Atack., D., Heitner, C., Stationwala, M. I., Svensk Papperstid. 81(5):164 (1978).

26. Atack., D., Heitner, C., Karnis, A., Svensk Papperstid. 83(5):133 (1980).

27. Heitner, C., Atack, D., Karnis, A., Svensk Paperstid. 85(12):R78 (1982).

28. Sinkey, J. D., Appita 36(4):301 (1983).

29. Åkerlund, G. and Jackson, M., "Manufacture and end-use application of CTMP and CMP from softwoods and hardwoods," EUCEPA 1984 Chemical Processes in Pulp and Paper Technology Proceedeings, EUCEPA, Torremolinos, p. 171.

30. Webb, M., Paper Tech. Ind. 26(6):281 (1985).

31. Gummerus, M., "Studies on upgrading mechanical pulp through chemical modification of the long fiber fraction or screen rejects," Doctoral thesis, Helsinki University of Technology, Department of Forest Products Technology, Espoo 1987, 35 p. + app.

32. Bohn, W. and Sferrazza, M., "Alkaline peroxide mechanical pulping, a revolution in mechanical pulping," 1989 Mechanical Pulping Conference Proceedings, KCL, Helsinki, p. 184.

33. Heimburger, S., Quick, T., Sabourin, M., Tremblay, S., Shaw, G., TAPPI J. 79(8):139 (1996).

34. Sherbaniuk, R., Tappi J. 75(1):61 (1992).

35. Heitner, C., Argyropoulos, D. S., Miles, K. B., Karnis, A., Kerr, R. B., J. Pulp Paper Sci. 19(2):J58 (1993).

36. Sundholm, J., unpublished KCL results, 1997.

CHAPTER 9

Screening and cleaning

1	**Efficiency calculations**	**252**
2	**Pressure screening**	**255**
2.1	Mechanisms in slot and hole screening	256
	2.1.1 Flows in the screen	257
	2.1.2 Fiber mat	258
	2.1.3 Pressure difference over the screen plate	258
	2.1.4 Pulses	259
	2.1.5 Higher consistency screening	259
	2.1.6 Passing speed	260
2.2	Screening variables	260
	2.2.1 Screen plates and rotors	264
2.3	Screen types	268
	2.3.1 Tangential and axial screens	268
	2.3.2 Medium-consistency screens	269
	2.3.3 Multistage screens	269
3	**Screening Systems**	**270**
3.1	Main line screening of GW and PGW	271
3.2	Main line screening of TMP and CTMP	273
3.3	Screening of GW and PGW rejects	274
3.4	Screening of TMP and CTMP rejects	274
3.5	Future screening systems	274
4	**Control of Screening**	**275**
4.1	Control of a single screen	275
4.2	Control of parallel screens and screens in series	277
4.3	Control of two-stage screening	279
5	**Cleaning and sand removal**	**279**
5.1	Theory	279
5.2	Cleaner types	281
5.3	Cleaning systems	282
5.4	Sand removal	282
5.5	Control of cleaning	284
	References	286

CHAPTER 9

Jouko Hautala, Ismo Hourula, Tero Jussila, Markku Pitkänen

Screening and cleaning

Today the screening room is one of the key elements for top quality mechanical pulp and good runnability at the paper machine. Pulp and paper quality is highly dependent on the function of the screening room, and a good performance in the screening room is an absolute necessity when manufacturing value added paper grades. One single harmful shive in the accept pulp can cause runnability or quality problems at the paper machine.

The mechanical pulping screening room includes several subprocesses, most notably screening, cleaning, and reject handling. Each of them has its own specific task in pulp processing, such as separation and/or retreatment of various impurities like sand, bark, shives, and coarse fibers. The cleaner's primary task is to remove inorganic debris, mainly sand, from the system. Today's screens usually are slotted pressure screens, and their task is to separate wood originating impurities from fibers and to divide the feed pulp into different fractions, which will be further processed separately. The main functional difference between screening and cleaning is that screens reject debris on the basis of fiber dimensional properties (length and width), whereas cleaners mainly do it based on density and specific surface area[1]. Separated undefibrated wood components and undeveloped (unfibrillated) fibers will be refined in the reject handling until suitable for further processing. Screening and cleaning technologies are both subject to complex fluid mechanics.

The best configuration for the screening room depends on the desired accept pulp quality, on the raw material, and on the pulping process. For example, the use of cleaners is questionable if the feed pulp is sufficiently clean from sand. If there are cleaners in the process, they are coupled together in 2–6 stages and are usually situated after the fine screening or in the reject line. The fine screening with pressure screens is typically done in 2–3 stages, and the accepts from each stage are combined. The screens rejects are further treated in the reject refiner, after which the refined reject pulp is screened in a reject screen from which the accept is usually led forward and rejects are led back to the reject handling feed. With modern equipment, there is no need for cascade connections in the screening room; screening efficiency has improved remarkably during the last decade[2,3].

Quality improvements of mechanical pulp can be done mainly in two ways: Either by changing the process conditions in grinding or refining or by improving the undeveloped fibers separation efficiency in screening and cleaning. An improved efficiency in screening means that the screening will be more selective and the reject refining can be directed only at those components, which need more treatment. The efficiency of screening and cleaning depends on the design of the screening room.

CHAPTER 9

The screening room has become a more important part of the pulping processes, and its importance tends to grow. The biggest challenges for the screening room equipment and design will in the future be higher screening consistencies for cost reduction reasons and higher fractionation efficiency for pulp quality reasons. For more information on screening, see the volume "Chemical Pulping" in this textbook series.

1 Efficiency calculations

Screening efficiency calculations utilize different analysis and testing methods. The most common methods which are used to describe the properties of the mechanical pulp and the efficiency of each process stages are consistency, drainability, shive content, and fiber length distribution (e.g., Bauer McNett fraction).

Wood originating rejectable material (shives and undeveloped fibers) can be divided into four classes: shives, minishives, chops, and coarse fibers. A shive is often defined as a fiber bundle that is at least 3 mm long and 0.10–0.15 mm in width. A minishive is often under 3 mm long and over 0.08 mm in width. Chops are cubical debris with dimensions of 0.25–1 mm^4. Because of undeveloped testing methods, coarse fibers are often defined as the BMcN + 14 fraction. Table 1 gives typical shive testing analysis. Figure1 presents relative dimensions between fibers and screen plates.

Table 1. Typical shive testing methods.

Shive Analysis	Operational Method
Somerville shive	Screen, normal slot size #0.15 mm
Pulmac shive	Screen, normal slot size #0.10 mm
PFI minishive	Screen, normal slot size #0.08 mm
PQM	Optical

A common efficiency calculation, which describes the efficiency of screening, fractionation, or cleaning, is the shive removal efficiency (SRE). It is based on shive content of different flows and reject rate by weight. It illustrates the proportion of how much debris is removed from the feed flow. The SRE formula can be used for calculation of every contaminant removal efficiency. On the other hand, the cleanliness efficiency represents the improvement in cleanliness of the processed pulp, thus comparing the cleanliness of feed and accept pulp. The long fiber yield tells the effectiveness of a screen to accept long fibers. The long fiber fraction is usually defined as the BMcN +28 fraction, excluding or including the coarse fiber (BMcN +14) fraction. The thickening ratio tells the ratio between reject and feed consistencies. This is due to water having higher probability of acceptance than fibers; therefore, reject consistency will increase. Tables 2 and 3 summarize the most important efficiency calculations, symbols, and formulas which are used to describe the screening efficiency.

Screening and cleaning

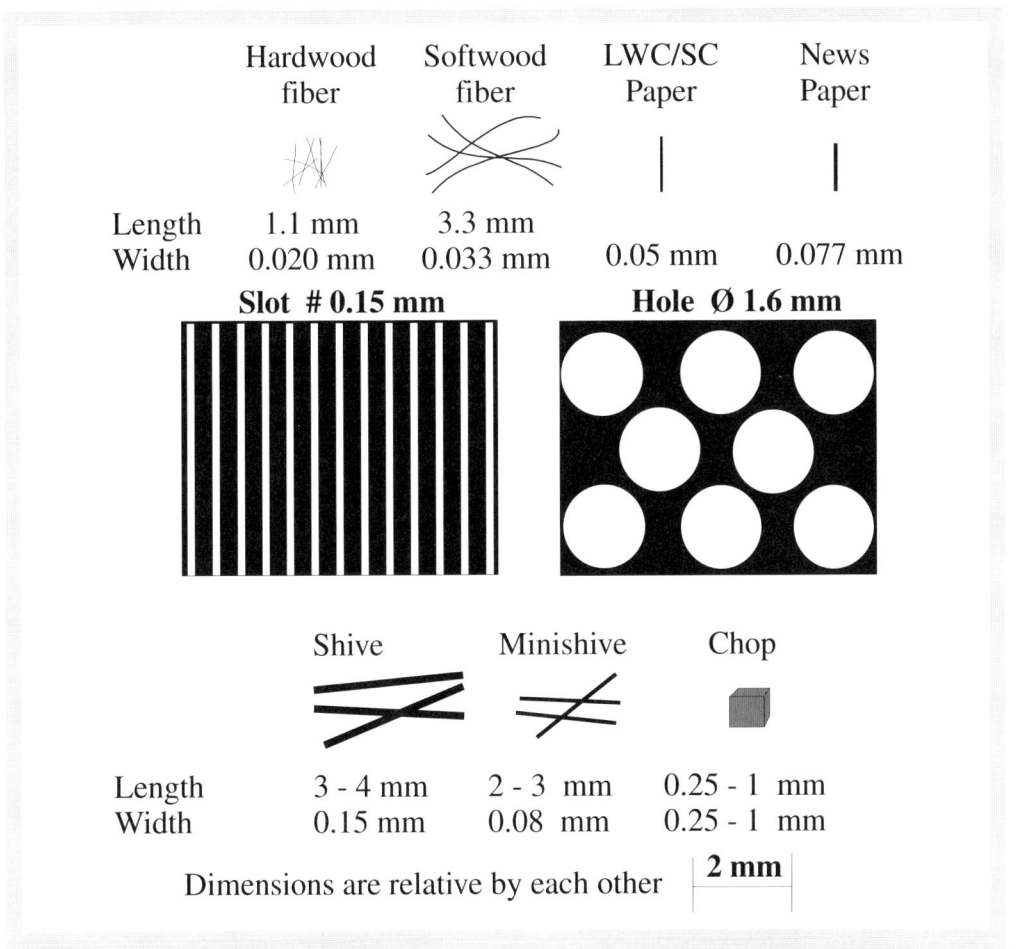

Figure 1. Dimensional relations in screening.

	Hardwood fiber	Softwood fiber	LWC/SC Paper	News Paper
Length	1.1 mm	3.3 mm		
Width	0.020 mm	0.033 mm	0.05 mm	0.077 mm

Slot # 0.15 mm   Hole Ø 1.6 mm

	Shive	Minishive	Chop
Length	3 - 4 mm	2 - 3 mm	0.25 - 1 mm
Width	0.15 mm	0.08 mm	0.25 - 1 mm

Dimensions are relative by each other   2 mm

Table 2. Screening room symbols.

Quantity	Symbol	Unit
Feed	F	
Accept	A	
Reject	R	
Volume flow	V	l/s
Consistency	c	%
Shive content (Pulmac, Somerville, PFI)	S	%
Long fiber content (General fraction content)	L	%
Coarse fiber content	C	%

CHAPTER 9

Table 3. Screening room efficiency and property calculations.

Quantity	Symbol	Unit	Formula
Mass flow	m	l/s	$m_i = V_i \times c_i \qquad i = F, A, R$
Reject rate by volume (Volumetric reject flow)	RR_V	%	$RR_V = 100 \times \dfrac{V_R}{V_F}$
Reject rate by weight (Mass reject rate, reject ratio)	RR_m	%	$RR_m = 100 \times \dfrac{m_R}{m_F}$
Shive removal efficiency (General removal efficiency)	SRE	%	$SRE = \left[1 - \left(1 - \dfrac{RR_m}{100}\right) \times \dfrac{S_A}{S_F}\right] \times 100$
Coarse fiber reduction efficiency	CRE	%	$CRE = \left[1 - \left(1 - \dfrac{RR_m}{100}\right) \times \dfrac{C_A}{C_F}\right] \times 100$
Long fiber yield (General fraction yield)	LY	%	$LY = \left[\left(1 - \dfrac{RR_m}{100}\right) \times \dfrac{L_A}{L_F}\right] \times 100$
Drainability drop	CSF	%	$\Delta CSF = \left[\dfrac{CSF_F - CSF_A}{CSF_F}\right] \times 100$
Thickening ratio	T_R		$T_R = \dfrac{c_R}{c_F}$
Shive removal index	Q		$Q = \dfrac{m_F \times (S_F - S_A)}{m_F S_F - m_A S_A}$
Cleanliness ratio (General cleanliness ratio)	E_C		$E_c = \dfrac{S_F - S_A}{S_F}$
Long fiber loss index	T		$T = \dfrac{m_A L_A}{m_F L_F - m_A L_A}$
Mass balance (Shive balance)			$c_F = \left(1 - \dfrac{RR_m}{100}\right) \times c_A + \dfrac{RR_m}{100} \times c_R$
Drainage balance (when $CSF_i < 300$ ml)			$\ln CSF_F = \left(1 - \dfrac{RR_m}{100}\right) \times \ln CSF_A + \dfrac{RR_m}{100} \times \ln CSF_R$

Screening and cleaning

With the very low shive contents that we have today, there are often problems with analytical accuracy because it is possible to reach shive levels so low that they cannot easily be confidently measured. It should not be forgotten that, in addition to this, there are also deviations in sampling, consistency measurements, and in the process itself. It is thus important to ensure that screening analysis results are correct, and it is recommended that shive and mass balances are calculated. Both balances can be calculated with the formulas that are presented in Table 3. Moreover, in such cases the drainage balance can also be checked and calculated[5].

The continuous development in fractionation will increase the need of better definitions for fractionating efficiency. One way to do this is based on fiber length analysis, in which each fiber length distribution percentage of the accepts is divided by the corresponding feed distribution. It illustrates how much of each fiber length is in the accept pulp. It is also possible to define fractionation efficiency by fibers width or flexibility.

2 Pressure screening

The pressure screen is the most common type of screening and fractionation equipment in mechanical pulp processing. Only in atmospheric groundwood lines is pulp usually prescreened with bull screens (vibrating flat screens) because of big slivers. A typical pressure screen, as the one in Fig. 2, has a cylindrical screen basket through which the accepted pulp fibers pass while the remaining particles are rejected and removed under the screen basket.

Current screening technology uses wedge wire slot screen baskets and new rotors. In the 1980s, hole screen baskets were used. When compared, a significant increase in pulp cleanliness was observed when screening was done over machined slot screen baskets. The introduction of wedge-shaped wire slot screen baskets and new rotors additionally improved both accept pulp cleanliness and throughput. This has enabled all the fine screening stage accepts to be combined and led forward. Cascade connections in the screening stages have been reduced, which means fewer and smaller screens, fewer screening stages, better runnability, and better pulp quality.

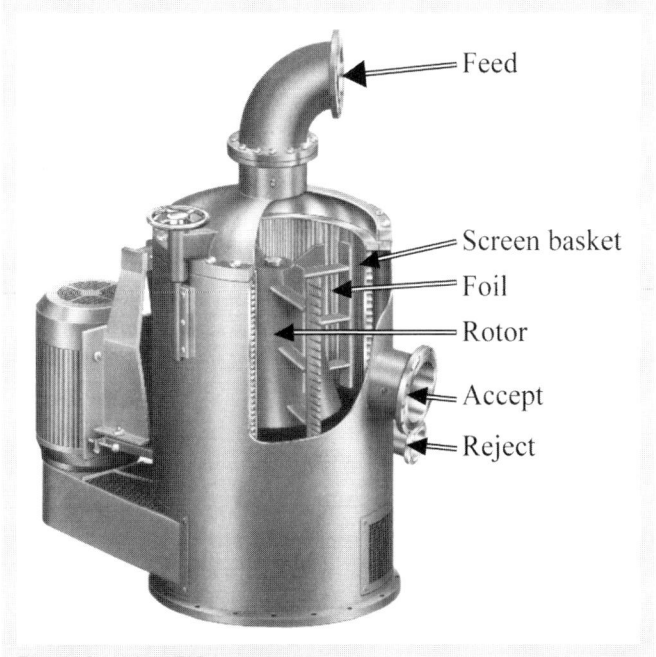

Figure 2. Valmet TAP pressure screen.

Also the energy consumption has decreased. The energy consumption of a pressure screen unit is dependent on the feed rate, the consistency, and – to a very high degree – the rotor speed. The specific energy consumption varies today from 10 to 20 kWh/t for a single screen. In the mid-1980s, the value was correspondingly 15–35 kWh/t[2]. We have also seen the introduction of the medium consistency screening technology, which operates at a consistency rate of 3%–5%. Higher feed consistency reduces pulp and dilution water volumes, meaning less pumping and thickening, and smaller storage chests.

Pressure screens accept and reject qualities and pulp accepting probability are mainly controlled by flow conditions inside the screen basket. Flow conditions can be changed by operational and design parameters[6]. Main operating parameters are consistency, rotation speed, and reject rate; the main design parameters are properties of the screen, screen basket, the design of the rotor, and foils. These parameters have a significant influence on the flow conditions and accordingly on the performance of the pressure screen. Today the narrowest industrially used slots have a 0.06 mm width, but usually in mechanical pulping the slot width range is 0.10–0.30 mm. The height of the profile in the screen basket are varied typically up to 1.5 mm. Nowadays hole screen baskets or plates are used only in coarse screening stages when it is necessary to protect the slot screen baskets, for example, from big slivers.

Today pressure screens are increasingly used also for pulp fractionation, which divides a fiber flow into two different fractions, usually one containing short and the other long fibers. These pulp fractions can be further treated in different ways or the fractions can be used for various purposes, such as in layers of composite board or in entire separate products. New pulp fractionation applications are continuously being developed, and these will play a more significant role in the pulp and paper industry in the near future[7]. In the future, the development of the pressure screening will concentrate on fiber fractionation and medium consistency screening; utilizing advanced measurements, control, and simulation techniques; and naturally on pulp quality improvements.

2.1 Mechanisms in slot and hole screening

The mechanisms in slot and hole screening are rather complicated because they depend on fluid flow dynamics, acceptance probability, fiber mat formation, fiber orientation, fiber properties, and flow condition and suspension flows in the screen basket. Most pulp screening takes place by probability screening[8]. The fibers and shives are of similar form, both being long and slender, but their flexibility and fibrillation differ from each other. The screen passage dimensions are generally equal to or larger than the diameters of fibers and shives, but smaller than their lengths. The shives are able to pass through easily, but their movement is inhibited by other factors such as their orientation to the screen plate and their interaction with the fiber mat[9, 10]. The very large shives are removed by barrier screening, where the shives are simply too big to squeeze through the passages of the basket.

2.1.1 Flows in the screen

The main flows inside the screen are the tangential flow induced by the rotor, the axial flow induced by reject removal, and the radial movement toward the screen basket caused by removal of the accept pulp. In addition to these primary flows, there is also movement between the solid matter and the water caused by centrifugal force. The centrifugal force and the accept removals are forces that drive the fibers toward the screen plate. The opposing force that moves particles away from the screen plate is the mixing induced by the rotor and foils[11, 12]. The pulp moves through the screen plate into the accept chamber only when the pressure inside the surface of the screen basket is higher than that in the accept chamber. Figure 3 presents flow directions.

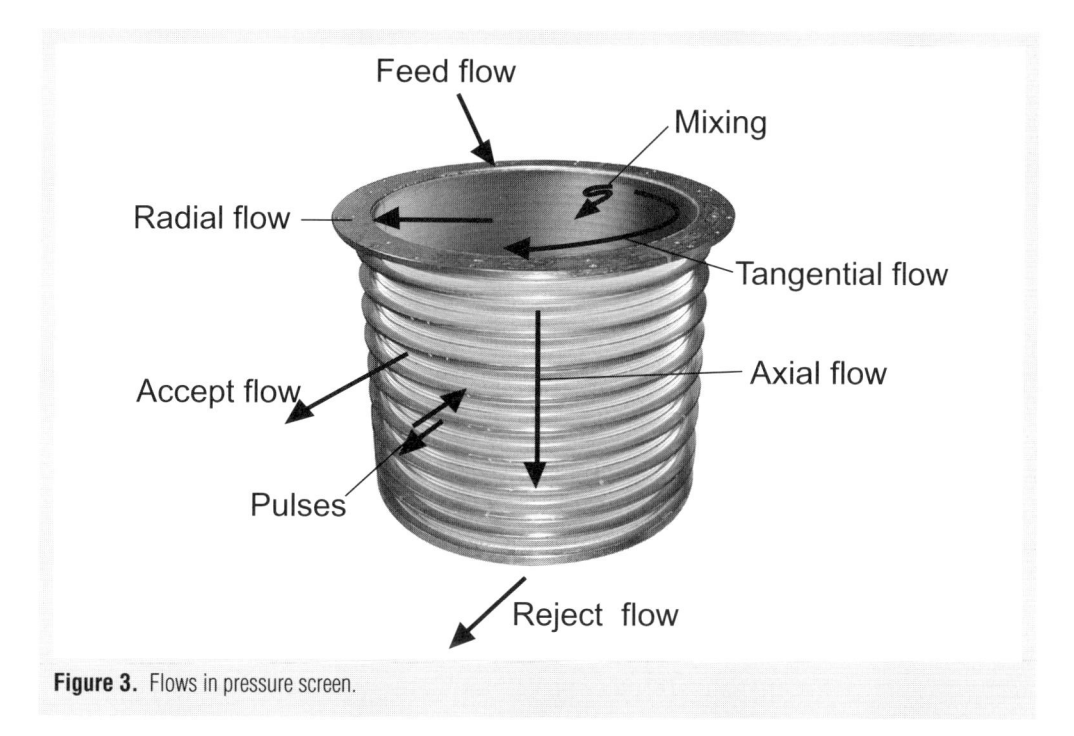

Figure 3. Flows in pressure screen.

Flow conditions inside the screen basket are mainly dominated by the tangential flow field induced by the rotor. When a cylindrical rotor body is used, the fiber suspension flows along a narrow channel between the screen basket and the rotor, where the flow has been shown to relate to a mathematical plug flow model[13]. However, in the case of the conical rotor body types, the mixing conditions inside the screen basket are not related to this model, as here the rotor creates mainly a tangential flow pattern inside the screen basket.

The difference between rotor and suspension flow velocity at the screen basket surface is one key parameter in achieving a good screening result. A higher velocity difference means higher turbulence at the screen basket surface, which means that longer fibers can turn and pass through the screen. Thus, the flow velocity ratio between rotor and suspension has the same effect as the screen basket profile height has. Velocity ratio can be controlled by changing the ratio of rotor baffle blades and foil diameter. When the baffle blades diameter is much smaller than the foils diameter, the suspension velocity is much slower than that of the foils and turbulence at the screen basket surface is high, which means that the shives acceptance probability is higher. When the baffle blades diameter and the foils diameter are almost the same, the suspension velocity is close to the rotor rotation velocity. The time for a fiber to turn into a slot becomes shorter in the case of increasing pulp rotation velocity, and this hinders penetration of long fibers and shives through a screen plate.

In slot screening, acceptance is greatly affected by pulsation and velocity of the rotors foils, as well as by screen basket properties. Screen basket properties are slot size, profile height, wire width, and design. Today the rotors and foils are designed just to keep the screening surface open. The screen basket profile height together with rotor pulsations causes microturbulence that fluidizes the fiber suspension at the screen surface. A higher profile means greater turbulence, which means that longer fibers can turn and pass through the screen. On the other hand, a high profile can be more or less compensated by rotation speed.

2.1.2 Fiber mat

Each pulp type has different fiber network strengths at the same pulp consistency, depending notably on average fiber length, fiber length distribution, and fluid apparent viscosity. The fiber mat has a higher consistency area near the screen plate that tends to form because the screen basket acts as a barrier that prevents over-sized particles from contaminating the accept stream. In addition to individual over-sized particles, fiber flocs can also form a mat if the flow conditions at the screen surface allow the flocs to remain coherent. Over-sized particles and flocs are transferred onto the screen plate by the influence of the flow field and the centrifugal effect, and the resulting mat alters the properties of the screen surface and tends to plug the screen. The flow through the screen plate leads to formation of the fiber mat on the screen surface, and the foils disrupt this mat. Nevertheless, the foils hinder the formation of a fiber mat by mixing the impurities back upstream or onward to the reject removal. Moreover the foils suction pulse sucks fines and water back inside the screen simultaneously washing the screen plate surface.

2.1.3 Pressure difference over the screen plate

The pressure difference over the screen plate is the pressure difference between the accept chamber and the inside surface of the screen plate. It is not the same thing as the pressure difference over the screen itself, which can be measured as a difference between accept and feed line pressures or accept and reject line pressures. Depending on the inlet, this difference can be either positive or negative. Over the screen plate, there is always a pressure drop. The tendency of screen plugging will increase if the pressure difference over the screen plate increases too much.

A higher arial load on the screen plate surface means a greater pressure difference over the screen basket. The pressure difference is also dependent on the structural properties of the screen plate, the apparent viscosity of the fluid, and probably also on the manner of formation of the fiber mat.

2.1.4 Pulses

The rotor and the foils induce suction pulses and microturbulence, which disrupts the fiber mat that tends to form on the screen surface[14]. The rotation frequency accordingly prescribes the cleanliness grade of the screen surface and thereby affects the screening results. The throughput rate defines the pressure difference over the screen basket and retention time, during which pulp is affected by rotor forces. Rotor frequency and throughput rate therefore determine the rotation velocity of pulp. The time for a fiber to turn into a slot becomes shorter in the case of increasing rotational pulp velocity, and this hinders penetration of long fibers and shives through a screen plate. It has been shown that, with different rotor speed and reject rate combinations, it is possible to achieve different ratios of the P14 and P28R14 fiber fractions in the accept pulp[23].

2.1.5 Higher consistency screening

Pulp screens have a low- and high-consistency limitation as described in Fig. 4. The low-consistency limitation is the increased hydraulic load, as a result of dilution down to lower consistencies. The high consistency limitation is the strength of the fiber network, which increases rapidly at higher consistencies. In order to make the screen work, the pulp in the boundary layer on the screen plate surface has to be fluidized, or deflocculated, for the separation process to take place. In a fluidized state, the individual particles are free to move independently of each other[15].

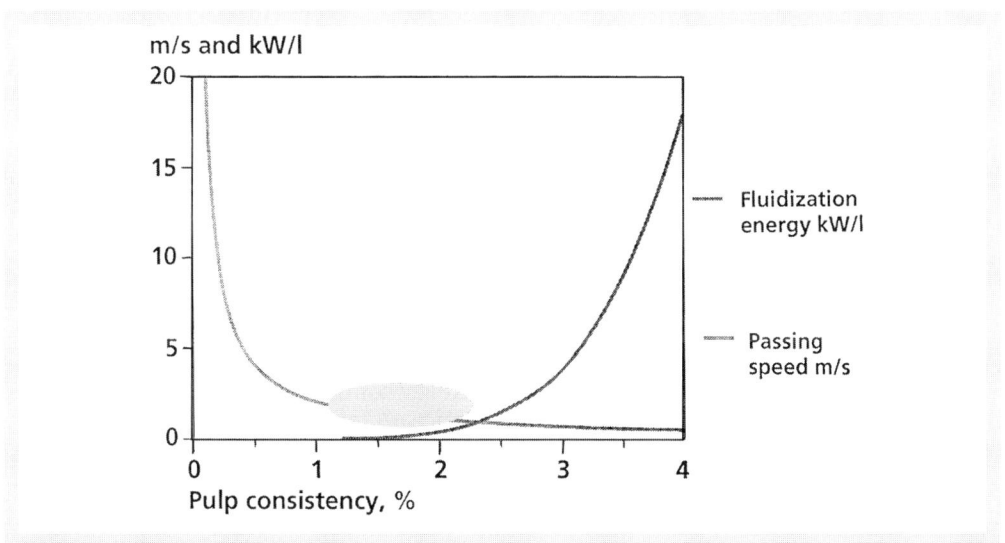

Figure 4. The consistency limits for screening of pulp.

CHAPTER 9

Thickening between feed and reject pulp occurs normally in the screening. With high feed consistency, thickening can be a problem; the reject consistency can be too high to allow reject removal from the screen. One way to prevent thickening is to use wider foils, which suck the filtrate from the accept side of the screen basket by a longer suction pulse. The amount of filtrate recovered should ideally be the same amount as was lost due to thickening.

2.1.6 Passing speed

Typically the slot passing speed (V_S) is calculated from the accept flow rate divided by the screen open surface as follows:

$$V_S = \frac{F_A}{A_S} \tag{1}$$

where F_A is accept flow
         A_S screen open area and is calculated for wedge wire slot baskets as follows:

$$A_S = \frac{S}{S+W} \cdot A_B \tag{2}$$

where S is slot width
         W wire width
         A_B screen basket area.

This calculation will give the average flow speed through the slot, but it does not tell anything about the speed variations. When the rotor suction pulses are very strong, a remarkable amount of suspension is sucked back inside the screen basket, and that also has to pass the screen. Therefore the real passing speed is a function of pulsation. However, the average passing speed is one design parameter that is connected to the screen capacity.

2.2 Screening variables

Pressure screening characteristics such as throughput capacity and pulp quality are defined by the operating and design parameters. Operating parameters such as reject rate and feed consistency are factors that can be varied in order to optimize the screening process. Changes in design parameters such as the screen basket or rotor type are needed if a considerable upgrade in product quality or in throughput rate is required. The reasons for initiating optimization can be problems in further papermaking process stages, such as breaks at the paper machine or coater. Below is one way to classify screen functional parameters:

Design parameters

    Screen basket properties

        1. Slot or hole basket

        2. Size of the slot (or hole)

        3. Wire size/width and design

        4. Basket's profile height

        5. Effective screening area

    Rotor properties

        1. Rotor body design

        2. Foils design

        3. Foils angle

    Screen itself and its design and properties

Operating parameters

    Screen properties

        1. Reject rates by weight and volume

        2. Rotation speed of the rotor

    Feed properties

        1. Feed consistency

        2. Feed flow

        3. Temperature

Functional optimization of a pressure screen requires a good knowledge of the screening process. The behavior of each parameter must be known in general so that its influence can be tested rationally without a high risk of unwanted incidents. An understanding of the flow conditions in the screen basket also provides a good basis for understanding the effect of each operating parameter on the screen performance[12, 16–19].

CHAPTER 9

The reject rate affects the screening efficiency, but in a different way with slot and hole baskets. Figure 5 presents screening efficiency as a function of reject rate. Figure 6 shows the freeness drop over a single slot screen as a function of reject rate. Increasing rotor rotation speed causes a larger difference between pulp and rotor velocity and therefore more turbulence. Figure 7 presents this influence and also illustrates how much accept properties can be changed with rotor design. Figure 8 shows the effects of slot size and optional screening parameters on SRE. Figure 9 illustrates the differences in SRE, LY, and cleanliness with slots and holes and with different profile heights. It can also be seen that slot size and profile height influence screening efficiency in different ways. The influence of screening parameters are more or less case sensitive, and screening efficiency must therefore always be optimized separately.

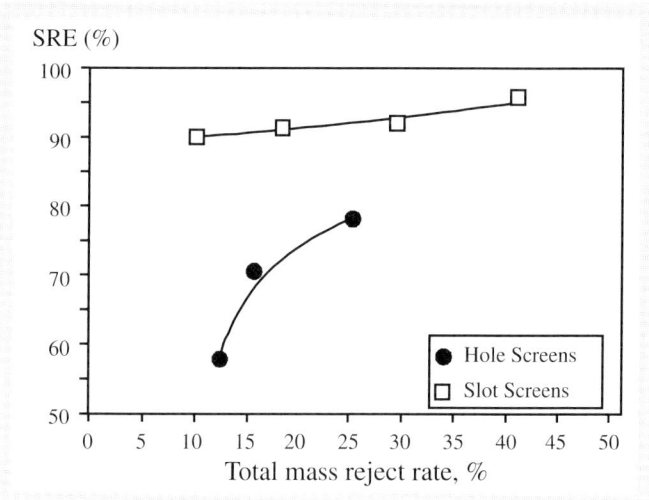

Figure 5. Two-stage screening shive removal efficiency as a function of total reject rate with hole and slot screen baskets. GW pulp[20].

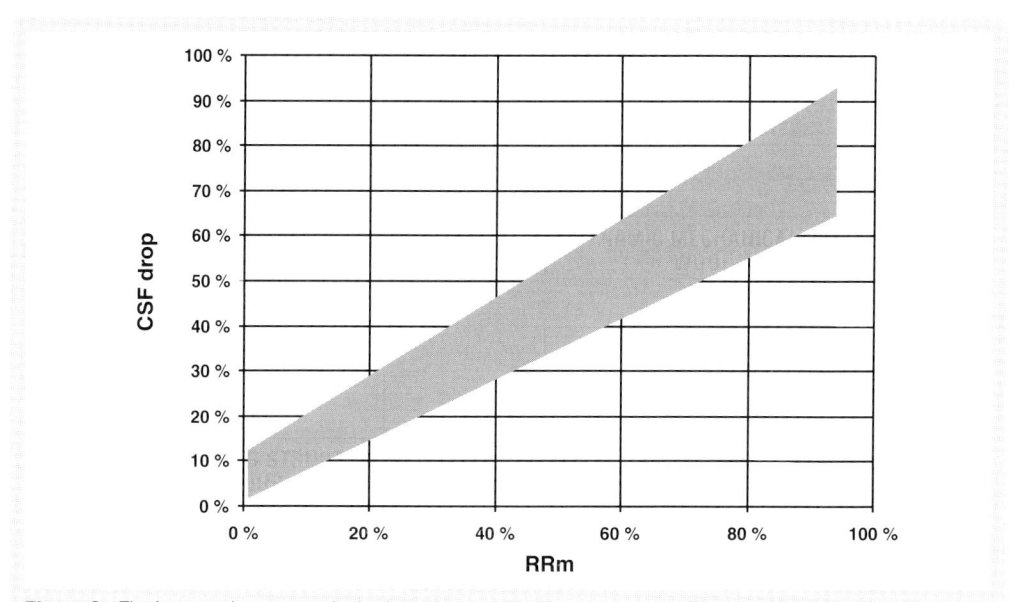

Figure 6. The freeness drop over a single slot screen as a function of reject rate. Typical values for a 100–150 ml freeness mechanical pulp.

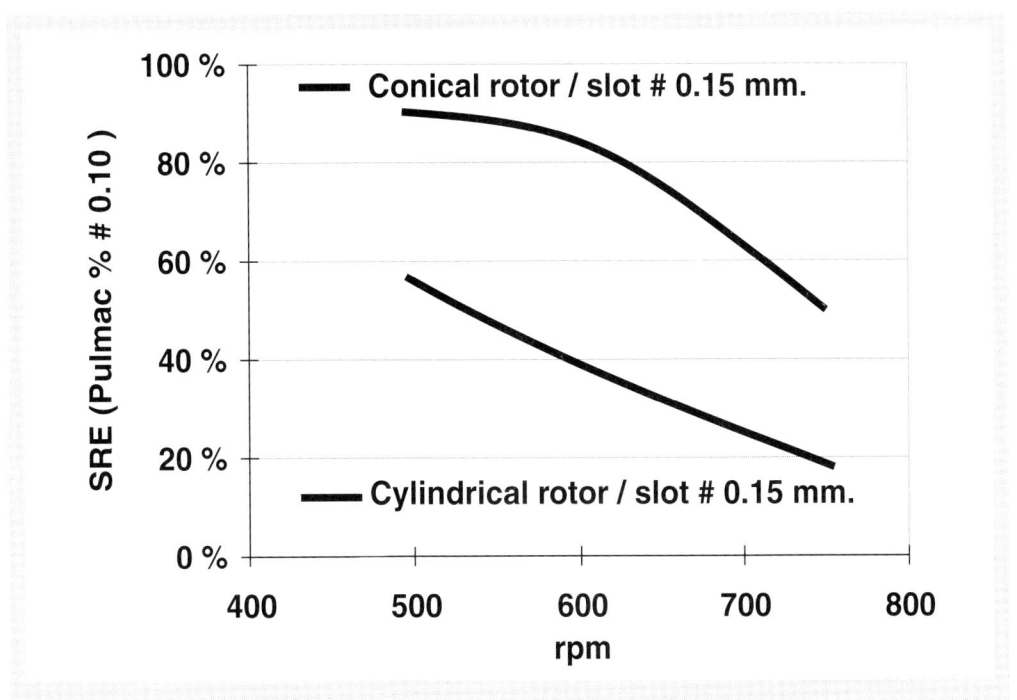

Figure 7. Rotation speed and rotor influence on SRE. PGW pulp.

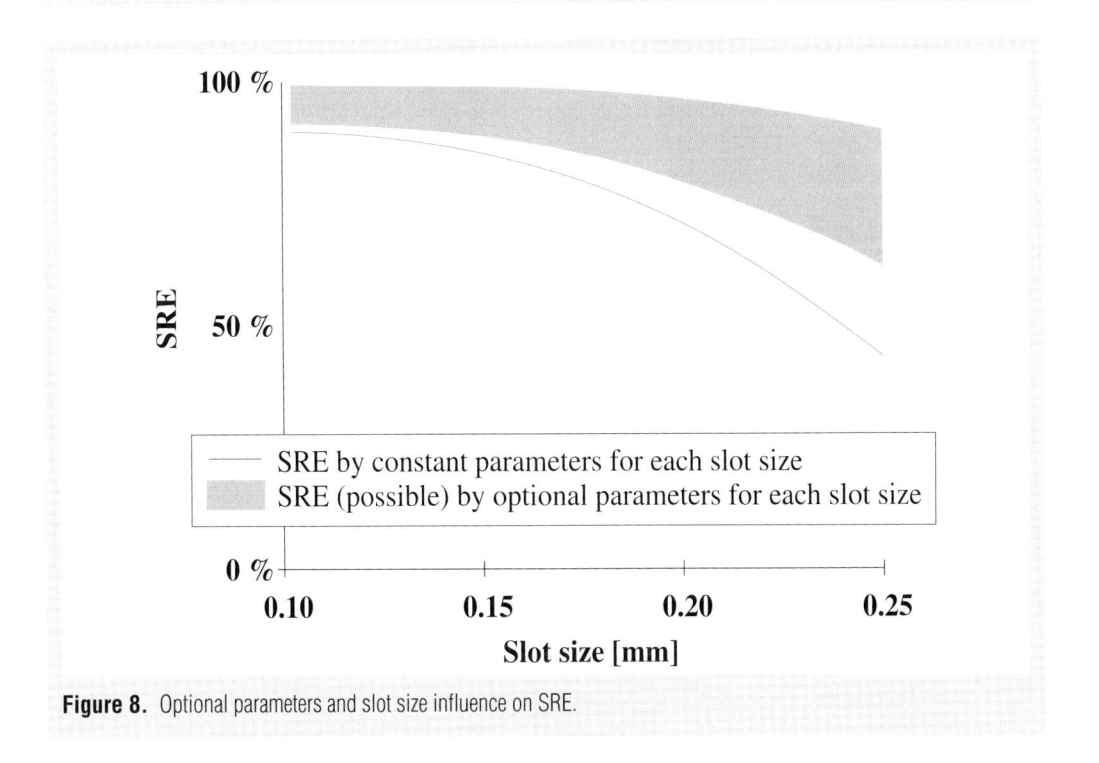

Figure 8. Optional parameters and slot size influence on SRE.

CHAPTER 9

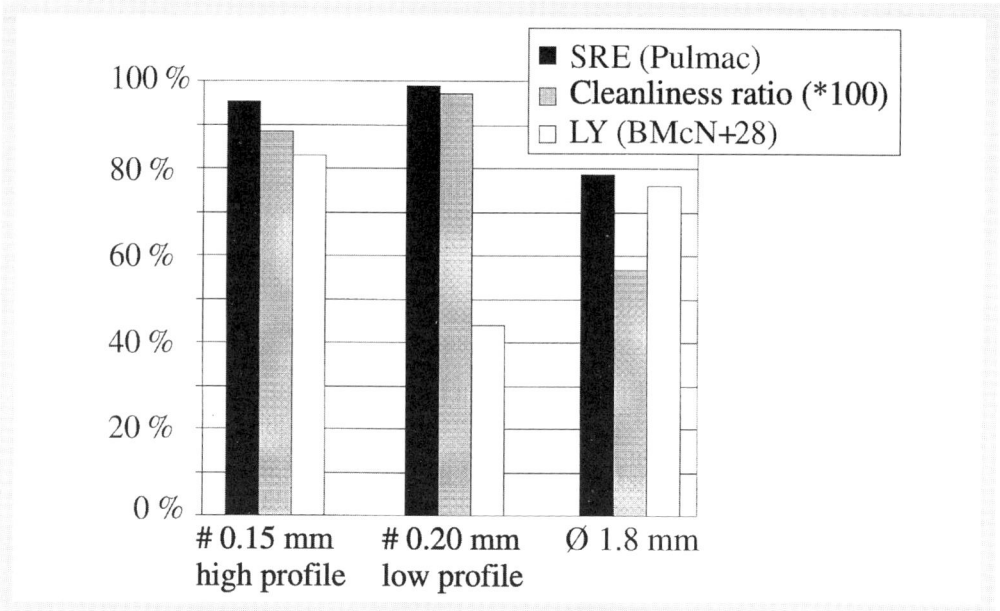

Figure 9. A comparison of SRE, LY, and cleanliness using different profile heights and slot and hole baskets. TMP pulp[21].

2.2.1 Screen plates and rotors

Traditionally, pulp was screened using screen plates with plain surfaces, and specifications were limited to the size of holes and slots and open area of the screen plate. During the early 1980s, contour screen plates were developed in which the surface of the screen plate facing the feed flow was milled to provide various configurations of ridges and grooves. With machined slot baskets, pulp cleanliness was improved remarkably as compared to hole baskets. However, with the continuous slot wedge wire screen baskets and new rotor concepts, accept pulp cleanliness has additionally improved. Considering the development of narrow slot screening, the success of the wedge wire screen baskets led to the development of new rotor concepts. This is due to the fact that the rotor and screen plate work in combination and therefore need to be optimized together.

Screen plates

There are numerous different screen plates. A rough division can be made between hole and slot screen baskets. Hole baskets can be drilled or perforated and contoured or plain. Slot baskets can be machined, contoured, laser machined, or wedge wire screen baskets.

Hole screening technology has largely given way to slotted screen baskets. Today hole screen baskets are typically used only in the coarse screening. Figure 10 and 11 show pictures of hole and slot baskets, and Fig. 12 presents some slot basket parameters.

Screen Baskets for Pressure Screens

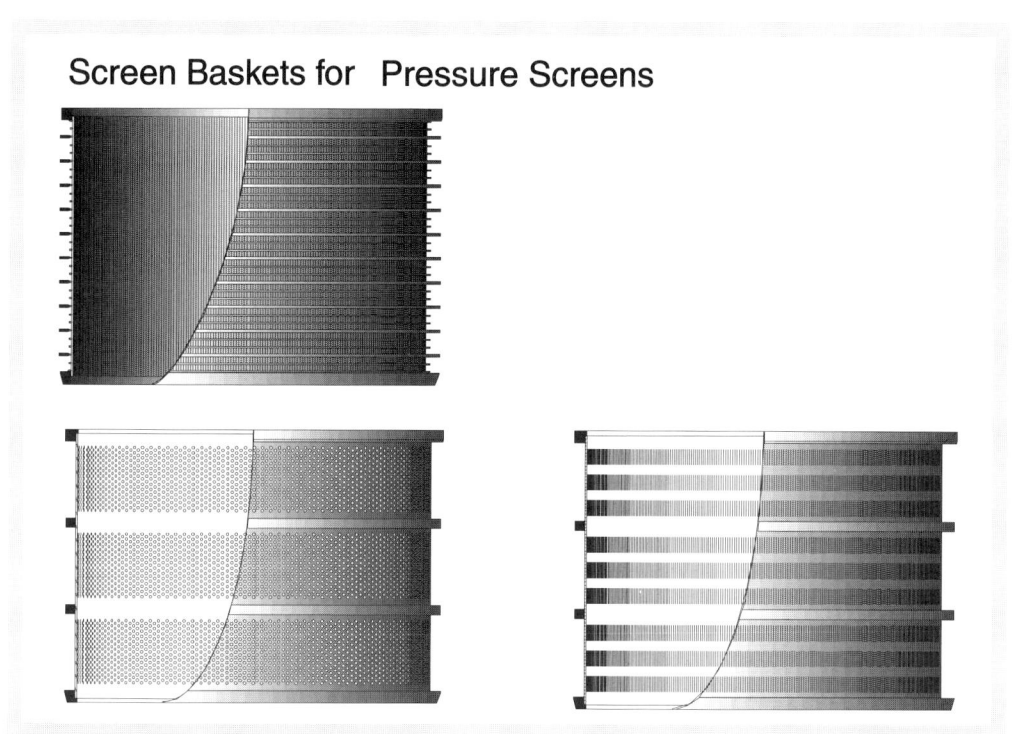

Figure 10. Different hole and screen baskets (Valmet): (a) continuous wedge wire basket, (b) hole basket, and (c) machined slot basket.

Figure 11. Continuous wedge wire basket (Ahlström) and machined multiprofile basket (CAE ScreenPlates).

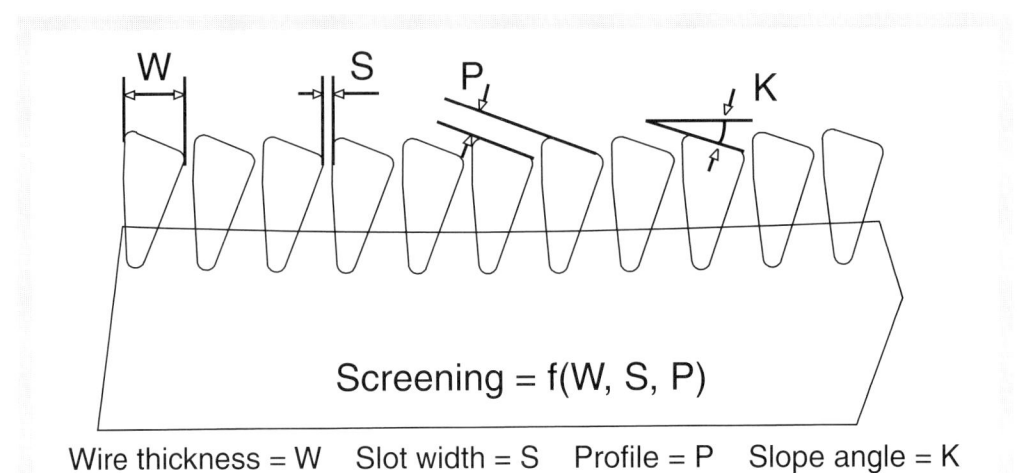

Figure 12. Wedge wire screen basket parameters. The parameters should be selected according to type of pulp and desired characteristics of end product.

Wire thickness = W   Slot width = S   Profile = P   Slope angle = K

Screening = f(W, S, P)

The basket's profile – measured from the highest point of one wire face to the bottom edge of the adjacent wire as shown in Fig. 12 – can be adjusted in manufacturing to give the degree of screening or fractionation required. A high profile generates more turbulence which means that more long or coarse fibers pass through. A flatter profile limits the turbulence so that long, coarse fibers have fewer chances to pass through the slots.

Screen rotors and foils

Despite the differences in design, all rotors serve two main functions. They accelerate and keep the pulp suspension in a tangential velocity, and they create suction pulses by foils. The tangential velocity, together with the axial flow, establishes the necessary flow field adjacent to the screen plate and induces microturbulence at the screen surface. The suction pulses prevent effective plugging of the screen.

Rotors can be classified as conical and cylindrical type rotors, as presented in Fig. 13. The shape of the rotor body affects the screening result, a conical rotor being more advantageous because it ensures a uniform flow in the axial direction of the pressure screen. With the conical rotor, the purpose is to create an even pressure toward the screening surface.

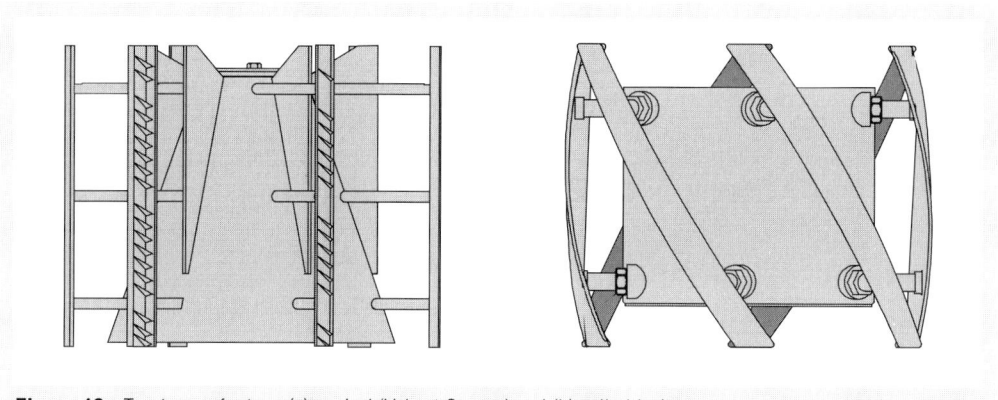

Figure 13. Two types of rotors: (a) conical (Valmet C-rotor) and (b) cylindrical rotor.

Just as it used to be important to design the foils to give a high-pressure pulsation, today it is important to keep pulsation low. Earlier, a high-pressure pulse made both fines and coarse particles, such as coarse fibers and shives, to flow through the slots to the accept side. Additionally an inverted pressure pulse sucked the fines back to the inside of the basket and to the rejects outlet. With rotor foil design, the pulsation on the screening surface has been diminished to 40% of that with former foil designs. This is just high enough to keep the screening surface open. Figure 14 presents a comparison of two pulsation curves.

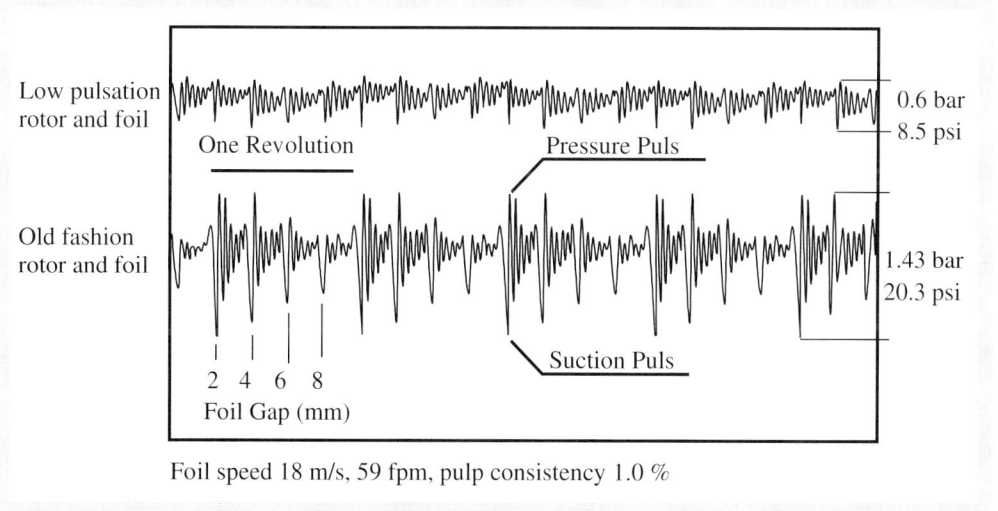

Foil speed 18 m/s, 59 fpm, pulp consistency 1.0 %

Figure 14. A comparison of the pulsations from two different rotors.

CHAPTER 9

2.3 Screen types

Pressure screens can be divided into different classes by their feed direction, operational principles, operation consistency, and combined function. The feed direction can be either tangential or axial. The flow operational principle can be centrifugal, centripetal, or a combination of these two. The rotation operational principle can either use a rotor or a screen basket that rotates. Today the main operational function is centrifugal with rotor. Typical screening consistency is 1%–1.5%, while medium consistency screens can operate at 3%–5% consistency. Multistage screens, which include several screening stages in a single screen, were presented in late 1990s.

2.3.1 Tangential and axial screens

Figure 15 illustrates the two ways of feeding a pressure screen. One can note that the screening areas are the same, but the screen sizes are different. Figure 16 illustrates another tangential application.

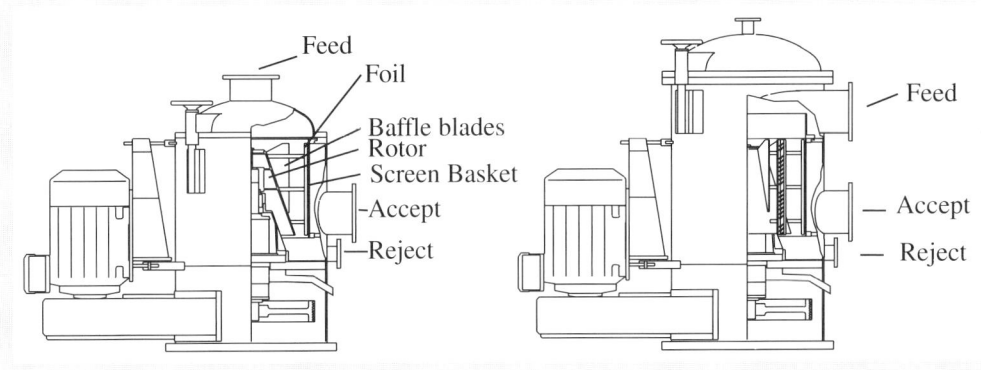

Figure 15. Screens with (a) axial feed (Valmet TAP screen) and with (b) tangential feed (Valmet TAS screen).

The feed flow direction affects the pressure difference over the screen. Axial feed produces a positive pressure difference over the screen while, with tangential feed, the pressure difference over the screen is negative. This means that, with axial feed, it is possible to reduce demands of process pumping requirements[19, 22].

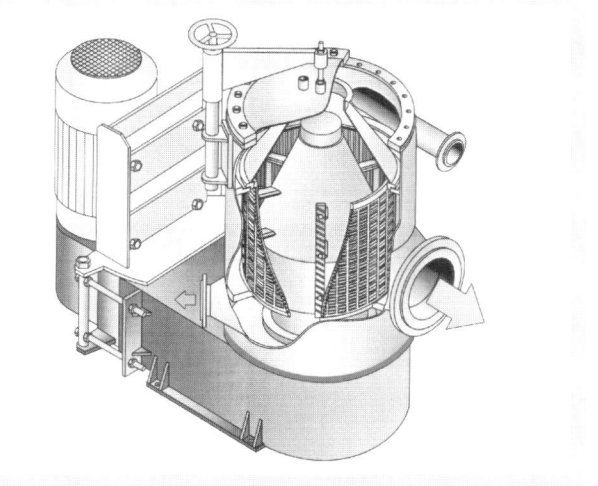

Figure 16. Tangential feed screen (Ahlstrom ModuScreen F).

Screening and cleaning

2.3.2 Medium-consistency screens

Medium-consistency screening is operating at the 3%–5% consistency. Usually this consistency range is classified in pulp processing as low-consistency but, while it is higher than normal screening consistency, it is better to separate these two consistency ranges. Figure 17 presents a medium-consistency screen.

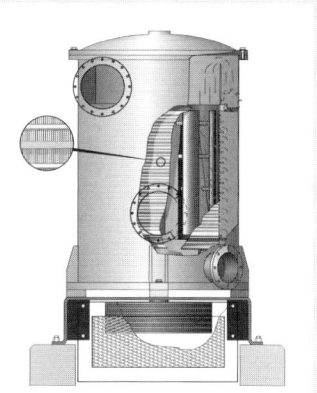

Figure 17. Sunds Defibrator Delta Screen.

2.3.3 Multistage screens

Multistage pressure screens are screens in which several screening stages are run in one screening apparatus. The benefits of this kind of screens are savings notably in energy and space. Figures 18 and 19 present a hole and slot screen combination. The hole plate (Fig. 18) on top of the screen is a pre-screen through which pulp goes into a screen basket. Figure 19 shows a multistage screen which actually has four different screening stages in one single screen. It is composed of a pre-screen and a three-stage fine screen.

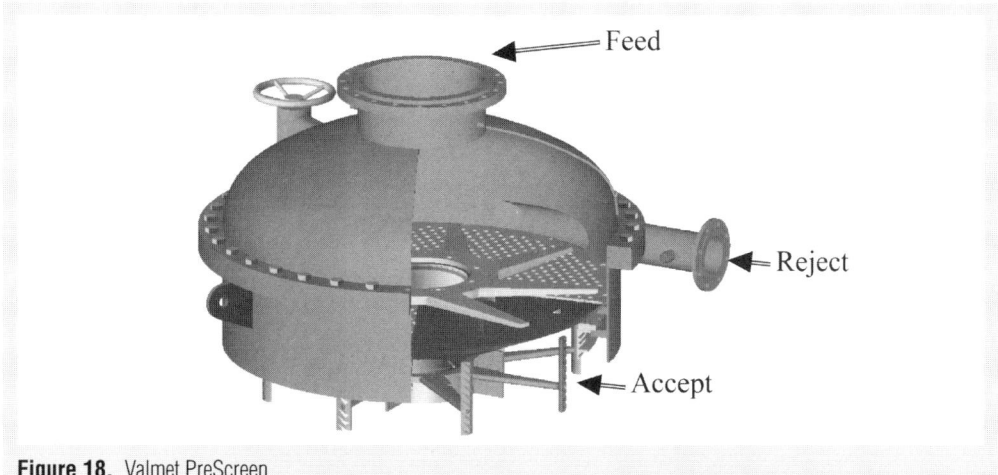

Figure 18. Valmet PreScreen.

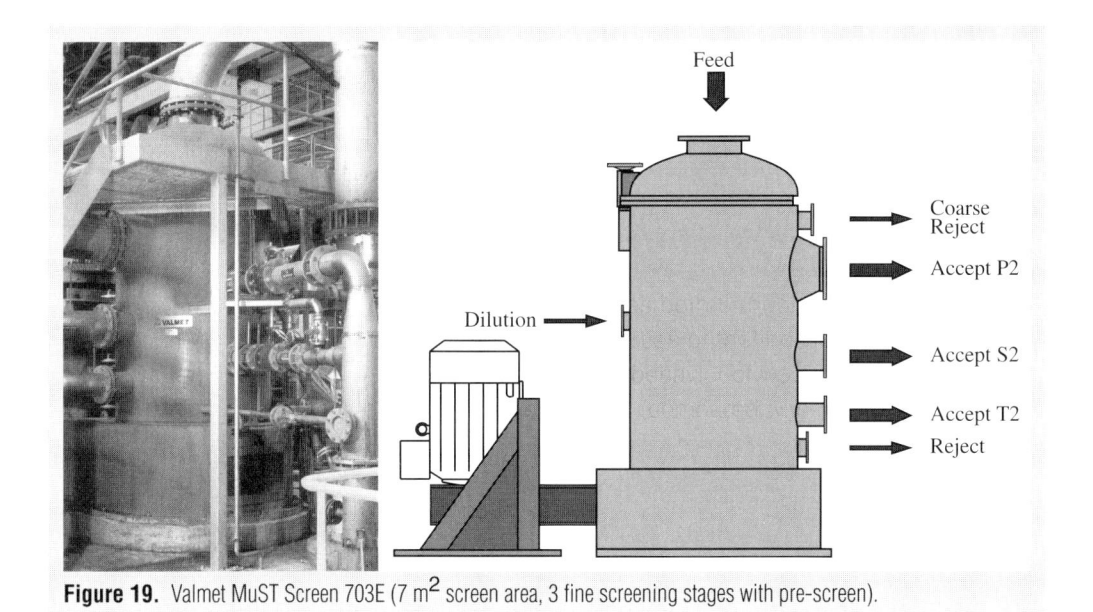

Figure 19. Valmet MuST Screen 703E (7 m^2 screen area, 3 fine screening stages with pre-screen).

3 Screening Systems

Screening connections are chosen according to required pulp and paper properties. Different pulp types contain various amounts of impurities and fractions, which must be removed or retreated. A screening process can be very simple and inexpensive if there are no high requirements for pulp quality. Also the reject treatment can be simple, and the total reject rate can be low. When higher requirements for pulp quality (fiber fractions, freeness control, shive content or strength properties) are set, the screening concept gets more complicated. In those cases, pulp will be fractionated or screened more efficiently, and quality requirements can be met by using narrow slots, high reject rates, and high specific energy consumption in the reject refining.

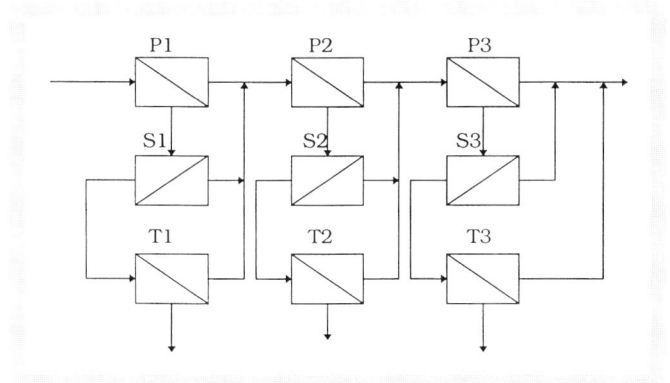

Figure 20. Screen symbols for different screen positions.

Figure 20 shows the naming principle of different screen positions. Screens in the main line are called primary screens (P1, P2, P3). Primary screen rejects are led to secondary screens (S1, S2, S3), and secondary screen rejects are led to tertiary screens (T1, T2, T3).

Screening and cleaning

3.1 Main line screening of GW and PGW

In GW lines, the pulp from the grinders is normally pre-screened using vibrational screens because of big slivers in the pit pulp. The accept pulp from these vibrational screens goes to pressure screening, and the reject goes to the reject treatment. In PGW lines and sometimes also in GW lines, pulp is shredded before screening.

The first-stage screen in pressure screening is called primary one screen (P1). Accept from the P1 screen is led to the primary two screen (P2). The primary one screen (P1) is equipped with a hole basket. Hole screens remove big shives before slot screening. The primary two screen (P2) is equipped with a slot basket. The final pulp quality is determined by slot screens. In older systems, a primary three screen (P3) also can be used, equipped with narrower slots than the P2 screen. Figure 21 shows this type of screen room layout. Basket type (hole size, slot size, profile) is chosen according to the paper grade and position of the screen.

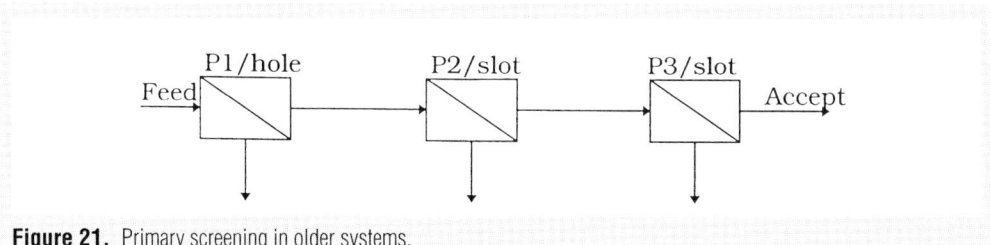

Figure 21. Primary screening in older systems.

The reject of the P1 screen goes to the reject treatment. Rejects from P2 and P3 screens are screened with the secondary two screen (S2). In this type of system, the secondary screen accept is led back to the feed of the system (P1 feed). The reject from the S2 screen goes to reject treatment. Figure 22 shows such a flow sheet.

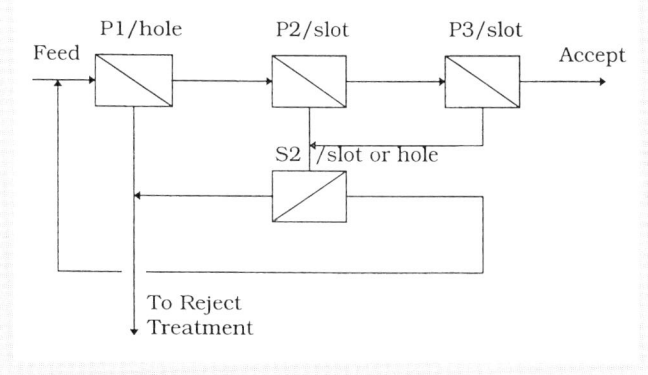

Figure 22. Main line screening in older systems.

CHAPTER 9

There are only P1, P2, and S2 screens in modern systems (Fig. 23). The P1 reject is refined, and the P2 reject is screened by the S2 screen. The S2 reject goes to the reject treatment. With the modern wedge wire screening technology, both P2 and S2 accepts can be combined and led forward while maintaining highquality accept pulp.

For high production rates, parallel screening lines can be installed. Figure 24 shows a screening system with two parallel P1 and P2 screens.

In some cases when reject treatment capacity is limited compared to the groundwood mill production, a tertiary (T2) screen can be used (Fig. 25). The idea of the tertiary screen is to minimize the reject amount or to direct only the most coarse reject to refining.

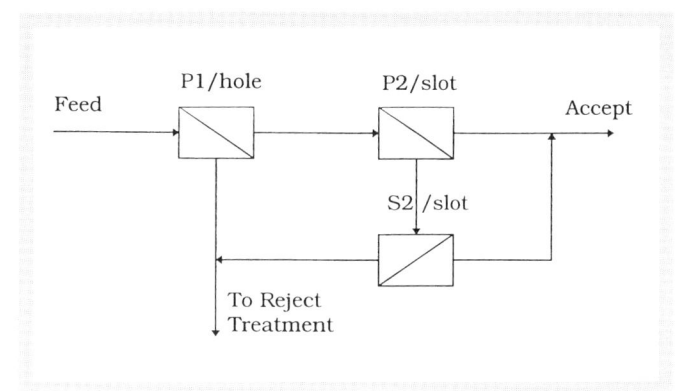

Figure 23. Modern main line screening.

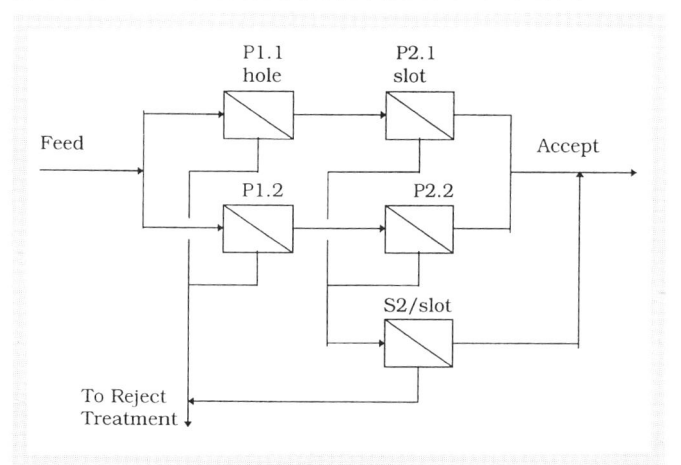

Figure 24. Screening system with parallel screens.

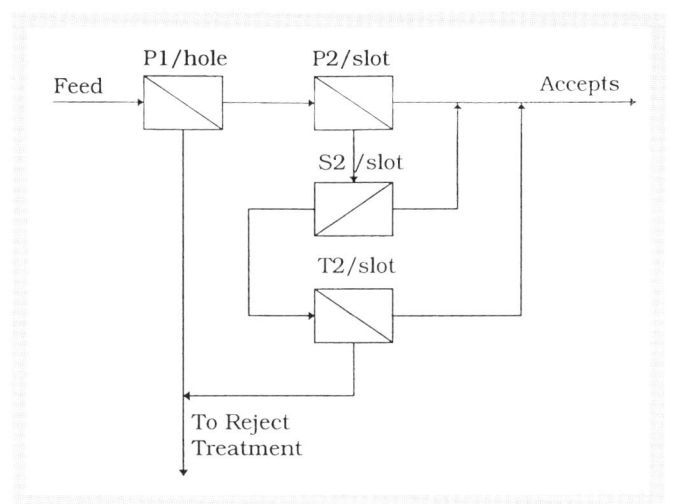

Figure 25. Screening system with the tertiary screen.

3.2 Main line screening of TMP and CTMP

Screening of TMP can be performed using primary one (P1) and primary two (P2) screens in series. The P1 screen can be equipped with a hole or slot basket while the P2 is equipped with a slot basket. The reject from both primary screens is led to the secondary two (S2) screen, which is equipped with a slot basket (Fig. 26).

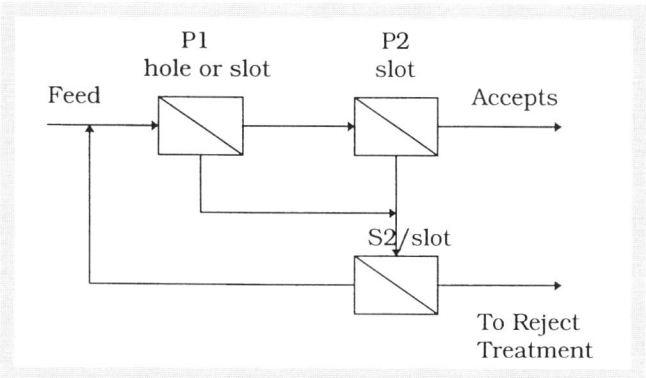

Figure 26. Screening system with P1 and P2 screens.

In the modern TMP screening room, it is not necessary to have mainline screens in series or to lead secondary screen accept backward. Depending on the situation, there can be two- (P1 and S1) or three-stage (P1, S1 and T1) screening (Fig. 27). That kind of connection has become possible because slot screening with the wedge wire technology has improved pulp quality. All screens are equipped with slot baskets.

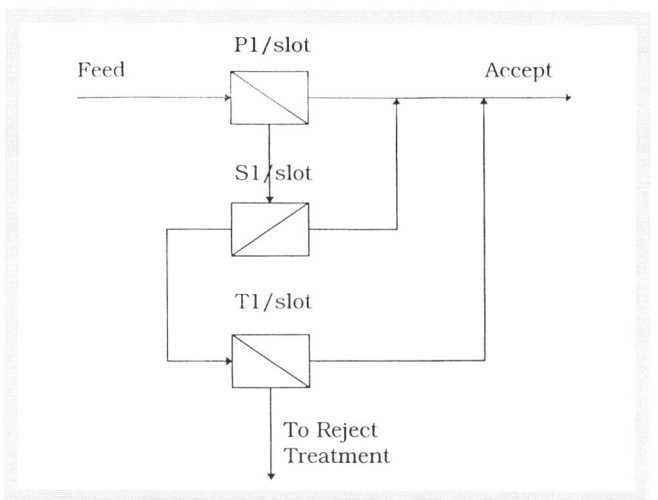

Figure 27. Modern TMP screening system.

The screening principles for CTMP are quite the same as for screening of TMP. For high freeness CTMP, no tertiary screen is needed (Fig. 28).

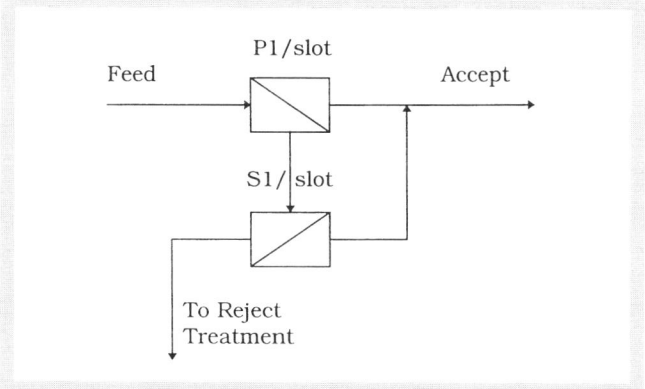

Figure 28. Board grade CTMP screening system.

CHAPTER 9

3.3 Screening of GW and PGW rejects

In some cases, the refined reject pulp is simply returned to the beginning of the process (the primary one screen feed). In such a case, there is no need for a separate reject screening, but normally the refined reject is screened separately. Figure 29 shows a typical flowsheet. Reject screen one (R1) is equipped with a hole basket to protect reject screen two (R2) against big shives and possible disturbances in reject refining. The R2 screen is equipped with a slot basket.

The reject screening can also be done with one screen (R1) if reject refining is well controlled and efficient and if there is no risk that unrefined shives can get through the refiner. In that case, the screen is equipped with a slot basket.

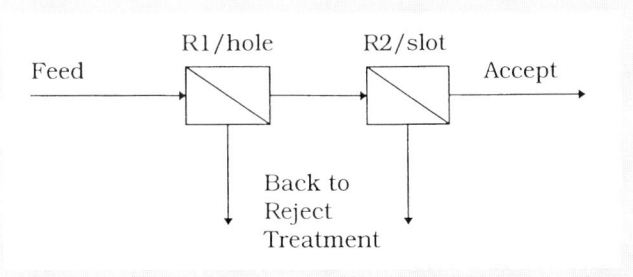

Figure 29. Screening system for refined groundwood rejects.

3.4 Screening of TMP and CTMP rejects

In TMP and CTMP lines, the screening of refined reject is typically made in one stage (R1), which is equipped with a slot basket. A secondary reject screen (RS1) can be used if the amount of reject to be returned to refining needs to be limited (Fig. 30). All screens are equipped with slot baskets. This kind of screening arrangement is also used with groundwood pulps.

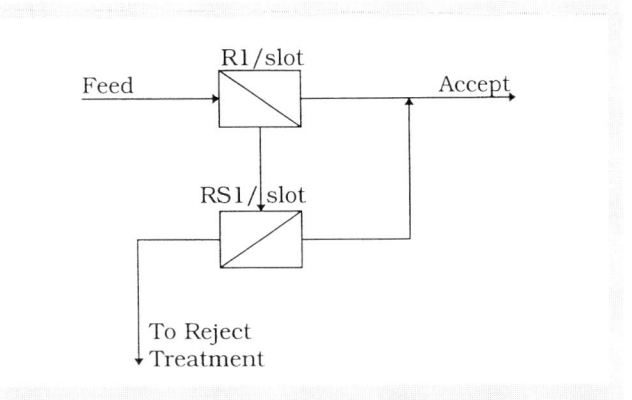

Figure 30. Screening of refined TMP and CTMP rejects.

3.5 Future screening systems

In the future, the screening systems will be largely simplified by the use of the new multistage screens. A multistage screen is a screen with several screening stages incorporated into a single screen body. The multistage screen in Fig.19 can be used to replace the four screens in Fig. 25, as is shown in Fig. 31.

Screening and cleaning

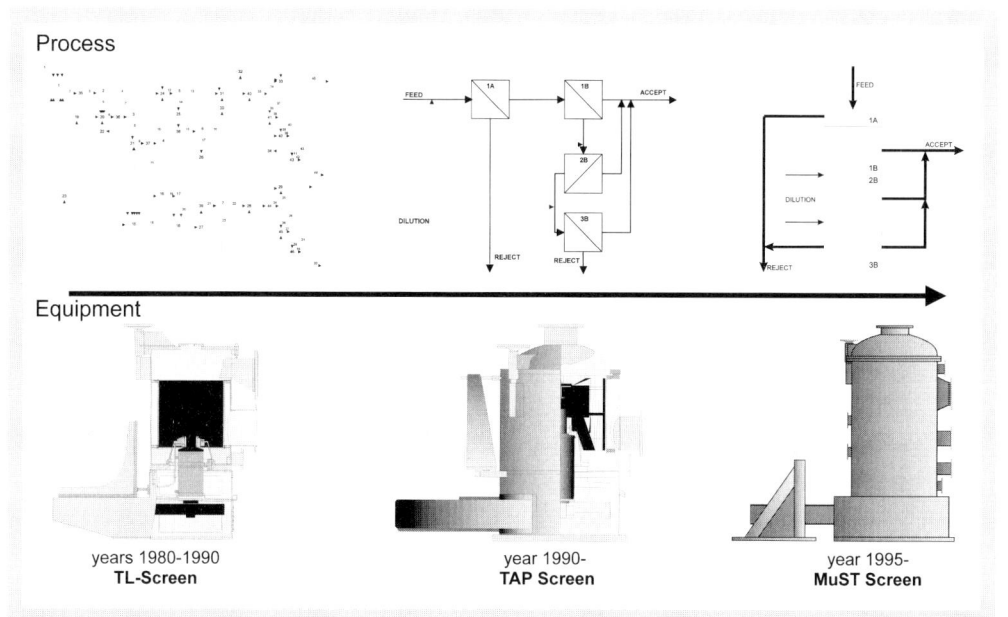

Figure 31. Replacement of a five-screen system with one single multistage screen (Valmet MuST-screen).

Equipment labels (left to right):
- years 1980-1990 **TL-Screen**
- year 1990- **TAP Screen**
- year 1995- **MuST Screen**

4 Control of Screening

The high pulp quality requirements of today set high demands on the screening control, the purpose of which is to ensure a stable process and a constant pulp quality. The control is based on process measurements. Pressure in the feed and accept side of each screens is measured. Based on those, the pressure drop over the screen can be calculated. The plugging tendency will increase if the pressure drop over the screen basket increases. Normally accept and reject flows are measured and feed flow can be calculated. Flows need to be known to maintain good runnability of the screening process and constant reject rates for the stability of the pulp quality. Feed consistency of the screen is measured and controlled with dilution water to maintain optimal conditions in the screening system. Dilution water into the screen can be used to prevent pulp from thickening too much and to ensure reject removal to the next process stage. The screen can plug if pulp consistency becomes too high. Dilution water can be fed to one or several places of the screen, e.g., middle or bottom of the screen.

4.1 Control of a single screen

Figure 32 shows an example of controlling a single pressure screen. The controlling principle for parallel screens and screens in series is identical to that.

CHAPTER 9

Figure 32. Control of a single screen.

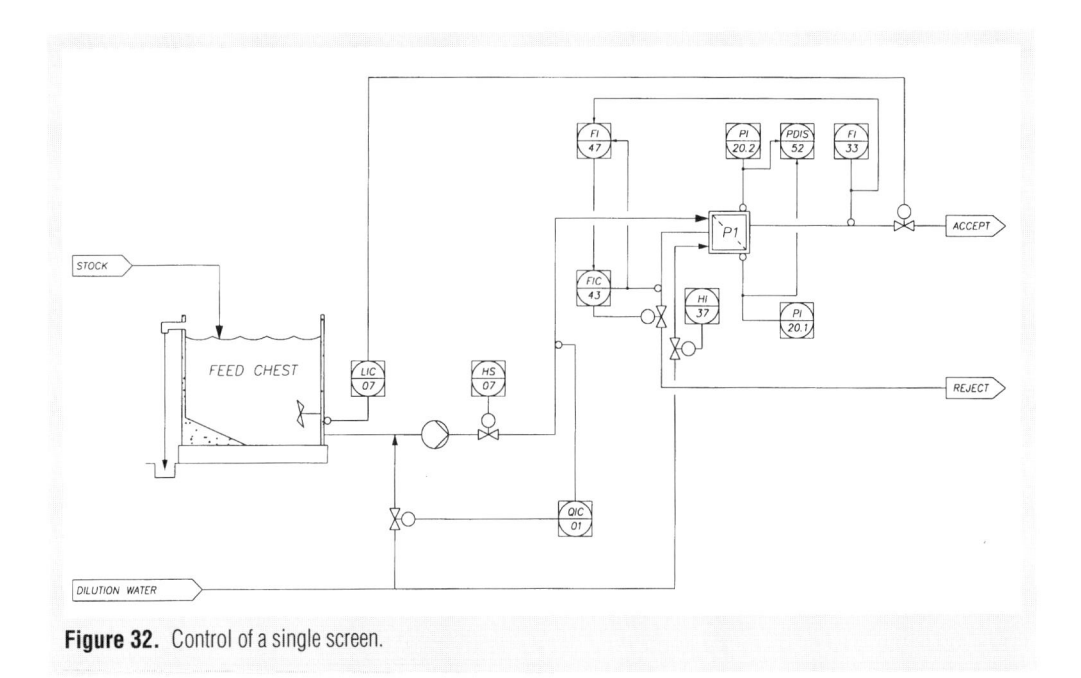

The feed consistency of a screen is measured and controlled (QIC-01 in Fig. 32) by dilution water. Accept flow can be controlled by the level of feed chest (LIC-07) as shown in Fig. 32. Depending on process and situation, accept valve opening can also be controlled by accept flow (FIC), accept pressure (PIC), or pressure difference (PDIC). Two of the main flows feed – accept and reject – need to be measured. In Fig. 32, the accept flow (FIC-33) and the reject flow (FIC-43) are measured. Feed flow is calculated (FI-47) based on measured accept and reject flows.

In this case, feed (PI-20.1) and accept pressure (PI-20.2) of the screen are measured, and pressure difference (PDIS-52) over the screen is calculated. Pressure difference can also be calculated based on reject line pressure instead of feed line pressure, as an indicator of the pressure drop over the screen basket. Dilution water flow into the screen is controlled by the HI-37 valve. The flow can also be measured and controlled according to flow (FIC). HS-06 is an ON-OFF valve, which is the startup valve for a feed pump.

The feed to the screen can also come from the white water chest (TMP screening). Figure 33 shows that kind of controlling principle. All controls are identical to the single-screen situation except for the feed line controls. Pump speed (SHI-05) is controlled by consistency measurement (QIC-04). This loop ensures constant feed consistency to the screen. Stock flow to suction side of the screen feed pump is coming from latency chest. Consistency is measured and controlled (QIC-01). Stock flow is controlled by the level of the latency chest. It can also be controlled by flow (FIC).

Screening and cleaning

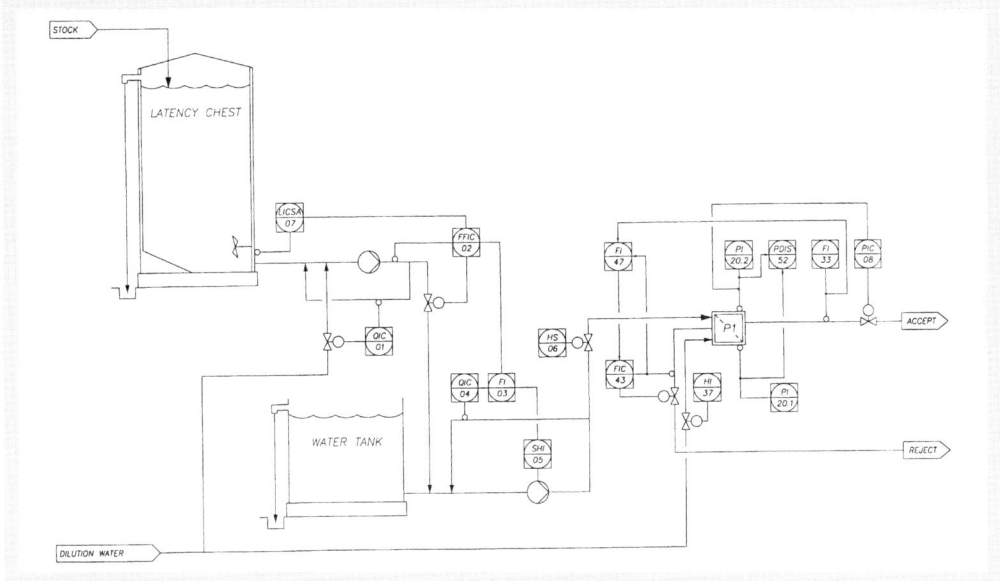

Figure 33. Control of a single screen with feed from the white water tank.

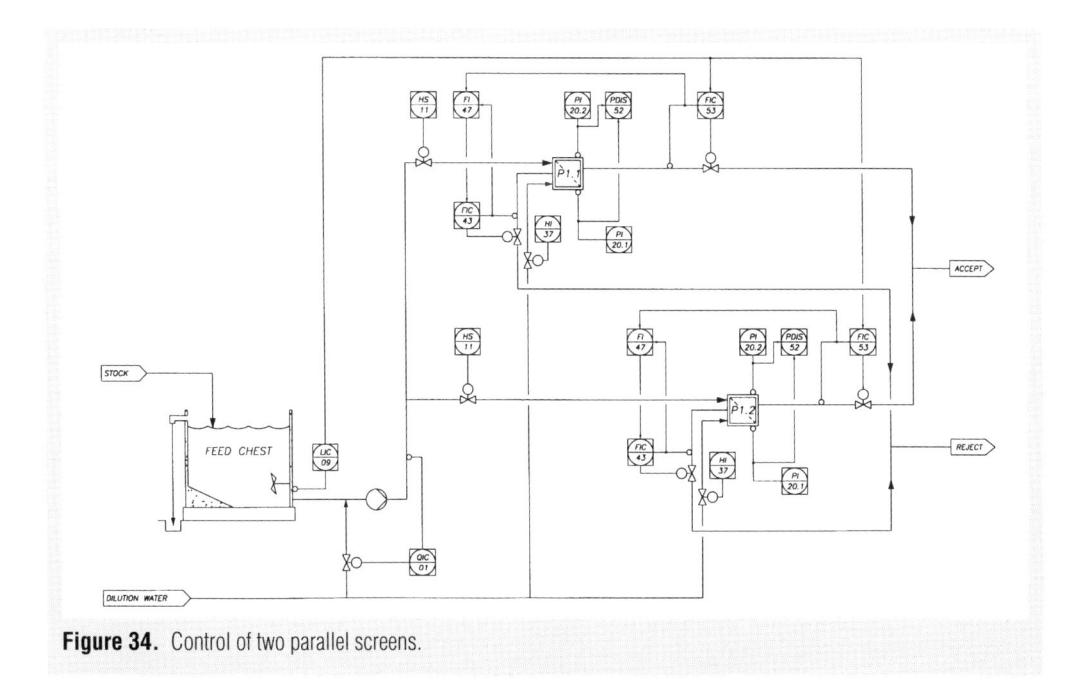

4.2 Control of parallel screens and screens in series

Control of a single screen is not affected by screening configuration. Figure 34 shows control of parallel screens. Two separate screening lines can be selected by ON-OFF valves (HS-11). This valve can also be used in a single-screen process for a smooth startup.

Figure 34. Control of two parallel screens.

CHAPTER 9

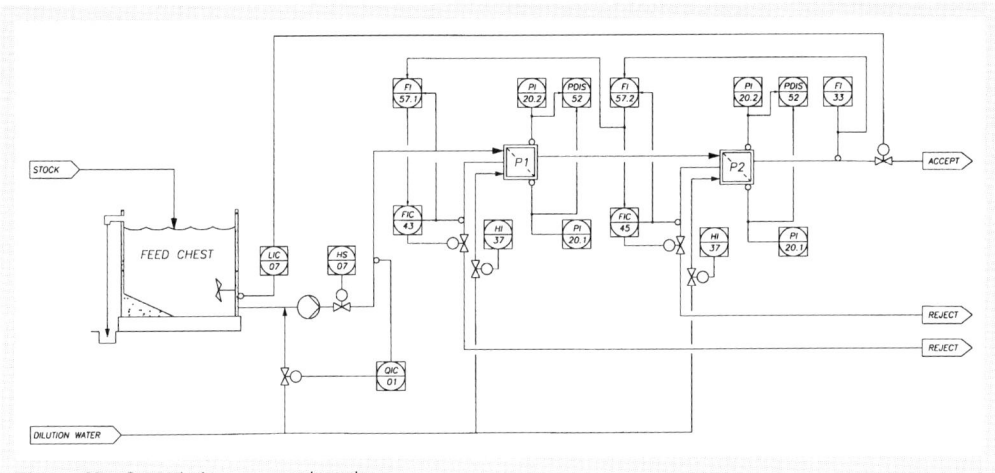

Figure 35. Control of two screens in series.

Figure 35 shows the control principle of screens in a series. The primary one (P1) screen has no accept valve in a system with screens in series. Control of accept flow can be accomplished by controlling accept and reject flows of the primary two screen (P2). The reject flow (FIC-45) of the P2 screen is controlled by P2 screen feed flow (FI-57.2). Feed flow is calculated by measuring P2 screen reject (FIC-45) and accept (FI-33) flows. P1 screen reject rate is controlled likewise as the P2 screen reject rate. Feed of P1 screen (F1-57.1) is calculated (P2 accept + P2 reject + P1 reject). All other controls are analogous to those of a single screen.

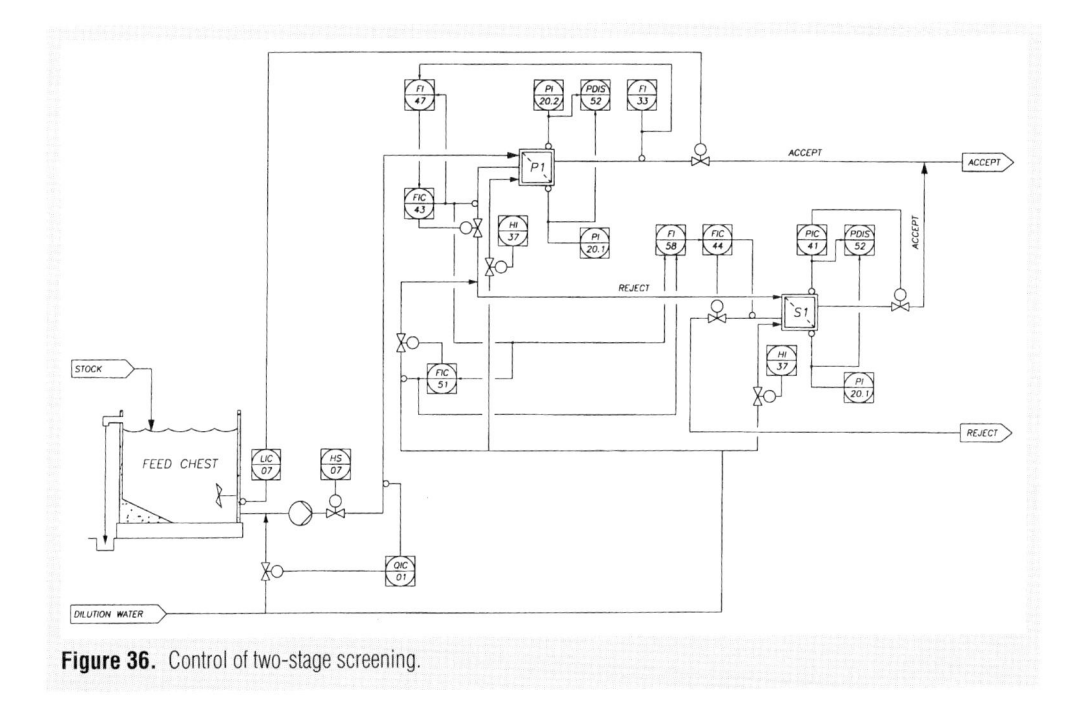

Figure 36. Control of two-stage screening.

Screening and cleaning

4.3 Control of two-stage screening

Figure 36 shows a two-stage screening configuration. Dilution water (FIC-51) is used to control feed consistency to the secondary screen, and it is proportional to the P1 reject flow. Feed flow of the secondary screen (FI-58) is calculated from the primary screen reject flow (FI-43) and dilution water flow (FI-51). All other controls are identical to single-screen controls.

5 Cleaning and sand removal

5.1 Theory

Hydrocyclones or cleaners have been used for a long time in the pulp and paper processes. Hydrocyclones are used in pulp lines as well as in front of the paper machines. The main purposes of a cleaner include separating bark, sand, metal, or other contaminants from the pulp slurry and separating shives and undeveloped fibers which are returned to the process for further treatment.

In the hydrocyclone, the low-consistency pulp is injected tangentially under pressure into a stationary cylindrical and conical vessel, i.e., it is brought into rotation. Thus, a cyclone is a device which utilizes fluid pressure energy to create a rotational fluid motion. Figure 37 displays a hydrocylone of the most common design. A core of air or gas travels in the center of the cyclone. Such a core appears in practically every cyclone.

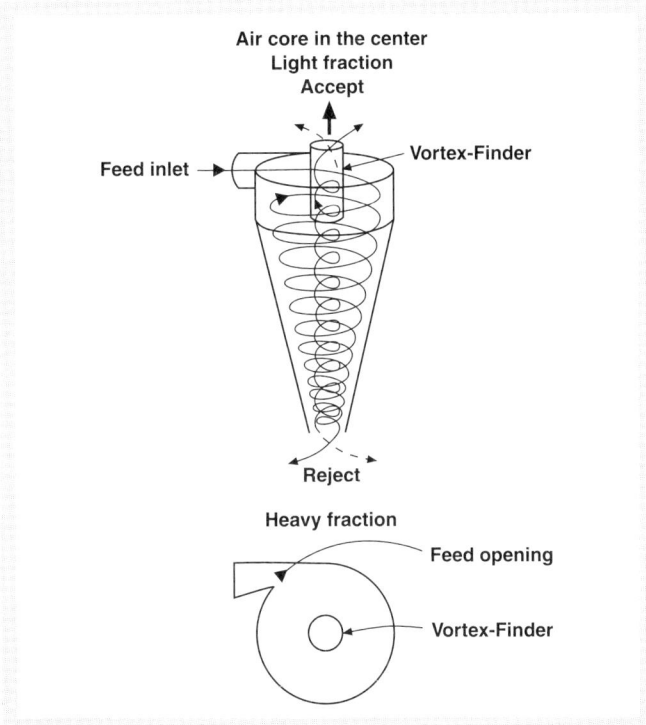

Figure 37. A hydrocyclone of common design. A core of air or gas travels in the center of the cyclone.

There are four forces which cause separation within a cleaner as can be seen from Fig. 38. They are as follows:

The centrifugal force	A result of the slurry rotating.
The drag force	Resulting from liquid moving past an object.
The buoyancy force	The same force as exerted on a submerged beachball.
The lift force	The force that makes a spinning baseball curve as it passes through the air.

CHAPTER 9

There is a peculiarity connected to the lift force. The flow pattern in the outer vortex zone is a free vortex, while in the inner zone it is a forced vortex. This causes the lift force to change directions. In the outer vortex, the lift force is directed outward while, in the inner vortex, it is directed inward. The lift force is significant for objects which have a high specific surface area. The net effect of these four forces determines whether a particle of a given size, shape, and density is accepted or rejected by the cleaner. The arrows in Fig. 38 indicate the relative directions and strengths of the forces directed at a single particle.

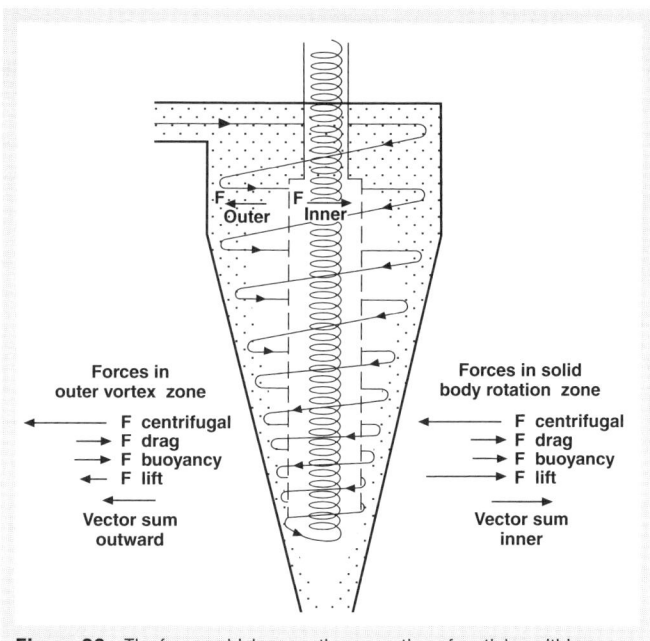

Figure 38. The forces which cause the separation of particles within a cleaner.

The physical properties of the particles that are being handled will have a strong effect on how they behave in the cyclone, i.e., whether the particles end up in the accept or the reject zone. The data given in the table below has to be understood as relative values[24]:

Characteristic of the particle	With high likelihood into the accept	With high likelihood into the reject
Gravity	Light	Heavy
Shape 1	Short	Long
Shape 2	Thin	Thick
Surface area	Large	Small
Size	Small	Large

The larger and heavier a particle is, the bigger the certainty is for it to end up in the reject flow. Smaller particles remain further away from the walls but, if they get close to the flow directed toward the apex of the cone, they will be rejected. If a heavy particle has a greater size than the thickness of the vertical flow directed toward the apex, the particle will be thrust from the vertical flow into the faster flow directed toward the accept outlet at the base of the cone only to be thrown back into the vertical flow due to its greater rotation diameter and tangential speed. This can continue until the particle has worn down enough to fit into the reject flow section and disappears through the reject. A particle that remains spinning in the cone, due to its gravity and shape, can have a rather circular rotational motion, so the weardown caused by the rubbing against the walls is centralized on a specific small area of the cone. This can cause the cone to deteriorate rapidly.

As a particle travels through fluid, its motion is restricted by a friction force which is in proportion to the projection surface area of the particle. Therefore a particle with a greater projection surface will end up in the accept zone more likely than one with a small projection surface. Based on this, for example, barkspecs that are weightwise in the same class as fibers but have a smaller projection surface get rejected. In the same way, the effect of fibrillation to the likelihood of separation is thereby explained. The more the fiber is fibrillated, the greater its projection surface is and the bigger its chance to end up in the accept zone. Shives usually get rejected for that same reason.

The length and shape of a particle has a dual effect on the separation process. First, a particle situated between two flows with different tangential speeds will be swept away by the faster flow or, in other words, toward the center region of the cleaner. Long fibers end up in the accept zone easier than short fibers. The same happens to shives with the same projection surface but different lengths.

The second mechanism is dependent on both the length and the stiffness of the particle. The section toward the apex of the cone gets thinner the closer to the apex it gets. In order for the particle to fit this reject flow section, it must neither be too long nor too thick in size.

Based on this information, short and "fluffy" shives will be rejected more effectively than long shives. The larger the cleaner's diameter is, the better it can deal with removing longer and bigger shives.

5.2 Cleaner types

Cleaners are mainly used for removal of heavy particles. However, because of an increased use of secondary fibers, there are many cleaner types also removing plastic, rubber, wax, and other lighter particles.

There are four different groups of cleaners:

- The coarse/high-density/heavyweight cleaner
- The medium-density cleaner
- The fine cleaner
- The lightweight cleaner.

Heavy- and medium-weight cleaners act mainly as protection equipment removing heavy debris such as sand, staples, and other larger impurities. They are large in diameter, i.e., 250–2500 mm and used under a low-pressure drop between the inlet and the accept 15–120 kPa. The feed consistency ranges from 1.0 to 3.5%.

CHAPTER 9

Fine cleaners enhance the visual appearance of the end product and reduce abrasive wear of the system components by removing impurities such as sand, metal, bark, shives, and pitch. These cleaners have a small diameter, i.e., 50–250 mm, and run with a medium- to high-pressure drop, between 100 and 200 kPa. The feed consistency ranges between 0.5 and 0.8%. Cleaners with a diameter of 100–150 mm are mainly used for the centricleaning of mechanical pulp. The smooth inner cone of these cleaners improves the separation process of shive-like particles. Figure 39 shows an example of a fine cleaner.

Lightweight cleaners increase the quality of the end product and protect paper machine clothing and dryer cylinders by removing contaminants such as air, resin, wax, pitch, stickies, and plastic. The cleaners are small in diameter, i.e., 50–250 mm, and are used under a medium- to high-pressure drop, between 100 and 210 kPa.

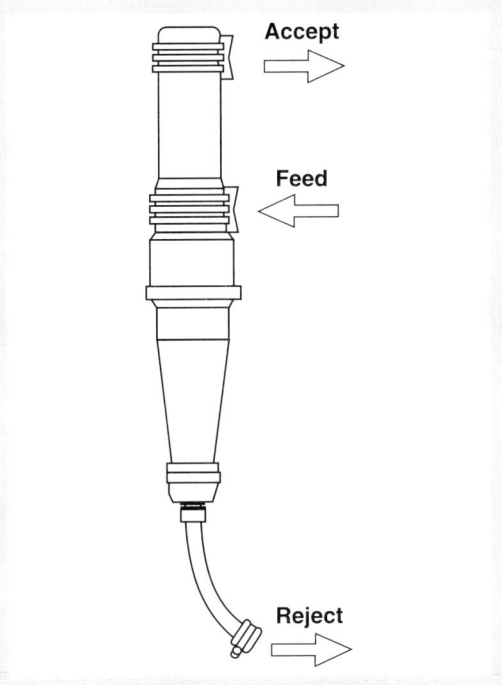

Figure 39. A fine cleaner (Celleco). This cleaner type is manufactured in a variety of cones (smooth, steprelease); it has a small pressure drop (100–120 kPa) and a high efficiency on small particles.

5.3 Cleaning systems

A cleaner plant in the pulp and paper industry consists normally of 2–6 stages which means that, in most cases, the coupling is a cascade (Fig.40). The accepts from the latter stages can be taken away for refining in order to avoid consistency levels increasing too much, which reduces cleaning efficiency.

5.4 Sand removal

The centrifugal force has the biggest effect on the separation of small and heavy particles such as sand (quarts) particles. This means that even a small pressure drop will cause the sand particle to end up in the outer flow section and thus toward the reject. The bigger the pressure drop being used (between inlet feed and accept) is, the smaller are the particles which will be separated. The outward directed force, owing to a strong centrifugal force, enables the separation to take place also in higher consistencies. A good sand separation efficiency can be achieved up to consistencies of 1.2% (Fig.41). Using even higher consistencies raises the problem of the reject flow not being able to leave the cyclone due to the outlet clogging up. The reject consistency is normally 1.5–3.0 times higher than that of the inlet.

Screening and cleaning

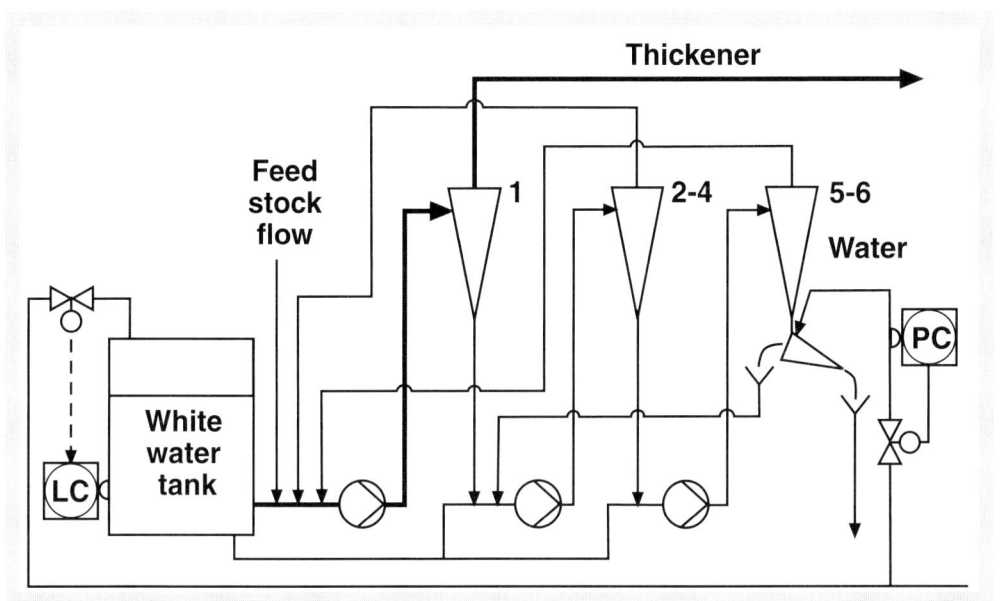

Figure 40. A cleaner plant situated in the pulp line.

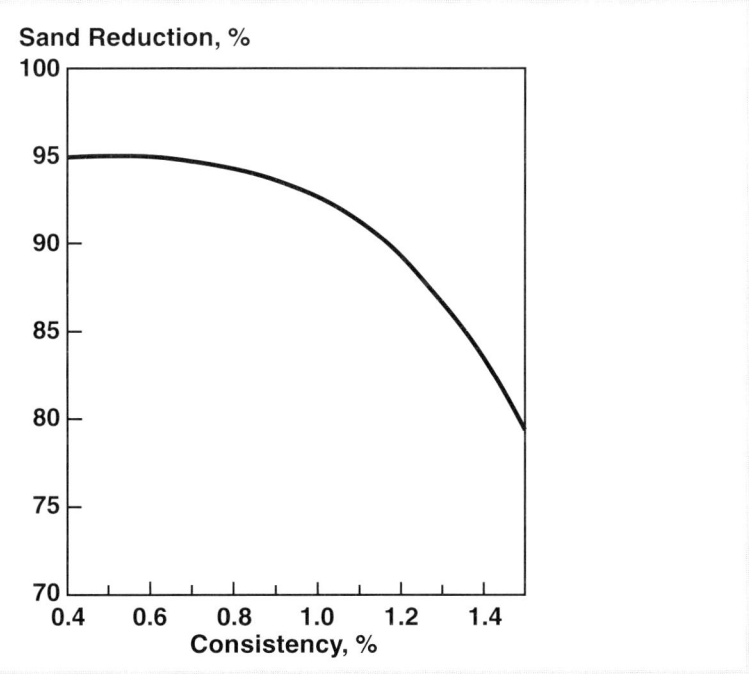

Figure 41. The sand removal efficiency in a cleaner. The cleaner in question is a 6-in. cleaner, and the pressure drop between the inlet and accept is 150 kPa.

CHAPTER 9

5.5 Control of cleaning

The following operating circumstances have an effect on the separation efficiency of the hydrocyclone:

- Pressure drop from inlet feed to accept
- Separation consistency
- Reject rates
- Temperature.

Increasing the pressure drop (inlet feed/accept) results in higher throughflowing capacity, incoming speed of the pulp, and rotation speed of the whirl, as shown in Fig. 42. Increasing the pressure drop will increase the separation efficiency to a certain point at which the flow motion in the cleaner becomes turbulent.

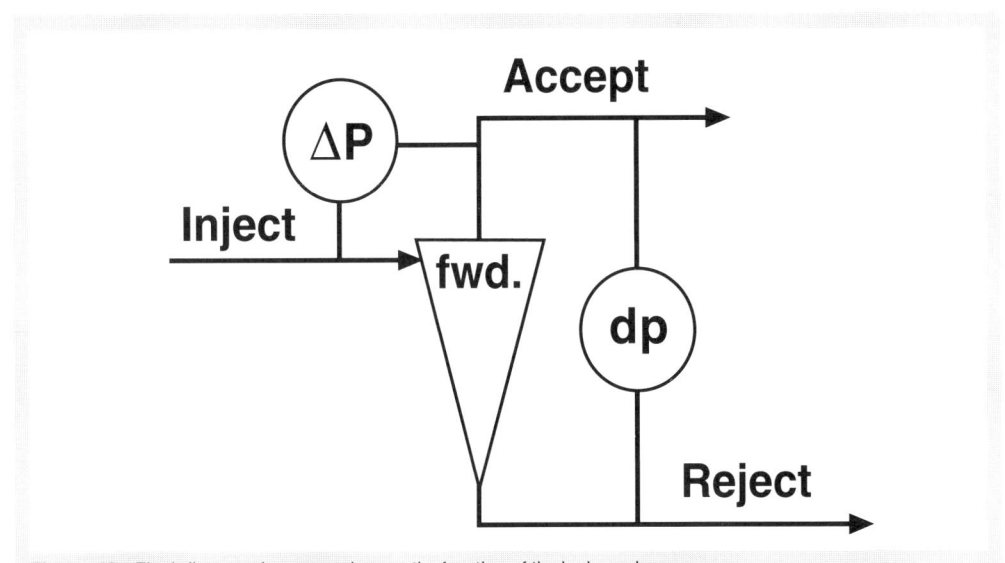

Figure 42. The influence of pressure drop on the function of the hydrocyclone:
A higher ΔP: higher efficiency and higher capacity.
A higher dp: higher reject flow and and higher consistency in the following stage.

The pressure drop between the inlet feed and the accept varies normally between 100 and 200 kPa. By improving the structure and dimensions of the cleaner, it has become possible to lower the necessary pressure drop and thus reduce the consumption of energy. The feed, accept, and reject pressures are regularly checked in a centricleaning plant. By control of these, the right centricleaning efficiency and reject quantities can be maintained.

Generally the process itself will determine the temperature of the pulp and, in return, this determines the temperature of the pulp to be centricleaned. The temperature has been proven to have a slight effect on the classification ability. As a general rule it can be said that, the higher the temperature is, the better the separation efficiency will be. A higher temperature also increases the reject flow consistency; therefore, higher reject rates must be selected when calculating for the balance sheet.

When requiring a cleaning system that is reliable and has good classification capacity, the following dimensioning practice can be used. When transferring between the first and the last stage, the inlet feed consistency should get lower and the reject rates per stage higher. In connection with the startup of the plant, a series of samples is taken from the various stages to establish the consistency. Calculations and fine tuning of reject quantities are made based on these samples. By taking these samples on a regular basis, the correct method of running the plant can be ensured. The temperature's effect on the measures also has to be taken into consideration. If the later stages have been wrongly dimensioned, they will become a problem area in the production line. The inlet consistencies are high, and the separation capacity is low.

CHAPTER 9

References

1. Williamson, P. N., Pulp Paper Can. 95(4):9(1994).
2. Ahnger, A., Appita 1995 Annual General Conference Proceedings, Appita, Carlton, p. 549.
3. Kleinhappel, S., 1995 International Mechanical Pulping Conference (Ottawa) Proceedings, CPPA, Montreal, p. 267.
4. Leask, R. and Kocurek, M., Pulp and Paper Manufacture, TAPPI/CPPA Joint Textbook Committee, Atlanta, 1987.
5. Paulapuro, H., Paper and Timber 59(4):297 (1977).
6. Niinimäki, J., Dahl, O., Hautala, J., Tirri, T., Kuopanportti, H., TAPPI 1996 Pulping Conference Proceedings, TAPPI PRESS, Atlanta, p. 761.
7. Niinimäki, J., Heikkinen, P., Kortela, U., IASTED 1997 International Conference on Modelling, Identification and Control Proceedings, IASTED, Innsbruck, p. 280.
8. Gooding, R. W. and Kerekes R.J., Can. J. Chem. Eng. 67(10):801 (1989).
9. Dulude, B. K., Pulp Paper Can. 96(5):40 (1995).
10. Kumar, A., "Passage of fibres through screen apetures," Ph.D. thesis, University of British Columbia, Vancouver, 1991.
11. Gooding, R. W. and Kerekes, R. J., J. Pulp Paper Sci.15(2):J59 (1989).
12. Yu, C. J. and DeFoe, R. J., Tappi J. 77(9):119 (1994).
13. Gooding, R.W. and Kerekes, R.J., Tappi J. 75(11):109 (1992).
14. Yu, C. J. and DeFoe, R.J., Tappi J. 77(8):219 (1994).
15. Frediksson, B., "New, efficient HC screening technology for recycled fiber pulp," Paper Recycling '94 Proceedings, PPI, San Francisco.
16. Baehr, T. and Rienecker, R., "A new generation of pressure screens for stock preparation systems," TAPPI 1991 Pulping Conference Proceedings, TAPPI PRESS, Atlanta, p. 1065.
17. Bliss, T., Pulp and Paper Manufacture: Stock Preparation, 3rd edn., TAPPI PRESS, Atlanta, Vol. 6, 1992.
18. Gallagher, B. J., TAPPI 1991 Improving Screening and Cleaning Efficiencies Short Course Notes, TAPPI PRESS, Atlanta.

19. Hourula, I., Systems Engineering Laboratory Series A: Reports 17-18, University of Oulu, Oulu, 1996.

20. Ora, M., Saarinen, A., Hautala, J., Paulapuro, H., Paper and Timber 75(5):330 (1993).

21. Ahnger, A., Haikkala, P., Hautala, J., Nerg, H., TAPPI Mechanical Pulping Committee Meeting Notes, October 1993, Atlanta.

22. Hourula, I., Kortela, U., Ahnger, A., Hautala, J., TAPPI 1996 Recycling Symposium Proceedings, TAPPI PRESS, Atlanta, p. 435.

23. Repo, K. and Sundholm, J., 1995 International Mechanical Pulping Conference Proceedings, TAPPI PRESS, Atlanta, p. 271.

24. Fryckhult, R., Celleco's centricleaners, internal training material, Alfa Laval, Stockholm.

CHAPTER 10

Refining of shives and coarse fibers

1	**Equipment and systems used in refining of shives and coarse fibers**	**289**
1.1	Thickening of reject	289
	1.1.1 The bow screen	290
	1.1.2 The screw press	290
	1.1.3 The twin-roll press	292
	1.1.4 The twin-wire press	293
1.2	Reject refiners	294
1.3	Reject refining systems	296
2	**Operation and control of reject refining**	**300**
2.1	Key variables of reject refining	300
2.2	Process conditions and intensity in reject refining	302
2.3	Disturbances in reject refining	303
3	**Effect of reject refining on pulp and paper properties**	**304**
3.1	Development of fiber properties	304
3.2	Development of pulp and paper properties	308
	References	310

CHAPTER 10

Erkki Huusari

Refining of shives and coarse fibers

The mainline refining or grinding defiberizes the wood matrix into particles of different size and properties. *Fiber development* can be used as a term to indicate the changes of size and properties of fibers which occur in different stages of mechanical pulping processes. Fiber development is important in determining the papermaking potential of mechanical pulp.

After the grinding or the mainline refining, the quality of the coarse fractions of the mechanical pulp is poor and further refining is needed to develop the properties to a level needed in high-quality paper grades. The task of screening and cleaning is to remove impurities (sand, metal, etc.) and to separate fiber bundles (shives) and undeveloped coarse fibers from the prime quality pulp for further refining. The original target of this reject refining was to make usable pulp out of the stone groundwood shives but, nowadays, especially with TMP, in addition to shives removal the treatment of poorly developed long fibers has become a major objective of reject refining. Accordingly, this chapter is named "Refining of shives and coarse fibers" and not "Reject refining."

1 Equipment and systems used in refining of shives and coarse fibers

Reject refining can be made at low consistency (3%–5%, LC), at medium consistency (10%–15%, MC), or at high consistency (30%–50%, HC). Medium-consistency refining develops the pulp properties most likely in a similar way as low-consistency refining. Benefits of medium-consistency when compared to low-consistency refining are obviously improved loadability and lower idling power. However, medium-consistency refining is seldom used; therefore, it will not be discussed here. Low-consistency refining does not require expensive thickening equipment. Investment costs for low-consistency reject refining are low, and it also has a good shive reduction efficiency. But, although the high consistency reject refining is the most expensive reject refining system, it is the one commonly used due to the good development of fiber properties in HC refining. High-consistency refining enables the use of high specific energy, which results in good development of fiber properties in single-pass refining. However, there are still a number of LC reject refining operations, especially for LWC-groundwood.

1.1 Thickening of reject

Nowadays bow screens, called also hydrasieves or side hill screens, are commonly used as prethickeners for increasing consistency to 3%–5% before final thickening to

CHAPTER 10

30% or higher consistency. In addition to thickening, a bow screen also washes fines out from the reject. This increases the freeness of the reject, which in turn helps to dewater the pulp in the main thickeners. The removal of fines also improves loadability of reject refiners, which makes the treatment of coarse fibers more effective.

There are three main types of thickeners: the screw press, the twin-roll press, and the twin-wire press. The screw press is most commonly used due to a lower investment cost. The energy consumption of a screw press is higher than the energy consumption of a twin-roll or a twin-wire press.

1.1.1 The bow screen

The screen plate of the bow screen is constructed of wedge bars forming a bowed surface with horizontal slots between the bars, the normal slot width being 150 microns. The pulp is fed through a nozzle which distributes the flow tangentially to the full screen width by means of an adjustable slice. As the suspension flows along the screen plate, thin layers of water are sliced off by the screen bars. Fibers in the suspension orient themselves in the direction of the flow. Thus the majority of fibers longer than the slot width will be retained on the screen surface and slide down to discharge at a consistency of 3%–5%. In case plugging of the screen plate occurs, it can be easily cleaned with a water hose. If needed, the screen can be equipped with an oscillating cleaning device. If the pulp suspension contains abrasive particles, the upper edge of the bars will eventually be rounded. In such cases, the turning of the screen plate inside the screen housing will increase the lifetime of the screen plate.

1.1.2 The screw press

The main field of application for the screw press is the high-freeness range as is the case in the dewatering of the reject before high-consistency refining. The pulp suspension is pushed into the press by pumping pressure or by the backwater height of the suspension level in the headbox. This initial pressure is required to fill the first section of the screw and to obtain some degree of dewatering. As the screw shaft revolves, usually at a speed of 20–30 revolutions per minute, the suspension is compressed more and more and is dewatered through the screen baskets in the shell. The holes in the inner screen basket are always step-drilled or internally tapered. These outwardly tapering perforations are required to prevent clogging. In the low-pressure zone, the required stability can be obtained with the screen plates alone; however, supporting structures of increasing strength are required in the subsequent medium and high pressure zones.

Dewatering to a high consistency requires higher pressure at the end of the screw. The so-called counterpressure device is used in the startup process and for precision adjustments of the pressure. The main part of the counterpressure is generated in the final section of the shaft which has no flighting. In this part of the press, the dry pulp cake has to be pushed out against the friction occurring along the length of the shaft and against the conical shape of the outer shell. This process provides the major part of the pressure required for dewatering. The pressure applied by the counterpres-

Refining of shives and coarse fibers

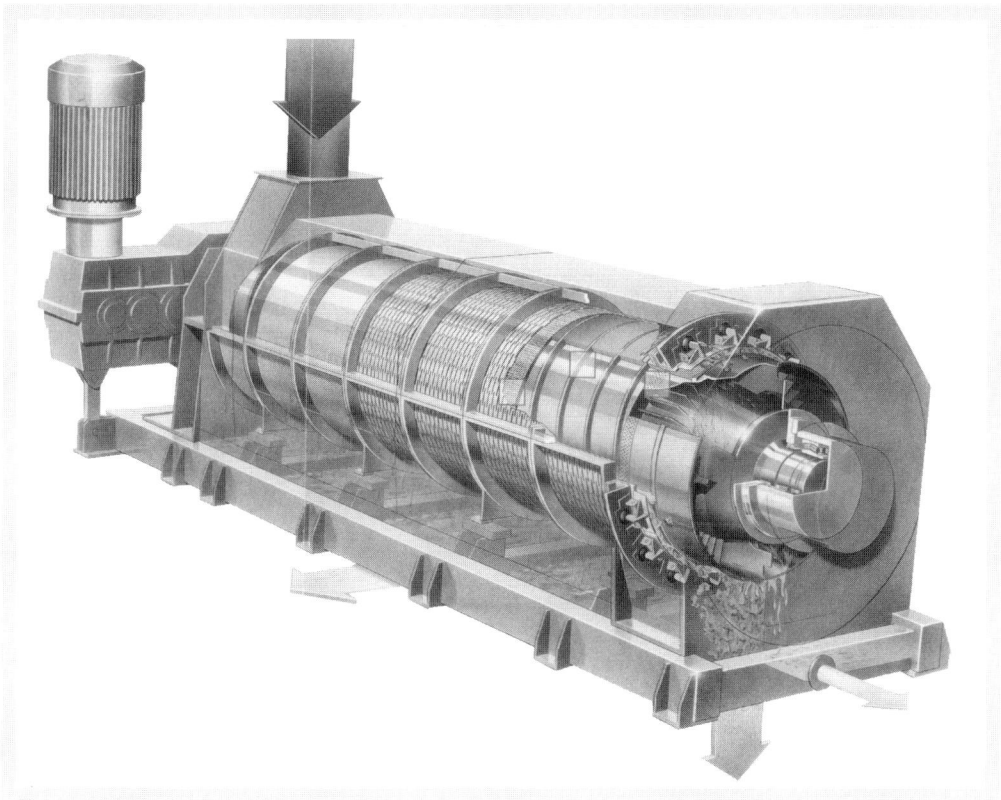

Figure 1. Screw press (Andritz).

sure device has only relatively little influence on the final dryness because the major resistance comes from the plug in the section without flighting. The pulp consistency after the press can be controlled by means of the torque setting. The second factor affecting the consistency is the throughput. If the throughput drops, the speed also drops if the torque remains the same, and then the pulp has more time available to dewater and the dryness increases. Thus a higher discharge consistency can be obtained by using a larger size of screw press.

Screw presses operate primarily with cake thicknesses of 20–30 mm. At high consistency, there is already considerable resistance to filtrate draining from the area around the shaft to the outer screen. In order to boost dryness, modern screw presses have shaft dewatering in the final section where there is no flighting, and this allows the filtrate to drain off through a screen plate mounted on the shaft.

Although the basic design of a screw press is simple and in general supervision of operation and maintenance of the screw press is reduced in minimum, high pressure and torque will cause wear at the high-pressure section of the press. Wear can require excessive maintenance if the pulp contains abrasive particles such as grits from pulp stones. The two main manufacturers of screw presses are Kvaerner Eureka (Thune) and Andritz.

CHAPTER 10

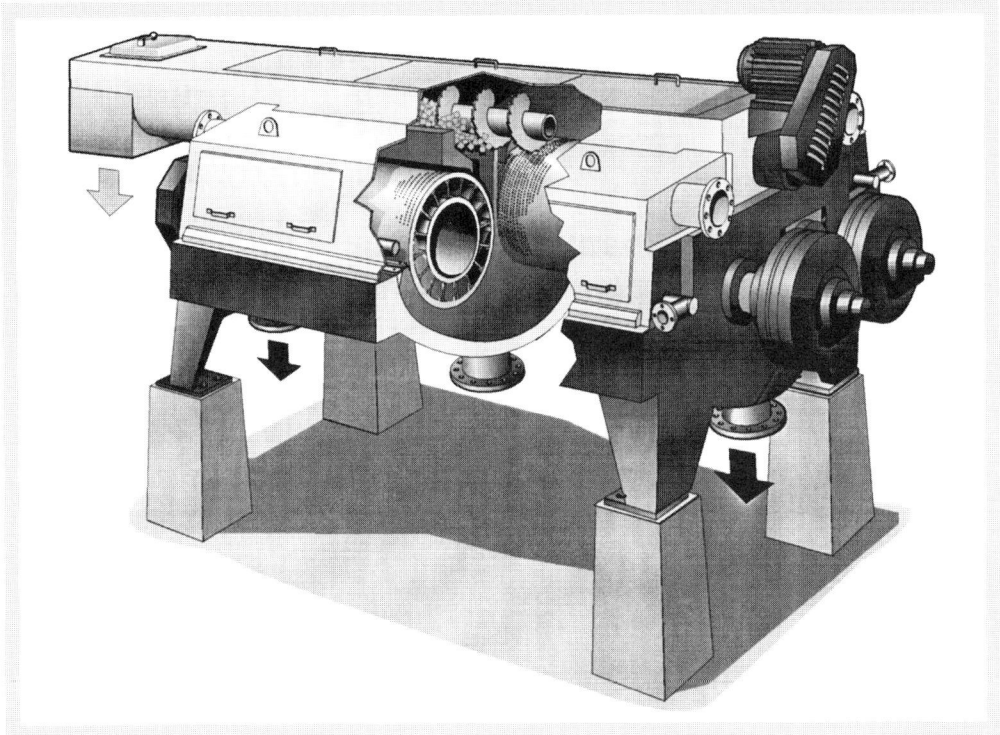

Figure 2. Sunds Defibrator twin-roll press.

1.1.3 The twin-roll press

The twin-roll press is capable of dewatering pulp with an inlet consistency of 3%–10% to a dryness of 35%–45%, depending on the type of press.

When the pulp suspension enters the pressurized vat, the fluid is drained through the holes in the roll face and a mat of fibers is formed on each roll. The fiber mat is pressed to an increased consistency in the nip between the rolls. After the nip, the pulp is taken off by a shredder screw to the outlet of the press. The filtrate flows through the openings at the end of rolls and is discharged from outlets at the bottom of the vat.

Dryness after the press depends on the freeness, the flow, the consistency, and the temperature of the feed pulp and the vat pressure. At varying operation conditions, a constant nip pressure control helps to stabilize dryness variations of the discharged pulp.

The dewatering rolls are synchronously counter rotating. One roll is in a fixed position, and the position of the other roll is adjusted by means of hydraulic cylinders. A hydraulic gap relief control protects the rolls against overloading. A hydraulic motor drives each roll, which provides constant torque at a variable speed. This is important for an even discharge consistency. The flow of the oil controls the speed of the rolls, and the oil pressure determines the torque. In order to reduce the fines content of the filtrate, the dewatering press can be equipped with wire covered rolls but, in reject thickening, wires are not needed. In fact, fines removal from pulp is favorable for a stable dryness of

Refining of shives and coarse fibers

the discharge pulp, which improves operation of the reject refiner. The even dryness of the pulp and the compact design of the press are benefits of the twin-roll press.

1.1.4 The twin-wire press

The twin-wire press is capable of covering the entire range of potential freenesses from 30 to 600 ml CSF, and it also can be used over a very wide range of inlet consistencies. The twin-wire press requires more space than other types of presses.

Figure 3. Andritz twin-wire press.

The headbox distributes the suspension over the entire working width of the twin-wire press. A twin-wire press with a low-consistency headbox is used with consistencies up to around 6%. The pulp is pumped from below into the headbox, which widens gradually in such a way that the pulp is fed in evenly over the entire width of the machine. When the pulp suspension is brought into the wedge zone between the two revolving wires, it has a slight overpressure in order to cause the pulp to dewater at the beginning of this zone. Due to the wedge formed by the wires, the filtrate is displaced continuously and pushed through the two wires and through the perforated wedge plates supporting the wires in this part of the machine. Since the pulp suspension is still free-flowing at this stage, the wedge zone must be well sealed at the sides to prevent the pulp from running out. At the end of this zone, the pulp has a consistency of 12%–15%.

An alternative type of headbox and wedge zone can be used for medium inlet consistencies of 6%–12%. In this case, the full length of the wedge zone is no longer needed. The top section is shortened and a medium-consistency headbox is used instead with pulp feed from above, which distributes the very high-viscosity (thixotrope) pulp suspension over the working width. Although there is no real need for a wedge at inlet consistencies of 10%–12%, a short wedge zone allows the press to be operated under disturbed plant conditions, e.g., low inlet consistencies to the press.

After the wedge zone, the mat is then further dewatered to 20%–25% by applying the area pressure in the next zone. This area pressure is generated by wrapping the wires around rolls. The smaller the roll diameter and the higher the wire tension are, the greater is the pressure generated. The larger the angle of the wire wrap around the roll and the larger the roll itself, the longer is the dewatering time.

After this area pressure dewatering stage follows a press section, which contains one to five press nips, depending on the desired final dry content. At the beginning of this zone, the linear pressure applied is low; however, it increases toward the end of the machine and can be as much as 250 N/mm. This is equivalent to a pressure of 25 t over a width of 1 m. Due to high pressures, a final dryness of over 50% is possible. In applications for dewatering rejects, however, lighter models with lower line pressures and fewer nips are in use because dryness of 30%–35% is sufficient.

1.2 Reject refiners

The single disc (SD) refiners are most commonly used in reject refining. The mechanical construction of the modern pressurized SD refiners might vary, but the basic operation principle is the same: one disc is rotating and the other disc is stationary. Thickened reject is fed by means of a screw feeder through an opening in the stator disc into the eye of the refiner. At the refiner inlet, rotor wings make the pulp rotate and the centrifugal force feeds pulp into the gap between the segments of the refiner disc. The load of the refiner motor is set by adjusting the gap clearance between the rotor and stator segments. From the gap, the refined pulp enters the refiner casing and, from there, pulp is blown by means of steam generated in refining to a cyclone where steam and pulp are separated.

In addition to single disc refiners also double disc refiners and in large TMP plants also CD- and Twin-refiners may be used for reject refining as well. In addition to high consistency refiners also low consistency (LC) refiners may be used for reject refining. The LC refiners as well as other than the SD- high consistency refiners are described elsewhere in this book (see chapters on thermomechanical pulping and post refining).

Sunds Defibrator RGP 200 series refiners

The Sunds Defibrator RGP 200 series refiners are compact single-disc refiners for high-consistency refining. The RGP 200 type refiner combines a sturdy construction and easy operation with modern hydraulic and instrument systems. The refiner frame is a heavy iron casting and refiner housing of cast stainless steel. Design of the frame and housing are symmetrical so that thermal expansion and pressure forces do not affect parallelism of stator and rotor discs. The front cover is of solid stainless steel plate. The refiner has a large feed opening through the front cover for uniform feeding and extraction of the blow back steam. In the pressurized high-consistency refining, material to be refined is fed by means of a plug screw and ribbon screw feeders into the eye of

Refining of shives and coarse fibers

the refiner. Refined pulp is discharged through the blow line connection flanges. For atmospheric applications, the refined pulp is discharged through a large bottom opening. The rotating assembly is a rugged ready-to-mount cartridge unit, complete with rotor, shaft seal, bearings, and hydraulic rotor positioning system. The large-diameter shaft is supported by spherical roller bearings which eliminate the need for a separate thrust bearing. For atmospheric application, the shaft sealing is a packing box of traditional design. For pressurized applications, a mechanical shaft sealing is used.

Figure 4. Sunds Defibrator RGP 268 single disc refiner.

The refiner gap is controlled by changing position of the rotor assembly. This is made by means of a hydraulic system, which makes it possible to control the refiner gap with high precision. The refiner can be furnished with the True Disc Clearance (TDC) sensor for a precise control of the refining gap. Since the hydraulic gap positioning system is counterbalanced by the pressure between the refining discs, a loss of raw material feed would cause the refining discs to clash. In order to protect the refining discs, the feed guard unit instantly activates the hydraulic disc-positioning system for rapid opening of the gap in case of such feed disturbances. The rotor as well as the stator refining segments are premounted on segment holders, forming a ready-to-mount assembly. For changing segments, the top half of the refiner will be removed.

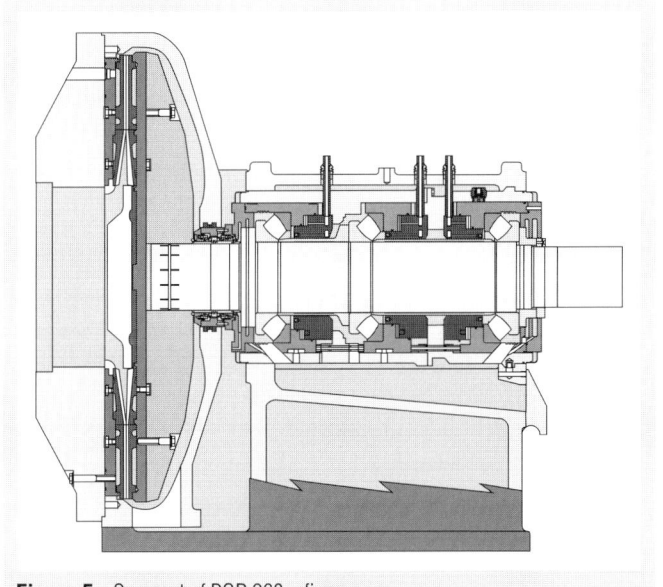

Figure 5. Crosscut of RGP 268 refiner.

CHAPTER 10

The disc diameter of the RGP 200 series refiners depends on required refining power and varies from 44 to 68 inches (RGP 244–RGP 268). RGP 268 is designed for a maximum thrust load of 90 tons and max. rated power of 15 MW. Design steam pressure of the RGP 200 series refiners is 1.4 MPa.

Andritz SB 150 and 170 refiners

The Andritz SB150 and 170 single-disc refiners are designed for high-consistency mechanical pulping and reject refining applications.

Material is fed into the refiner by means of a side entry feeder, which feeds the material to the large-diameter, high-speed, shaft-driven ribbon feeder. Large open area and high centrifugal force improves steam and fiber separation at the refiner feed. The rotor disc is mounted between the bearings. This arrangement provides a high degree of dynamic stability. The refiner design enables a short and compact thrust path by pulling the rotating disc against the stationary disc, which guarantees a good parallelism of the refiner gap. A plate gap monitor provides rapid feedback of gap clearance variations, which can be used for process control adjustments as well as to prevent plate contacts. A servo-based hydraulic system has the capability to operate in constant pressure or fixed position mode. A large refiner casing opening provides an easy change of the refiner plate segments. The compact design minimizes the floor space requirements. The maximum rated power of the SB 150 and 170 refiners are 12 and 15 MW respectively.

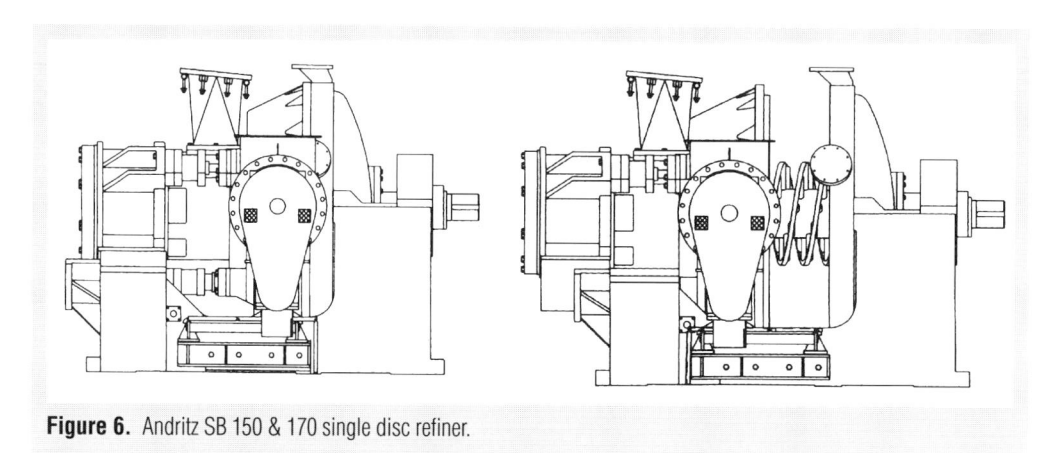

Figure 6. Andritz SB 150 & 170 single disc refiner.

1.3 Reject refining systems

Constant conditions at feed of the reject refiners is needed for good and stable refiner operation. Thickening equipment should be dimensioned for the lowest consistency and freeness and highest production in order to eliminate excessive variations due to lack of thickening capacity. Sometimes speed-controlled screws are used for proportioning thickened reject from common transportation screws to one or more reject refiners, but more common is a system where all the reject refining lines have their own separate thickening.

Refining of shives and coarse fibers

Different kinds of reject refining systems can be used:

- Reject refining in the TMP mainline refiners
- Atmospheric one-stage refining
- Pressurized refining in one or two stages
- Atmospheric or pressurized refining in two stages with screening between the stages
- LC-refining
- Combinations of HC- and LC-refining.

Reject refining in mainline refiners is used sometimes to reduce the investment cost of the reject refining, but the refining efficiency of the coarse fibers as well as the controllability of the process are not comparable with a separate reject refining system[1].

Atmospheric reject refining is generally used when the need of the refining energy is relatively low. When higher SEC is needed, large volumetric flow of steam at the feed of an atmospheric refiner can result in feeding problems. Elimination of these feeding problems by excessive use of dilution water results in low refining consistency and reduced pulp quality.

Pressurized reject refining systems are used when higher energy input is needed. Nowadays most of the new reject refining installations are pressurized. Figure 7 presents a typical pressurized reject refining system. A reject refining system might be only slightly pressurized in order to transport the refined pulp by blowing it by means of the steam generated in refining. At a constant level of SEC, the difference in pulp quality between pressurized and atmospheric refining is small. The main benefit of a fully pressurized system (approximately 3 bar) is the improved loadability of pressurized refiners. This enables the use of higher specific energy which, in turn, results in improved development of the fiber properties. In the pressurized systems, refiners are equipped with plug screw feeders and with pressurized cyclones for separation of steam and pulp. Figure 8 presents typical control loops of pressurized reject refining.

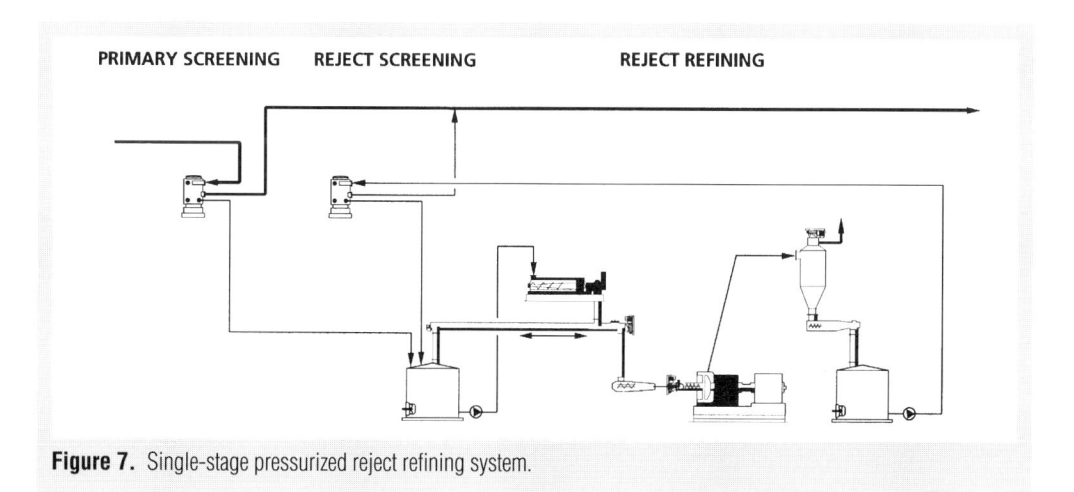

Figure 7. Single-stage pressurized reject refining system.

CHAPTER 10

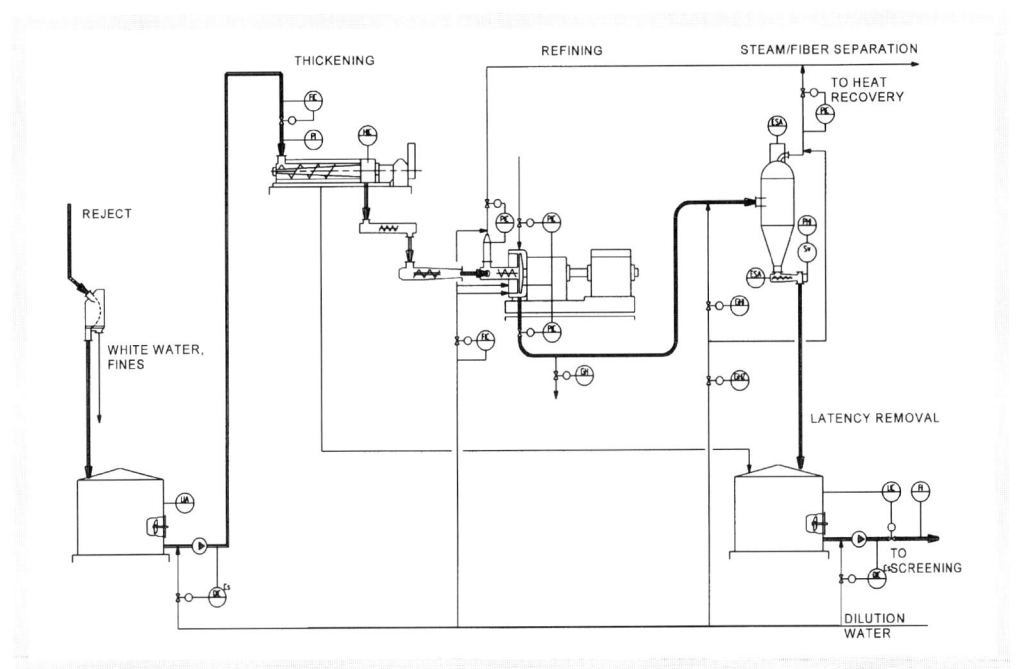

Figure 8. Control loops of a single-stage pressurized reject refining system.

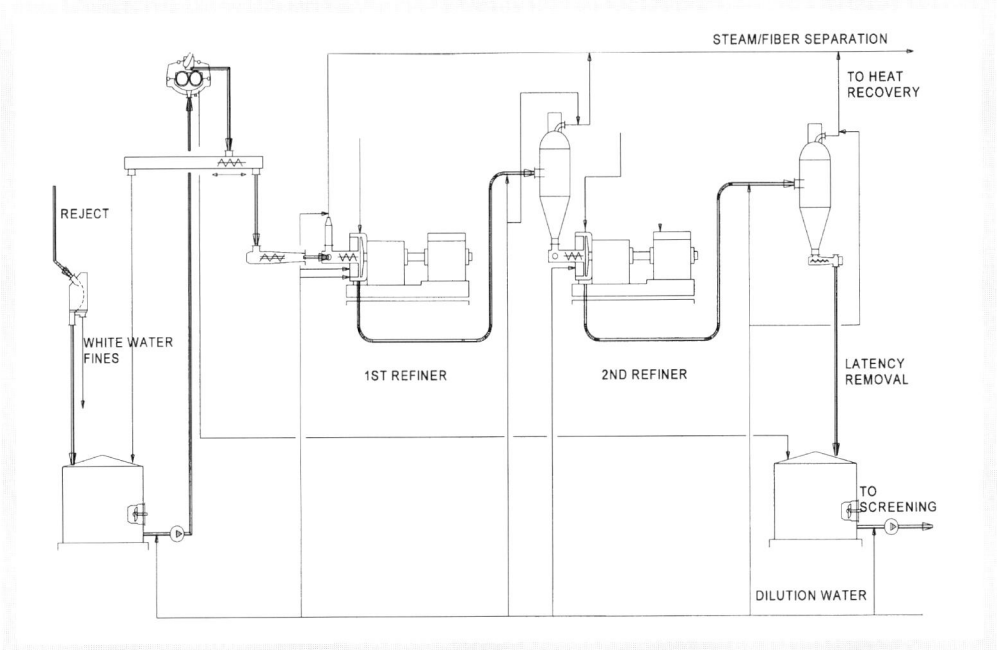

Figure 9. Two-stage pressurized reject refining system.

Refining of shives and coarse fibers

Two-stage pressurized reject refining is used when the need of specific refining energy is very high. This is usually the case when long fiber TMP reject is refined to meet the demands of a high-quality printing paper. In two-stage refining, it is easier to maintain the fiber length when high specific energy is used in refining. Properties in single- and two-stage refining are almost equal. Shive reduction in single-stage refining might be even better than in two-stage refining because, at a constant SEC, the plate gap clearance of single-stage refining is smaller than in two-stage refining.

Screening or fractionation between the reject refining stages is used to improve the separation and refining of long, coarse fibers. In this way, by using more energy, properties of those fibers can be improved.

The Thermopulp® reject refining process[2] has a very high temperature and pressure in refining (6–7 bar, 165°C–170°C). Figure 10 presents a flowsheet of the Thermopulp reject refining system. Due to compression of steam at high pressure, the Thermopulp refiner operates at a reduced gap clearance, which increases the refining intensity and reduces the specific energy consumption. Fiber cutting is avoided by improved flexibility of fibers at increased temperature.

Low-consistency reject refining has the lowest investment cost. LC refining effectively reduces the amount of large shives, and it has some potential to improve bonding ability of fibers with less energy than high-consistency refining. In spite of lower capital expidentures, LC refining is not commonly used in the refining of mechanical pulp screen reject because with maintained fiber length only relatively small amounts of refining energy can be used in LC refining. In the conical type LC refiners, higher SEC can be used and improved fiber development can be reached with less cutting than in disc-type LC refiners[3]. Regardless of the LC refiner type, the low energy input in one or two passes of LC refining is not sufficient for a proper development of the coarse fibers

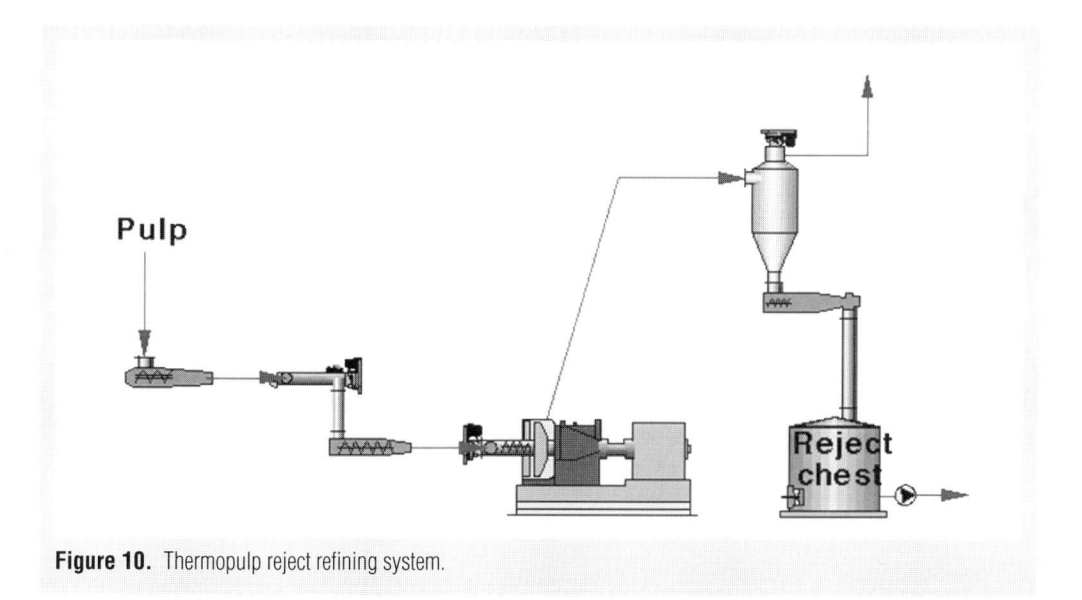

Figure 10. Thermopulp reject refining system.

CHAPTER 10

of TMP reject[4]. Low-consistency refining of groundwood reject is successfully applied in some mills producing groundwood-based LWC paper.

A suitable combination of HC and LC refining of reject could result in improved treatment of the most critical fiber fractions. In this way, a part of expensive thickening equipment can also be eliminated and the good energy efficiency of LC refining can be utilized. The position of the LC refiner in the process depends on the targets of LC refining: When additional refining of the whole pulp after screening and reject refining is needed, then traditional LC post refining should be used (see the chapter on post refining); when the target of LC refining is an improved treatment of long coarse fibers, then the LC refiner should be placed in the reject line after HC reject refining.

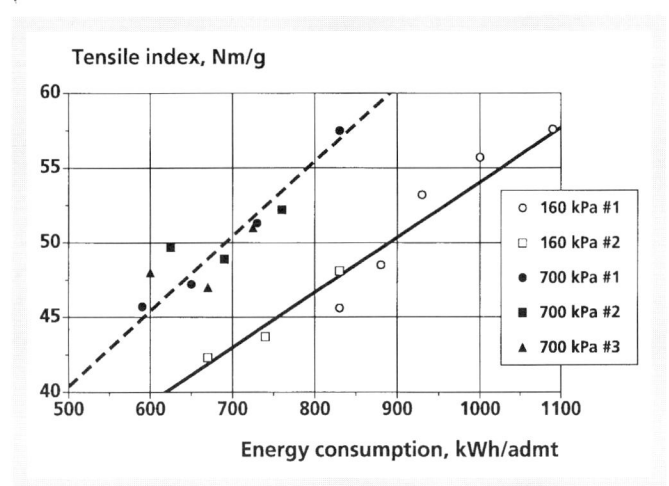

Figure 11. Development of tensile index in Thermopulp (700 kPa) and in conventional (160 kPa) reject refining. Results of mill trial in RGP 65 DD refiner.

The following general rules for a well operating reject refining system are worth mentioning:

- Good fractionation efficiency of shives and coarse fibers in the screening and cleaning
- Reject refining in a separate reject refining system
- Sufficient dimensioning of equipment used in fractionation, thickening, and reject refining
- Sufficient SEC in refining of coarse pulp fractions
- Individual fractionation, reject refining, and circulation water systems for each paper machine.

2 Operation and control of reject refining

2.1 Key variables of reject refining

Although extensive studies[5, 9] have been made, gap phenomena of high-consistency refining are not completely known. The key variables of reject refining are: the properties of the feed pulp, the specific energy consumption, and the intensity of refining.

Refining of shives and coarse fibers

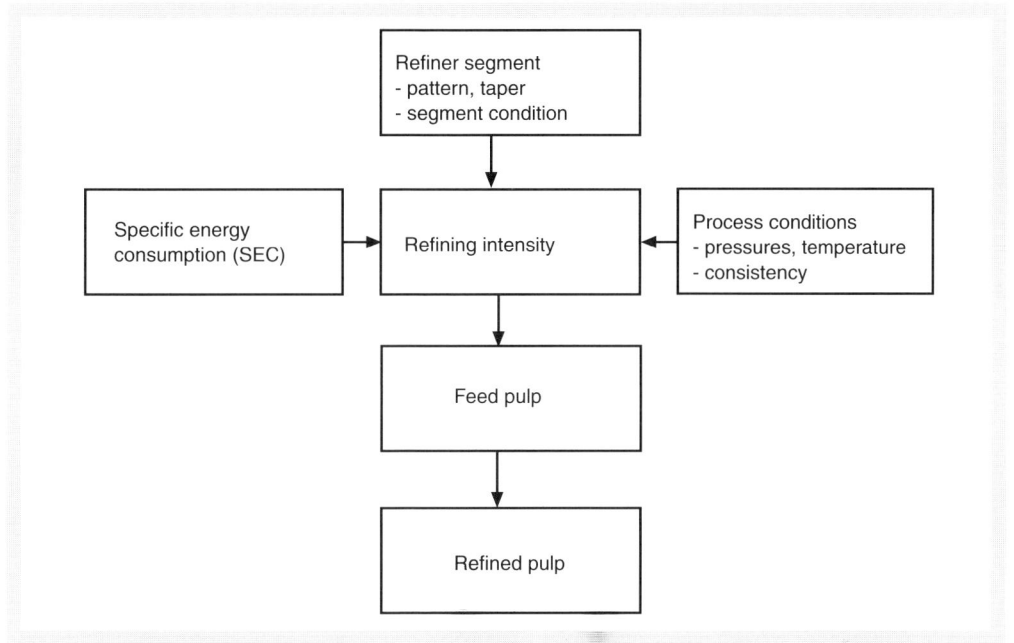

Figure 12. Key variables of reject refining.

The properties of the feed pulp determine the need of the refining energy. Stone groundwood and short-fiber DD-TMP consume less energy to reach a certain freeness than a long-fiber SD-TMP. Long-fiber rejects tolerate the use of higher SEC and intensity than short-fiber rejects without an excessive shortening of fiber length. Removal of fines from the feed pulp increases the freeness and consistency, which improves the loadability of the reject refiner.

The properties of feed pulp can also be changed by increasing the temperature of fibers as is made in the Thermopulp reject refining.

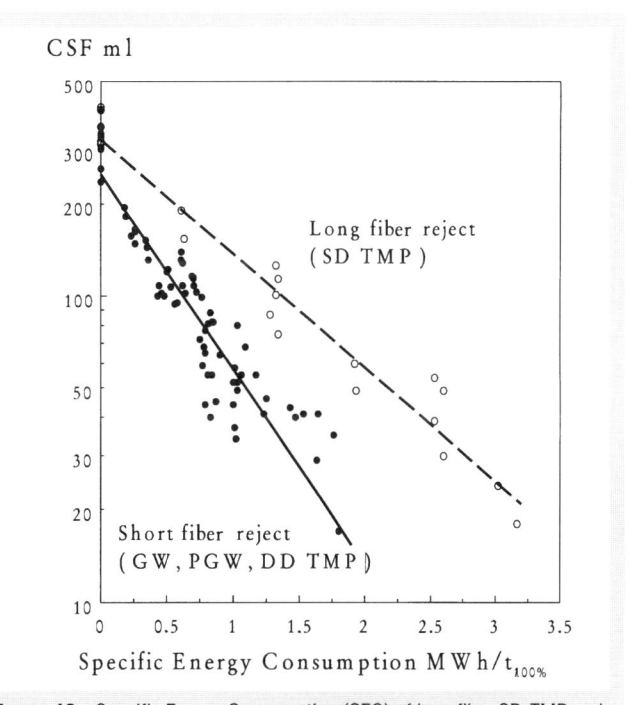

Figure 13. Specific Energy Consumption (SEC) of long fiber SD-TMP and short fiber GW-, PGW- and DD-TMP reject.

CHAPTER 10

This makes fibers flexible to tolerate higher refining intensities. In some cases, chemical treatment (sulfonation, alkaline peroxide, etc.) is used to improve the bonding ability of fibers. Chemical treatment also results in essential reduction of refining energy consumption. In order to avoid excessive reduction of optical properties and increased effluent load, only a small coarse fraction of reject should be treated with chemicals. For a more detailed description of the chemical treatments of coarse fibers, see the chapter on chemimechanical pulping.

Specific energy consumption (SEC, MWh/t) is the main variable of reject refining. In most refiners, the amount of refining energy is controlled by moving the position of the rotor disc in relation to the stator disc. Motor load is increased when the clearance between the refiner discs is reduced. When refining energy is increased, the freeness is reduced and the bonding ability of fibers is improved in relation to the specific energy consumption (see Figs. 11 and 13).

The refining intensity is the specific energy consumption and refining power per bar impact[5]. The refining intensity depends on the flow conditions of the pulp and the steam in the refiner gap. A short residence time of fibers in the gap reduces the amount of pulp in the gap, which in turn results in reduced gap clear ance and increased refining intensity. The change of the refining intensity changes the nature of the refining: Lowintensity refining can be described as a fibrillating type of refining and very high-intensity refining as a cutting type of refining. There is an optimum for the refining intensity – too low refining intensity increases the specific energy needed for a certain development of fibers, and too high intensity deteriorates the pulp quality by excessive reduction of the fiber length. Cutting-type of refining should be avoided due to deterioration of the bonding ability of fibers and fiber fractions.

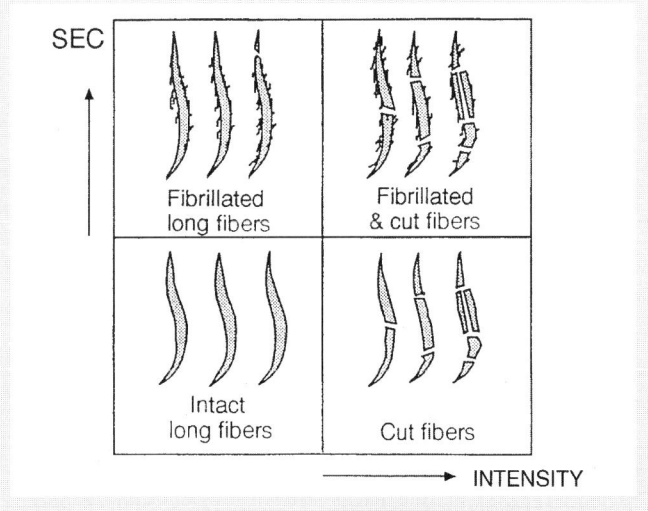

Figure 14. Increased refining energy (SEC) develops surface properties (fibrillation) of fibers and increased refining intensity has a tendency to reduce length of fibers (cutting).

2.2 Process conditions and intensity in reject refining

The refining intensity is determined through the refiner gap clearance. By closing the refiner gap, both SEC and intensity of refining are increased. By means of the loading device, the refiner gap clearance can be opened or tightened, but the exact size of the gap is determined by the flow conditions of fibers and steam in the gap. The process conditions affecting the steam and pulp flow in the gap are: the refiner throughput, the refining consistency, the steam pressures at the feed, the casing of the refiner, and the design of the refiner segments.

Reject refiner throughput should be kept as constant as possible. Uncontrolled variations in throughput change SEC and pulp quality as well. The refiner throughput is calculated based on the pulp flow and consistency after the latency chest of refined reject. Inaccuracy of consistency measurements makes the calculated throughput, however, unrealiable for the exact control of the SEC in reject refining.

Refining consistency is determined by measuring the consistency of samples taken at the discharge of the reject refiner. Consistency affects the refining intensity, e.g., a decrease of the refining consistency reduces the gap clearance by speeding up the pulp flow through the gap (higher centrifugal force) and by reducing the steam pressure in the gap. In general, refining consistency is kept as constant as possible by avoiding variations in reject refining conditions. Nowadays it is also possible to control refining consistency based on the measurement of the blowline consistency.

Steam pressures in the feed and inside the casing of the refiner have an influence on the steam and the pulp flow conditions in the gap. Increased differential pressure between the casing and the feed turns the steam to flow more toward the refiner feed. This increases the recirculation of pulp in the gap, which in turn increases the gap clearance and decreases the refining intensity. Increased steam pressure level of refining results in reduced gap clearance and increased refining intensity due to the compression of steam to a smaller volume. Compared to atmospheric refining, the reduced volume of steam makes it easier to use higher SEC in pressurized refiners. Although increased pressure in refining reduces the gap clearance, fiber cutting in a pressurized refiner is not increased because higher pressure/temperature makes fibers more flexible to tolerate higher refining intensities. This phenomenom has been utilized in the Thermopulp refining system[2].

Refiner segments play the key role in the runnability, in the pulp quality development, and in the energy consumption of reject refining. As far as runnability is concerned, efficient steam removal from the gap is important. Selective segment plate patterns with specific grooves for steam removal are used to improve runnability of segments (see chapter 7). Refiner segment design has a great impact on the intensity of refining. The segment design parameters increasing the refining intensity (reducing the gap clearance) are: the increased groove volume, the reduced number and height of dams, the reduced width of bars, the shorter length of refining (parallel) zone, and the forward pumping angle of rotor bars, etc. Each type of refiner segment has its specific curve according to which pulp properties are developed when the specific refining energy is increased. Test runs are needed to find out the best combination of refiner segment design and refining conditions. For more information on refiner segment patterns, see the chapter on thermomechanical pulping.

2.3 Disturbances in reject refining

During normal steady operation, a target is to keep all the variables of the reject refining as constant as possible. However, due to different disturbances in normal mill conditions, this is not possible. Typical disturbances are uncontrolled variations of throughput, refining consistency, refining pressures, and condition of the refiner segments[6].

Reject refiner throughput variations reflect in the changes of the SEC and the quality of refined reject. Changes in the mainline refining will change conditions of screening which in turn change the reject rate and the throughput of the reject refiner. A poor operation of the screw presses might also be a reason for the throughput variations. Due to the lack of reliable measurements, the amount of the feed pulp is not exactly known. Throughput calculations based on the flow- and consistency measurement often give erratic results due to the sensitiveness of the measuring equipment to temperature, pulp quality, amount of air, etc.

Refining consistency can vary due to poor operation of thickening equipment. The throughput and the freeness variations or the lack of the dewatering capacity are the normal reasons of the consistency variations. Therefore, it is difficult to keep the main variables under control, when both the amount of the feed pulp and the refining consistency are varying. Sampling in high consistency at the refiner discharge will not give reliable results because part of the steam is condensated into the pulp.

Steam pressure variations at the refiner feed or the refiner casing affect the gap clearance by changing the steam flow conditions in the gap as well as by changing the balance of the axial forces in the refiner. A good control of pressure variations is difficult due to the slow operation of the control valves. A separate steam separation cyclone for each reject refiner helps to keep steam pressures under control. Steam pressure fluctuations can result in considerable changes of the refiner motor load and SEC in refining.

Condition of the refiner segments has an influence on pulp quality. Worn segments should be replaced in order to maintain the pulp quality. Under controlled conditions, the lifetime of the reject refiner segments is 2000–3000 hours. If the feed pulp contains abrasive material such as sand, the lifetime of the segments will reduce drastically. Sand removal from the groundwood reject is essential for a good lifetime of the reject refiner segments. Wear problems caused by sand have become more common when using modern screen baskets with narrow slots, tight enough to guide the sand particles into the reject flow of the screens.

3 Effect of reject refining on pulp and paper properties

3.1 Development of fiber properties

The quality of the mechani-cal pulp is determined mainly by the size distribution and by the bonding ability of different fiber fractions of the pulp[7, 8]. Poorly developed fibers are separated in screening and cleaning and treated in the reject refiners to meet the demands of the end product.

Reject rate, or the amount of fibers to be taken into refining, depends in addition to the end product requirements on the separation efficiency of fractionation and also on the distribution of the refining energy between the mainline refining and grinding and the reject refining. Excessive refining in the mainline refiners should be avoided because the cutting type of refining occurs more easily in the mainline refining (Fig. 15) than in reject refining. SEC of mainline refining should be limited to a level where severe shortening of fiber length will not yet occur[9]. Presumably longer fibers of reject form a stronger mat of fibers in the gap, which tolerates higher refining intensities without excessive shortening of long fibers.

Specific refining energy needed for a certain development of fiber properties depends, in addition to wood and fiber characteristics, on the defibration process and on the specific energy used in defiberizing and refining. High intensity in the reject refining reduces the consumption of the refining energy, but this is mostly reached at the cost of the length and the bonding ability of fibers. The shive content and the light scattering, however, can be improved by means of increased refining intensity. A coarse refiner segment pattern increases the refining intensity. Table 1 shows effects of fine and coarse segment patterns on pulp properties. A coarse pattern (high-refining intensity) reduces SEC, bonding (tensile strength), and shive content, but light scattering increases.

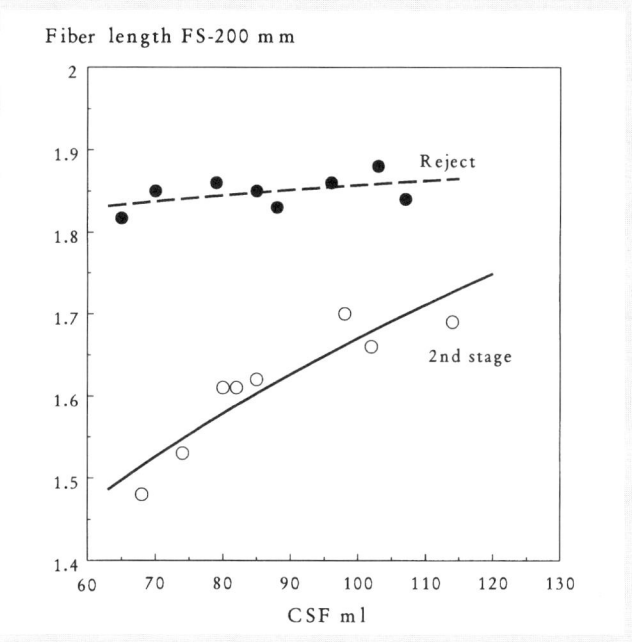

Figure 15. Freeness and fiber length after mainline and reject refining of TMP.

Table 1. The effects of fine and coarse segment patterns on pulp properties.

	Segment 1 Fine Pattern	Segment 2 Coarse Pattern	Difference
CSF, ml			
feed	523	508	
refined	118	126	
Tensile index, Nm/g	66	61	− 8%
Tear index, mNm^2/g	8.3	8.2	
Light scattering, m^2/g	42.8	45.9	+ 7%
Bauer McNett fractions, %			
+16	53.6	37.3	
+28	10.5	13.0	
+48	17.9	26.6	
+200	6.0	8.9	
−200	12.0	14.2	
Shives, %	0.09	0.03	
SEC, MWh/t	1.43	1.11	−22%

CHAPTER 10

Although in reject refining defiberizing of stone groundwood shives requires less refining energy than TMP shives, refined TMP reject at a certain freeness usually has less shives than stone groundwood reject (Fig. 16).

Shive content is reduced almost linearly with decrease of the refiner gap clearance. However, if the refining intensity is too high (too small gap), fibers as well as shives are cut down to smaller particles which have poor bonding ability. This reduces strength properties and can cause other problems such as linting in offset printing. An LC refiner usually operates with a smaller disc clearance than an HC refiner. LC refining is usually quite efficient in removing large- and medium-sized shives and at constant SEC even better than in HC refining in this respect. On the other hand, due to poor loadability of LC refining, more refining stages are needed to reach similar development of fiber properties as are found in HC refining.

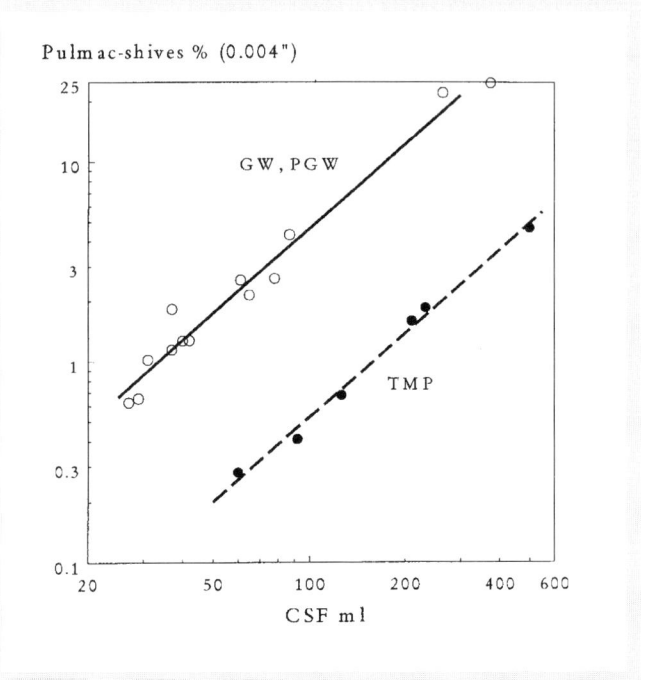

Figure 16. The reduction of shive content at a different degree of refining is linear with logarithmic scales of shive content and freeness.

Shive removal efficiency (SRE) of a reject refiner is calculated as follows:

$$SRE = (shives_{feed} - shives_{refined})/shives_{feed} \times 100$$

Fiber length is more or less reduced in refining: The higher the refining intensity is, the more the length of fibers is reduced in refining. At the very beginning of refining, the amount of long fibers might even increase when the shives are reduced to long fibers. An excessive shortening of fibers should be avoided in refining because the quality of middle and fine fractions produced by cutting the long fibers is poor.

The changes of the fiber properties in refining are described as an external and internal fibrillation of fibers. During refining, the delamination of the fiber wall reduces the stiffness and the coarseness of fibers decreases as the outer layers of the fiber wall are peeled off[10, 13]. Bonding ability of long fibers is increased linearly with the increase of the refining energy. Meanwhile the tear index at first increases, but then at higher SEC tear decreases due to reduced strength of single fibers.

The strength properties of the whole reject pulp develop basically in a similar way as the properties of long fiber fraction. Due to improved flexibility and bonding ability of long fibers, sheet consolidation increases in reject refining. Especially the development of stiff, thick-walled latewood fibers is important for the development of paper properties. It is obvious that stiff, thick-walled latewood fibers react differently to refining than do the slender earlywood fibers. Some researches report that the amount of the thick-walled latewood long fibers is reduced in reject refining without noticeable changes in bonding ability[11, 12]. Some other studies report about improved bonding ability of latewood fibers[13]. It might be that the behavior of thick-walled latewood fibers in refining depends on the conditions used in refining. Heavier intensities result in the shortening of stiff latewood fibers, and lower intensities (as well as higher temperatures) develop properties of latewood fibers instead of cutting.

Due to their large specific surface, the amount and quality of fines have a significant effect on the properties of mechanical pulp. The amount of fines is increased in reject refining by gradual external fibrillation of fibers. Refining determines the nature of fiber breakage: At lower instensities, coarseness of fibers is reduced and about the same amount of fines (P 200 mesh) will be generated. When refining intensity is increased by loading up the refiner over a certain limit, generation of coarse fines (R 200 mesh) instead of P 200 mesh fines increases rapidly. If the refiner load is further increased, long fibers are cut and middle fractions with poor bonding ability will be produced. Properties of fines varies depending on the process: Average lengths of the fines particles increase from GW to PGW and to TMP. The fines from reject refiners contain more long fibrillar particles than primary fines originating from defiberizing. Long fibrillar secondary fines improve bonding of the sheets while the light scattering is influenced mainly by the amount of the fines[14].

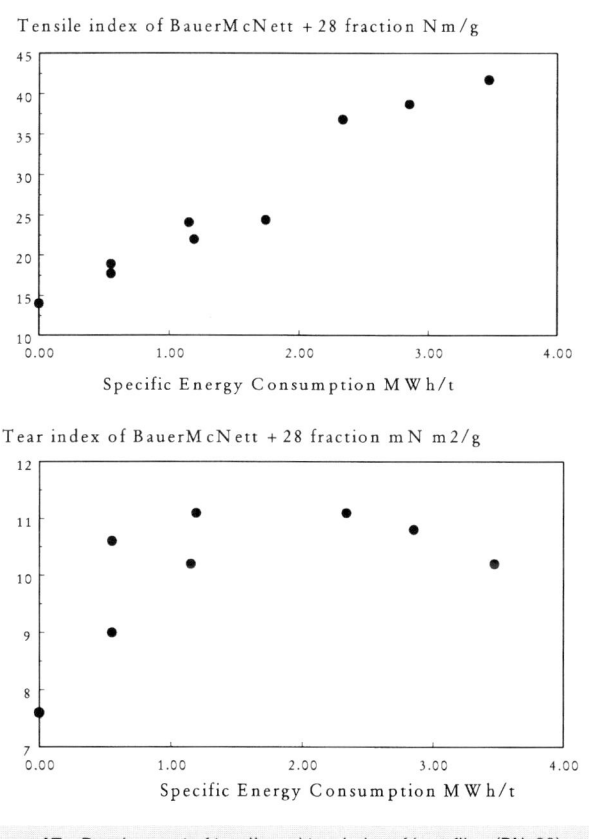

Figure 17. Development of tensile and tear index of long fiber (BN+28) fraction of TMP in reject refining.

CHAPTER 10

The main targets of the reject refining are:

- To eliminate fiber bundles (shives and minishives)
- To improve properties of fibers and fiber fractions and specifically properties of long and stiff fibers
- To increase amount of fines
- To maintain fiber length
- To control dewatering properties of final pulp.

3.2 Development of pulp and paper properties

The amount and quality of different fiber fractions determine the strength and surface properties of paper. Treatment of small amounts of reject is sufficient to reduce the shive content to a level which is acceptable, e.g., for interior layers of board. Higher reject rate and SEC are needed to eliminate problems caused by linting and surface roughening (fiber puffing) of LWC offset paper.

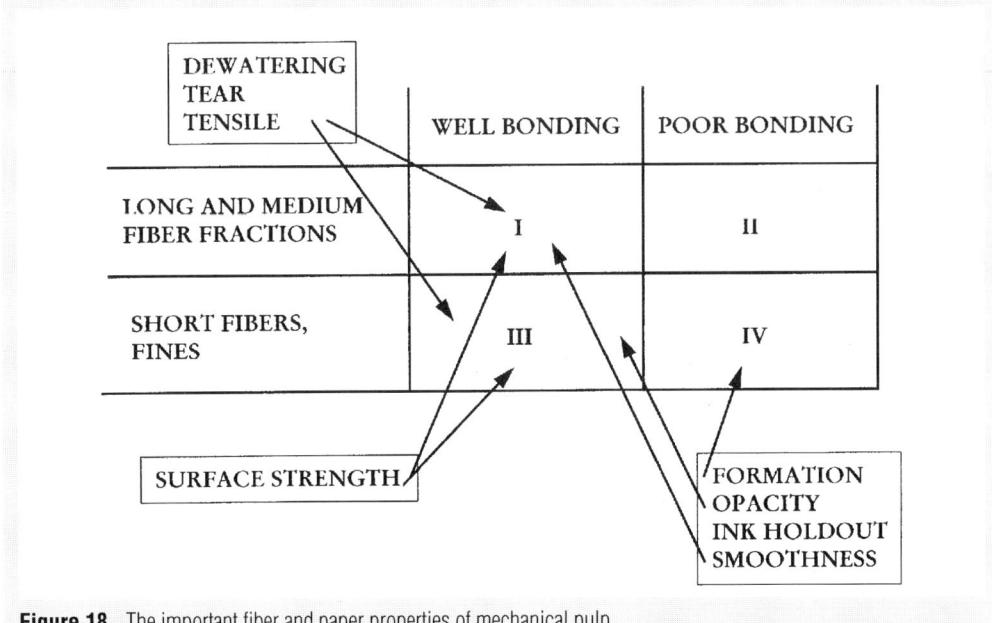

Figure 18. The important fiber and paper properties of mechanical pulp.

Good surface properties are needed for good printability of paper. In offset printing, potential problems of linting, surface roughening, and blistering are reduced or eliminated by efficient refining of all the fiber fractions, although treatment of the long coarse fibers is most critical. Good printability in offset is reached by sufficient development of fibers before screening and cleaning and by the efficient separation and refining of the coarse fibers. Surface smoothness of paper is improved by reducing the amount of the long fibers or by the effective development of the long fiber fraction separated in

Refining of shives and coarse fibers

screening. In rotogravure printing, a long fiber pulp with well developed long fiber fraction can have a good combination of printing smoothness and bulk[16].

Good opacity is most important for the printability of paper. The amount of fines and properties of fibers developed in mainline and reject refining determine the optical as well as the strength properties of mechanical pulp. The optical properties are very much influenced by the specific surface of the pulp, e.g., for a pulp containing about 25% fines (P 200 mesh), about half of the light scattering is obtained from the fines. Good development of coarse fibers reduces the need of calendering, which reflects also in an improved opacity of paper. Good strength improves the opacity by reducing the need of chemical pulp in the paper furnish.

Good runnability in papermaking as well as in printing with a minimum addition of expensive chemical pulp is important for the economy of papermaking. Mechanical pulp with long, slender fibers with good bonding ability is favored for the best combination of runnability and printability of

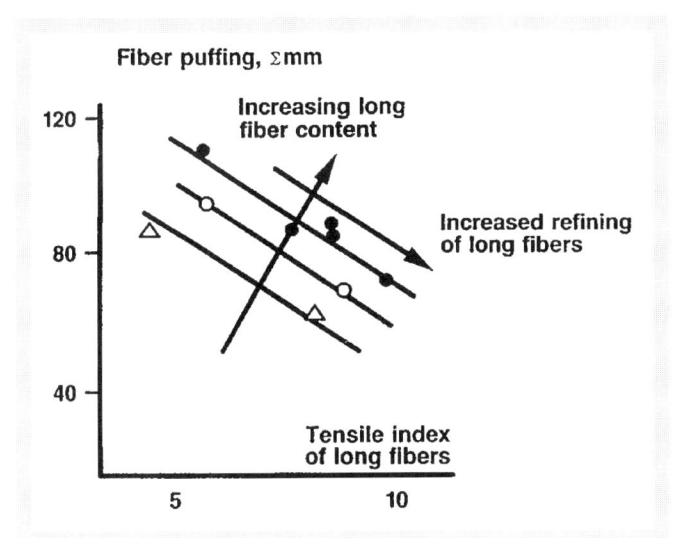

Figure 19. Fiber roughening as a function of long fiber content and bonding ability of long fibers of TMP[15].

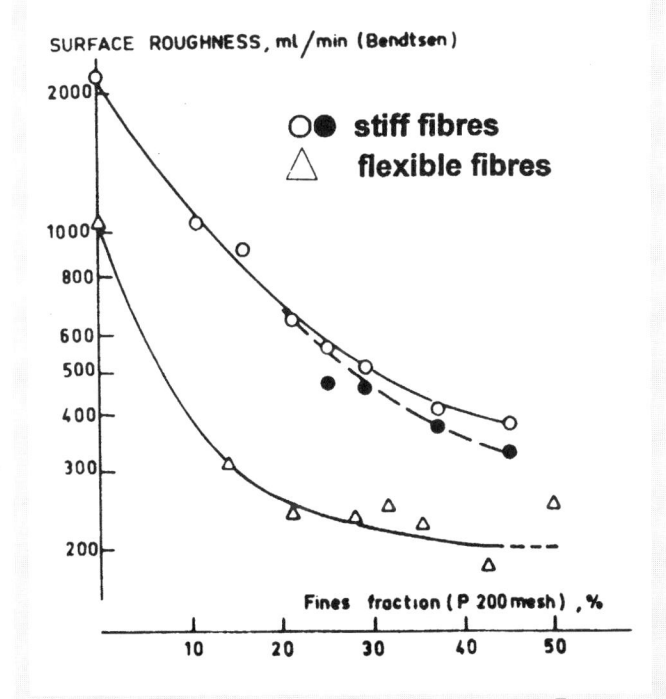

Figure 20. Effect of long fiber properties on surface roughness[7].

paper. Well controlled selective refining of mechanical pulp reject improves both runnability and printability of paper by concentrating the refining energy on the undeveloped long fiber fractions of mechanical pulp.

References

1. Corson, S., Appita J. 49(5):309 (1995).

2. Höglund, H., Bäck, R., Falk, B., Jackson, M. "Thermopulp® – A new energy efficient mechanical pulping process," 1995 International Mechanical Pulping Conference (Ottawa) Proceedings, CPPA, Montreal, p. 213.

3. Lumiainen, J., "Comparison of the mode of operation between conical and disc refiners," 1997 International Refining Conference Proceedings, PIRA, Leatherhead, p. 14:1.

4. Viljakainen, E., Huhtanen, M., Heikkurinen A., "Efficient papermachine operation with tailormade mechanical pulps," 1997 International Mechanical Pulping Conference Proceedings, SPCI, Stockholm, p. 173.

5. Miles, K. B., and May, W. D., J. Pulp Paper Sci. 16(2):J63 (1990).

6. Kortelainen, J. and Nystedt, H., unpublished project report of Sustainable Paper Energy Technology Research Programme 1993(98.

7. Mohlin, U. -B., "Properties of TMP fractions and their importance for the qualitly of printing papers," 1979 International Mechanical Pulping Conference (Toronto) Proceedings, CPPA, Montreal, p. 57.

8. Strand, W. B., Mokvist, A., Jackson, M., "Improving the reliability of pulp quality data through factor analysis and data reconciliation," 1989 International Mechanical Pulping Conference Proceedings, KCL, Helsinki, p. 362.

9. Härkönen, E., Ruottu, S., Ruottu, A., Johansson, O., "A Theoretical Model for a TMP-refiner," 1997 International Mechanical Pulping Conference Proceedings, SPCI, Stockholm, p. 95.

10. Karnis, A., "The mechanism of fiber development in mechanical pulping," 1993 International Mechanical Pulping Conference Proceedings, PTF, Oslo, p. 268.

11. Koljonen, T. and Heikkurinen, A., "Delamination of stiff fibers," 1995 International Mechanical Pulping Conference (Ottawa) Proceedings, CPPA, Montreal, p. 79.

12. Corson, S. R. and Ekstam, E. I., Paperi ja Puu – Paper and Timber 76(5):334 (1994).

13. Mohlin, U. -B., J. Pulp Paper Sci. 23(1):J28 (1997).

14. Heikkurinen, A. and Hattula, T., "Mechanical pulp fines – characterization and simplications for defibration mechanisms," 1993 International Mechanical Pulping Conference Proceedings, PTF, Oslo, p. 294.

15. Hoc, M., Tappi J. 72(4):165 (1989).

16. Forseth, T. and Helle, T., J. Pulp Paper Sci. 23(3):J95 (1997).

CHAPTER 11

Bleaching

1	**Bleaching or brightening**	**313**
2	**Factors affecting the brightness of high-yield pulps**	**313**
3	**Principles of bleaching high-yield pulps**	**318**
4	**Peroxide bleaching**	**319**
4.1	Reactions	319
4.2	Process variables	320
	4.2.1 The peroxide dose	320
	4.2.2 Removal of transition metal ions	320
	4.2.3 Pulp pH	321
	4.2.4 Additives	322
	4.2.5 Consistency	322
	4.2.6 Temperature	324
	4.2.7 Retention time	324
	4.2.8 Acidification	324
4.3	Bleach plant design and operation	324
	4.3.1 One-stage medium-consistency bleaching	325
	4.3.2 High-consistency bleaching	326
	4.3.3 Recycling of peroxide	329
	4.3.4 Refiner bleaching	329
	4.3.5 Flash dryer bleaching	329
5	**Dithionite (hydrosulfite) bleaching**	**329**
5.1	Reactions	329
5.2	Process variables	331
	5.2.1 Dithionite (hydrosulfite) dose	331
	5.2.2 Chelating agents	331
	5.2.3 Pulp pH	331
	5.2.4 Consistency	331
	5.2.5 Temperature	332
	5.2.6 Retention time	332
5.3	Bleach plant design and operation	332
	5.3.1 Tower bleaching	332
	5.3.2 Two-stage bleaching with peroxide and dithionite (hydrosulfite)	332
	5.3.3 Chest bleaching	334
	5.3.4 Refiner and grinder bleaching	334

6	**Other bleaching chemicals**	**334**
6.1	Potential bleaching chemicals	334
6.2	Sodium bisulfite	335
6.3	Formamidine sulfinic acid	336
7	**Bleaching of chemimechanical pulps**	**336**
8	**Brightness reversion**	**337**
8.1	General	337
8.2	Determination of brightness reversion	338
8.3	Thermal brightness reversion	339
8.4	Light-induced brightness reversion	340
9	**Effect on yield and papermaking properties**	**340**
9.1	Dithionite (hydrosulfite) bleaching	340
9.2	Peroxide bleaching	340
	References	342

CHAPTER 11

Carl-Anders Lindholm

Bleaching

1 Bleaching or brightening

The objective of bleaching mechanical and chemimechanical pulps is the same as that of bleaching chemical pulp: to increase pulp brightness. The means, however, are different. As chemical pulp bleaching mainly is based on lignin removal, mechanical pulp bleaching is based on elimination of colored groups in lignin. For that reason, bleaching of mechanical pulp is often referred to as lignin-retaining bleaching or brightening, to distinguish it from the principle applied in bleaching of chemical pulp, i.e., lignin-removing bleaching.

2 Factors affecting the brightness of high-yield pulps

Pulp brightness is determined by the light-absorption and light-scattering ability of a sheet made from the pulp. The optical properties of pulp can be examined with the aid of the Kubelka-Munk theory[1]. This theory determines the relation between the properties in the following way:

$$R_\infty = 1 + (k/s) - \sqrt{(k/s)^2 + 2(k/s)} \tag{1}$$

where R_∞ is brightness
       k light-absorption coefficient
       s light-scattering coefficient.

If the light-absorption and light-scattering coefficients are related to the brightness value, they should be determined at the same wavelength as brightness (457 nm; SCAN-C 11:75), and not at the wavelength normally used for determining the light-scattering coefficient (557 nm; SCAN-M 7:76). Figure 1 graphically illustrates the relation between the light-absorption and scattering-coefficients and pulp brightness. The light-absorption coefficient is a measure of the quantity of colored matter in the pulp. The light-absorption coefficient of mechanical pulps is mainly dependent on the wood raw material, although some changes can take place during the pulping stage. The light-scattering coeffcient is dependent on the pulping method.

CHAPTER 11

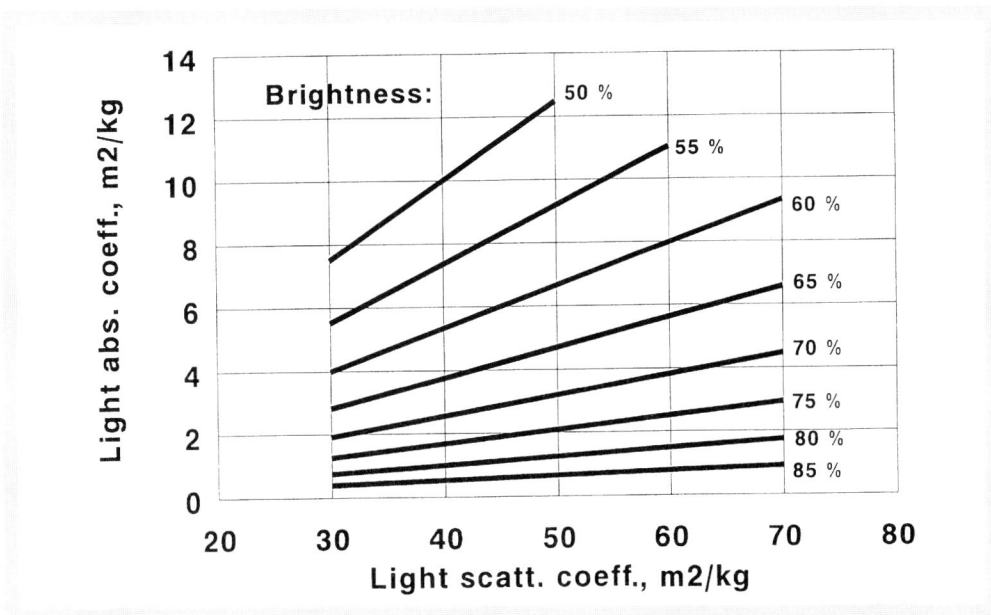

Figure 1. The relationships between pulp brightness and the light-absorption and light-scattering coefficients.

The brightness of high-yield pulps is close to that of the wood raw material. The wood substance is always colored, but the color varies between different wood species. Table 1 shows typical differences in brightness of groundwood pulps produced from various wood species. Quite large deviations from these typical values can occur.

Table 1. Brightness of unbleached groundwood pulps produced from various wood species.

Name	Wood species	Brightness, %
Europe[2]		
Norway spruce	*Picea abies*	65
Scots pine	*Pinus silvestris*	65
Aspen	*Populus tremula*	70
Western Canada[3]		
Western hemlock	*Tsuga heterophylla*	48
Spruce	*Picea sitchensis*	56
Balsam	*Abies amabalis*	50
Cottonwood	*Populus triocharpa*	60
Eastern Canada[3]		
Spruce	*Picea glauca*	61
Balsam	*Abies balsamea*	60
Pine	*Pinus banksiana*	55
Poplar	*Populus grandidentata*	48
New Zealand[4]		
Radiata pine	*Pinus radiata*	63

Of the main wood components, cellulose and the hemicelluloses are practically uncolored. Lignin is one of the main sources of color in wood. Lignin is not homogeneously colored, but it has certain functional groups, called chromophores, that absorb light and make the lignin colored. Coniferylaldehyde and α-carbonyl groups and various quinoidic structures are the most common chromophores in lignin (Fig. 2). In addition, phenolic groups that are not colored in wood can condense at elevated temperature into colored structures. Part of the differences in brightness between wood species can also be caused by the extractives; the amount and color of the extractives vary within broad ranges, especially in North American softwood species.

Brightness values of various components are not directly additive, but the light-absorption coefficients are almost additive. The contribution of the chemical components of wood to the light absorption coefficient can thus be expressed by an equation of the type:

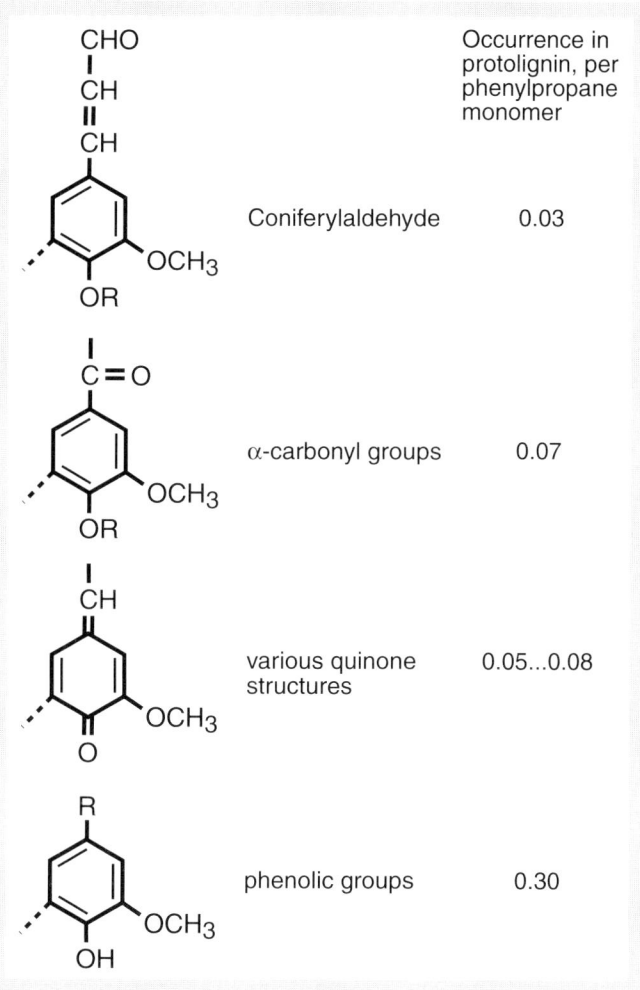

Figure 2. Various groups in lignin contributing to the color of lignin or taking part in reactions affecting the color[5].

Group	Occurrence in protolignin, per phenylpropane monomer
Coniferylaldehyde	0.03
α-carbonyl groups	0.07
various quinone structures	0.05...0.08
phenolic groups	0.30

$$k_p = c_C \cdot k_C + c_L \cdot k_L + c_E \cdot k_E \tag{2}$$

where k_P is the light-absorption coefficient of the pulp
       $c_C, c_L,$ and c_E the relative amounts of carbohydrates, lignin, and extractives, respectively, in the pulp
       $k_C, k_L,$ and k_E the light-absorption coefficients of carbohydrates, lignin, and extractives.

CHAPTER 11

Based on this equation, the contribution of the various chemical components to the light-absorption coefficient can be estimated as shown in Table 2. The example shows that more than 90% of the colored matter in Norway spruce (*Picea abies*) originates from lignin. The light-absorption coefficient of the extractives is also relatively high, but the contribution to the light-absorption coefficient of the pulp is limited because of the the low content of extractives in wood.

Table 2. Calculations of the contribution of the various components of Norway spruce (*Picea abies*) to the light-absorption coefficient.

	Relative amount, c_X	Light-absorption coefficient, k_X, m^2/kg	Contribution to the light-absorption coefficient of the pulp $c_X \cdot k_X$, m^2/kg
Carbohydrates	0.70	0.35	0.25
Lignin	0.28	20	5.6
Extractives	0.02	7.5	0.15
Whole pulp	1.0		6.0

Norway spruce (*Picea abies*) does not contain any significant amounts of colored extractives. The light-absorption coefficient has been determined to be 4.0–6.0 m^2/kg[6]. Under the assumptions that the light-absorption coefficient is not changed in mechanical defibration and that the light-scattering coefficient is 50–70 m^2/kg, this means that the brightness would be 62%–71% ISO. Mechanical pulps from Norway spruce are normally somewhat darker than this, due to some raw material- and process-related factors.

Storage of the wood raw material rapidly decreases pulp brightness. One-year storage can reduce the brightness by 7% ISO[7]. It therefore is important to minimize the storage time when striving to maximum brightness. The bark content of the pulp can also have a clear impact upon pulp brightness. Bark contents of 3% have been shown to reduce pulp brightness by up to 10% ISO[8].

In the pulping process, the elevated temperatures in the defibering stage can cause some condensation of the lignin structures, and metal ions in the process water can initiate darkening reactions (cf. Fig. 3). Table 3 presents typical brightness levels of unbleached mechanical and chemimechanical pulps from Norway spruce.

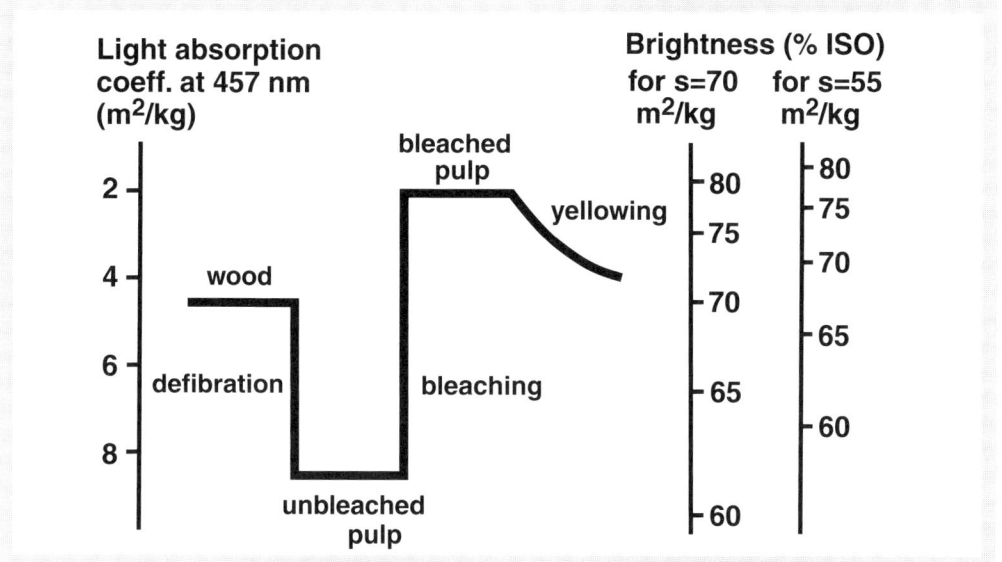

Figure 3. Development of the light absorption coefficient in mechanical defibration, bleaching, and in brightness reversion (yellowing). The corresponding brightness level depends on the light scattering coefficient[6].

Table 3. Typical brightness values of unbleached pulps produced from Norway spruce (*Picea abies*) with various mechanical and chemimechanical processes[9].

Pulp type		Brightness, % ISO
Stone groundwood	SGW	60–65
Pressure groundwood	PGW	60–63
Refiner mechanical pulp	RMP	60–62
Thermomechanical pulp	TMP	57–60
Chemithermomechanical pulp	CTMP	60–67
Chemimechanical pulp	CMP	45–55

The differences in brightness between the various pulp types can be due to the following factors:

- The high temperatures used in some of the processes, e.g., PGW, TMP, and CMP, can give rise to condensation and other reactions increasing the amount of colored structures.

- Pulps produced from wood chips as raw material can have a higher bark content than groundwood pulps.

- The pretreatment chemicals applied in chemithermomechanical and chemimechanical pulping can have either a bleaching (e.g., bisulfite) or darkening effect (e.g., sodium hydroxide) on lignin.

- Differences in light-scattering ability can have a noticeable effect on pulp brightness. This can explain, for example, the low brightness level of CMP pulps because they can have fairly low light-scattering ability.

3 Principles of bleaching high-yield pulps

Bleaching of high-yield pulps can have several objectives: increasing pulp brightness, reduction of the extractives content, and – in some cases – increasing pulp strength and fiber bonding. The primary objective normally, however, is to increase pulp brightness.

Table 2 shows that, like in chemical pulps, lignin is the main source of colored material in mechanical pulp. Chemical pulps are bleached by removal of almost all residual lignin from the pulp. In the case of mechanical and chemimechanical pulps, this approach cannot be used because the lignin content is very high. Removal of all lignin would need severe chemical treatments and would result in low yield, i.e., the benefits of mechanical pulping would be lost. Therefore, mechanical pulp bleaching is based on destruction of some of the colored groups in lignin only (Fig. 4).

By elimination of chromophores, it is possible to greatly reduce the light absorption of lignin and thus increase pulp brightness. The chemicals used in lignin-retaining bleaching, however, are not able to eliminate all colored structures. As a result, the maximum attainable brightness is lower than in chemical pulp bleaching. For example, mechanical pulps produced from Norway spruce (*Picea abies*) can be bleached to brightness levels of about 80% ISO. Darker softwood pulps can have lower maximum brightness. Some poplar pulps can be bleached to brightness levels of about 85% ISO.

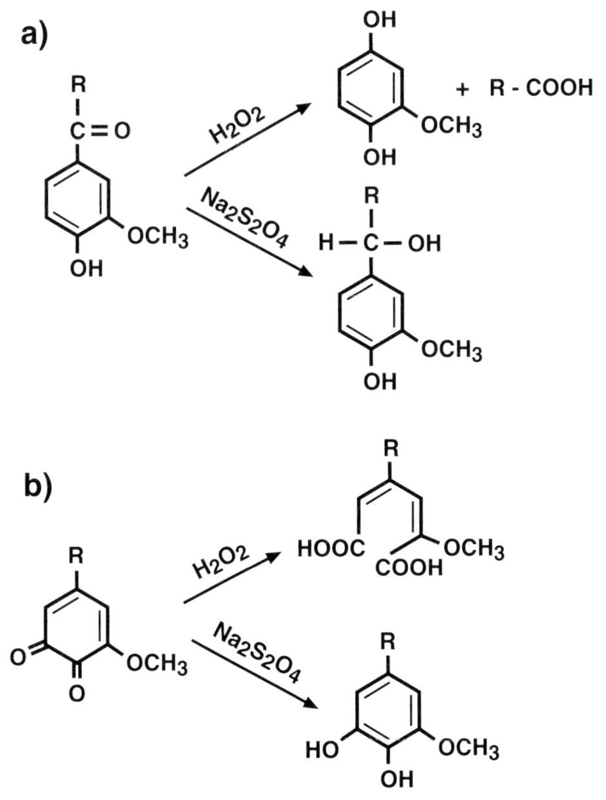

Contrary to chemical pulps, mechanical pulps are not always bleached to maximum attainable brightness,

Figure 4. Examples of chromophore elimination with hydrogen peroxide or sodium dithionite.

but the target brightness is determined by the end product. The reason for this is that a reasonable increase in brightness often can be attained with relatively low chemical charges, whereas high brightness levels can demand very high doses.

4 Peroxide bleaching

4.1 Reactions

Peroxide bleaching today is normally performed with hydrogen peroxide, H_2O_2; in earlier times, sodium peroxide, Na_2O_2, was also used. Hydrogen peroxide is a colorless liquid with a slightly acidic odor. It is completely miscible with water. It is delivered to the pulp mill as 35%–70% solutions in water (normally 50%) and is diluted before use.

It is generally accepted that the active mechanism in chromophore elimination with hydrogen peroxide involves the perhydroxyl ion OOH⁻. Hydrogen peroxide bleaching therefore is performed in alkaline systems to produce the active ion:

$$H_2O_2 + OH^- \leftrightarrow OOH^- + H_2O \tag{3}$$

The formation of the perhydroxyl anion can be enhanced by increasing the pH or by increasing the temperature.

Hydrogen peroxide readily decomposes under bleaching conditions according to the reaction:

$$2H_2O_2 \rightarrow O_2 + 2H_2O \tag{4}$$

Sodium silicate and Epsom salt (MgSO$_4$ • 7H$_2$O) is normally added to the bleach liquor in order to stabilize peroxide.

Transition metal ions like iron, manganese, and copper catalyse peroxide decomposition through chain reactions of the type[10]:

$$H_2O_2 + M \rightarrow M^+ + HO^- + HO\cdot \tag{5}$$

$$HOO^- + M^+ + HO^- \rightarrow M + H_2O + O_2^-\cdot \tag{6}$$

$$M^+ + O_2^-\cdot \rightarrow O_2 + M \tag{7}$$

$$O_2^-\cdot + HO\cdot \rightarrow O_2 + HO^- \tag{8}$$

CHAPTER 11

Metal ions enter the system with the wood raw material and the process water. Some metal can also dissolve from the process equipment. The transition metal-induced decomposition reactions waste hydrogen peroxide because they lead to a loss of perhydroxyl ion. The concentration of metal ions in the pulps therefore has to be reduced ahead of a peroxide stage by treatment with acid or chelating agents.

Hydrogen peroxide is believed to oxidize at least para and ortho quinone groups into colorless structures and decompose coniferyl aldehyde groups and conjugated double bond structures. The radicals probably also react with phenolic structures changing them into quinoidic structures, thus creating new chromophores.

4.2 Process variables

4.2.1 The peroxide dose

The peroxide dose is the most important process variable. Already a peroxide dose of 1% on pulp has a clear impact on pulp brightness (Fig. 5). Doses of this order of magnitude increase pulp brightness by 6%–8% ISO. Increasing the dose up to 4% further increases pulp brightness, even if the specific effect decreases with increasing doses. Under optimum conditions, pulp brightness can be improved by 15%–20% ISO in a peroxide stage.

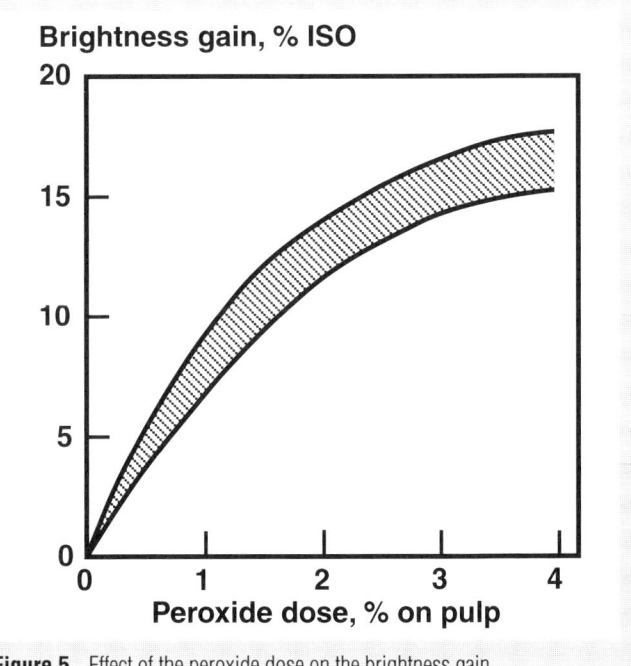

Figure 5. Effect of the peroxide dose on the brightness gain.

4.2.2 Removal of transition metal ions

Contrary to hydrogen peroxide bleaching of chemical pulp, metal-initiated delignification reactions are not desirable in mechanical pulp bleaching. Because transition metals decompose peroxide, it is important to remove them before a peroxide stage.

Transition metals can be removed either by acidifying the pulp to pH levels of 2–3 or by treatment of the pulp with a chelating agent (normally ethylenediaminetetraacetic acid, EDTA, or diethylenetetraminepentaacetic acid, DTPA). Acidification and the chelating agent releases the metal ions from the fibers, but they still have to be washed out of the system in a washer ahead of the bleaching stage[11].

4.2.3 Pulp pH

Hydrogen peroxide is performed at alkaline conditions because presence of hydroxyl ions promotes the dissociation reaction [Formula 3]. Too high pH levels, on the other hand, should be avoided because they lead to decomposition reactions. Optimum pH is to some extent dependent on the overall conditions, but an initial pH of 11.5 is considered suitable.

It is difficult to determine the initial pH directly because the pH rapidly decreases immediately after addition of the alkaline peroxide liquor to the pulp. It therefore has been suggested[12] that the initial pH (pH_i) should be defined as the pH of the total bleaching liquor at 24°C with no pulp present. The total bleaching liquor also includes the amount of water normally held by the wet pulp. Hence a separate liquor sample has to be prepared for determination of the initial pH.

The amount of alkali needed to reach the optimum pH depends on the peroxide dose. The higher the peroxide dose, the more alkali is needed. Figure 6 shows an approximate relationship between the peroxide dose and the alkali dose needed for an initial pH value of 11.5.

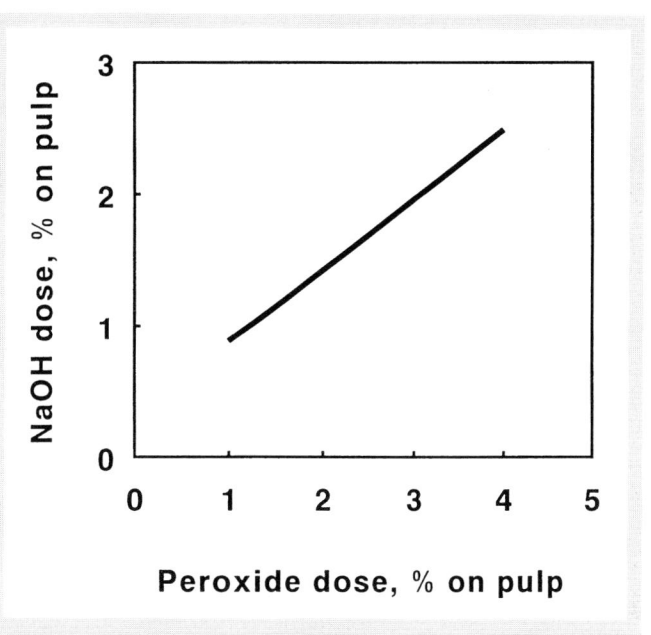

Figure 6. The NaOH dose needed for an initial pH level of 11.5 as a function of the peroxide dose.

Sodium silicate that is normally used as an additive in peroxide bleaching contributes to the total alkalininty of the bleach liquor. A 41 °Bé solution of sodium silicate contains approximately 11.5 wt % NaOH, so it contributes to the alkalinity in the following way:

% Total alkali = % Sodium hydroxide + 0.115% Sodium silicate.

4.2.4 Additives

Since the beginning of peroxide bleaching of mechanical pulps, sodium silicate (Na$_2$O · 3–4 SiO$_2$) has been used as an additive. Sodium silicate is normally applied as a 41 °Bé solution. It is evident that sodium silicate can improve the bleaching result significantly (Fig. 7). Former doses of up to 5% were not uncommon, but because silicate has a disposition to form buildups in closed water loops and to disturb the wet-end chemistry of paper machines, it is normally tried to minimize the dose. Today doses of 3% or less are therefore more common.

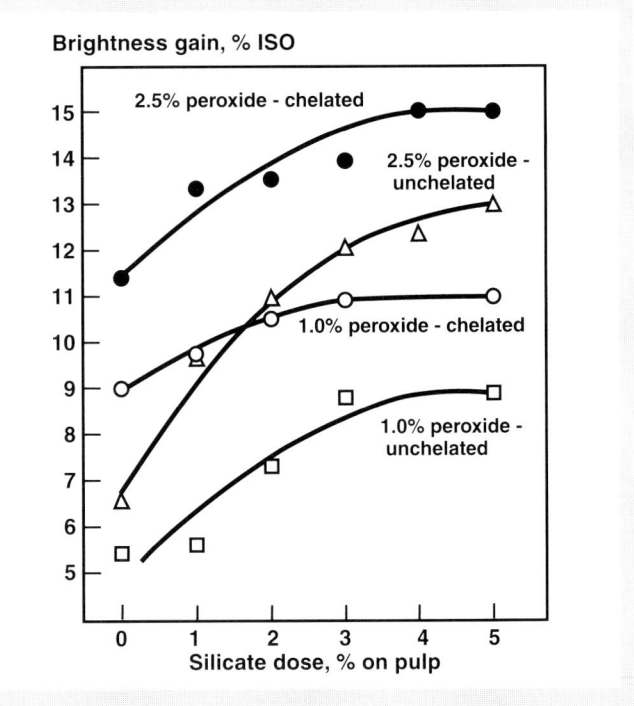

Figure 7. Effect of the sodium silicate dose on the brightness increase in peroxide bleaching of unchelated and chelated groundwood pulp [13].

The exact mechanism of the effect of silicate in peroxide bleaching is not known. Several theories have been proposed:

- The silicate has a stabilizing effect on peroxide.
- Sodium silicate acts as a buffer.
- Silicate has the ability to form metal complexes together with magnesium and thus prevent transition metal ions to catalyze peroxide decomposition.

None of these mechanisms alone, however, can explain the effect of silicate. The lack of knowledge of the exact mechanism has also been a difficulty in searching for substitutes for silicate.

Small amounts (0.05%–0.1%) of magnesium sulfate (usually in the form of Epsom salt, MgSO$_4$ · 7H$_2$O) are commonly included in the peroxide bleach liquor. Magnesium acts as a stabilizer of peroxide, probably together with sodium silicate. In some mills, sufficient amounts of magnesium are brought into the system either with the wood raw material or with the process water, and no additional benefits are then obtained with addition of magnesium salts.

4.2.5 Consistency

As a general rule, higher consistency improves bleaching effectiveness. This is revealed both as faster bleaching reactions (Fig. 8) and as higher attainable final brightness

(Fig. 9). One explanation of this is the increased concentration of bleaching chemicals in the liquid phase. The reduced quantity of water-soluble impurities (e.g., transition metals) present in the bleaching stage can also play an important role.

Although increasing the consistency up to levels of 35%–40% improves the bleaching result (Fig. 9), it is common to perform peroxide bleaching at clearly lower consistency, i.e., in the range of 15%–20%. The reasons are equipment-related: for a long time no efficient dewatering systems or high-consistency chemical mixers were commercially available. Within the last few years, high-consistency (HC) technology has won ground in peroxide bleaching allowing the use of higher consistencies[16].

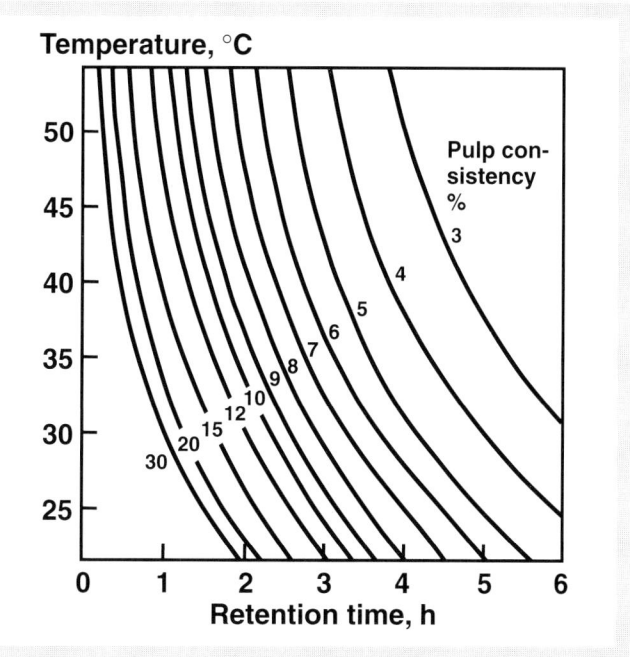

Figure 8. Equi-level curves showing various combinations of pulp consistency, bleaching temperature, and retention time resulting in the same bleaching response in peroxide bleaching[14].

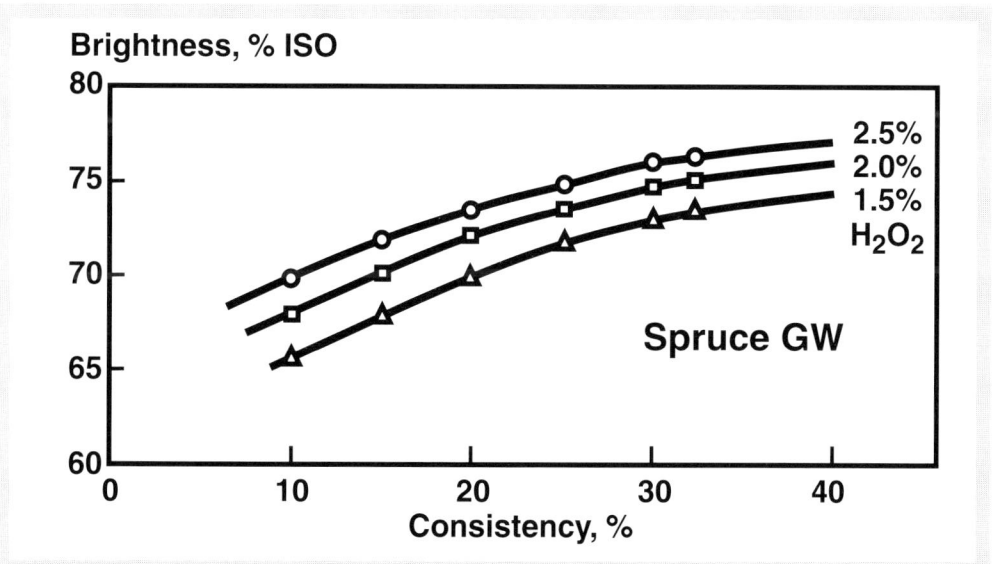

Figure 9. Effect of pulp consistency on the brightness level attained in peroxide bleaching with various peroxide doses[15].

4.2.6 Temperature

An increase in the temperature accelerates the reactions in peroxide bleaching[17]. This means, however, that not only the bleaching reaction becomes faster, but also peroxide decomposition and reactions between alkali and pulp.

At high temperatures (> 80°C), the brightness development is fast and maximum brightness can be attained in less than 30 minutes[18]. Longer retention times will darken the pulp. At lower temperatures, the brightness development is slower, but the risk for darkening reactions at extended retention times is small. The temperature in tower bleaching therfore is normally in the 40°C–70°C range.

Because the reactions between alkali and pulp accelerate at high temperatures, the alkali dose has to be reduced for maximum brightness.

4.2.7 Retention time

The combination of temperature and retention time has to be chosen so that a reasonable portion of the charged peroxide is consumed. However, all peroxide must not be consumed, but there should be a small residue at the end of the bleaching preventing alkali darkening of the pulp.

The temperature and retention time are thus closely interrelated and an increase in the temperature allows the use of a shorter retention time and vice versa (Fig. 8).

4.2.8 Acidification

Although peroxide bleaching is performed at highly alkaline conditions, the presence of peroxide prevents the pulp from alkali darkening. At the end of the bleaching stage, however, pulp pH has to be reduced before all peroxide is consumed in order to avoid darkening reactions. The pulp is normally acidified to pH 5–6 with sulfuric acid, bisulfite, or sulfur dioxide water. Acidification with bisulfite or sulfur dioxide, in addition to reducing pulp pH, also decomposes residual hydrogen peroxide according to the following reaction:

$$H_2O_2 + H_2SO_3 \rightarrow H_2O + H_2SO_4 \tag{9}$$

Destruction of residual peroxide is essential if the peroxide stage is followed by a reducing bleaching stage. If not destroyed, the residual peroxide consumes bleaching chemicals in the reducing stage.

4.3 Bleach plant design and operation

The most common commercial method for peroxide bleaching of mechanical pulp is bleaching in a tower. Tower bleaching requires equipment for dewatering (and metal removal) ahead of the bleaching stage itself, mixers for mixing steam and chemicals with the pulp, a bleaching tower providing a proper retention time, dilution at the bottom of the tower, and acidification of the bleached pulp.

Conventionally, tower bleaching with peroxide is performed in one stage at medium consistency (15%–20%). During the past few years, various modifications of the tower bleaching process have been developed. Examples of such modifications are recirculation of residual peroxide, two-stage peroxide bleaching, and high-consistency (HC) bleaching. In striving to high brightness levels, two stage bleaching with peroxide and dithionite can also be used.

In addition to tower bleaching, some methods based on more simple equipment are in use in some mills. Examples of such systems are refiner bleaching and flash dryer bleaching.

4.3.1 One-stage medium-consistency bleaching

The process used most frequently in commercial practice is bleaching in one stage in a bleaching tower (Fig. 10). Conventional bleach plants use a decker or a disc filter for dewatering the pulp ahead of the bleaching tower. The dewatering capacity of such equipment delimits the bleaching consistency to the medium-consistency range, 12%–20%.

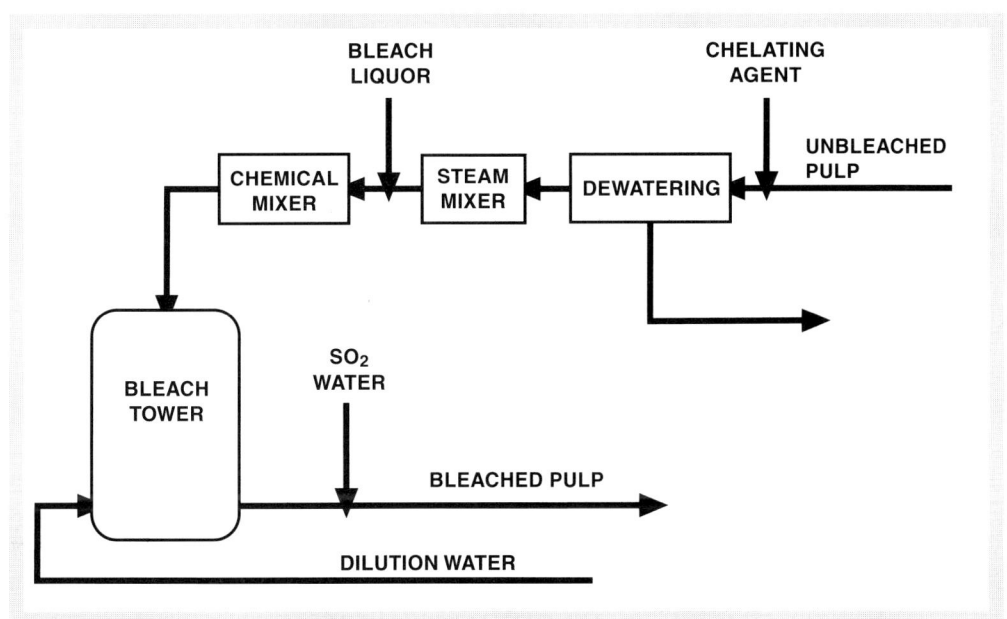

Figure 10. Principle of conventional medium-consistency tower bleaching with peroxide.

Some chelating agent is added to the pulp ahead of the decker or filter in order to bind transition metals. These are then removed together with the filtrate when dewatering the pulp. After dewatering, the bleach liquor is mixed with the pulp and the pulp is heated with steam in peg or paddle mixers.

From the mixer, the pulp is fed to the bleaching tower that normally is of the downflow type. This type of tower allows to vary the retention time. The retention time can vary from one to four hours. Most systems include acidification of the pulp in connection with the bottom dilution.

CHAPTER 11

A conventional peroxide bleaching system does not include washing of the bleached pulp like in bleaching of chemical pulp because the concentrations of residual chemicals and dissolved material are low.

Today, fewer mills operate at medium consistency because of brightness gain limitations and the trend is toward high-consistency bleaching.

4.3.2 High-consistency bleaching

The requirement to bleach mechanical pulp to high brightness led to the development of high-consistency peroxide bleaching in the beginning of the 1980s. In these systems, the bleaching stage is carried out at consistencies of 30%–40%. Special dewatering equipment such as twin-wire, twin-roll, or screw presses are used for dewatering the pulp to such high consistencies. In high-consistency bleaching, uniform mixing and retention times are essential. The lack of suitable large-scale mixers for mixing bleaching chemicals with pulp at high consistency has been a limiting factor in developing high-consistency bleaching systems, but these are now overcome. Figures 11 and 12 show two HC mixers, and Fig. 13 shows a discharge system from the HC tower.

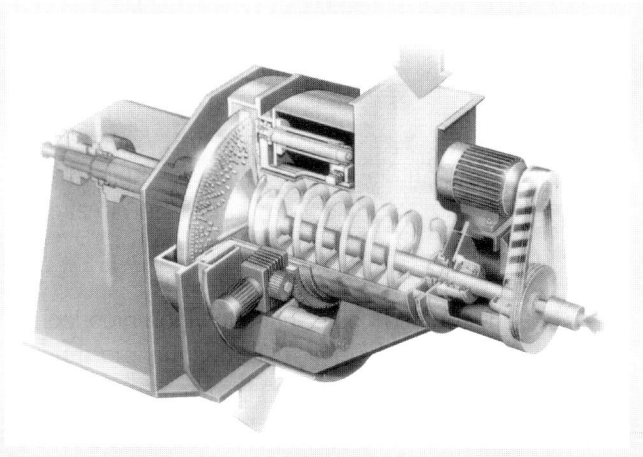

Figure 11. HC peroxide mixer (Andritz).

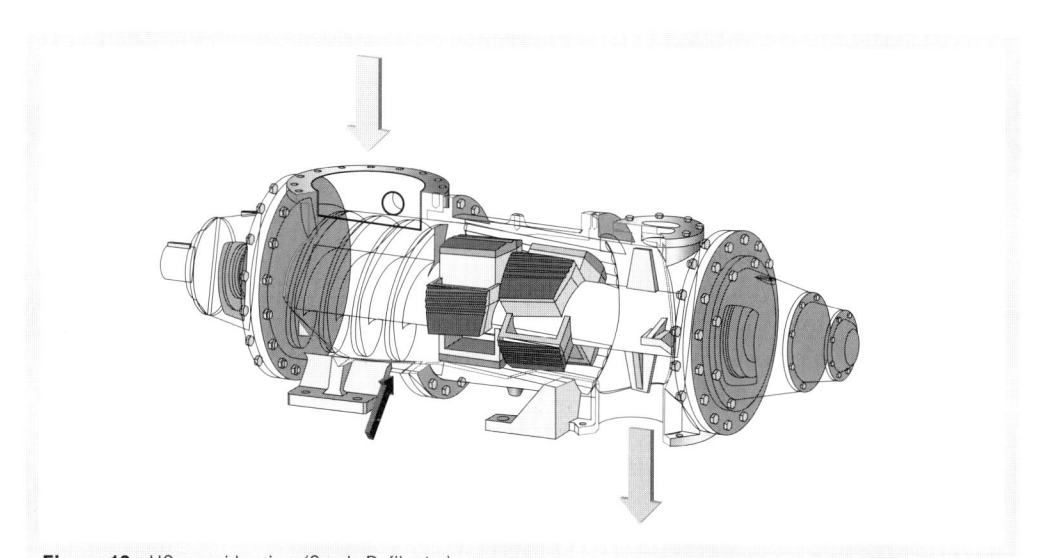

Figure 12. HC peroxide mixer (Sunds Defibrator).

Figure 13. Discharge system from HC bleaching tower (Andritz).

Besides the unconventional dewatering and mixing equipment, high-consistency bleaching systems (Fig. 14) are similar to medium-consistency bleaching systems.

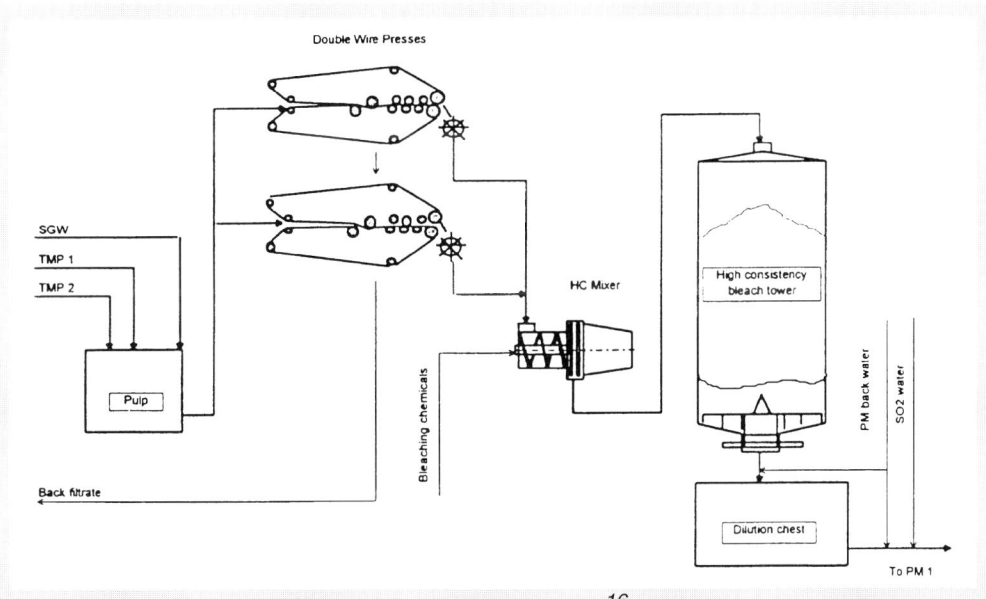

Figure 14. Principle of high consistency tower bleaching with peroxide[16].

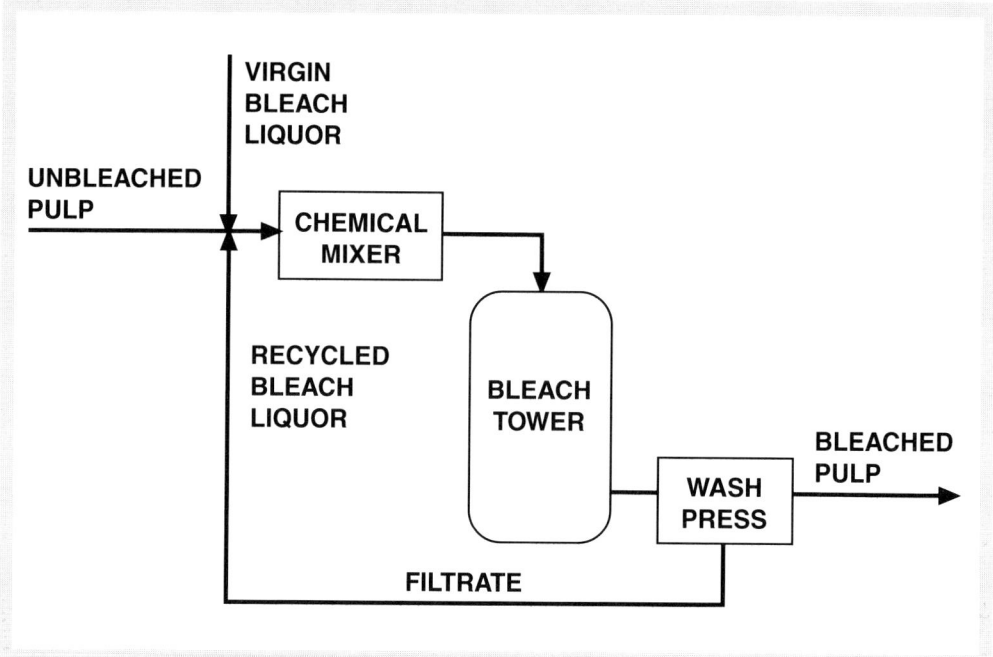

Figure 15. Principle of recirculation of residual peroxide.

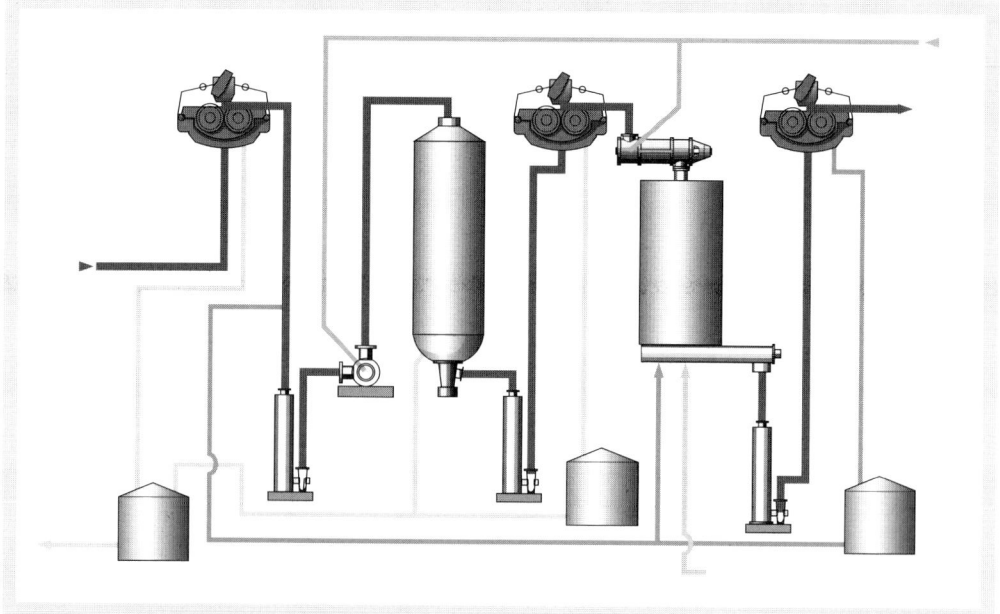

Figure 16. Example of two-stage medium consistency + HC peroxide bleaching with recirculation of residual chemicals (Sunds Defibrator).

Bleaching

4.3.3 Recycling of peroxide

When bleaching with low peroxide doses (about 1%), it is relatively easy to use combinations of temperatures and retention times that allow most of the peroxide charged to be consumed. When applying high peroxide doses, however, quite a great share of the peroxide dose can be left in the pulp after a reasonable retention time. For example, with peroxide doses of 4%, it is not uncommon that 50% or more of the peroxide is present unreacted in the pulp leaving the bleaching tower. A high peroxide dose, however, is needed as a driving force in the bleaching stage.

In cases like this, part of the residual peroxide can be removed from the bleached pulp in a washing press and recycled to a prebleaching or a first bleaching stage (Figs. 15 and 16).

4.3.4 Refiner bleaching

Although a high brightness is only achieved in tower bleaching, there are other peroxide bleaching methods that can be applied if very high brightness levels are not necessary. One such method is refiner bleaching. In this process, peroxide bleach liquor is introduced in the refiner zone of a TMP refiner. The very short retention time does not allow brightness increases as high as in tower bleaching, but brightness increases of about 10% ISO can be attained.

One benefit of refiner bleaching is the possibility of avoiding the costs of building and operating a separate bleach plant. In addition, it has been claimed that refiner bleaching can reduce the energy consumption and improve strength properties of the pulp.

4.3.5 Flash dryer bleaching

Another bleaching method that is attractive from the low capital requirement is bleaching with peroxide in connection with flash drying of market pulp. In this method, the peroxide bleach liquor is sprayed on the pulp that is fed to the flash dryer. After drying, the bleach reaction is completed in the stored bale. In this process, the pulp cannot be acidified in the normal way, so it is important that the alkali dose is adjusted so that the pulp is neutral or slightly acid by the time all peroxide has been consumed to avoid alkali darkening.

5 Dithionite (hydrosulfite) bleaching

5.1 Reactions

The active chemical in dithionite bleaching (also known as hydrosulfite bleaching) is either sodium dithionite (sodium hydrosulfite, $Na_2S_2O_4$) or zinc dithionite (zinc hydrosulfite, ZnS_2O_4). Currently very little of zinc dithionite is used, but sodium dithionite is dominating.

Sodium dithionite is a white crystalline solid. It can be supplied to the pulp mills in a number of different forms. It can be purchased as a liquid or powder, but it can also be produced on-site from commercial solutions containing sodium borohydride and sodium hydroxide. Such solutions typically contain 12% sodium borohydride, 40% sodium hydroxide, and 48% water (the molar ratio is 3.2 mole NaOH to 1 mole NaBH$_4$). When making dithionite, more sodium hydroxide and sulfur dioxide or sodium bisulfite is added. Sodium bisulfite is then reduced by borohydride and one atom of borohydride reduces eight sulfur atoms. The complete reaction is as follows:

$$(NaBH_4 + 3.2NaOH) + 4.8NaOH + 8SO_2 \rightarrow 4NaS_2O_4 + NaBO_2 + 6H_2O \quad (10)$$

Sodium dithionite produced in this way behaves in the same way as conventional dithionite.

In water solutions, dithionite can hydrolyze in various ways[13]:

$$S_2O_4^{2-} + 2H_2O \rightarrow 2HSO_3^- + 2H^+ + 2e^- \quad (11)$$

$$S_2O_4^{2-} + 2H_2O \rightarrow S_2O_6^{2-} + 4H^+ + 4e^- \quad (12)$$

$$S_2O_4^{2-} + 3H_2O \rightarrow HSO_3^- + HSO_4^- + 4H^+ + 4e^- \quad (13)$$

$$S_2O_4^{2-} + 4H_2O \rightarrow 2HSO_4^- + 6H^+ + 6e^- \quad (14)$$

$$2S_2O_4^{2-} + 2H_2O \rightarrow S_2O_3^{2-} + 2HSO_3^- \quad (15)$$

$$S_2O_4^{2-} + H_2O + O_2 \rightarrow HSO_3^- + HSO_4^- \quad (16)$$

The reactions presented in Eqs. 11–14 act in a reducing way and are thus active bleaching reactions. A decrease in pH accelerates the decomposition reaction shown in Eq. 15. Decomposition can be observed already at pH 5 and, at pH 4, the decomposition is very fast. This can be avoided by proper pH control. Presence of air leads to the decomposition reaction shown in Eq. 16.

The reactions of dithionite with the chromophores in mechanical pulps have not been as comprehensively studied as the reactions of peroxide bleaching. It is generally considered, however, that the bleaching effect of ditionite is due to:

- Reduction of quinoidic groups into hydroquinones
- Reduction of α-carbonyl groups
- Reduction of coniferyl aldehyde groups
- Probably also reduction of colored Fe^{3+} compounds into less colored Fe^{2+} compounds.

5.2 Process variables

5.2.1 Dithionite (hydrosulfite) dose

The most important variable is, of course, the dithionite dose. Normally doses of 0.5%–1.0% $Na_2S_2O_4$ on pulp are applied. In dithionite bleaching, there is a ceiling in the brightness gain that can be achieved by increasing the dose. Doses in excess of 1.0% have a very limited effect on brightness (Fig. 17). Under optimum conditions, the brightness can be improved by about 10% ISO in a dithionite stage.

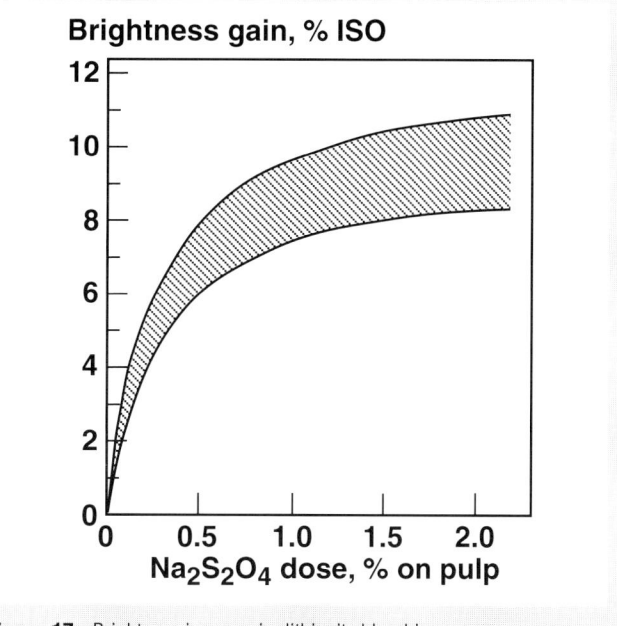

Figure 17. Brightness increase in dithionite bleaching.

5.2.2 Chelating agents

Some transition metals in pulp can reduce the bleaching response due to catalyzed decomposition reactions of dithionite and the re-oxidation of reduced chromophore groups. Most commercial sodium dithionite blends contain sufficient chelating agents to control the retarding effect of small amounts of metals. If the concentration of metals in the pulp is high, additional amounts of chelating agents might be required.

5.2.3 Pulp pH

The optimum pH range in zinc dithionite bleaching is 4–6 and, in sodium dithionite bleaching, is 6–6.5. High pH levels darken the pulp; at lower pH levels the dithionite starts to decompose. At pH levels below 4, immediate decomposition occurs. The pH is reduced by 0.3–1.0 units during the bleaching stage. The pH can be controlled by addition of sodium carbonate, sodium tripolyphosphate (STPP), or sodium hydroxide.

5.2.4 Consistency

Dithionite bleaching is normally performed at low consistency, in order to avoid entrainment of air into the pulp. Air (i.e., oxygen) in the bleaching system decomposes dithionite. Oxygen can also react with the reduced chromophores and re-oxidize them to their colored form. Within the range of 3%–5%, pulp consistency has a minimum effect on the brightness response.

New equipment, i.e., MC mixers, have made it possible to perform dithionite bleaching also in the 8%–12% consistency range. Medium-consistency bleaching can improve the brightness response by 1–2 points. It will also reduce the size of the bleaching tower.

CHAPTER 11

5.2.5 Temperature

The optimum temperature in dithionite bleaching is dependent on various factors. Generally, increasing the temperature accelerates the bleaching reactions, but also brightness reversion. The bleaching temperature is typically in the range of 50°C–70°C. Above 75°C, the brightness tends to drop slightly.

5.2.6 Retention time

The retention time in dithionite bleaching is not critical. Most bleaching normally occurs within the first 10–20 minutes. Somewhat longer retention times, 30–60 minutes, are normally used in order to fully utilize the bleaching effect.

The normal conditions in dithionite bleaching are thus:

Dithionite application:	0.5%–1% on pulp
Chelating agents:	added if the metal concentration in the pulp is high
Pulp pH:	6–6.5
Consistency:	3%–5% (8%–12%)
Temperature:	50°C–70°C
Retention time:	30–60 min

5.3 Bleach plant design and operation

5.3.1 Tower bleaching

Dithionite bleaching is conventionally performed in a tower. The equipment has to be designed to prevent contact between pulp and air that would result in harmful oxidizing reactions. Attention has to be paid to the properties of the bleach liquor prior to addition to the pulp and to the pulp both during the mixing and the retention period. Dithionite bleaching towers therefore are normally of the upflow type and the pulp moves through the tower as a plug flow. Figure 18 shows a typical layout of dithionite bleaching system.

Because of the relatively low chemical charges and the limited amount of wood material dissolved in a dithionite bleaching stage, washing of the pulp after the bleaching tower is not necessary.

5.3.2 Two-stage bleaching with peroxide and dithionite (hydrosulfite)

The bleaching effect of dithionite and peroxide is partially additive. Sequential bleaching with these two chemicals can thus be used when aiming at high brightness levels. A two-stage bleaching concept consists of the following steps (Fig. 19):

- A peroxide tower bleaching stage
- Destruction of residual peroxide in the pulp by acidification with sulfur dioxide or sodium bisulfite (cf. Eq. 9)
- A dithionite tower bleaching stage.

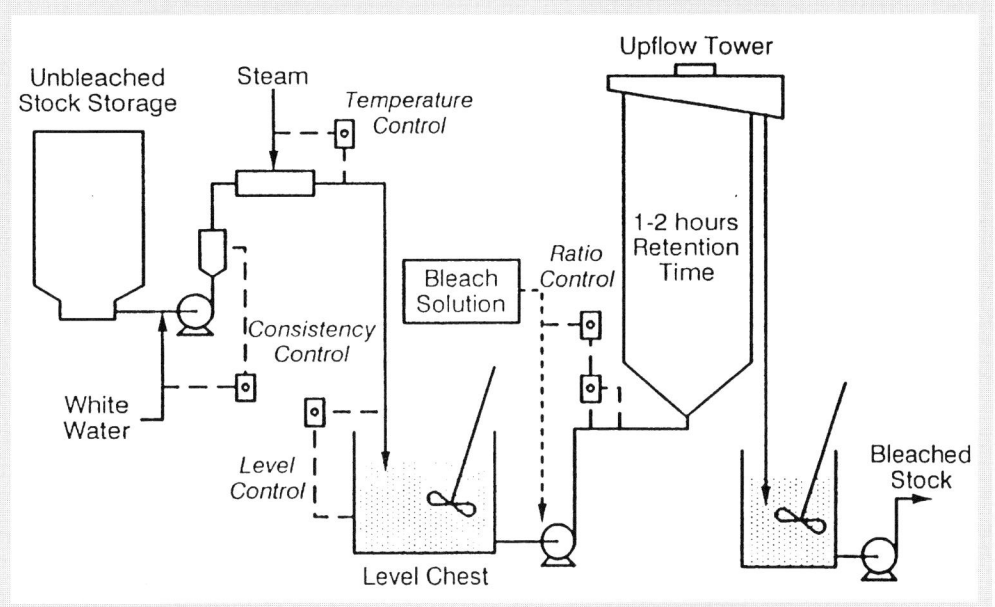

Figure 18. Typical layout of tower bleaching with dithionite [19].

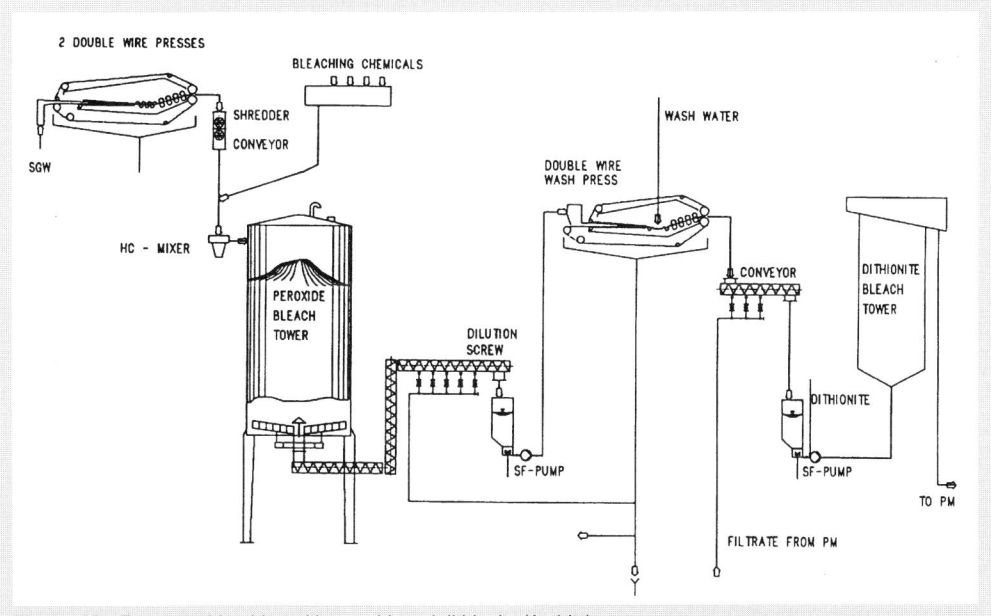

Figure 19. Two-stage bleaching with peroxide and dithionite (Andritz).

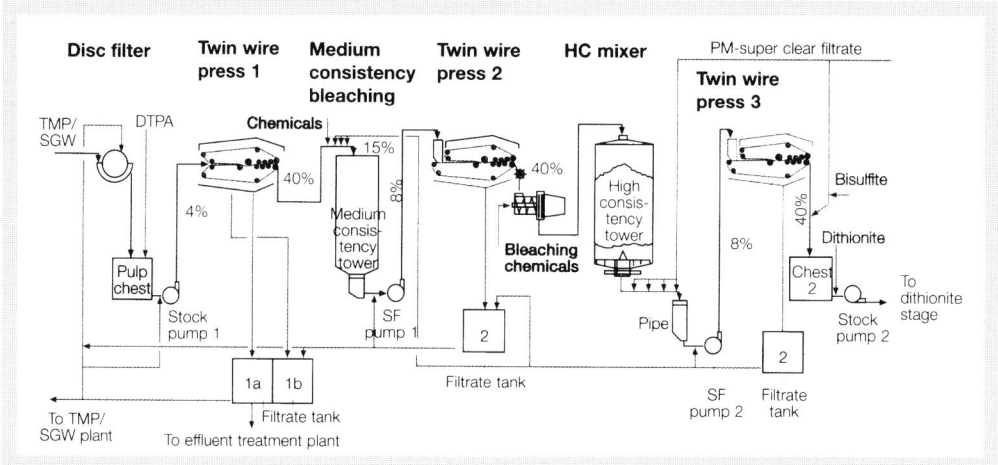

Figure 20. Three-stage bleaching: Medium consistency peroxide + HC peroxide + Medium consistency dithionite (Andritz)[33].

Brightness gains of 15–20 units can be attained in two-stage bleaching with peroxide and dithionite. For even higher brightness gains a three-stage bleaching system might be considered (Fig. 20).

5.3.3 Chest bleaching

Instead of bleaching towers, stock chests can be used for holding the pulp during the retention period. In chest bleaching, it is not possible to control the retention time as properly as in tower bleaching. In addition, there is more exposure of the pulp to air in a chest. As a result, the brightness gain is somewhat lower than in tower bleaching.

5.3.4 Refiner and grinder bleaching

In refiner-based mechanical pulping processes, an option is to perform bleaching with dithionite in the refiner. Refiner bleaching can be realized in a relatively simple way. All that is needed is a bleach solution line with suitable control valves to the dilution water line at the refiner.

Dithionite can also be fed into the shower water in the groundwood process, thus increasing the brightness of the pulp by approximately 5% ISO[19].

Refiner or grinder bleaching can be combined with subsequent bleaching in a tower. Such a system can provide either higher brightness gain or the desired brightness gain at reduced costs compared with tower bleaching alone. As a two-stage system, it can also be more easily controlled.

6 Other bleaching chemicals

6.1 Potential bleaching chemicals

In addition to hydrogen peroxide and sodium dithionite, there are some other oxidizing or reducing chemicals that might be used for bleaching mechanical pulps. Ozone and peracetic acid have been proposed as oxidizing chemicals[20, 21]. Ozone, O_3, has been found to have a slight brightening effect on hardwood pulps, but cannot be used for bleaching softwood pulps. Peracetic acid, CH_3COOOH, can be used for pulp bleaching, but the high price has so far restricted its use.

Sodium borohydride, $NaBH_4$, has been found to be an effective chemical for reductive bleaching of mechanical pulps, but the high price has restricted the use of this chemical, too. Still borohydride is used indirectly as alkaline borohydride solutions for on-site production of dithionite. Other reducing chemicals of some importance are bisulfite and formamidine sulfinic acid.

6.2 Sodium bisulfite

Bleaching of mechanical pulps with sodium bisulfite, $NaHSO_3$, is an old method, but it is not very widely used. The bleaching effect of bisulfite is based on reducing reactions of the type shown in Fig. 21a. Sulfite can also react with conjugated carbonyl groups forming sulfonic acid groups, Fig. 21b.

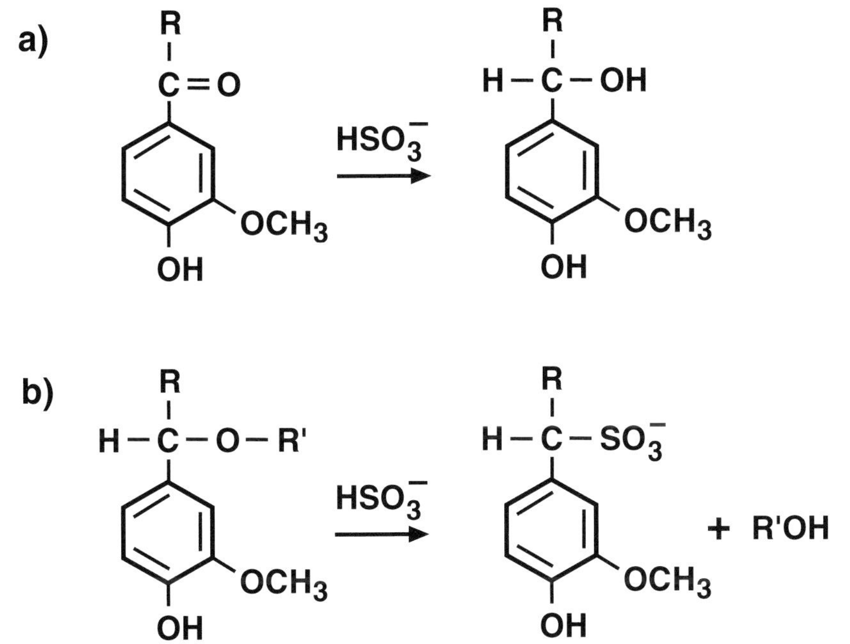

Figure 21. Bleaching reactions in bisulfite bleaching.

The maximum brightness increase in bisulfite bleaching with a chemical charge of 1% is 3%–4% ISO. Brightness gains of 6%–7% ISO can be attained with substantially higher doses, but this is hardly any economical alternative for mechanical pulp bleaching[22].

CHAPTER 11

Sodium bisulfite is also frequently used as a pretreatment chemical in production of chemimechanical pulps. In this case, reactions of the type indicated in Fig. 21 can destruct chromophores already in the pretreatment stage, thus reducing the light absorption coefficient of the pulp. Even if a small reduction of the light scattering coefficient is considered, such a reduction in the light absorption ability can improve pulp brightness by 3%–4% ISO.

6.3 Formamidine sulfinic acid

Formamidine sulfininc acid or FAS (HO_2S-CNH-NH_2) is a reducing agent that has about the same bleaching effect on mechanical pulps as sodium dithionite[23]. Formamidine sulfinic acid is more commonly applied in bleaching of recycled fibers than in bleaching of mechanical pulps.

7 Bleaching of chemimechanical pulps

Bleaching of high-yield chemimechanical pulp has to be performed with lignin-retaining chemicals like bleaching of mechanical pulps because of the high lignin content of these pulps. The main chemical applied in bleaching chemimechanical pulps is hydrogen peroxide.

In bleaching chemimechanical pulps with peroxide, is it important to wash the pulp well ahead of the bleaching stage. Carry-over of sulfite might consume peroxide and clearly impair the bleaching response (Fig. 22).

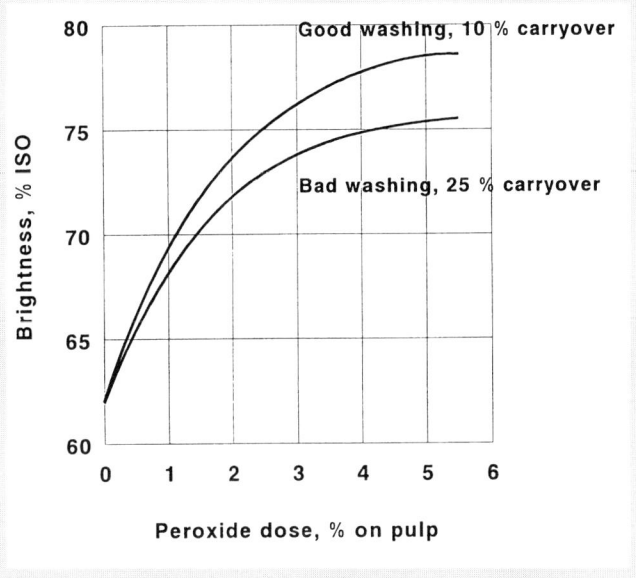

Figure 22. Effect of carryover from a CTMP process on the bleaching reponse in peroxide bleaching[24].

If the pulp is properly washed, the consumption of peroxide in bleaching chemimechanical pulp can be lower compared with mechanical pulp. Regardless, the bleaching response is of the same order of magnitude or even higher (Table 4).

Table 4. Effect of pulp type on the peroxide consumption and bleaching response in peroxide bleaching with 2.5% H_2O_2 on pulp[25].

Pulp type	Sulfonate content, % on pulp	Peroxide consumption, % on pulp	Brightness increase, % ISO
TMP	–	2.2	12.3
CTMP	0.4	1.7	15.0
CMP 1	1.3	1.0	11.7
CMP 2	1.6	1.2	11.4

8 Brightness reversion

8.1 General

Although mechanical pulps are relatively bright compared with unbleached kraft pulps and can be further bleached with oxidative and reducing agents, they have a significant drawback: They are relatively prone to diminish in brightness with age. The reason for this is that lignin is not removed in the type of bleaching applied for mechanical pulps. Reduced chromophores can be reoxidized to their colored form, and new chromophoric groups can be created.

Brightness reversion or yellowing is a common drawback of all high-yield pulps: The brightness drop is of the same order of magnitude for groundwood, thermomechanical, and chemithermomechanical pulp. Bleaching does not remove the yellowing tendency either, but the brightness reversion can be higher for bleached pulps (Fig. 23).

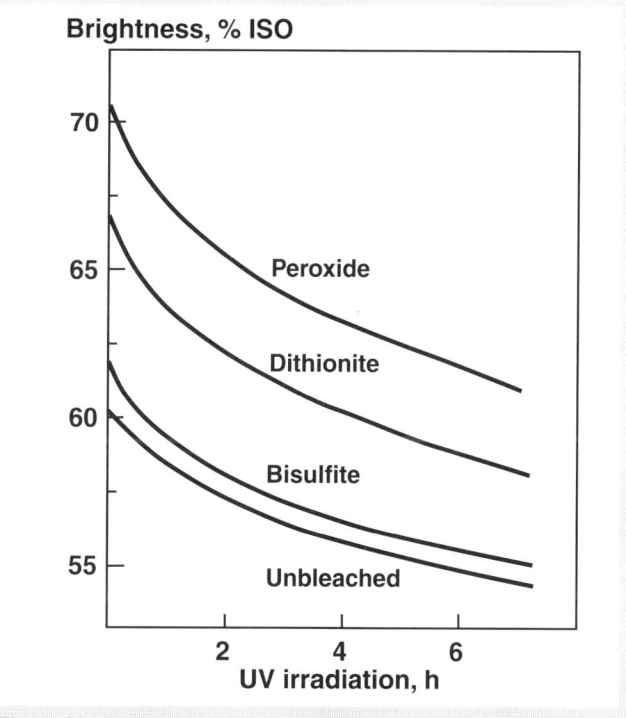

Figure 23. Example of brightness reversion of differently bleached pulps on UV irradiation.

CHAPTER 11

Brightness reversion is affected by the storage conditions such as temperature, relative humidity, and irradiation. Depending on the conditions, various mechanisms can be responsible for formation of chromophores in the pulp. Brightness reversion is normally classified according to the following two processes:

- Thermal brightness reversion caused by long-time storage of paper
- Light-induced brightness reversion caused by exposure to daylight.

This paragraph discusses determination of brightness reversion and some main factors affecting brighness reversion. Brightness reversion is discussed in more detail elsewhere in the volume "Forest products chemistry" (P. Stenius, Ed.) of this textbook series.

8.2 Determination of brightness reversion

Determination of brightness reversion is based on brightness measurements. The most simple way of expressing the magnitude of brightness reversion is to report the brightness drop as brightness units (% ISO). This way of expressing the brightness reversion might be suitable in practical operation, but it is not necessarily fitted for more scientific reflections because the brightness drop is not directly proportional to the change in light-absorption capacity, as exemplified in Table 5.

Table 5. An aging treatment that increases the light-absorption coefficient with 2 m^2/kg has a different impact on the reduction in brightness at different brightness levels. The light scattering coefficient is assumed to be 65 m^2/kg.

Brightness before aging, % ISO	Light abs. coefficient before aging, m^2/kg	Assumed effect of aging on the light-abs. coefficient, m^2/kg	Brightness after aging, % ISO	Reduction in brightness, % ISO	pc number
80	1.6	1.6 + 2 = 3.6	71.7	8.3	3.1
70	4.2	4.2 + 2 = 6.2	64.9	5.1	3.1
60	8.7	8.7 + 2 = 10.7	56.8	3.2	3.1

If it is assumed that the light-scattering coefficient is not changed at aging, the change in the k/s value is directly proportional to the change in the light absorption coefficient k. The k/s value is more easy to determine than the light absorption coefficient because it can be calculated directly from the brightness value according to the Kubelka-Munk theory[1]:

$$k/s = (1 - R_\infty)^2 / 2R_\infty \qquad (17)$$

Bleaching

Based on this fact, Giertz[26] proposed the use of a pc number (post color number) for expressing the magnitude of brightness reversion:

$$pc\ number = 100 \times [(k/s)_a - (k/s)_b] \tag{18}$$

where $(k/s)_a$ and $(k/s)_b$ are values determined after (a) and before (b) brightness reversion.

Brightness reversion of pulp, especially if the pulp is not exposed to light, is a relatively slow process. Various accelerated aging treatments therefore have been developed to make it possible to predict the disposition of pulps to yellow. Some of these methods are based on storage at elevated temperatures and, in some cases, also at specified relative humidity. A wide variety of conditions has been proposed[27]. Dry heat treatment at 105°C (ISO Standard 5630/1 and TAPPI Useful method 200) is a common method. Other methods might use different temperature levels (ISO 5630/4) or moist heat treatments at specified relative humidity (ISO5630/2 and ISO 5630/3).

Various methods have been proposed for accelerated light-induced aging, too. These methods are mostly based on irradiation with UV-rich light.

8.3 Thermal brightness reversion

Thermal brightness reversion is caused by storage in the dark under warm and moist conditions. The wood species can have a clear effect on the brightness drop, partly due to differences in extractives composition (Fig. 24). The pulping method can also affect the disposition of the pulp to yellow. Chemi-mechanical pulps are more prone to yellow than mechanical pulps. Pulp properties like pH and the ferric ion content can also affect thermal brightness reversion.

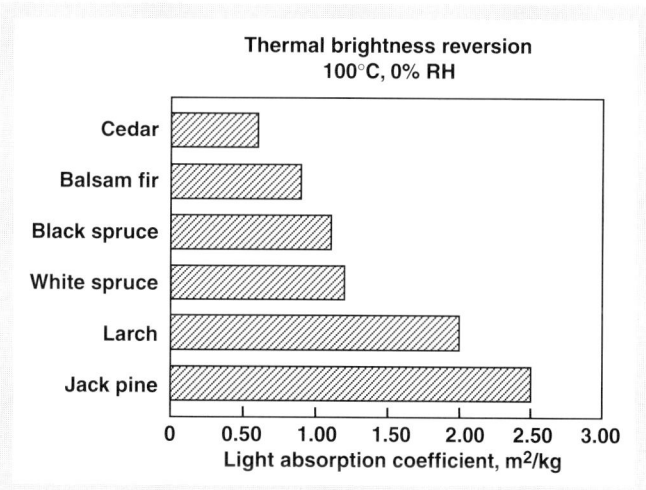

Figure 24. Thermal brightness reversion of mechanical pulps produced from various Canadian softwood species[28].

Bleaching has a significant effect on thermal brightness reversion. Both peroxide and dithionite bleaching increases yellowing compared with unbleached pulp[29]. The yellowing tendency of pulp bleached with oxidative agents like peroxide is clearly lower than that of pulps bleached with reducing agents. Peroxide bleaching, in some cases, can even reduce the brightness reversion expressed as the pc number[34]. The reason is that reductive bleaching agents brighten mechanical pulps by reducing colored groups

CHAPTER 11

into their uncolored form. These reduced groups, however, may be oxidized back to their colored form upon storage. The bleaching reactions of peroxide, on the contrary, are irreversible (cf. Fig. 4).

8.4 Light-induced brightness reversion

Light-induced yellowing can reduce the brightness of mechanical and chemimechanical pulps by some twenty to thirty units. Light-induced brightness reversion is due to oxidation reactions intiated by UV-irradiation. Oxygen is necessary for these reactions; no yellowing takes place in the absence of oxygen. On the contrary, light-induced brightness reversion is less dependent on the moisture and the pH of the pulp than thermal reversion.

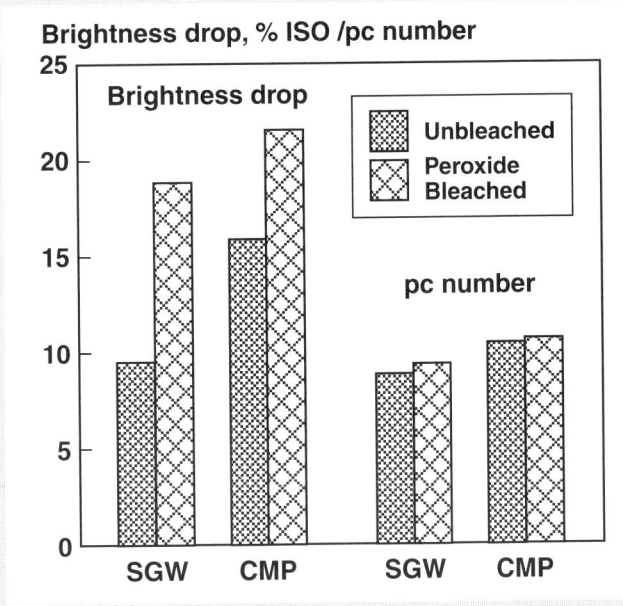

Figure 25. Effect of peroxide bleaching on light-induced brightness reversion of groundwood and chemimechanical pulp[34].

Bleaching with reductive bleaching agents has been found to increase light-induced brightness reversion[29], but the effect of oxidative bleaching is less clear. A recent investigation showed that peroxide bleaching of groundwood and chemimechanical pulp clearly increased the brightness reversion expressed as the brightness drop, but only marginally changed the pc number (Fig. 25).

9 Effect on yield and papermaking properties

9.1 Dithionite (hydrosulfite) bleaching

Dithionite bleaching is performed at relatively mild conditions and does not significantly affect the freeness or the papermaking potential of the pulp. The yield of dithionite bleaching is 98%–100%.

9.2 Peroxide bleaching

Peroxide bleaching can affect the properties of mechanical pulp in many ways. The alkaline conditions of peroxide bleaching remove extractives and can improve fiber flexibility. Extractives can cover the fiber surfaces and reduce the bonding capacity. Partial removal of the extractives in a peroxide stage can improve pulp strength.

Partly contradictory results, however, have been reported for the effect of peroxide bleaching on the physical properties of mechanical pulps. Some investigations have shown an increase in pulp freeness in peroxide bleaching, whereas others have indicated a small reduction in freeness. Similar differences have been reported for the strength properties. Several investigations have not found any significant impact of pulp strength, whereas others report either slightly reduced (e.g.,[30]) or somewhat improved (e.g.,[31]) strength properties.

One probable reason for these contradictions is that the conditions in peroxide bleaching can vary considerably. The chemical charges applied, for example, have been shown to affect pulp strength. Moldenius[32] showed that neither high NaOH or high H_2O_2 doses alone affect pulp strength properties significantly, but a combination of high pH (i.e., high NaOH doses) and high peroxide doses can have a significant effect. An initial pH of 13 combined with peroxide doses of 4%–6% on pulp can increase sheet density and tensile strength by 40–60%.

The bleaching yield in peroxide bleaching with moderate doses (1%–2% H_2O_2 on pulp) is 97%–98%, but bleaching with high peroxide (and hydroxide) doses can clearly reduce pulp yield even more.

CHAPTER 11

References

1. Kubelka, P. and Munk, F., Z. tech. Physik 12(11a):593 (1931).
2. Unpublished KCL results, 1997.
3. Rapson, W. H., Wayman, M., Andersson, C.B., Pulp Paper Mag. Can. 66(5):T255 (1965).
4. Allison, R. W., Appita J. 36(5):362 (1983).
5. Rydholm, S., Pulping Processes, Interscience Publishers, New York/London/Sydney, 1965, p. 896.
6. Norrström, H., Svensk Papperstid. 72(2):25 (1969).
7. Nyblom, I., "Puuraaka-aineen vaikutus mekaanisen massan ominaisuuksiin." Publication 112-83 Päällystämättömien offset-papereiden valmistus, laatu ja käyttö, Insinöörijärjestöjen koulutuskeskus, Helsinki, 1983.
8. Lorås, V., Tappi 59(11):99 (1976).
9. Paulapuro, H., Vaarasalo, J., Mannström, B., "Mekaanisen massan valmistus," in Puumassan valmistus (N.-E. Virkola, Ed.) 2nd edn., SPIY, TTA, Turku, 1983, p. 621.
10. Colodette, J. L., Rothenberg, S., Dence, C. W., J. Pulp Paper Sci.14(6):J126 (1988).
11. Janson, J. and Forsskåhl, I., Nordic Pulp Paper Res. J. 4(3):197 (1989).
12. Moldenius, S., Svensk Papperstid. 85(15):R116 (1982).
13. Ali, T., McArthur, D., Stott, D., Fairbank, M., Whitling, P., J. Pulp Paper Sci.12(6):J166 (1986).
14. Gavelin, G., Paper Trade J.150(26):42 (1966).
15. Soteland, N., "Modern bleaching of high yield pulps," 1993 International Mechanical Pulping Conference Proceedings, PTF, Oslo, p. 373.
16. Bräuer, P., Kappel, J., Bjerke, P., "Operating experiences with a high consistency bleach plant for TMP," CPPA 1995 International Mechanical Pulping Conference (Ottawa) Proceedings, CPPA, Montreal, p. 277.
17. Dence, C. W. and Omori, S., Tappi J. 69(10):120 (1986).
18. Kappel, J. and Sbaschnigg, J., Pulp Paper Can. 92(9):60 (1991).
19. Theodorescu, G., Wand, D., L. -K., Forsberg, P., Haikkala, P., "Hydrosulfite grinder bleaching of pressure groundwood pulp," TAPPI 1991 Pulping Conference Proceedings, TAPPI PRESS, Atlanta, p. 959.

20. Soteland, N., Pulp Paper Mag. Can. 78(7):T157 (1977).
21. Wayman, M., Anderson, C. B., Rapson, W. H., Tappi 48(2):113 (1965).
22. Kuys, K. and Abbott, J., Appita 49(4):269 (1996).
23. Daneault, C. and Leduc, C., Tappi J. 78(7):153 (1995).
24. Hägglund, T. -Å. and Lindström, L. -Å., Tappi J. 68(10):82 (1985).
25. Kouk, R. S., Meyrant, P., Dodson, M. G., J. Pulp Paper Sci. 15(4):J151 (1989).
26. Giertz, H. W., Svensk Papperstid. 48(13):317 (1945).
27. Neimo, L., Paperi ja Puu 46(1):7 (1964).
28. Lee, J., Balatinez, J. J., Whitling, P., "The optical properties of eight Canadian wood species. IV. Brightness stability," TAPPI 1989 International Symposium of Wood and Pulping Chemistry Notes, TAPPI PRESS, Atlanta, p. 529.
29. Jansen, O. and Lorås, V., Norsk Skogindustri 22(10):342 (1968).
30. Brecht, W. and Meltzer, K. -P., Wochenblatt für Papierfabrikation 95(17):677 (1967).
31. Lindahl. A. and Norberg, P. H., Pulp Paper Can. 81(6):T138 (1980).
32. Moldenius, S., J. Pulp Paper Sci.10(6):J172 (1984).
33. Bräuer, P., Fiber Spectrum (1):6 (1998).
34. Forsskåhl, I. and Jansson, J., Nordic Pulp Paper Res. J. 6(3):118 (1991).

CHAPTER 12

Thickening, storage, and post refining

1	**Thickening**	**345**
1.1	Disc filters	345
1.2	Broke deckers	347
1.3	Bow screens	348
	1.3.1 Screw presses, twin-roll presses, and twin-wire presses	348
2	**Storage of mechanical pulps**	**348**
2.1	Storage systems	348
2.2	Effect of storage on pulp properties	350
3	**Post refining**	**351**
3.1	Targets of post refining	351
3.2	Equipment used in post refining	352
3.3	Post refining systems	355
3.4	Operation of a post refiner	355
3.5	Development of fiber and pulp properties in post refining	357
	References	362

CHAPTER 12

Markku Pitkänen, Markus Mannström, Jorma Lumiainen

Thickening, storage, and post refining

1 Thickening

1.1 Disc filters

Mechanical pulps are usually thickened with disc filters (Fig.1). The disc filter has shaft mounted discs that are composed of segments. The pulp that is to be thickened is directed into the filter vat at a consistency of 0.5%–1.2%. As the segment is descending under the pulp level, pulp starts to filter onto the segment surface. A wire cloth on the

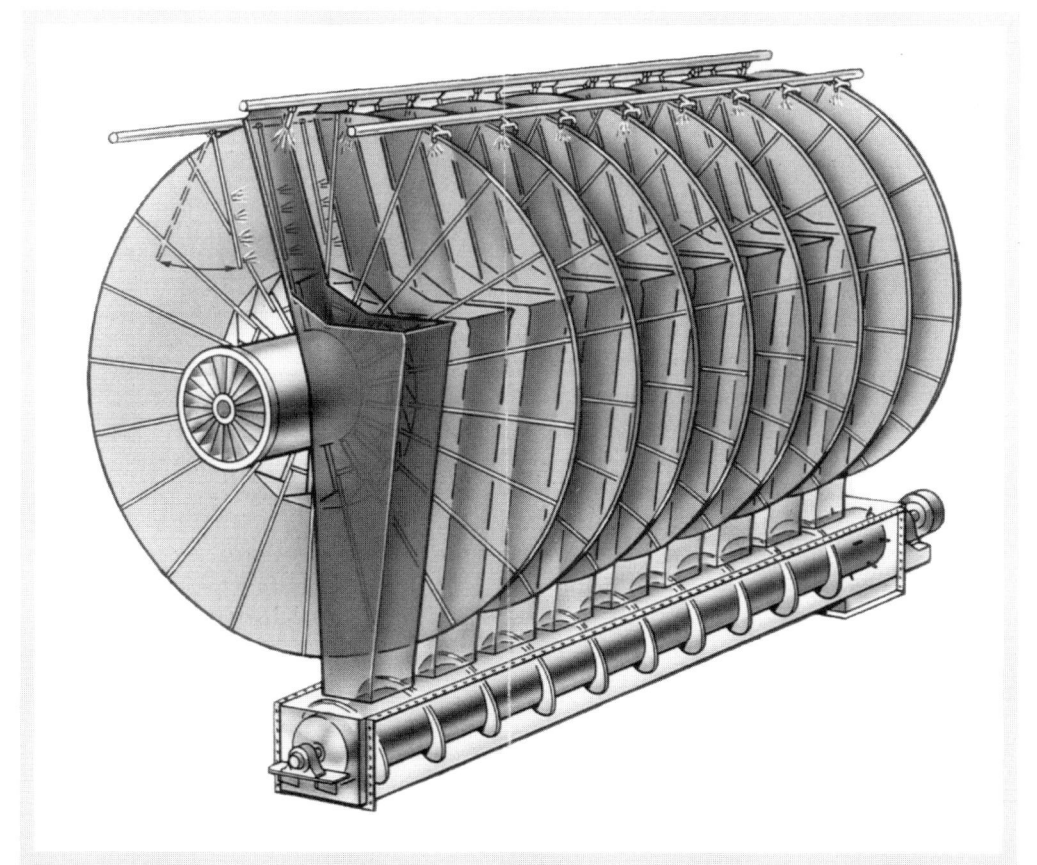

Figure 1. Inside principle of a disc filter (Alfa Laval Celleco).

surface of the segment prevents the pulp from entering directly into the segment. The water or filtrate that has filtered into the segment is diverted into the shaft. The filtrate flows into the valve at the gable through narrow channels within the shaft. From the valve, the filtrate continues to the suction leg, where a vacuum is created by the difference in height, 2–4 m. The suction improves the thickening process on the segment's surface and thus a fiber mat of higher consistency can be achieved. A thicker fiber mat means cleaner filtrate. Normally the cleaner filtrate is diverted from the valve with a suction pipe into its own tank, the clear filtrate tank. The solids content in the clear fitrate is about 50–100 mg/l. The dirtier filtrate from the early stages goes under the name of cloudy filtrate. When there is use for it in the process, even a third filtrate, called superclear filtrate, can be separated from the disc filter. This filtrate can be used, for example, as spray water, as its solid content is less than 50 mg/l.

When the segment rises above the pulp surface, the fiber mat is removed from the segment with a specific discharge shower. The fiber mat drops into the collection chutes, and from there it continues to the screw conveyor. In the screw conveyor, the pulp consistency is 8%–15% depending on the pulp freeness, the temperature, the rotation speed of the disc filter, etc. After the removal of the fiber mats, the segments are cleaned with a wire cleaning shower in order to keep the wire cloth on the segment's surface clean.

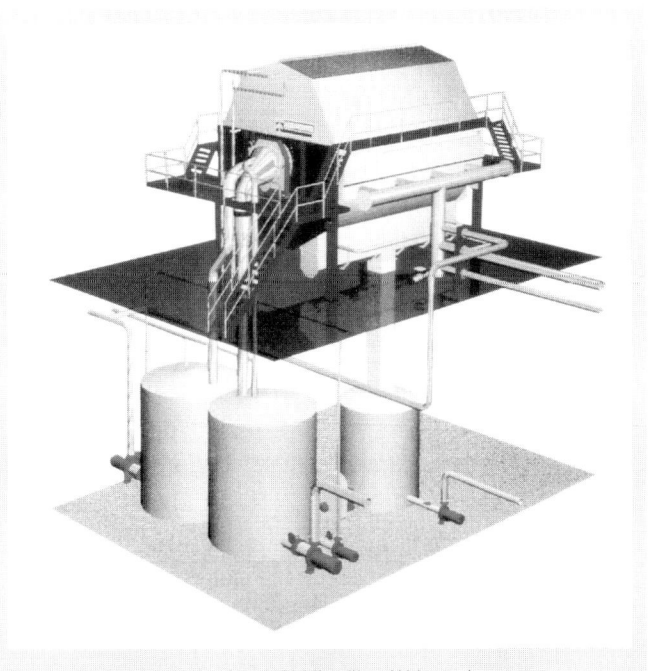

Figure 2. Process layout for an Ahldisc filter (Ahlström).

There are disc filters with variable diameters. The smallest are 3.0 meters and the largest 5.5 meters. The maximum amount of discs in a single filter is 30 pcs, which means that one filter can contain over 1000 m^2 of filtration surfaces. Figure 2 shows a typical layout. In the PGW "hot loop," the disc filter is pressurized, which means that filtration water with a temperature of over 100°C can be retained (Fig. 3).

Thickening, storage and post refining

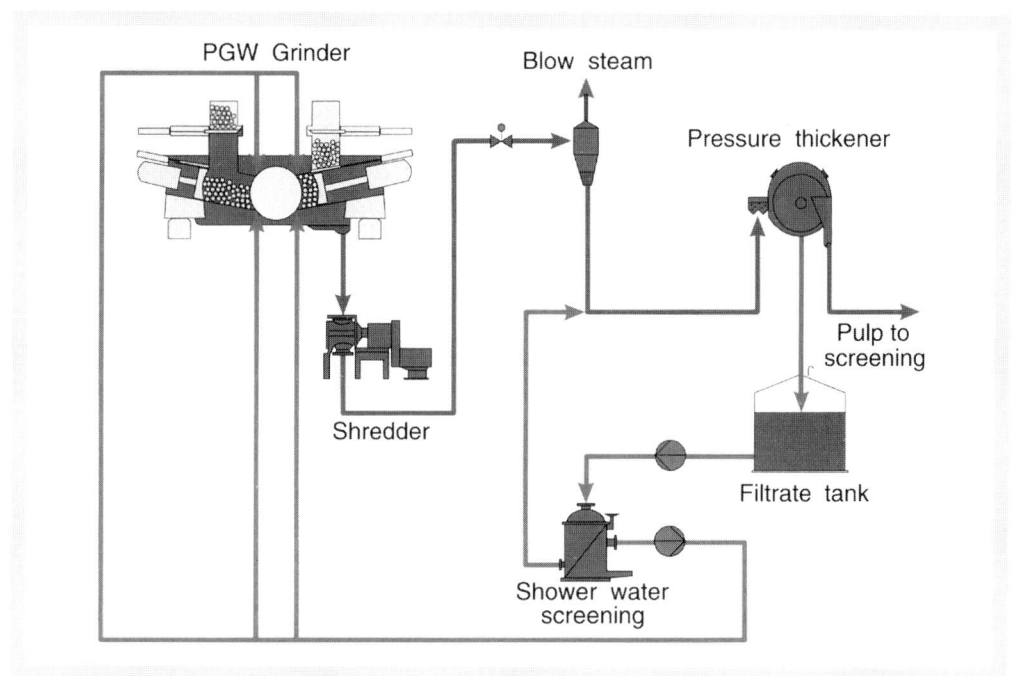

Figure 3. The coupling of a pressurized disc filter in the PGW "hot loop."

1.2 Broke deckers

Also the broke decker (Fig. 4) can be used for the thickening of mechanical pulps. In a broke decker, the wire cloth is fastened onto the surface of the drum. The pulp is directed into a vat where it is filtered while the filtrate enters and exits the drum. As there is no suction in the drum, the consistency of the outgoing pulp is 5%–8%. Broke deckers also have a low capacity. Therefore they are suited only for small pulp lines. Furthermore the solid content of broke decker filtrates is 500–800 mg/l.

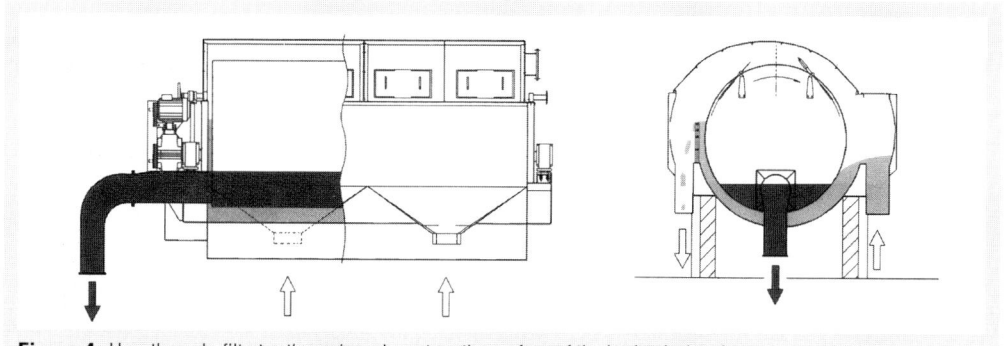

Figure 4. How the pulp filtrates through a wire set on the surface of the broke decker drum.

1.3 Bow screens

In the reject line, the pulp is often thickened with a bow screen prior to the refining process. The feed consistency of a bow screen is around 1% and the outlet consistency is 3%–4%. Figure 5 shows the operation principle. The chapter "Refining of shives and coarse fibers" discusses the operation of bow screens in more detail.

1.3.1 Screw presses, twin-roll presses, and twin-wire presses

Screw presses, twin-roll presses, and twin-wire presses are used for dewatering before reject refining and bleaching (see Fig. 16 in the chapter on bleaching). The chapter "Refining of shives and coarse fibers" describes these presses. A new design of roll press is the Sunds Defibrator PreRoll™ press, which is developed for dewatering of pulp suspensions with feed consistency of 1%–4% to a discharge consistency of 4%–15%. Figure 6 presents the operating principle of the PreRoll press. It has a compact design and little floor space is needed in relation to the dewatering capacity. The first two presses of this kind are in operation at Union Bruk in Norway (see flow sheet in Fig. 29 in the chapter on TMP).

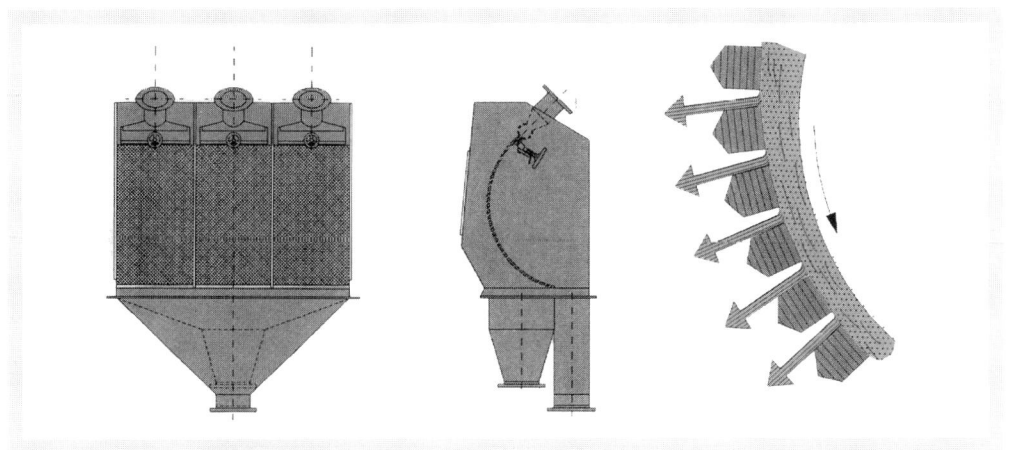

Figure 5. The operation principle of a bow screen: The pulp is directed along the bowlike screen plate and, when the bending plate surface forces the pulp flow to change direction, the water is filtrated through the screen plate.

2 Storage of mechanical pulps

2.1 Storage systems

For storage of mechanical pulp, the most common solution is to have a storage tower between the mechanical pulp plant and the paper machine. The size of the tower is determined by:

- The time-based need for storage
- Storage consistency.

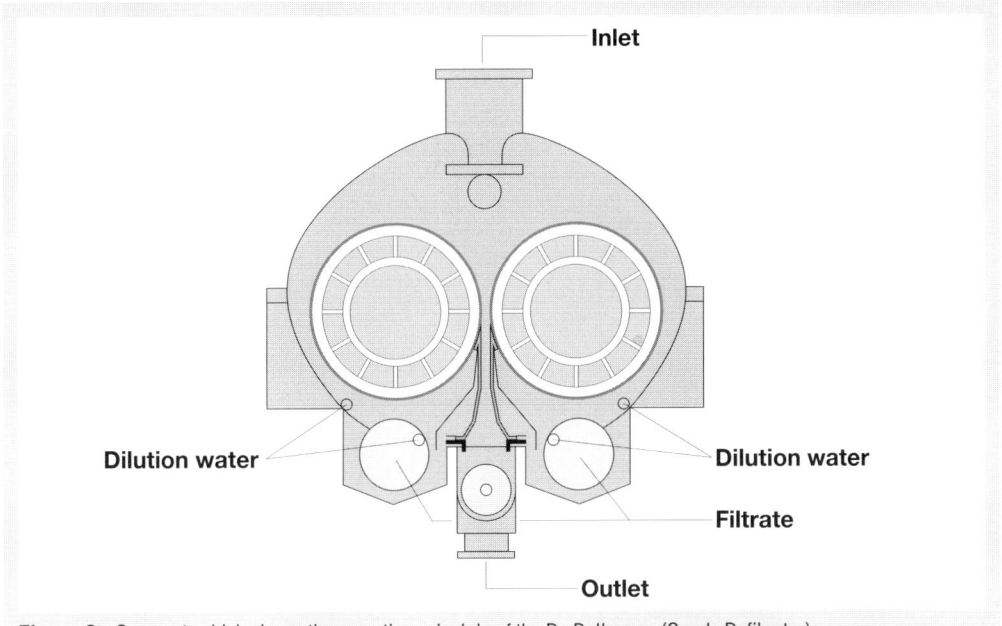

Figure 6. Crosscut, which shows the operating principle of the PreRoll press (Sunds Defibrator).

The time-based storage capacity is usually in proportion to the shutdown time at the mechanical pulp plant for service and maintenance. The time-based storage capacity is typically 8–12 h when medium-consistency storage is used.

The storage consistency is determined by the process solutions after the pulp bleaching. If a wash press is used after the bleach tower, the pulp is normally pumped in medium-consistency to an medium-consistency tower for storage. In older mills, one can see solutions where there is no washing after the bleaching stage and the pulp is diluted to 4%–5% consistency for storage. For low-consistency towers, the storage time cannot be as long as for medium-consistency towers because the size of the tower would be too big.

In some cases, the storage tower can be used as a second-stage bleaching tower in a peroxide + dithionite (hydrosulfite) bleaching process. A storage tower can also be used for reducing quality variations in the pulp. If a storage tower is used for reducing quality variations, sufficient agitation must take place. Also the size of the tower should be selected based on the cycle of the variation.

Today mechanical pulp is typically stored at medium consistency, e.g., 12%–13%. The tower is usually made of concrete and the shape is somewhat conical. The upper part of the tower is used for storage. In the lower (conical) part, there is a dilution and sufficient agitation. The pulp is diluted in the lower part in order to be pumped further in the process with a centrifugal pump at 3.5%–4.5% consistency. The agitators can be of either top or side entry type. In large towers, two side entry agitators are preferable. The pulp is diluted in the lower part of the tower with PM white water (clear filtrate). Figure 7 shows a typical process for mechanical pulp storage.

CHAPTER 12

For detailed information concerning the agitation of pulp, reference is made to two textbooks[1,2].

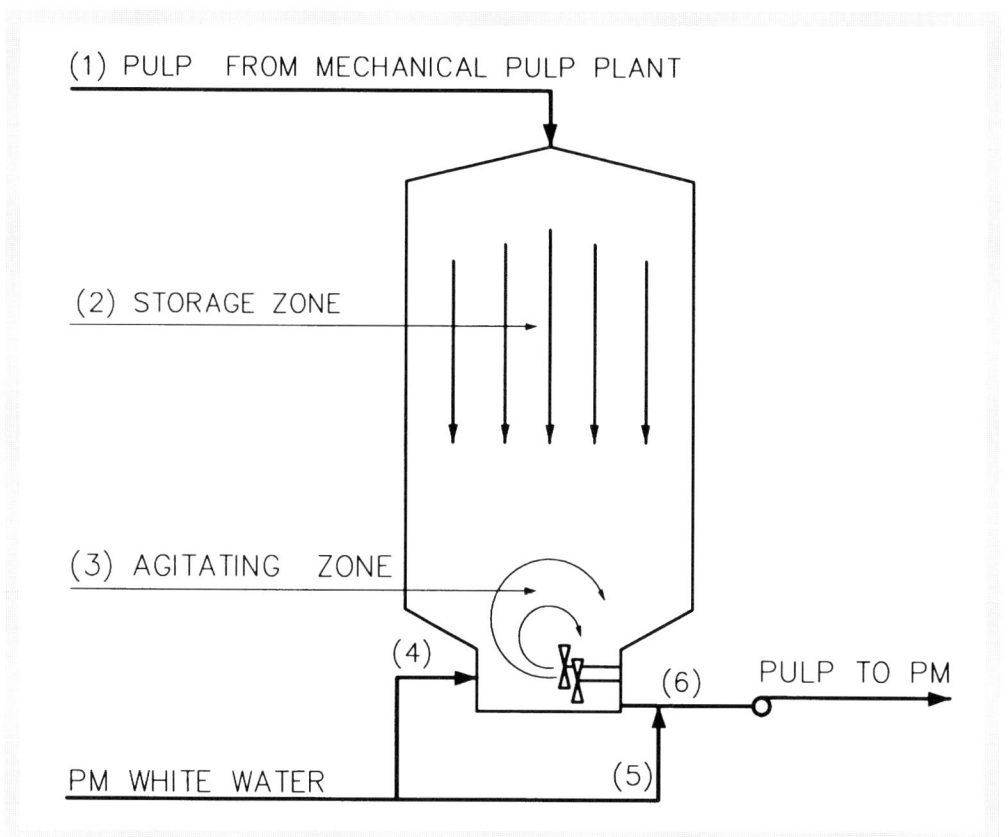

Figure 7. Typical mechanical pulp storage tower. The pulp is entering from the top of the tower (1). The upper part of the tower is used for storage (2) and the lower part of the tower is used for agitation (3). The consistency of the incoming pulp is 12%. The pulp in the lower part of the tower is diluted with PM white water (4) to a consistency where agitation with two agitators is optimized in terms of energy consumption. The consistency of the pulp is adjusted with PM white water (5) on the suction side of the pump (6) which is pumping the pulp further to the process.

2.2 Effect of storage on pulp properties

Especially when using neutral papermaking as more and more is the case in the production of coated mechanical papers, there is a danger for deterioration of pulp properties during pulp storage. In a laboratory trial[6], dithionite bleached mill PGW was stored after dilution with PM white water. The results (Table 1) showed that pH decreased, COD increased, and brightness decreased.

Table 1. Laboratory storage trial with dithionite (hydrosulfite) bleached PGW at temperature 55°C in a closed tank.

Time, h	pH	Turbidity, FTU	COD, mg/l	Brightness, %
0	6.81	80	947	75.4
1	6.73	158	1031	75.3
3	6.66	205	1201	75.2
5	6.54	266	1239	74.1
7	6.44	267	1280	73.3
9	6.16	216	1295	72.8
11	5.97	232	1305	72.7

3 Post refining

Low-consistency post refining often is an important unit operation when producing mechanical pulp for high-quality mechanical printing papers. Typically post refining is performed immediately before mechanical pulp fibers enter the paper machine. It is a final fiber upgrading phase before the paper machine[3].

3.1 Targets of post refining

Post refining serves several purposes, which vary from mill to mill. The main purpose is to improve or restore the bonding ability of fibers. Briefly, the objectives are to improve runnability, to improve printability, and to reduce paper manufacturing costs. Those objectives are achieved by developing fibers so that they are more suitable for papermaking. Post refining can improve the bonding ability of fibers by fibrillating and also perhaps by collapsing long and stiff fibers. Post refining removes latency by straightening curly and kinky fibers.

Runnability is improved by improving bonding ability (tensile strength) of fibers, and by reducing the shives content. Printability is improved in many ways. Fibrillation of fibers improves bonding ability, which reduces dusting or linting on printing presses. Fiber collapsing reduces thickness of coarse fibers, which improves surface smoothness, and fiber shortening by cutting reduces fiber rising to some extent. Post refining reduces porosity by reducing pinholes, thereby improving printability due to reduced print-through. Paper manufacturing costs might be reduced in two ways. At first, if the final freeness drop is performed in post refining, the total energy consumption in pulp manufacturing is reduced or capacity of pulp manufacturing is increased. Secondly, the better bonding ability of fibers might reduce the need for expensive reinforcement kraft pulp.

Sometimes, post refining is used to eliminate harmful shives. This has been very typical in the past, but today shive content usually is not a big problem because of efficient screening and reject refining. One typical post refining application is the final free-

CHAPTER 12

ness control at multimachine paper mills. The pulp mill produces constant freeness pulp, but the paper machines can have various freeness requirements since they produce different paper grades. In this case, post refining is used for trimming of freeness to suit specific requirements.

3.2 Equipment used in post refining

In the past, only single- or double-disc refiners were able to perform correct post refining. The segments of other refiner types (conical refiners) were too coarse and not suitable for post refining. Today, there are three refiner types which are able to perform correct post refining: namely the Conflo medium angle conical type refiner (Fig. 8), the conventional double-disc type refiner (Fig. 9), and the Multi-Disk type refiner (Fig. 10).

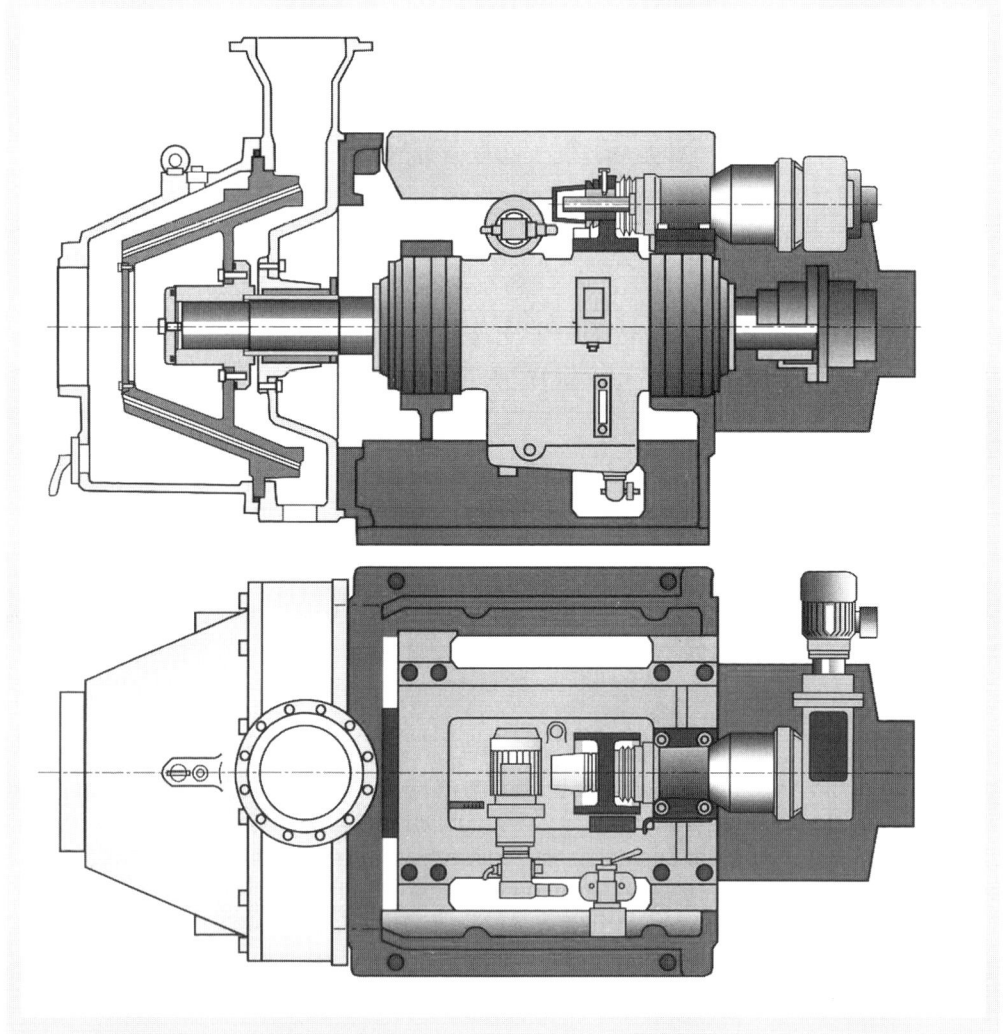

Figure 8. Sunds Defibrator Conflo® refiner.

Thickening, storage and post refining

The Sunds Defibrator Conflo refiner series comprises six different sizes. The connected maximum power range is 110–3500 kW and the gap clearance adjustment is electromechanical. Typical data for post refining installations are:

Capacity range	50–600 bdmt/d
Bar width of segments	2.0–3.0 mm
Refining intensity	0.5–1.5 J/m or 250–500 J/m^2
Refining consistency	4.0%–6.0%

As an example of several double-disc type refiners on the market, figures for the Andritz Sprout-Bauer Twin Flo III double-disc refiners are presented. These refiners are available in six different sizes ranging from 12 in. (305 mm) to 58 in. (1473 mm) disc diameter. Plate adjustment at a power range of 90 to 3000 kW is accomplished either by an electromechanical positioning mechanism or hydraulically. Typical data for post refining installations are:

Capacity range	20–800 bdmt/d
Bar width of segments	1.3–2.0 mm
Refining intensity	0.3–1.0 J/m or 200–350 J/m^2
Refining consistency	4.0%–6.0%

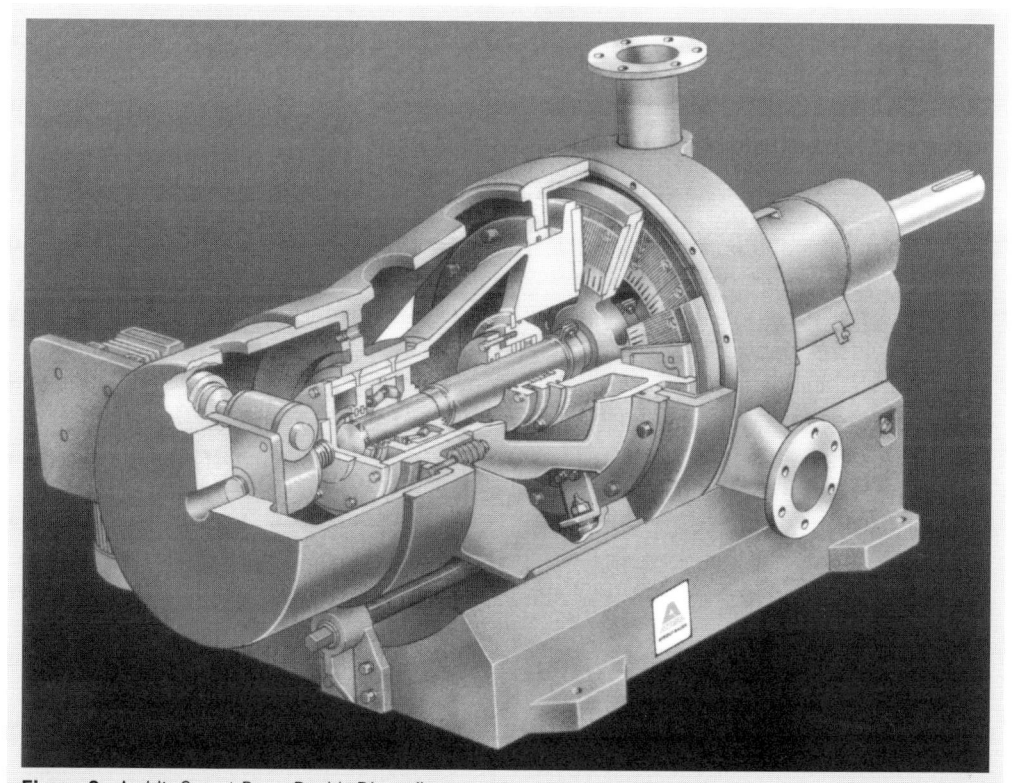

Figure 9. Andritz Sprout-Bauer Double Disc refiner.

CHAPTER 12

The Beloit Multi-Disk Series MD4000 comprises six different sizes covering disc diameters from 20 in. (508 mm) to 42 in. (1067 mm) and power ranging from 350 kW to 1800 kW. MD refiners have three or four rotors and can be converted to double-disc (one rotor) operation, 20 in. (508 mm) to 46 in. (1168 mm) diameter. Typical data for post refining installations are:

Capacity range	50–400 bdmt/d
Bar width of plates	2.0–2.5 mm
Refining intensity	0.2–1.0 J/m or 90–350 J/m^2
Refining consistency	4.0%–5.0%

With every refiner type, the actual power, capacity, segment type, refining intensity, and consistency figures depend on the refining resistance of fibers and on the targeted refining result. In general, TMP fibers are longer and have a higher refining resistance than groundwood fiber. Accordingly, TMP fibers also require more energy and coarser segments than groundwood fibers[5].

Figure 10. Beloit Multi-Disk refiner.

Thickening, storage and post refining

3.3 Post refining systems

A typical post refining system comprises a refiner feed chest, pump, and one or two refiners in series (Fig. 11). Pulp is fed from the refiner directly to the paper machine blending chest. Depending on the type of refiner and on the capacity variations, there might be a circulating line after the last refiner back to the pump suction. The purpose of this circulation line is to ensure sufficient fiber flow through the refiners in all conditions.

It is important that the post refining stage is very close to the paper machine. If it is located too early in the process, fibers in a pulp storage tower lose their bonding ability and regain latency due to high temperature and consistency. It should also be noted that, if a final freeness drop is carried out in the paper mill, the thickening before the storage tower is easier due to higher freeness.

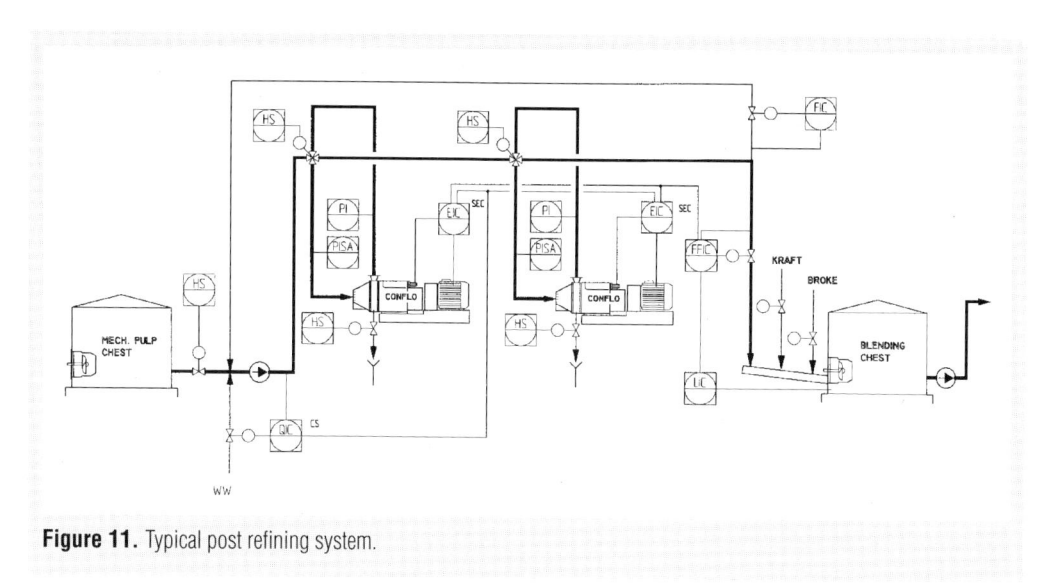

Figure 11. Typical post refining system.

3.4 Operation of a post refiner

There are a great number of variables describing refining conditions, but the most important are the nature or intensity of the refining, the amount of net energy input, consistency, and pH.

The nature of refining is described by using various refining intensity figures. The most common one has been specific edge load, SEL (J/m), which describes the amount of net energy for each meter of bar crossing:

$$Specific\ edge\ load\ SEL\ (J/m) = \frac{Net\ refining\ power\ (kW)}{Cutting\ speed\ (km/s)} \qquad (18)$$

Net refining power is the total absorbed refiner load minus no load power. The no load power is measured with water flowing through the running refiner, and the gap clearance is the narrowest possible without plates touching each other or any substantial increase in power.

CHAPTER 12

Recently, the new specific surface load theory (SSL theory) has replaced the old specific edge load theory. The specific surface load theory also takes the width of refiner bars into consideration and describes the refining intensity through two factors, namely, the length of refining impacts (m) and the specific surface load (J/m²). The new specific surface load value is obtained by dividing the specific edge load with the bar width factor:

$$\text{Specific surface load } SSL(J/m^2) = \frac{\text{Specific edge load SEL (J/m)}}{\text{Bar width factor, IL (m)}} \quad (19)$$

in which the bar width factor,

$$IL(\text{impact length}) = \frac{Wr + Wst}{2} \times \frac{1}{\cos \alpha / 2} \quad (20)$$

in which Wr is Width of rotor bars
Wst Width of stator bars
α Average intersecting angle.

The refining intensity figures for mechanical pulp, as for any pulp, depend on the refining resistance and on the physical dimension of fibers. In general, too low an intensity is not able to treat fibers effectively. On the other hand, too high an intensity destroys fiber flocs on the bar edges, and "peeled off" fibers flow back into the grooves without receiving any refining treatment. In the event that the fiber flocs are not broken, a higher intensity increases fiber cutting.

Sufficient refining intensity and width of bars also depend on the type of refiner. Typically, the disc-type refiners have slightly narrower bars and operate at lower intensity than Conflo refiners[4]. As guidelines, the following intensity figures can be given for post-refining applications:

Multi-Disk	0.2–0.6 J/m or 200–400 J/m²
Double Disc	0.3–1.0 J/m or 200–350 J/m²
Conflo	0.5–1.5 J/m or 250–500 J/m²

The amount of refining is determined by means of the net energy input or the amount of net energy transferred to fibers. This is the best way for evaluating the refining conditions inside the refiner. However, the total energy consumption for obtaining correct refining conditions should also be considered since it determines the energy costs. There also are limitations on how much energy can be transferred to fibers in one pass. Depending on the refining resistance of fibers, this varies from 30 to 100 net kWh/bdmt ton in one pass. If more is required, there must be several refiners in series.

$$\text{Net refining energy (kWh/bdmt)} = \frac{\text{Net refining power (kW)}}{\text{Fiber mass flow (bdmt/h)}} \quad (21)$$

Consistency should not be considered as an independent figure since the coarseness of fillings has an effect on it. In general, coarser segments with wider grooves require higher consistency than finer segments with narrower grooves. Typically, consistency in post refining should be approximately 4.0%–5.0%.

The pH value has an influence on the water penetration into fibers. The best pH area is close to neutral, since too low pH prevents water penetration and too high pH makes fibers slippery (soapy).

Refiner load is controlled by adjusting the gap clearance and the most typical control system is the net energy (net kWh/bdmt) control, in which fiber consistency is controlled before refining and fiber flow is controlled prior to the blending chest. Consistency and fiber flow determine fiber mass flow expressed as bdmt/hour. When that is multiplied by net kWh/bdmt, the result is required net kW for the refiner. Total refiner load is then obtained so that no load power in kW is added to net power in kW. No power load is obtained with only water flowing through the refiner with discs just a little bit open. The basic set value is the net kWh/bdmt and that must be determined on site, case by case, so that required fiber development is obtained. After setting the net kWh/bdmt figure, the control system automatically follows flow and consistency values and controls refiner load so that correct net energy is obtained.

Refiner segments or plates are manufactured by casting, typically from martensitic stainless steel. The basic design parameters of the refiner segments are width of bars and grooves, height of bars, number, placement and design of dams, and angle of bars from the radial direction. There are various material compositions available and the heat treatment before final finishing determines the final characteristics of segments or plates.

Typical Conflo refiner segments for post-refining applications have 2.0–3.0 mm wide and 7.0 mm high bars. Grooves are 1.0 mm wider than bars and average intersecting angle between rotor and stator bars is approximately 20 degrees.

3.5 Development of fiber and pulp properties in post refining

Post refining develops fibers in many ways and the development, in addition to refining conditions, greatly depends on the net energy input. In general, an increase in the refining energy has the following effects on the fiber development:

- Freeness is reduced.
- Tensile and burst strength are improved at first, but will decrease after prolonged refining.
- Tear remains or slightly increases at first, but will soon decrease after prolonged refining.
- Fibers are shortened, but sometimes when refining curly fibers, fiber length can slightly increase.
- Typically coarse and long fiber fractions are reduced and medium fiber fractions are increased

CHAPTER 12

- There is no significant increase in the fines fraction, unless refining conditions are selected for that purpose.
- Shives content is reduced.
- Air permeability is reduced or air resistance is increased.
- Bulk is reduced or density is increased.
- Absorbency is reduced.
- Opacity and light scattering remain unchanged or are slightly changed (typically reduced when tensile strength is improved).
- Smoothness is improved.

The following illustrations show typical fiber development in post refining. The curves are average curves for Norway Spruce pulps and are based on serial refining in recommended refining conditions so that for TMP a maximum of 75 kWh/bdmt is applied in one stage. The corresponding maximum 50 kWh/bdmt is applied in one stage when refining groundwood pulp fibers. Depending on the pulp quality the development can vary, so the curves must be considered as indicative curves only. It should be noted that PGW pulp normally is between conventional groundwood and TMP.

Figure 12 indicates the freeness reduction as a function of the net refining energy input. The higher the initial freeness the larger is the freeness drop in ml. It

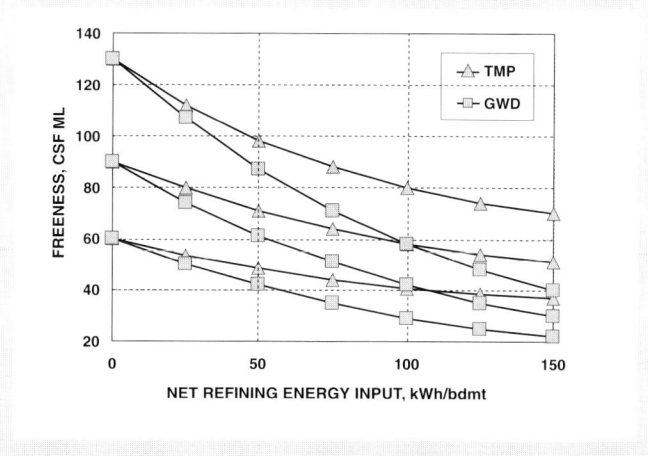

Figure 12. Freeness vs. net energy in post refining.

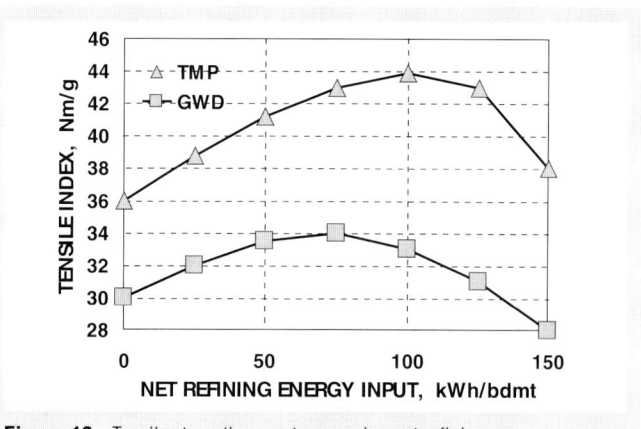

Figure 13. Tensile strength vs. net energy in post refining.

also can be seen that groundwood pulp fibers require less energy for a given freeness drop than TMP pulp fibers. An increased refining intensity would slightly, but not significantly, fasten the freeness reduction.

Figure 13 shows tensile strength development. Too high an intensity would decrease tensile strength increase with both pulp types. When fibers are post refined correctly, there will only be a slight reduction in the tearing strength or in the fiber length as indicated in Figs. 14 and 15. Typically, tearing strength is improved at first since improved bonding ability makes it more difficult to pull fibers out of the sheet in the tear test.

Too high a refining intensity has a definite negative effect on fiber properties since it involves more fiber cutting, and the tensile and tearing strength does not develop as favorably as in correct post refining. The disc-type refiner, if run at similar intensity as a Conflo-type refiner, is somewhat more fiber shortening. The following curves show fiber development with a Norway Spruce TMP pulp when a multipass refining has been

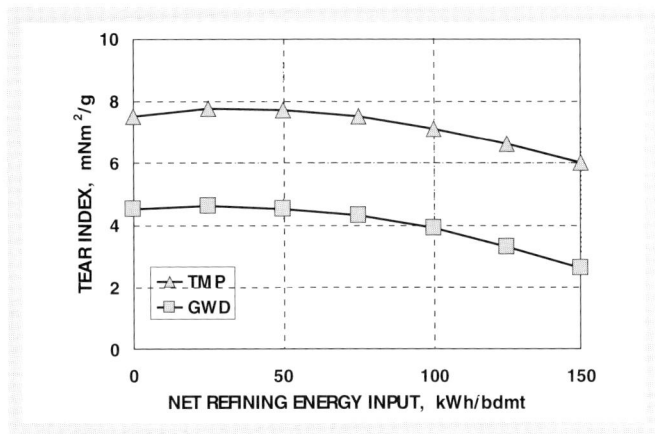

Figure 14. Tear vs. net energy in post refining.

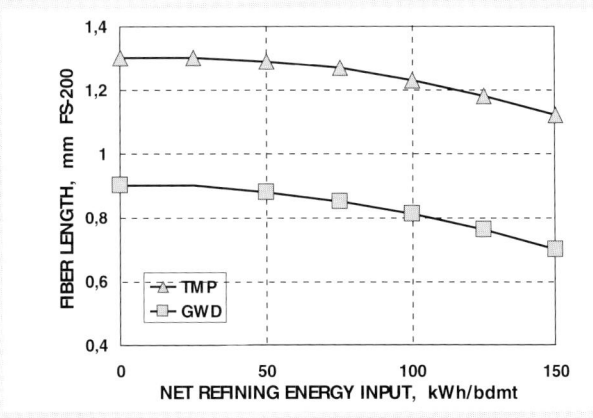

Figure 15. Fiber length vs. net energy in post refining.

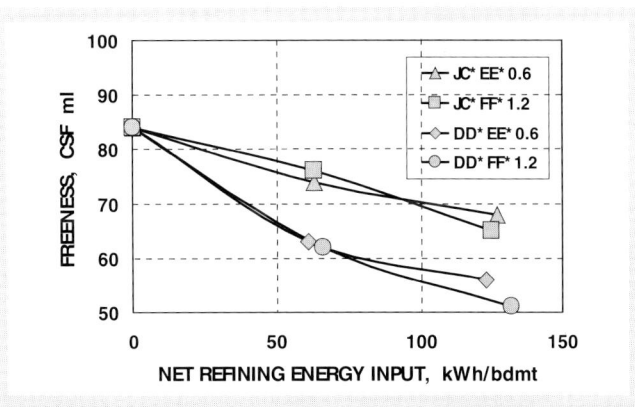

Figure 16. Freeness vs. net energy in multipass post refining with conical and disc refiners.

used. The first two lines with open symbols and JC marking are valid for the Conflo, and the last two lines with closed symbols and DD marking refer to the double-disc refiner. With both refiners, the lower intensity – 0.6 J/m – was obtained with 2 mm wide bars and the higher intensity – 1.2 J/m – with 3 mm wide bars. Correspondingly, the specific surface load figures were approximately 300 J/m^2 and 400 J/m^2. Refining consistency was 3.8%. An increase in refining intensity slightly increased freeness reduction (Fig. 16) but, at high intensity, the post refining did not improve tensile strength (Fig. 17). The effect of the type of the refiner was greater than the effect of the intensity.

A sufficient low-refining intensity with Conflo has at first improved tearing strength. Conflo refining at a higher intensity as well as all double-disc refiner results showed a declining tendency from the very beginning (Fig. 18). Lower refining intensity does not shorten fibers as much as the higher refining intensity (Fig. 19).

Post refining effectively reduces the shives content, and the higher the initial shives content is the larger is the shives reduction. Not only large shives

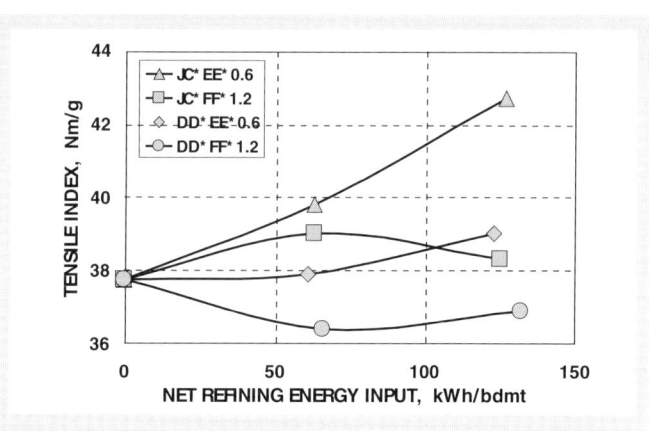

Figure 17. Tensile vs. net energy in multipass post refining with conical and disc refiners.

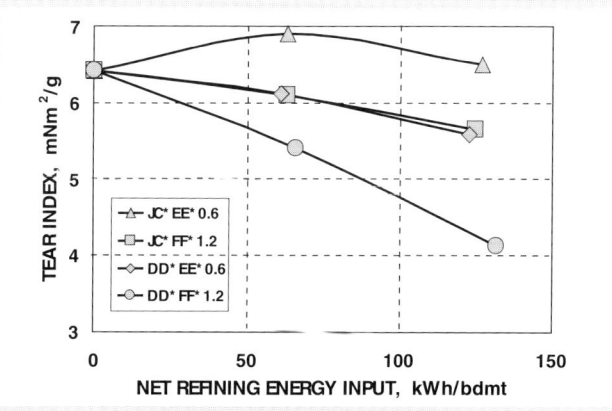

Figure 18. Tear vs. net energy in multipass post refining with conical and disc refiners.

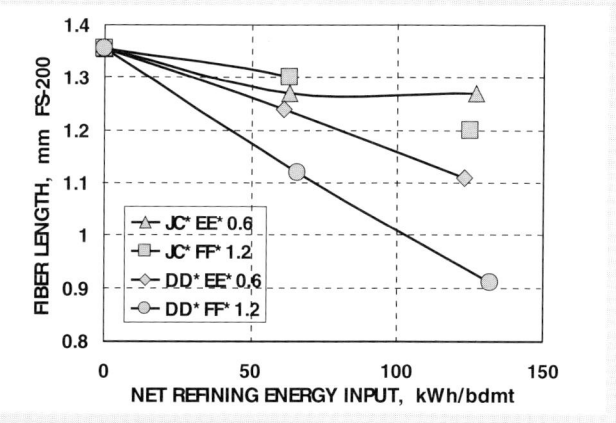

Figure 19. Fiber length vs. net energy in multipass post refining with conical and disc refiners.

but also medium-sized shives (PFI minishives) are reduced by post refining (Fig. 20). Conflo refining at the lower intensity shows practically no shives reduction; on the other hand, this gives the best tensile improvement and has the least effect on the tear or on the fiber length. Typically, post refining reduces light scattering, but the higher refining intensity with the double-disc refiner shows an increase in the light scattering (Fig. 21). However, in those conditions freeness, tensile, tear, and fiber length are all significantly reduced.

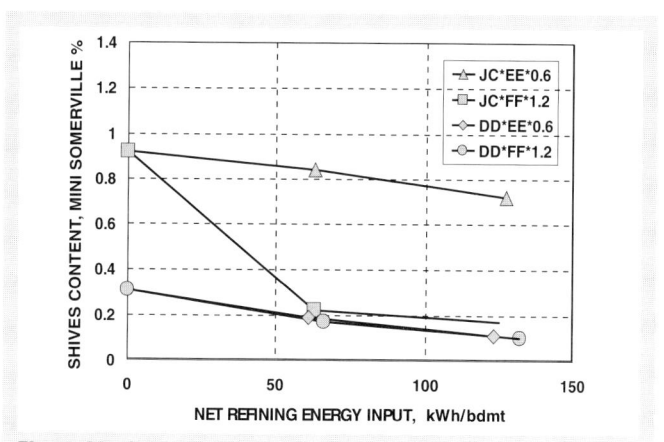

Figure 20. Shives content vs. net energy in multipass post refining with conical and disc refiners.

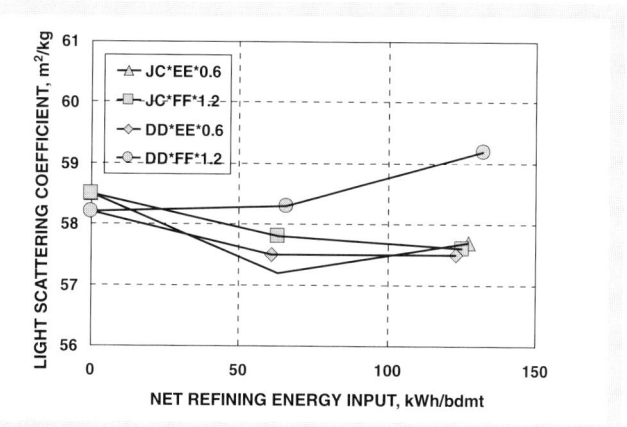

Figure 21. Light scattering vs. net energy in multipass post refining with conical and disc refiners.

CHAPTER 12

References

1. Handbok för elenergibesparing vid omröring inom massa- och pappersindustrin, SPCI meddelande nr 55, SPCI, Stockholm, 1989.

2. Yackel, D. C., Pulp and Paper Agitation, TAPPI PRESS, Atlanta, 1990.

3. Lumiainen, J., "Post refining of mechanical pulps," PIRA 1997 International Refining Conference Proceedings, PIRA, Leatherhead, p. 8:1.

4. Lumiainen, J., "Comparison of the mode of operation between conical and disc type refiners," PIRA 1997 International Refining Conference Proceedings, PIRA, Leatherhead, p. 14:1.

5. Demler, C. and Silveri, L., Appita 49(2):87 (1996).

6. Alhoniemi, O., diplomityö, Helsinki University of Technology, 1993.

Thickening, storage and post refining

CHAPTER 13

Flow sheets for various paper grades

1	**Groundwood and thermomechanical pulping lines**	**365**
1.1	Processing of GW and PGW pulps	367
1.2	Processing of TMP and CTMP	367
2	**Flow sheets for newsprint**	**367**
2.1	PGW pulps for newsprint	368
2.2	TMP for newsprint	369
3	**Flow sheets for SC and LWC**	**370**
3.1	GW and PGW pulps for SC and LWC papers	370
3.2	TMP for SC and LWC papers	371
4	**Flow sheets for board**	**372**
4.1	GW and PGW pulps for board	373
4.2	TMP and CTMP for boards	373

CHAPTER 13

Tero Jussila, Taisto Tienvieri, Jan Sundholm

Flow sheets for various paper grades

1 Groundwood and thermomechanical pulping lines

Figure 1 shows a simplified flow sheet or general overview of a pure mechanical pulping line. The process starts with the wood handling: for groundwood slashing and debarking and for TMP debarking and chipping or handling of saw mill chips. At the groundwood mill, the debarked logs are fed to the grinders by the use of belt conveyors. At the TMP mill, the chips are washed and preheated before fed to the refiners. After the grinding or refining, the simplified flow sheet looks the same for groundwood and TMP. The ground or refined pulp is screened, the rejected material is re-refined, and finally, if brightness requirements exceed those of newsprint, the pulp is bleached and washed. In some special cases, there is a separate post-refining stage just before the paper machine.

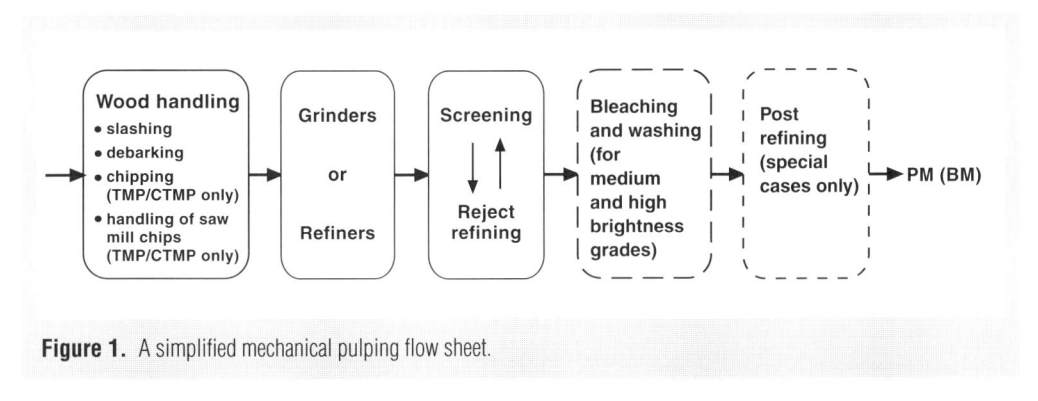

Figure 1. A simplified mechanical pulping flow sheet.

Even if the simplified flow sheet in Fig. 1 shows similarities between groundwood and TMP mills, a closer look at the mills will show many differences. Figures 2 and 3 show typical modern PGW and TMP installations.

CHAPTER 13

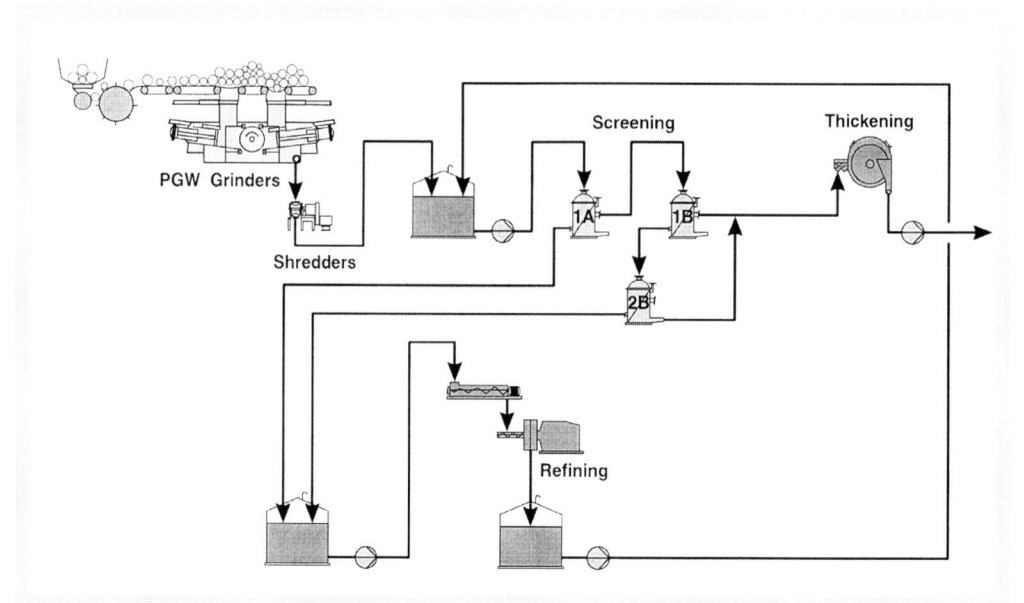

Figure 2. A pressure groundwood (PGW70) mill (Valmet). The subprocesses in the picture are from left to right: belt conveyors, pressure grinders, shredders, screening and reject refining, and dewatering in disc filter.

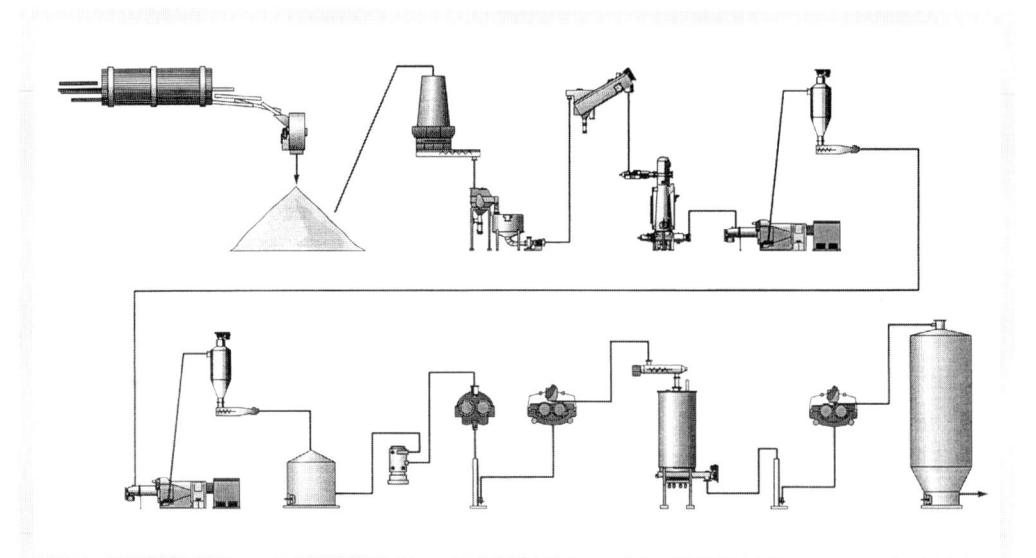

Figure 3. A TMP mill (Sunds Defibrator). The subprocesses in the picture are from left to right: debarking, chipping, presteaming, chipwashing and dewatering, preheating, first-stage refining, separation of steam, second-stage refining, separation of steam, removal of latency, screening and reject refining (here symbolized by one single screen), predewatering, pressing, high-consistency bleaching, pressing, and finally storage.

A groundwood mill for board pulp, for example, is also quite different from a groundwood mill producing magazine grade pulp. The largest differences in mechanical pulp mills producing pulps for different paper or board grades are found in the screening and also in bleaching and washing. This chapter, however, concentrates on the screen room flow sheets for various paper grades, and it starts with a look at the basic differences between the GW/PGW and the TMP pulp processing.

1.1 Processing of GW and PGW pulps

In PGW lines and also in modernized GW lines, the pit pulp is led through shredders, which break up slivers to big shives small enough for pressure screening. In GW lines, the pulp is normally pre-screened with vibration screens, which separate slivers to reject handling. The accept of the vibration screens goes to the pressure screening.

The first stage of the pressure screening is always equipped with hole baskets in groundwood lines. The purpose of hole screening is to separate big shives before slot screening and to lead those to reject treatment. Reject rates of these screens are as small as possible. The slot screening will determine the final quality of the pulp. The slot screening can be followed by a cleaner plant for removing additional shives and heavy particles like sand. Modern slot screening technology has increased screening efficiencies and pulp qualities so much that cleaner systems are usually not necessary in the pulp mills. However cleaners can be used for very demanding processes like high quality SC and LWC pulps. Groundwood processes contain some sand, which will be concentrated in the reject line. Therefore it is worthwhile to have cleaners in the reject line for sand separation. Cleaners should be positioned before the reject refiners.

The production rate of the mechanical pulping line can vary due to changes in paper machine production and paper grade. In such cases, it is essential that the screening system is constructed in such a way that the screening room can also function optimally when only a part of the screens are utilized.

1.2 Processing of TMP and CTMP

The refined mechanical pulp is usually first disintegrated in a pulper and then led to the latency chest and further on to screening. The screening system does not require screens with hole baskets as thermomechanical pulp does not contain slivers or big shives, which are typical for groundwood systems. On the other hand, TMP and CTMP pulps contain more long, coarse, and undeveloped fibers than groundwood pulps. Therefore reject rates and thus also the amounts of re-refined reject need to be larger compared to groundwood pulp lines. In TMP and CTMP mills, cleaner systems are usually even less necessary than in groundwood mills.

2 Flow sheets for newsprint

The main requirements for news quality mechanical pulp are sufficient strength properties and sufficient low shive content, which ensure good runnability in the paper machine and in the printing press. Newsprint is mainly printed in offset; thus, low linting tendency is an important property.

CHAPTER 13

The most suitable PGW procesess for newsprint is the PGW-S process with either 95 or 120°C shower water temperature, but any PGW or GW process can be used for newsprint. The pulpstone grit size is usually 54–60, and the pulp freeness after the grinder is usually in the range 140–180 ml. Newsprint TMP-mills usually have standard two-stage refining using SD-, CD-, or Twin refiners. Because more and more emphasis is put on low energy consumption, newsprint TMP mills with single-stage DD refining, the RTS process, or the Thermopulp process have been built during the last years. The pulp freeness after main line refining is in the same range as for groundwood pulp or somewhat higher.

2.1 PGW pulps for newsprint

Figure 4 shows a typical flow sheet for GW/PGW pulp screening for newspaper. The pulp is treated in a shredder (15–20 mm holes) before the primary one screen. The P1 screen is equipped with 1.6–2.0 mm holes, and the accept goes to the P2 screen and the reject to reject handling. The primary two and secondary two screens have slot baskets. Both accepts are led forward to the thickening, and the secondary screen reject is led to reject handling. The P2 and S2 screens are equipped with #0.15–0.20 mm slots.

The P1 and S2 rejects contain a lot of shives as well as stiff and coarse fibers. Those are treated in high-consistency reject refining, which develops pulp properties. The refined reject is screened with reject screen one (R1) equipped with a slot basket (#0.15–0.20 mm). The accept from the reject screen is led forward, and the reject goes back to reject refining. In some cases, the reject screening system contains two screens in series. Then the R1 screen is equipped with a hole basket to protect the slot screen (R2) against big shives. Normally 15%–30% of the groundwood production passes through the reject line.

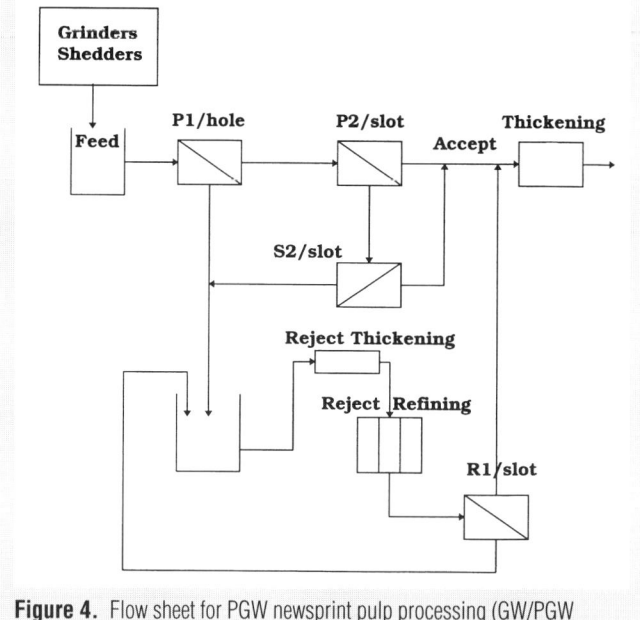

Figure 4. Flow sheet for PGW newsprint pulp processing (GW/PGW alternative 1).

Figure 5 shows an alternatively connection, where the refined reject is led back to the feed of the primary 1 screen. A disadvantage of that kind of connection is that all developed long fibers will not be accepted forward.

Flow sheets for various paper grades

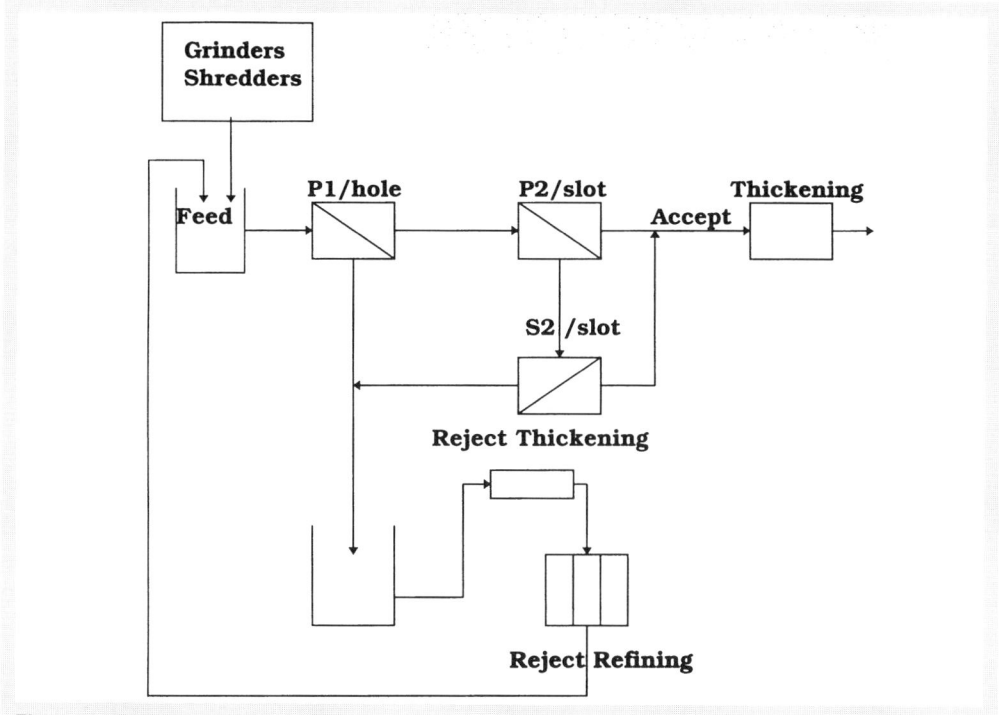

Figure 5. Flow sheet for PGW newsprint pulp processing (GW/PGW alternative 2).

2.2 TMP for newsprint

Fig. 6 shows a typical TMP screening system. The primary and secondary screens are equipped with #0.15–0.20 mm slot baskets and both accepts are led forward. The secondary screen reject goes to reject treatment. The refined reject is screened with the R1 screen, which also is equipped with #0.15–0.20 mm slots. The accept is led forward and the reject goes back to the reject handling. Some 25%–35% of main line refiner production passes through the reject line.

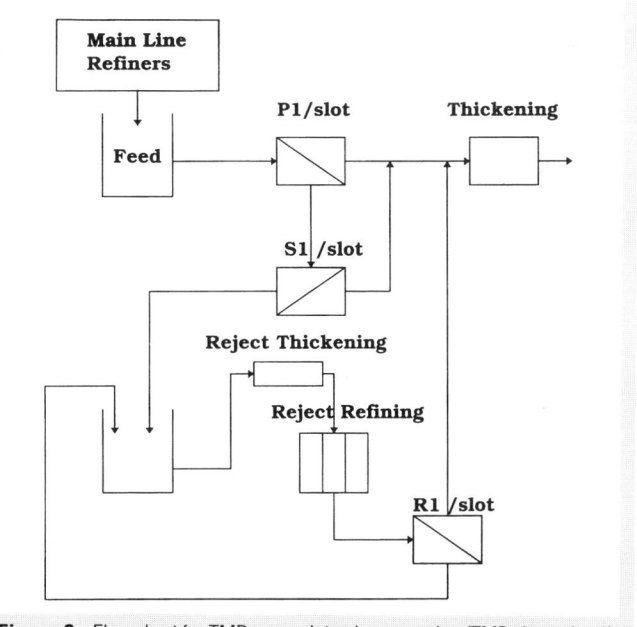

Figure 6. Flow sheet for TMP newsprint pulp processing (TMP alternative 1).

The mainline screening can also contain a third slot screen stage (T1), in which case all three accepts are led forward (Fig. 7). This kind of a system makes it possible to minimize the total reject rate of the process and to have higher reject rates for the separate screens. Sometimes reject refining is replaced by leading the reject back to the second stage refiners in the main line.

3 Flow sheets for SC and LWC papers

SC and LWC papers put roughly the same requirements on the mechanical pulping line and, thus, can here be treated together. Typically SC pulps have somewhat lower freeness than LWC pulps or an increased fines content. Good optical properties (brightness and opacity) are essential for SC and LWC paper pulps.

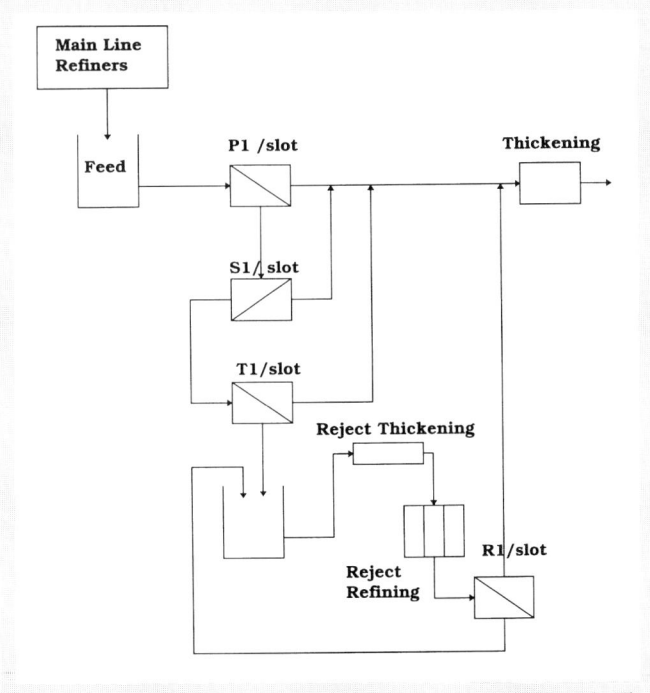

Figure 7. Flow sheet for TMP newsprint pulp processing (TMP alternative 2).

The targets for the SC pulp flow sheet are somewhat different for SC rotogravure and SC offset paper. The main quality requirements for SC rotogravure are high smoothness, high opacity, and good strength. For the offset grade, a low linting tendency and high gloss of printed area are important. Because of this, the freeness is often lower and the reject ratio higher when producing SC offset.

Also for LWC there are some different requirements on offset and rotogravure paper pulps. The rotogravure pulp should have high compressibility and high air resistance. The offset pulp should have high internal strength and more slender and fibrillated fibers together with a very low shive and coarse fiber content to prevent fiber roughening and blistering.

Of the groundwood processes, either normal GW or low-temperature PGW are considered most suitable for SC and LWC pulp production. The pulpstone grit size is usually 70 or 80, and the pulp freeness after grinding is normally in the range of 70–110 ml. For TMP, the main line refining is normally performed in 2–3 stages and the pulp freeness is in the range of 90–120 ml.

3.1 GW and PGW pulps for SC and LWC papers

The screening principles for SC and LWC groundwood are mainly the same as for newspaper, but the requirements for pulp quality are much higher. Some 30%–50% of the groundwood production passes through the reject refiner. Refining can be made either in high or low consistency. The trend is toward high-consistency refining because of its fibrillating nature.

Figure 4 shows a typical flow sheet for groundwood screening. The P1 screen is equipped in the case of SC and LWC papers with 1.4–2.0 mm hole baskets, and the P2, S2 and R1 screens are equipped with #0.10–0.15 mm slot baskets. The feed to the secondary screen is coarser than the feed to the primary screen. An equal pulp quality in both screening stages can be achieved by optimizing slot width and profile height of the screen baskets. The reject rate and screening consistency are also adjustment parameters for pulp quality.

The flow sheet in Fig. 8 shows a screening system with three-stage slot screening in the main line. All three accepts are led forward and the screens are equipped with #0.10–0.15 mm slot baskets. This kind of system provides a possibility to minimize the total reject rate of the process and to have higher reject rates for separate screens. This system also allows the use of higher SEC in reject refining. There are two screens in the reject line. The R1 screen is equipped with a 1.4–2.0 mm hole baskets and R2 screen with #0.10–0.25 slots. The reject of the R1 screen is led back to the reject handling, and the reject of the R2 screen is led to the feed to the S2 screen. A hole screen is not needed if reject refining is effective enough to eliminate all big shives (two reject refiners in series).

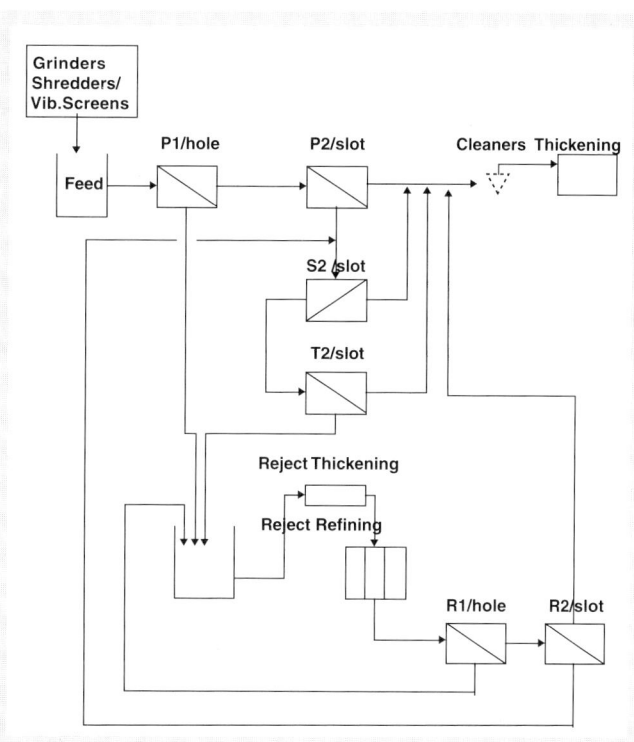

Figure 8. PGW pulp processing for SC and LWC papers (GW/PGW alternative 3).

3.2 TMP for SC and LWC papers

The processing principles for SC and LWC TMP are mainly the same as in the case of newsprint, but the pulp quality requirements are much higher. The amount of pulp taken to reject refining is 40%–60% of the main line refiner production. Typical flow sheets are shown in Figs. 6 and 7. All screens are equipped with slot baskets and the slot widths are #0.10–0.15 mm depending on the screen position.

For the most demanding processes, flow sheets of the type shown in Fig. 9 are used. The reject refining is performed in two stages in order to ensure effective refining of coarse and stiff fibers. All screens are equipped with slot baskets and all accepts are led forward. The secondary screen reject goes to the first-stage reject refining. The re-refined pulp is screened with the reject screen (RS1), and reject from that screen goes to the second-stage reject refining. The refined pulp is screened with the secondary and tertiary reject screens (RS1, RT1), and reject from those screens can be led either to the first- or the second-stage reject refiner. The RS1 and RT1 screens are used to minimize the amount of recirculated reject.

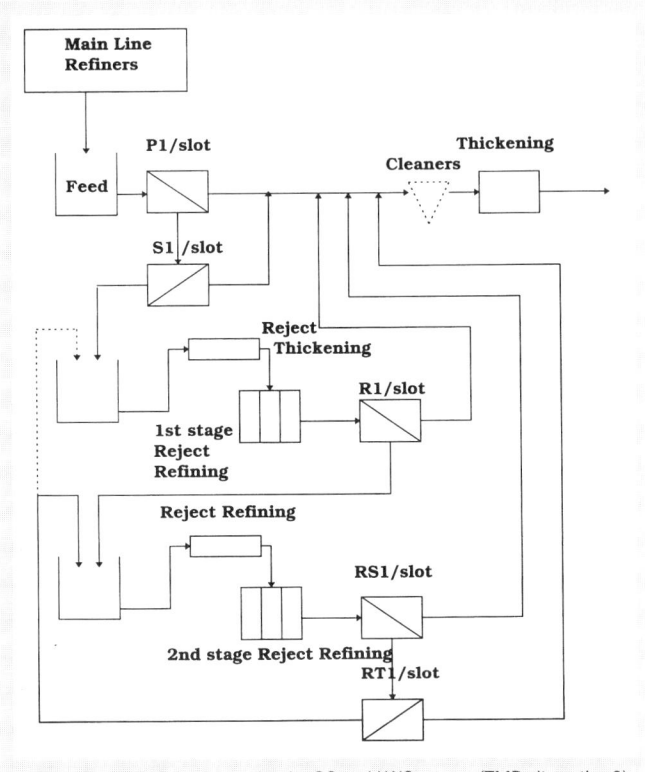

Figure 9. TMP Pulp processing for SC and LWC papers (TMP alternative 3).

4 Flow sheets for board

The main target for board pulp is a good combination of bulk and internal bond strength. The amount of shives and fibers without bonding ability need to be minimized to avoid linting.

Standard GW is still the most widely used raw material for folding boxboard, where it is used in the middle layers. Also low temperature PGW (PGW70 and PGW95) are very well suited for boardmaking. The grit size is normally 36–46, and the freeness before screening is in the range 350–450 ml. TMP is not widely used for board. Single- or two-stage refining is used to a lower freeness than in the groundwood case because of the low fines content in TMP, which causes poor internal bonding strength. CTMP is very suitable for use in the middle layer of liquid packaging board because of the high bulk compared with kraft-based solidboard and because of the low extractive content. In order to reach the required DCM content of < 0.1%, the CTMP must be washed 3–5 times. The CTMP freeness before screening is similar to or higher than that of board grade groundwood.

4.1 GW and PGW pulps for board

A typical flow sheet for GW or PGW board pulp processing is similar to that shown in Figs. 4 and 5. The P1 screen is equipped with 1.6–2.4 mm hole baskets, and the P2, S2, and R1 screens are equipped with #0.20–0.35 mm slot baskets. In older systems, the secondary screen accept is usually led backward to the feed of the P1 screen. In the reject line, a combination of hole and slot baskets can also be used (R1:1.6–2.4 mm holes, R2: #0.25–0.35 mm slots). Some 15%–25% of the groundwood production passes through the reject refiner. The main target in reject refining is to remove large shives and to retain bulk; therefore, LC refining – which is effective in these respects – is commonly used.

4.2 TMP and CTMP for boards

Similar screening systems can be used in the screening of TMP and CTMP grades for board grades. Figure 10 shows such a system. The shive content of CTMP is lower, and the required bonding properties are reached at a higher freeness. About 20%–35% of the main line refiner production passes through the reject refiner. The screens are typically equipped with #0.2–0.3 mm slot baskets.

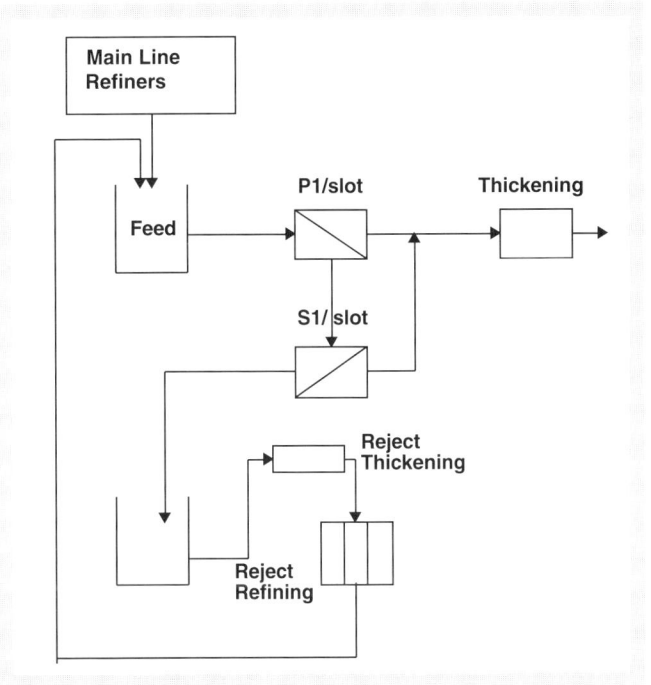

Figure 10. TMP and CTMP processing for board (bleaching and washing stages not shown).

CHAPTER 14

Environmental impacts of mechanical pulping

1	**Woodyard, debarking, pulping, and bleaching processes**	**375**
2	**Emissions from the processes**	**376**
3	**Wood and chip storage**	**377**
3.1	Trends in techniques	377
3.2	Changes in wood during storage	378
4	**Debarking and chipping**	**378**
4.1	Wet or dry techniques	378
4.2	Water circulation systems in debarking	379
4.3	Need for water	379
5	**Treatment of debarking effluent**	**380**
5.1	Composition and concentration of debarking circulation water	380
5.2	Effluent load	380
6	**Release of wood components into water in mechanical pulping and bleaching**	**381**
6.1	Discharge data for different pulping processes	381
6.2	Influence of bleaching stages	382
6.3	Chemical character of released wood components	382
6.4	Toxic components in effluents	385
7	**Factors influencing release of wood components into water**	**385**
7.1	Effect of pulping conditions	385
7.2	Effects of wood quality	387
8	**Evaporation of compounds in mechanical pulping**	**387**
8.1	Volatility of wood compounds	387
8.2	Contents of volatile compounds	388
9	**Water circulation systems in mechanical pulping plants**	**388**
9.1	Connection between pulp mill and paper mill	388
9.2	Washing stage	389
10	**Treatment of mechanical pulping effluents**	**390**
10.1	Purification methods	390
10.2	Outlook on internal cleaning stages in mechanical pulping plant	390
10.3	Zero effluent technology	391
	References	392

CHAPTER 14

Hannu Manner, Pekka Reponen, Bjarne Holmbom, Joseph A. Kurdin

Environmental impacts of mechanical pulping

1 Woodyard, debarking, pulping, and bleaching processes

Environmental impacts of mechanical pulping originate from various stages of the manufacturing. At many mill sites, the mechanical pulp mill is the only user of wood raw material. Forestry itself and logistics of wood cause certain impacts. A life-cycle approach should include these operations, but they are not considered here.

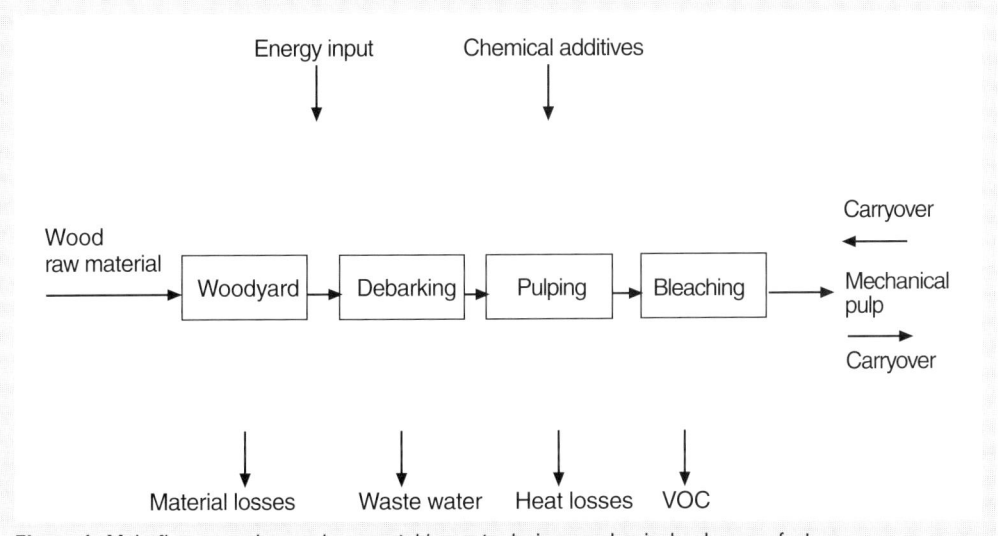

Figure 1. Main flows causing environmental impacts during mechanical pulp manufacture.

Defiberizing processes, either a variant of the grinding process or a variant of the refiner process, together with screening, reject treatments, thickenings, and dilutions, etc., are the main steps in mechanical pulping. They correspond to most of the energy input in the whole mill. They cause the greatest part of the wood fiber yield loss, all forms of which are subsequently found as environmental impact.

Residuals of chemical additives, such as residuals of treatment chemicals in CTMP manufacture or residuals of agents used in bleaching stages, can also form a noticeable part of the impacts. Moreover chemi-treatment or bleaching stages reduce yield. The wood material and the process chemicals which are released into process waters will end up in the mill effluents, if not carried away with the paper product or taken out as solid waste.

In this chapter, the emphasis is laid on water management and effluent formation in mechanical pulping and bleaching. Debarking effluents are also discussed, as well as release of volatile organic components in mechanical pulping. The volume "Environmental Control" in this textbook series deals with the treatment of effluents and solid waste.

2 Emissions from the processes

Most of the solid material that forms waste flows is created during debarking: bark, debarking loss of wood, sawdust, and chipping dust. A common practice is to use the solid wastes as fuel and thus utilize their energy value. A smaller part of the solid waste is composed of fiber losses during pulping.

The effluent discharge from debarking is significant. The effluent volume can be low compared to the volume flows from other parts of a mill, but its content of solutes can well correspond to one-third of the COD load of the whole paper mill. Dissolved and dispersed substances then mainly come from bark. Its biologically toxic nature is one reason why the effluent is kept separate from the water system of pulping and bleaching.

The dissolved and dispersed substances in the process waters of mechanical pulping systems originate from several sources:

- Components of wood dissolved and dispersed during the high-temperature conditions of the defiberization processes (hemicelluloses, pectins, lignin, extractives, and inorganic salts).

- Residuals of processing chemicals possibly applied in mechanical pulping and bleaching.

- Leakages from sealing and lubrication and dissolution of equipment materials in to the process.

- Carryover of residuals of papermaking chemicals with the circulation water that is circulated backward from papermaking to pulping.

- The former possibly including those organic and inorganic substances that came in with fresh water intake and purification.

- Carryover of dissolved and dispersed material from debarking. As the debarking process usually is relatively closed, the concentration of process water becomes high and contaminated water penetrates into wood.

The dissolved and dispersed substances in the process waters constitute the main environmental impact of mechanical pulping, and there is a great number of research reports dealing with these issues. There is, however, a considerable incompatibility between results of different studies. One obvious reason is that the amount and composition of dissolved and dispersed substances depend on the composition of the wood which varies largely and which, e.g., is different at different points of the annual growth cycle. Another obvious reason is that the water system of the mechanical pulping department is connected to the water system of papermaking which causes interference between the two systems. The water system of mechanical pulping also has internal circulations causing accumulation effects. Thus, the concentrations of substances in mill water are dependent not only on the wood and defiberizing conditions, but also on the structure of the water system.

Some material losses take place by evaporation. Those volatile compounds are removed with exhaust steam flows or condensed back into an liquid phase.

The final environmental emissions from mechanical pulp manufacture after all system closings, recovery, and purification steps can be listed as follows:

- Bark and other wood-based residues that are recovered as a valuable fuel, corresponding to about 0.8 MJ heat energy per ton of mechanical pulp produced.
- Concentrates from internal and external effluent purification. Biological purification is a common practice and the concentrate of external cleaning is biosludge, which is usually treated together with primary sludges.
- Residual contaminants in clarified water outlet from effluent purification, which contain oxygen-consuming organic substances, nutrients, and various other organic and inorganic impurities. Cooling and sealing waters remain uncontaminated.
- Air emissions, from different stages of the operation.
- Noise caused mainly by wood handling, outdoor conveyors, and improper traffic routing.
- Heat excess from defibration which is partly used for internal purposes of the process, partly recovered to be utilized for external purposes, and partly lost as exhaust steam or warm effluents.

Thus, the analysis of environmental impacts due to mechanical pulping should give an answer to what these items mean in a specific case and what the means are to influence them.

3 Wood and chip storage

3.1 Trends in techniques

Wood storage time and storage conditions can cause changes in wood quality leading to changes in solubility of components, which consequently can influence the environmental impact.

There are two important trends in managing wood operations: The delay times between wood harvesting and use at the mill are getting shorter due to intentional efforts to get wood for processing as fresh as possible. Therefore any quality impact of storage delay that might exist will become smaller. Another trend is that harvesting is more and more widely done by motorized harvesters which debark the wood to a great extent. The logs lose 30%–50% of bark, which means that drying and other influences in wood will be faster and short delays become more important. But this also means that dissolution of bark components during debarking becomes smaller.

Total storage times at a maximum of a few months are relevant when considering changes that might happen in wood during storage. Often the storage times are clearly shorter and the conditionings during storage reduce the effects even more.

3.2 Changes in wood during storage

Changes in wood quality and changes in fiber quality due to storage time and due to different storage methods have been an interesting topic for research all over the world. Less attention has been paid to possible yield loss and hence environmental impacts. An extensive study was carried out by the Finnish Pulp and Paper Research Institute and partner companies in the turn of the 1990s. An essential part of the results is unpublished. The study's main objective was to find the influences of storage variables on the quality potential of pulp. The main changes in wood during storage are due to bacterial and fungal activity and drying followed by accelerated oxidation. The main quality influence is a loss of brightness. A part of the change is probably due to the penetration of water-soluble bark extractives into the wood. Drying and temperature during storage have an influence. Earlier studies have reported loss of pulp strength, indicating decomposition of wood polymers. The study did not show any larger strength deterioration during a time of a few months, independently of conditioning during storage.

Other studies have reported a significant reduction in the amount of lipophilic wood extractives during storage. This is probably the case with wood species containing high amounts of extractives as evidenced by well known loss of volatile components and decrease of tall oil yield during longer storage. Studies on spruce wood storage have shown that a great part of the triglycerides are hydrolyzed to free fatty acids[1,2], while the total amount of extractives is not essentially affected during storage up to four months. As a consequence, there will be changes in the composition of extractives during harvesting and storage, but not in their total amount.

4 Debarking and chipping

4.1 Wet or dry techniques

Roundwood at mechanical pulp mills is almost always debarked by the aid of drum methods. In the wet technique, an ample amount of water is used over the whole drum delay to promote release and removal of the bark. In the dry technique, there is no closed section in the drum and only a small amount of water is used to assist debarking during the drum delay. Dry techniques use less water circulation through the drum stage. However, in frozen conditions, de-icing of the wood surface for debarking presumes that dry debarking is preceded by a separate smelting stage and here water is needed as heat transfer medium. Details of drum techniques can vary, resulting in different efficiency and in different wood loss vs. debarking degree ratios.

The debarking technique is often different for the wood chip raw material that arrives as sawmill waste. Knife techniques retain more bark residues than drum techniques, and these are then carried over to the refiner pulping process, influencing process stability and pulp quality.

During 1994–96, the Finnish Pulp and Paper Research Institute conducted an extensive study on the environmental impacts of debarking[3].

4.2 Water circulation systems in debarking

As a rule, water systems in debarking plants are kept very closed and consequently concentrations become very high. A cleaning stage is needed in the circulation to remove solid material from the circulation water. High concentrations of dissolved and dispersed substances will not essentially affect the operation of debarking but can cause carryover to pulping. Released bark is thickened to a high dry substance content by compressing. Filtrates of bark compression have the highest concentration. If there is a need to decrease concentrations in the circulation water, then a part of the bark compression filtrate can be separated to the waste water treatment. Additional water of lower concentration can be the circulation water from the pulp mill or paper mill or even purified waste water. In debarking, additional water is needed, e.g., for the washing showers.

4.3 Need for water

In wet debarking, circulation water is showered in the initial section of the drum, in the outlet of the drum, during washing, and during conveying. The bark remains under wet conditions for a long time and will be mechanically crushed into fine particles. Dry debarking needs circulation water only for log washing, for separation of extraneous material, and for dust control. Only a small part of the bark will remain in a wet state. There are solutions between the two extremes – wet debarking and dry debarking.

Intermediate forms of debarking techniques and individual mill practices have resulted in a certain variation in specific water consumption, as shown in Table 1.

Table 1. Specific water consumption in debarking plants of some Finnish mills (m^3 per ton dry wood).

	Type of debarking	Specific water consumption, m^3/ton wood Summer	Specific water consumption, m^3/ton wood Winter
Mill A	wet	4.5	5.7
Mill B	wet	0.9	1.0
Mill C	wet	6.0	6.0
Mill D	wet	5.1	4.8
Mill E	wet	1.2	2.7
Mill F	mixed	2.7	2.7
Mill G	dry	2.1	1.8

CHAPTER 14

5 Treatment of debarking effluent

5.1 Composition and concentration of debarking circulation water

The dissolved and dispersed substances in the circulation water of debarking mainly come from the bark. The influence of the source of the additional water is small. The chemical composition of bark is complex, and its composition is not even fully known. In chemical pulp mills and integrated mills, the situation is still more complex because water systems for softwood and hardwood species are mixed. Generally, bark contains high amounts of water-soluble substances, including extractives and nutrients rich in phosphorus and nitrogen. Some of the water-soluble compounds are toxic and some give a strong color to the water.

There are three important water flows to be considered in debarking: Circulation water, which is a mix of all flows; a part of it which is internally cleaned or clarified; and filtrate water of bark compression which is the most concentrated flow and which also can be used to take out water from debarking to the final effluent treatment.

Mill examples corresponding to Table 1 describing concentrations of debarking circulation waters are given in Table 2.

Table 2. COD_{Cr} concentrations of debarking process waters for some Finnish mills (summer conditions).

	Type of debarking	Water taken in	Circulation water cons., g/l	Water taken out cons., g/l
Mill A	wet	mech. pulping circulation	4.8	4.8
Mill B	wet	fresh water	n.a.	16.0
Mill C	wet	cleaned waste water	n.a.	3.0
Mill D	wet	cleaned waste water	13.0	3.9
Mill E	wet	PM clear filtrate	8.0	6.0
Mill F	mixed	PM clear filtrate, fresh water	2.8	2.8
Mill G	dry	cleaned waste water	2.3	2.3

5.2 Effluent load

Common practice is that the collected circulation water of debarking, or a part of it, is internally clarified, either by sedimentation or by flotation, possibly assisted by chemical flocculants. Then mainly solid material is removed. A modest decrease in dissolved substances is obtained due to adsorption on solids. A main part of the clarified water is reused for circulation, and a part is separated to the waste water treatment. The degree of opening is a means to control the concentration of circulation water.

As the share of debarking effluent load is so high compared to the total load of the papermaking process, further closing of debarking, utilization of dry methods, and internal cleaning have been of interest. An interesting option would be separation and purification of the most concentrated fraction: the filtrate of bark compression.

Different methods of internal purification have been proposed: MVR-evaporation, membrane filtration preceded by proper removal of solids, biological treatment, etc. None of these have proved superior to the others.

6 Release of wood components into water in mechanical pulping and bleaching

6.1 Discharge data for different pulping processes

Mechanical pulping processes differ from each other with respect to the temperature-time profile to which the wood and the resulting fibers will be subjected. Chemimechanical processes apply higher pH levels, commonly above 8. Consequently, different pulping processes cause largely differing dissolution of wood substances into the process waters.

When wood is processed into mechanical pulp by conventional grinding or refining, 2%–5% of the wood material is dissolved, or dispersed as colloidal particles, into the process water. Chemical Oxygen Demand (COD), determined by titration with dichromate, is commonly used as a sum measure of these dissolved and colloidal substances in process waters and effluents. More recently, Total Organic Carbon (TOC) has been introduced as an alternative to COD. TOC is determined by catalytic oxidation and subsequent determination of released carbon dioxide by infrared spectrometry in automated instruments. Sometimes Total Dissolved Solids (TDS) are also reported. Earlier, Biological Oxygen Demand (BOD) was the most used parameter. BOD is a measure of biodegradable organic compounds, whereas COD and TOC include essentially all organic substances. TDS includes both organic and inorganic nonvolatile substances. The discharge of nutrients, causing eutrophication in watercourses, is usually measured by determination of total phosphorus and nitrogen in effluents.

Twice the amount of wood material, measured as BOD, was reported to be released in TMP compared to SGW and PGW pulping[4]. Similar results were reported in other studies[5–10].

The dissolution and dispersion of wood substances into process waters is in practical mill conditions influenced by many process factors, such as quality of wood furnish, pulping process conditions, and quality of process water which depends, e.g., on the water system configuration of the whole paper mill and its degree of closure. Furthermore, pulp consistency, temperature, and salt concentration have an influence on the dissolution[8, 10]. Thus, much varying data can be found in the technical literature. Typical data for mechanical pulping of spruce wood in Nordic conditions are given in Table 3.

Table 3. Typical characteristics of waters from mechanical pulping of spruce in Nordic mills. BOD, COD, P and N, data from Franzén and Jantunen[11]. TOC data calculated by TOC = 0.35×COD.

Pulping Process	BOD$_7$ kg/ton	COD kg/ton	TOC kg/ton	Total P g/ton	Total N g/ton
GW	10–12	30–40	10–14	20–25	80–100
PGW	12–15	40–50	14–18	20–30	90–110
TMP	15–25	50–80	18–28	30–40	100–130
CTMP	20–35	60–100	21–35	35–45	110–140

6.2 Influence of bleaching stages

Bleaching with alkaline hydrogen peroxide causes an additional load of 5–15 kg BOD/ton and 15–40 kg COD/ton, which is of the same magnitude as in mechanical pulping itself[11]. The discharge of phosphorus increases by 20–35 g/ton and the discharge of nitrogen can increase to 400–500 g/ton.

Alkalinity strongly influences the dissolution of hemicelluloses and pectins[12]. Ekman et al.[14] determined the influence of alkaline peroxide bleaching on the wood extractives. Triglycerides, fatty acids, and sterols remained essentially unchanged, but abietic-type resin acids, lignans, and stilbenes were to a large extent degraded.

Sodium hydroxide and sodium silicate are used as alkali sources in peroxide bleaching. Since high alkalinity causes large dissolution of wood substances, other sources of alkali have been proposed[15].

Residuals of bleaching chemicals, especially anionic polymerized silicate, can interfere with the function of cationic retention aids in papermaking. The common chelating agents, EDTA and DTPA, are relatively stable against biological degradation and are discharged with purified effluents. Being nitrogen compounds and becoming later photochemically decomposed, they add to the nutrient load. For example, a dosage of 2 kg chelating agent per ton of pulp adds to the nitrogen discharge by at least 100 g/ton pulp, which is approximately at the same level as the amount of nitrogen compounds released from wood during pulping[16].

Dithionite bleaching, which normally is carried out in the pH range of 5–6 and at temperatures about 60°C, does not notably increase the amount of dissolved organic material. However, salts are introduced into the water system. Decomposition products of dithionite can cause a decrease in pH and thus cause disturbances in papermaking.

6.3 Chemical character of released wood components

The main components of the dissolved and colloidal substances are hemicelluloses, lipophilic extractives (wood resin), lignans, and lignin-related substances. Table 4 gives typical amounts of substances released from TMP of Norway spruce.

An acetylated galactoglucomannan (GM) with a low galactose content is the dominating hemicellulose type dissolved in mechanical pulping of Norway spruce (Fig. 2). The dissolved GM is composed of mannose (Man), glucose (Glc), and galactose (Gal) sugar units in a ratio of about 4.5:1:0.6 (Fig. 3). The GM has one acetyl group per 2–3 sugar units. Only 5%–10% of the GM in wood is dissolved during mechanical pulping.

An acidic arabinogalactan (AG) is released especially from the heartwood[17]. AG from spruce heartwood was found to be composed of galactose, arabinose (Ara), and glucuronic acid (GlcA) units in the approximate ratio Gal:Ara:GlcA of 4:1:1. AG is thus highly anionic and is the main contributor to so-called "anionic trash" in unbleached mechanical pulp suspensions. Only small amounts of arabinoglucuronoxylan are released in mechanical pulping and bleaching, although xylan is a major hemicellulose type in spruce wood.

Table 4. Typical amounts of dissolved and colloidal substances in 1% suspensions of mill-produced spruce (*Picea abies*) TMP. Bleaching of the TMP was made with alkaline hydrogen peroxide (3% on dry pulp basis) in the laboratory (based on data from Thornton and Holmbom et al.[12, 13, 18, 35]). The TOC-values of these waters were 214 and 288 mg/l, respectively.

Component group, kg/ton TMP	Unbleached TMP	Peroxide-bleached TMP
Hemicelluloses	18	8
Galactoglucomannan	16	4
Other hemicelluloses	2	5
Pectins (galacturonans)	2	4
Lignans	2	1
Other lignin-related substances	7	11
Lipophilic extractives (resin)	5	4
Acetic acid	1	20
Formic acid	0.1	4
Inorganic constituents	<1	5

Pectins, composed mainly of galacturonic acid units (GalA), occur only in small amounts in unbleached waters, but constitute an important group after alkaline peroxide bleaching, as seen from the large amount of GalA units after peroxide bleaching (Fig. 2). Most of the carboxyl groups in the pectins in spruce wood and in unbleached TMP waters are methylated, whereas the pectins dissolved in peroxide bleaching are demethylated and thus highly charged. The demethylated pectins are preferably called pectic acids. The galacturonic acid carboxyl groups are dissociated at normal papermaking conditions (pk_A ca 3.7). The pectic acids account for a major part of the cationic polymer demand in peroxide bleaching process waters.

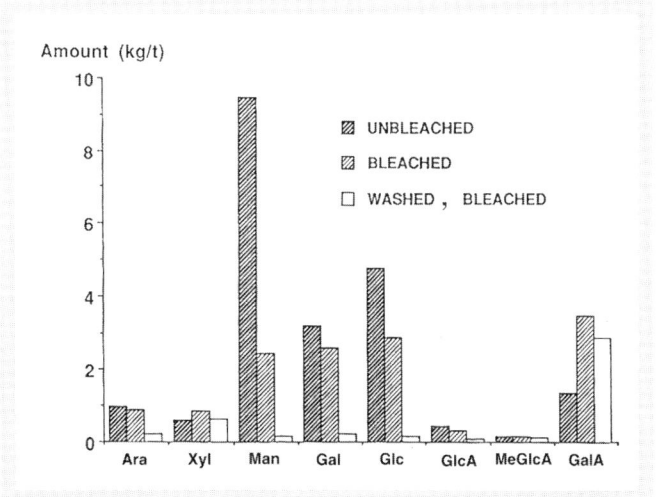

Figure 2. Amount and sugar composition of dissolved polysaccharides (hemicelluloses and pectins) in 1% TMP suspensions before and after alkaline peroxide bleaching[18].

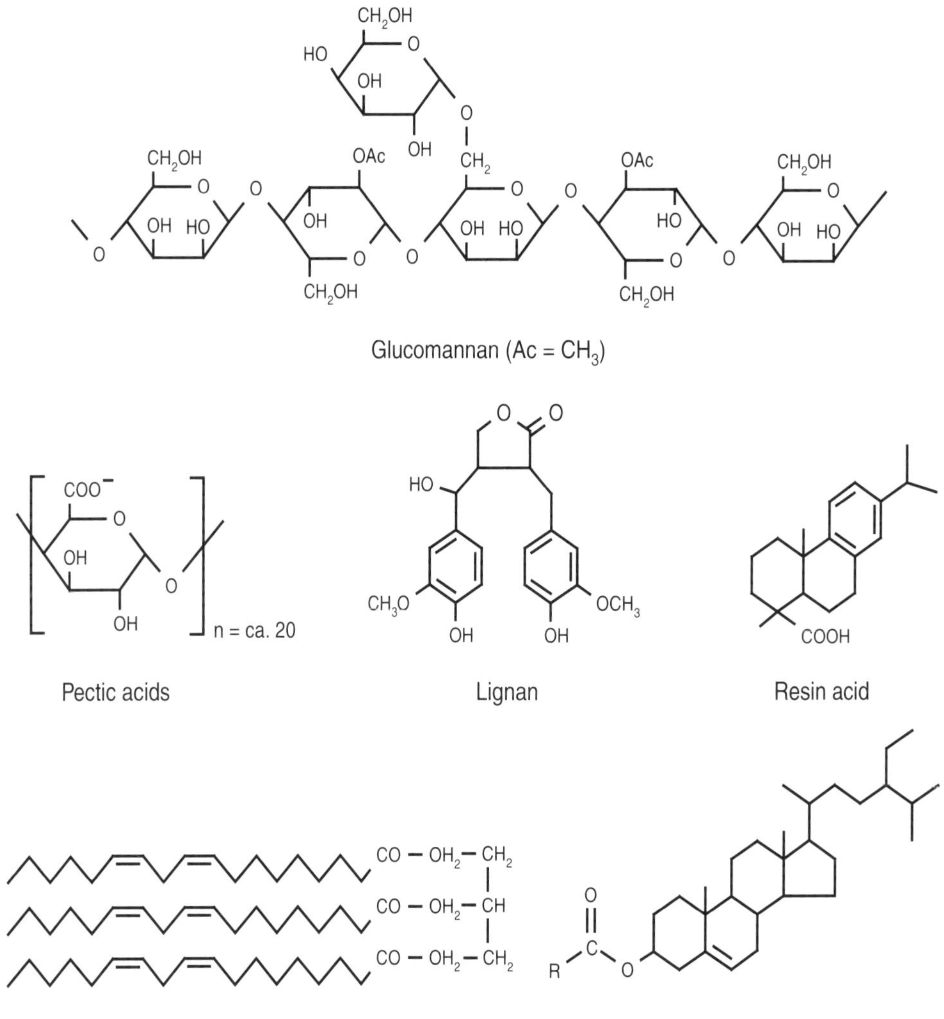

Figure 3. Structures of dissolved and colloidal substances in mechanical pulping effluents.

Another drastic change in peroxide bleaching is the decrease in dissolved GM (Fig. 2). This results from hydrolytic splitting of the acetyl groups in GM, resulting in a lower water solubility and consequently deposition of GM onto the fibers. Acetyl groups are split off also from GM in fibers, resulting in the release of about 20 kg acetic acid (acetate) per ton of mechanical pulp[19]. This yield loss is partly compensated for by the deposition of GM, by about 10 kg/ton pulp.

The lignans are typical of spruce heartwood. They comprise a complex mixture with two stereoisomers of hydroxymatairesinol as the main components. Most of the lignans are degraded by oxidation in peroxide bleaching[14]. The lignans are fairly water-soluble compounds that do not interact with cationic polymers or with fiber or filler surfaces. In highly-closed water systems, they can accumulate to high concentrations in the circulation waters and effluents.

The lignin dissolved in mechanical pulping is rather low-molecular and structurally very similar to native wood lignin[19]. During alkaline peroxide bleaching, more lignin is dissolved. This lignin contains carboxyl groups and has a slightly higher molar mass than the lignin in unbleached suspensions. In addition to true lignins, some lignin-like oligomers are found in mechanical pulp suspensions. The lignins are accumulated in the process waters by water system closure, and most probably contribute to a considerable part of the effluent color.

Wood resin forms the major part of the dispersed colloidal phase in mechanical pulp suspensions[20]. These resin droplets with diameters in the range of 0.1–2 µm have essentially the same composition as the resin in the wood raw material, when the pH is around 5. At a higher pH level, resin and fatty acid soaps are formed, and they are leached out of the droplets and the fibers into the water. At pH 8, most of the resin acids are dissolved in the water, but only part of the fatty acids. In fresh spruce wood, triglycerides dominate and the amount of free fatty acids is low. Storage of wood in humid or wet conditions leads to triglyceride hydrolysis by the action of lipase enzymes. Alkaline peroxide bleaching leads to oxidation of dissolved resin acids of the abietadiene type, but otherwise the resin is little affected.

Mechanical pulping and bleaching waters also contain colloidal size fiber fragments, called micro-fines[21]. The micro-fines have been demonstrated to play a role in flocculation phenomena, e.g., by promoting the aggregation of colloidal resin droplets with cationic polymers.

6.4 Toxic components in effluents

Mechanical pulping effluents are toxic to water organisms. Toxicity is usually determined with fish, water fleas (*Ceriodaphnia*), or special light-emitting bacteria (*Microtox*). The toxic components and the toxicity levels differ substantially between wood species[22]. Much of the toxicity in softwood pulping effluents is due to the resin acids[23]. Other diterpenoid compounds are also toxic. Dehydrojuvabione present in balsam fir is a particularly toxic component[22]. Well-operated biological treatment can eliminate most of the toxicity of mechanical pulping effluents[24].

7 Factors influencing release of wood components into water

7.1 Effect of pulping conditions

The difference between mechanical pulping processes regarding release of wood substances into water is mainly due to different process temperatures. The freeness level has practically no influence on the dissolution and dispersion[10, 25]. In two-stage

TMP refining, most of the wood material is released already in the first stage[36]. The shower water temperature in grinding has a strong influence on the release (Fig. 4).

It has generally been assumed that the composition of the organic material in process waters is the same for TMP and SGW, although more material is dissolved in TMP. However, the additional dissolution obtained by higher shower water temperature in laboratory grinding (PGW) was due mainly to increased dissolution of hemicelluloses and pectins[26]. Also lignin was released in slightly increasing amounts, whereas the dispersion of wood resin remained constant. Similar trends were seen in a laboratory study using ground spruce meal[27] where the treatment time was varied. The dissolution of lignin was strongly dependent on the treatment time.

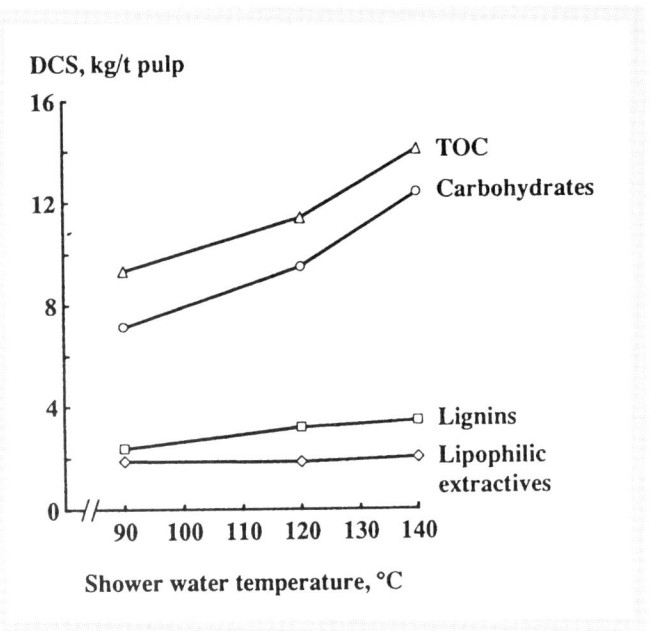

Figure 4. Dissolution and dispersion of spruce wood components into water at different grinder water temperatures[26].

Pulping at higher pH levels leads to more dissolved and colloidal substances in the waters. At pH above 8, two hydrolytical processes will occur in wood fibers; (i) deacetylation of glucomannans (in softwoods) or xylans (in hardwoods), and (ii) demethylation of pectins. The former reaction results in extensive dissolution of acetic acid, for softwoods up to 20 kg/ton of wood but, on the other hand, results in decreased dissolution of glucomannans and xylans. Demethylation of pectins leads to dissolution of both methanol and pectic acids.

Alkaline peroxide mechanical pulping (APMP) employs a pH range of 9–11 and most probably gives larger dissolution of wood material than ordinary mechanical pulping. However, no exact data have been found in the literature. In alkaline pressurized grinding in a laboratory grinder, the dissolved and colloidal substances increased from 8 kg/ton (measured as TOC) to 13–20 kg/ton when alkaline peroxide bleaching chemicals were added to the shower water[25]. The dissolution increased with the alkali and peroxide dose. The increased dissolution was due mainly to increased release of acetic acid, pectic acids, lignin, and wood resin. The amount of dissolved hemicelluloses, however, decreased due to deacetylation of glucomannans, which decreases their solubility.

Chemimechanical pulping with sulfite (CTMP) is usually made around pH 9. CTMP gives significantly higher dissolution of wood material than TMP. COD values of

60–100 kg/ton are typical for Nordic spruce CTMP, with an additional 20–30 kg COD/ton in the case of peroxide bleaching[11]. CTMP gives a higher dissolution, especially of lignin which is a result of lignin sulfonation, in addition to a higher dissolution of acetic acid and pectic acids than in TMP.

7.2 Effects of wood quality

Spruce sapwood and heartwood were compared in a laboratory study where finely ground spruce wood meal was vigorously mixed for extended times at 90°C[27]. Higher amounts of wood resin were dispersed from sapwood than from heartwood, reflecting the differing resin amount and composition. The dissolution of hemicelluloses and lignin was at the same level for sapwood and heartwood. Lignans were released only from heartwood.

Pine wood generally contains about twice the amount of wood resin than spruce and more resin is dispersed in pulping of pine wood. In a laboratory study with wood meal[28], it was found that pine wood *(Pinus silvestris)* released much more resin (especially the heartwood) than spruce wood *(Picea abies)*. Pine heartwood gave a much higher dissolution of arabinogalactans than spruce wood, and more pectins were also dissolved from pine wood than from spruce wood. Pine heartwood gave a considerable dissolution of pinosylvin.

Hardwoods are used only in chemimechanical pulping (in some cases in mechanical pulping, but then combined with peroxide bleaching). Hardwoods generally give a higher dissolution of wood substances than softwoods. This is due especially to larger dissolution of hemicelluloses and pectins.

8 Evaporation of compounds in mechanical pulping

8.1 Volatility of wood compounds

Due to the high temperature of defiberizing, a part of the wood extractives and fragmented wood polymers are evaporated during processing. There are also reports about significant evaporation of extractives in cases where the chip washing temperature is high[29].

Volatile organic compounds (VOCs) are a relatively new issue in mechanical pulping. Attention has been paid to them in connection with research of atmospheric changes, as the VOCs possibly take part in ozone formation[29, 30].

Earlier, the negative influence of evaporated compounds was seen as a drawback in heat recovery systems. Because of vapor condensation, they can accumulate in the water circulation and have a negative influence on equipment condition.

8.2 Contents of volatile compounds

Evaporated compounds from mechanical pulping are mainly monoterpenes from the oleoresin of softwoods and low-molecular-mass alcohols and organic acids derived from carbohydrates[30–32]. Reported research results are identified from TMP heat recovery condensate and pilot conditions. Boström *et al.* tried to estimate direct emissions (Table 5).

Table 5. Estimates of VOC emissions in a Swedish TMP plant[30].

	kg/ton dry pulp
Terpenes	0.9
Oxygenates	0.1
Aromatics	0.02
Alkanes and alkenes	0.01

The amount and composition of volatile compounds are greatly dependent on the resin content of wood. Recovery of volatile compounds is feasible in such cases where wood species with high extractive content are used. Volatile organic carbons are not the only air emission from mechanical pulping. Incineration of waste residues results in burning exhausts.

9 Water circulation systems in mechanical pulping plants

9.1 Connection between pulp mill and paper mill

Dissolved and dispersed compounds released into water in mechanical pulping follow the pulp flow to the paper machine system. A common practice is that there is a thick stock storage between the pulp mill and the paper mill. Before storage, pulp is thickened, e.g., to 10%–15%. Carryover is reduced because the filtrate of thickening is returned to the pulping process and pulp from the storage is diluted by paper machine water. As a consequence, the concentration of dissolved and dispersed compounds in the pulping system will raise due to accumulation. Thickening before storage does not reduce carryover unless the system is opened in the mechanical pulp mill and water is transferred to purification or, e.g., to debarking (Fig. 5).

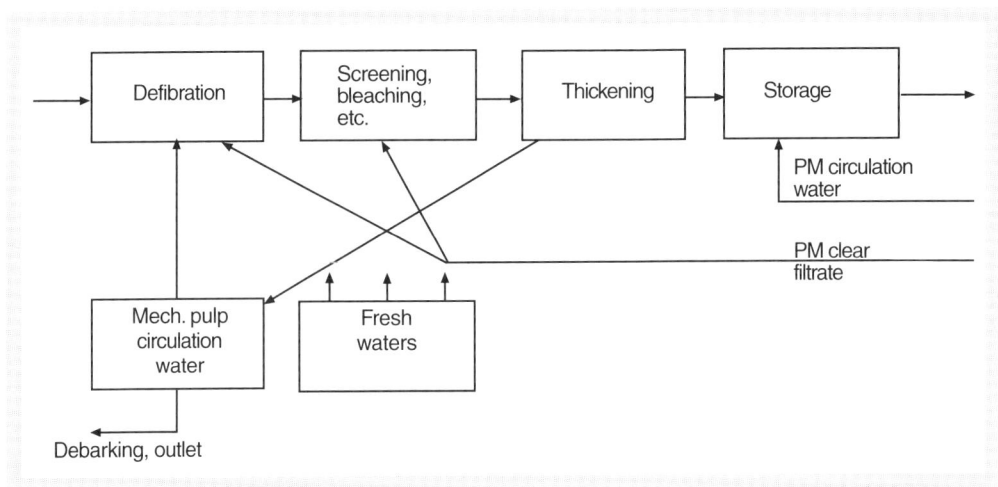

Figure 5. The main principle of the mechanical pulp mill water system.

Usually the clear filtrate from the paper machine is taken to purification. In that case, the cut-off effect of thickening between the pulp mill and the paper mill is weak. Instead there is a carryover of solutes from paper mill to mechanical pulping, as the pulp mill replaces the water shortage by taking clear filtrate from paper mill´s excess.

The diagram of Fig. 5 clarifies the main principle of the water circulation system in a mechanical pulping plant. In both grinding and refiner processes, the pulp is after defiberization diluted to low consistency. But in this dilution, circulation water and clear filtrate are used in different proportions. Refiner processes use clear filtrate for refiner dilution and circulation water for dilution before latency removal. Grinding processes use only little additional water to supplement their so-called hot circulation. The main principles are similar for other parts of the both types of processes. Individual solutions of the mills deviate much more from each other. Fresh water is needed for sealing, cooling, and some wash showers. Earlier, fresh water was used even for refiner dilution.

9.2 Washing stage

Recently, systems with a washing stage between mechanical pulping and paper mill have been adopted (Fig. 6). There are one or two thickening stages where consistency is raised, preferably up to 25%–40%, preceded by dilution using water that is cleaner than the filtrate of the last thickening stage. A great deal of dissolved and dispersed substances from pulping and bleaching are separated from the fiber flow and prevented to proceed to the paper mill system. Circulation water of the paper mill can be used as wash water, and in that case there is some carryover from the paper mill. Dilution after thickening is done by using circulation water from the paper machine. The filtrate from washing is returned to the pulp mill circulation. Some part of the filtrate is circulated in the washing circuit. Water circulation at the pulp mill side is to a certain extent opened in order to keep the accumulation at an acceptable level.

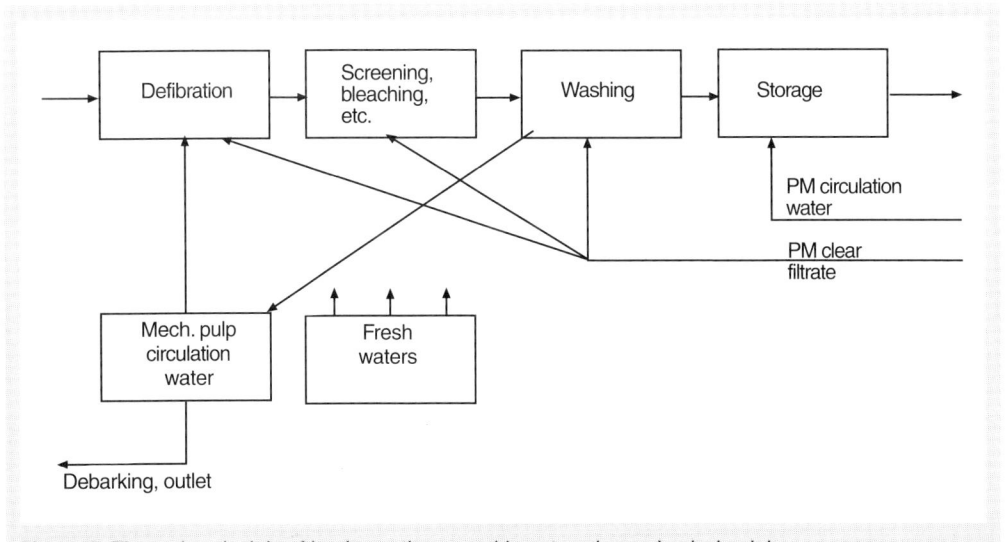

Figure 6. The main principle of implementing a washing stage in mechanical pulping.

Several high-consistency thickening techniques are available. The performance of washing in respect to its purifying effect can be influenced by controlling the conditions in thickening. The methods of thickening include filtering in disc or drum filters, wire or roll presses, or screw presses.

The removal efficiency of dissolved and colloidal material in the washing stage is mainly dependent on the dilution-thickening ratio because they, as a main rule, follow the water phase[33]. Temperature, pH, and other conditions have an influence, but efficiency is also dependent on solubility of materials and their different tendency to retain in the fiber mat at their different concentrations[34]. The wash water circulation arrangement in the the washing thickener has an influence on the purification.

There are some drawbacks to consider when applying a washing stage: A part of the fiber material is lost to the filtrate and returned to the pulp mill circulation. This might finally result in a significant loss of fiber material. The implementation of a washing stage also has a tendency to increase the use of water, as it changes the water balance of the mill.

A serious drawback of a washing stage is that it cuts off the heat flow from the pulp mill, where excess heat is generated, to the paper mill, where it is needed. A heat-exchanging step is necessary. Concentration of dissolved and dispersed substances in the mechanical pulp mill circulation will be higher when using a washing stage. This can have negative effects on the pulping process.

10 Treatment of mechanical pulping effluents

10.1 Purification methods

External purification of effluents is done together for all mill effluents which is a purposeful way to achieve as stable and economic operation as possible in the biological treatment stage. Debarking effluents could be separately treated externally. At least a separate pretreatment is needed for debarking effluents before the final treatment. Methods of purification are not further discussed here (see volume 19 of this textbook series).

10.2 Outlook on internal cleaning stages in mechanical pulping plant

A common practice in internal cleaning within debarking and mechanical pulping is that incinerable solid material is concentrated and recovered in debarking and solid nonfibrous impurities are separated and recovered in pulping.

There is an obvious opportunity to introduce internal cleaning stages in debarking to reduce the concentration of dissolved and dispersed compounds in the circulation water and process them to a removable form. The degree of closure in debarking can be kept very high without serious disadvantages. The most important disadvantages are corrosion, which can be controlled, and carryover of solutes to mechanical pulping which suggests internal cleaning.

There is also an obvious opportunity to introduce internal cleaning stages inside the pulping process. Washing stages offer an effective way to segregate concentration areas of dissolved and dispersed material and then to control their negative influence on papermaking. Internal cleaning stages to reduce concentration in inner circulation and to put concentrates in a removable form would promote maintaining of a good material balance, a stable operation in general and a good heat balance, and reduce the final environmental impact. The two important aspects are to avoid losses of useful fiber material and to utilize excess heat in the papermaking process.

10.3 Zero effluent technology

"Zero Effluent" technology was developed by Canadian market pulp mills. It means that no process effluent is discharged. All effluents from the mill are collected in storage ponds, providing many days of buffer to the operating mill. The suspended solids (2000 to 4000 ppm) are removed by mechanical clarifiers. The removed sludge is dewatered and burned in a waste fuel boiler.

The clarified effluent is concentrated by evaporation from about 2% to 40% solids. The concentrated solids contain 40% to 46% organics, which are fatty acids and some lignin. The rest of the solids are inorganic salts from the wood and the process chemicals. The concentrated solids have a heat value of about 10 MJ/kg, and are incinerated in a recovery boiler, or in a fluidized bed incinerator.

The smelt from the boiler contains (in the case of a CTMP or APMP mill) 25% sodium carbonate, while the rest are other inorganic salts like calcium, magnesium, silica, and some sulfur. 40% of the smelt can be reused by the mill after clarification to remove most of the iron and manganese contamination. This "green liquor" could provide the alkalinity for impregnating the wood chips in CTMP or APMP mills. The rest of the smelt is disposed by landfill.

Since the smelt is primarily sodium carbonate, it can be converted to sodium hydroxide by causticizing, a process well known to kraft mills. This would close up the chemical cycle, but to an additional cost. Zero effluent is an elegant way to eliminate effluent pollution, but it is expensive. The needed equipment will add about 25% to 30% to the capital of a CTMP market pulp mill.

CHAPTER 14

References

1. Ekman, R. and Hafizoglu. H., "Changes in spruce wood extractives due to log storage in water," 1993 International Symposium on Wood and Pulping Chemistry Notes, CTAPI, Beijing, p. 92.

2. Assarsson, A. and Åkerlund, G., Svensk Papperstid. 70(6):205 (1967).

3. Saunamäki, R. and Savolainen, M., "Effluent-free drum debarking," TAPPI 1998 International Environmental Conference (Vancouver), TAPPI PRESS, Atlanta, p. 247.

4. Mitchell, G. and Tapio, M., Pulp & Paper 64(6):97 (1990).

5. Stenberg, L.E. and Norberg, G., Paperi Puu 59(10):652 (1977).

6. Wong, A., Pulp Pap. Can. 78(6):T132 (1977).

7. Wong, A., Constantino, J., Breck, D., Herschmiller, D., Rae, J.G., Pulp Pap. Can. 80(1):T14 (1979).

8. Järvinen, R., Vahtila, M., Mannström, B., Sundholm, J., Pulp Pap. Can. 81(3):T67 (1980).

9. Auhorn, W., Wochenbl. Papierfabr. 112(2):37 (1984).

10. Wearing, J.T., Huang, S., Piggott, A., Ouchi, M.D. and Wong, A., Pulp Pap. Can. 86(5):T139 (1985).

11. Franzén, R. and Jantunen, E., "BAT 1999, Mechanical Pulping," in Environmental Issues within the Nordic Pulp and Paper Industry, Vol. 2, Rept. 1993:601, The Nordic Council of Ministers, Copenhagen, 1993.

12. Thornton, J., Eckerman, C., Ekman, R., "Effects of peroxide bleaching of spruce TMP on dissolved and colloidal organic substances," Proceedings of the 1991 International Symposium on Wood and Pulping Chemistry, Appita, Melbourne, p. 571.

13. Thornton, J., Ekman, R., Holmbom, B., Eckerman, C., Paperi Puu 75(6):426 (1993).

14. Ekman, R.and Holmbom, B., Nord. Pulp Pap. Res. J. 4(3):188 (1989).

15. Dionne, P., Seccombe, R., Vromen, M., Crowe, R., Paper Technol. 36(3):29 (1995).

16. Langi, A., Priha, M., Tapanila, T., FPPRI internal report, 1997.

17. Thornton, J., Ekman, R., Holmbom, B., Örså, F., J. Wood Chem. Technol. 14(2):159 (1994).

18. Holmbom, B., Ekman, R., Sjöholm, R., Eckerman, C., Thornton, J., Papier 45(10A):V16 (1991).

19. Pranovich, A. V, Sjöholm, R., Holmbom B., "Characterization of dissolved lignins in thermomechanical pulp suspensions," Proceedings of 1994 European Workshop on Lignocellulosics and Pulp, Royal Institute of Technology, Stockholm, p. 219.

20. Allen, L. H., Pulp Pap. Can. 76(5):T137 (1975).

21. Nylund, J., Lagus, O., Eckerman, C., Colloids Surf. A 85:81(1994).

22. O´Connor, B. I., Kovacs, T. G., Voss, R. H., Environ. Toxicol. Chem. 11:1259 (1992).

23. Lavallée, H. -C., Rouisse, L., Paradis, R., Pulp Pap. Can. 94(11):T393 (1993).

24. Gibbons, J. S., Kovacs, T. G., Voss, R. H., O´Connor, B. I., Dorica, J., Wat. Res. 26(11):1425 (1992).

25. Svedman, M., Lönnberg, B., Holmbom, B., Jäkärä, J., Paperi Puu 77(3):117 (1995).

26. Örså, F., Holmbom, B., Häärä, M., Paperi Puu 78(10):605 (1996).

27. Örså, F., Holmbom, B., Thornton, J., Wood Sci. Technol. 31(4):279 (1997).

28. Örså, F., "Utlösning och dispergering av komponenter ur granved i vatten," Tech. Lic. thesis, Åbo Akademi Univ., Åbo, 1995.

29. Sueiro, L. and Gill, J., "Control of VOC emissions from a TMP plant," TAPPI 1995 International Environmental Conference Proceedings, TAPPI PRESS, Atlanta, p. 925.

30. Boström, C -Å., Lindskog, A., Moldanova, J., Malmström, J., Andersson- Sköld, Y., Cooper, D., Berg, L., "VOC emissions from the Swedish Pulp and Paper Industry," TAPPI 1992 Environmental Conference Proceedings, TAPPI PRESS, Atlanta, p. 623.

31. Sayegh, N., Azarniouch, M., Prahacs, S., Pulp Pap. Can. 84(5):42 (1983).

32. Huusari, E.and Syrjänen, A., "Start-up experiences with the new Tandem TMP heat recovering system at Kaipola," 1981 International Mechanical Pulping Conference Proceedings, PTF, Oslo, p. VI:6

33. Lloyd, J., Deacon, N., Horne, C., Appita J. 43(6):429 (1990).

34. Ekman, R., Eckerman, C., Holmbom, B., Nordic Pulp Pap. Res. J. 5(2):96 (1990).

35. Thornton, J., "Dissolved and Colloidal Substances in the Production of Wood-Containing Paper," Ph.D. thesis, Åbo Akademi Univ., Åbo, 1993.

36. Wong, A., Breck, D., Constantino, J., Pulp Pap.Can. 81(3):T72 (1980).

CHAPTER 15

The character and properties of mechanical pulps

1	**Origin of mechanical pulp properties**	**395**
1.1	Wood fibers and pulp fibers	396
2	**Development of pulp properties**	**397**
2.1	Fiber wall thickness and structure	397
2.2	Pulp fractional composition and fiber properties	398
2.3	Latency in mechanical pulps	401
3	**Methods used to determine pulp suspension properties**	**402**
3.1	Method vs. property	402
3.2	Principles of drainage measurements	403
3.3	Principles of fiber classification methods	403
3.4	Principles of shive content measurements	405
3.5	Principles of optical fiber analyzers	405
4	**Fiber properties**	**406**
4.1	Flexibility and stiffness of fibers	406
4.2	Simons staining	407
4.3	Fiber wall thickness	407
4.4	Methods for the characterization of fines	408
5	**Handsheet properties and paper testing**	**409**
5.1	Preparation of handsheets	409
5.2	Testing of handsheet properties	410
5.3	Typical sheet properties of mechanical pulps	410
	References	412

CHAPTER 15

Annikki Heikkurinen, Leena Leskelä

The character and properties of mechanical pulps

1 Origin of mechanical pulp properties

The conditions of mechanical pulping affect the way the wood matrix is broken and the fibers are separated from each other. Shives, fibers, pieces of fibers, and fines are formed already during the fiber separation stage, and the conditions of the very initial defibration stage seem to determine the character of the fibers. Dimensions of the particles are changed during the course of defibration. Both the length distribution and cross-sectional dimensions are continuously changing as fibers are mechanically treated. Properties of fines and middle fraction are changed due to the new material formed during refining.

The grinding and refining processes separate the fibers in a different way. Figure 1 provides typical fiber length distributions of various pulp types. An increasing process temperature increases the fiber length in both grinding and refining.

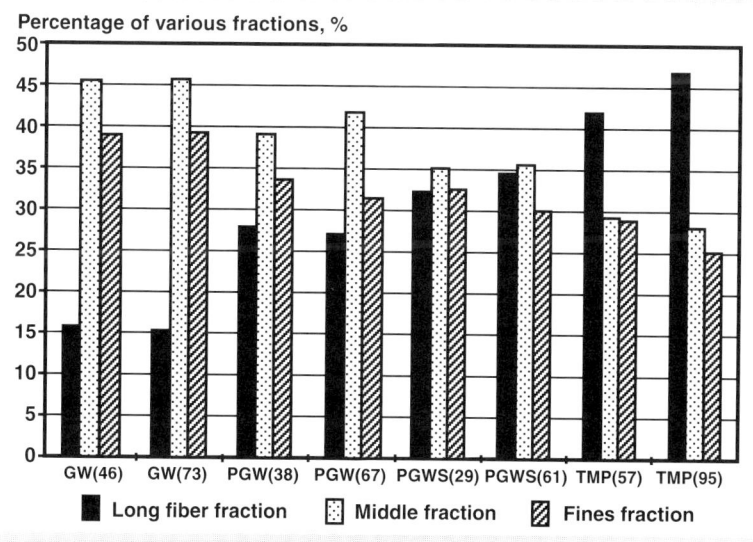

Figure 1. Typical fiber length distributions of mechanical pulps[1].

CHAPTER 15

1.1 Wood fibers and pulp fibers

The pulp fiber dimensions are to a great deal determined by the dimensions of wood fibers, but the properties of wood fibers also have an effect on how the wood is defiberized. The wood fiber properties vary, not only between different wood species, or growing sites, but also the age of the wood influences fiber properties greatly. Both the fiber length and the fiber wall thickness increase with wood age. The juvenile wood fibers are shorter, more narrow, and have thinner walls than mature wood. Typically the juvenile wood pulps have lower strength properties and higher light-scattering coefficients as shown by Tyrväinen who manufactured TMP in mill scale from first thinnings, regeneration cutting wood, and sawmill chips of Norway spruce[2]. The most important wood quality factors were fiber length, fiber cross sectional dimensions, and juvenile wood percentage. In several studies of *Pinus Radiata,* Corson has however shown that no simple trend could be found between wood properties, like basic density, and pulp handsheet properties, e.g., density, tensile strength, or light-scattering coefficient[3]. He suggests that there are compensating mechanisms in refining which balance out the relative contributions of thin and thick walled fibers and the fines formed from them. Hatton has investigated the behavior of second growth softwoods of the Canadian west coast pulps[4]. He found consistently higher light-scattering coefficients in pulps made of juvenile wood and that, at the same time, they also required more refining energy to reach the same Freeness level.

In addition to the age effect, the fiber wall thickness varies during the growing season so that the fibers formed in the beginning of the growing season (so called earlywood or springwood fibers) are thin walled and the fibers formed later (the latewood or summerwood fibers) are thick walled. According to Mork's rule, the fiber is classified as a latewood fiber when the thickness of a double fiber wall is higher than lumen width in the radial direction[5]. The latewood proportion is approximately 30% of fibers in wood. It is evident that, in addition to early and latewood fibers being so different, they also behave differently in defibration. Both Koljonen and Corson obtained results that showed how the proportion of latewood fibers in the long fiber fraction seemed to decrease with increasing energy in pulp refining[6,7]. These results mainly concerned reject refining because the specific energies used in both investigations were up to 5000 kWh/t. There are two possible explanations to these observations:

- As the fiber walls of latewood fibers become thinner, they are no longer classified as latewood fibers and the decrease in latewood fibers content is just an apparent phenomenon.

- Latewood fibers are broken down to smaller particles more often than earlywood fibers, and their relative proportion in long fiber fraction is really reducing.

Mohlin and Kure have presented opposite results of decreasing proportion of earlywood fibers. Mohlin concluded that the fiber shortening mainly concerned the thin walled earlywood fibers as the thick walled latewood fibers tended rather to unravel[8]. She measured the lumen perimeter and found the proportion of fibers with lumen perimeter larger than 60 μm were decreasing with decreasing R30 fiber fraction. Kure prepared TMP from thinnings and mature trees and measured dimensions of fiber cross sections from McNett + 50 fraction[9]. He ended up with the same conclusion as Mohlin. The main difference between the experimental procedures was that Mohlin and Kure investigated pulps prepared at an energy consumption range of 0–2000 kWh/t, but Koljonen and Corson used the refining energies up to 5000 kWh/t. Except for the different energy levels, the different ways of determining the summerwood proportion might be the reason for this discrepancy. In all these studies, it is agreed that the average fiber wall thickness is decreasing in high-consistency refining.

2 Development of pulp properties

2.1 Fiber wall thickness and structure

Karnis suggests that there are two competing mechanisms in fiber development depending on the process conditions[10]. The comminution-mechanism results in fiber breakage into smaller particles, and the peeling off mechanism results in gradual unraveling of the fiber wall from the fiber surface. Typical features for the development by peeling mechanism are

- Decreasing shive content
- Decreasing fiber coarseness
- The number of long fibers either remains constant or increases allthough the weight percentage of long fibers decreases
- Fiber wall thickness decreases
- Fiber collapsibility increases
- Fiber width either decreases or increases depending on the relative effect of peeling and fiber collapse.

With increasing energy, the coarseness of the long fiber fraction was reduced to that of chemical pulp. At the same coarseness, a chemical pulp was most flexible, and TMP was more flexible than RMP (Fig. 2). The difference between TMP and RMP indicates that the modulus of elasticity of the fiber wall might have decreased due to higher temperature during the TMP process. However, Karnis considers the decrease in fiber coarseness more important for the develop-

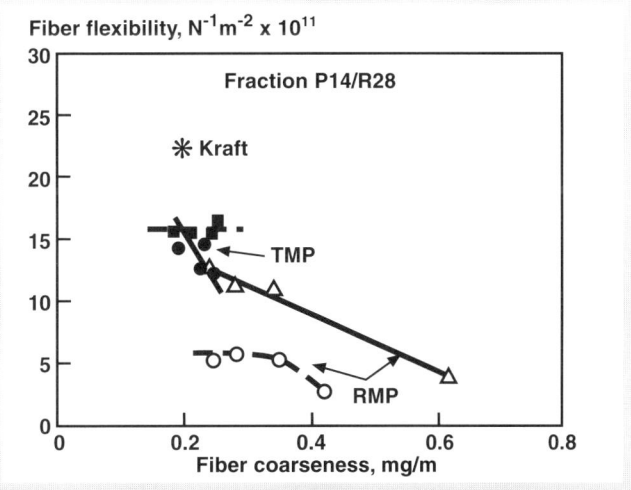

Figure 2. Fiber flexibilities and coarsenesses of various mechanical pulps according to Karnis[10]. There is no unique relationship between fiber coarseness and flexibility and, at the same coarseness, the fiber flexibility can vary.

ment of fibers than the flexibilization of fibers. Jang et al. have investigated how the fiber wall thickness and collapsibility have changed when refining intensity is changed[11]. Results show that the fiber wall thickness reduces more rapidly with high-intensity than low-intensity refining and that the fiber collapse appears to be further enhanced by a decrease in the rigidity of the fiber wall.

Lammi investigated the relationship between fiber flexibility and the moment of inertia by measuring the stiffness and the optical cross section of the same single fibers[12]. The fiber flexibility is a function of both the moment of inertia and the elastic modulus of the fiber. When the fiber wall thickness is reduced, the moment of inertia is also always decreasing, but it has been questioned whether there are such changes in the fiber wall structure that could affect the elastic modulus, too. Part of the fibers investigated were from coarse pulp that had been very slightly refined and part of them were highly refined. In both groups, the fiber wall flexibilization seemed to be an independent refining effect, and neither a decrease in fiber wall thickness nor a deformation of fiber shape was required to make the fibers more flexible. Figure 3 indicates how the stiff fibers could have either high or low moment of inertia and fiber wall thickness. This means that at the same moment of inertia the structure of the fiber wall varied, which is an indication of internal fibrillation, i.e., fiber wall flexibility had been increased.

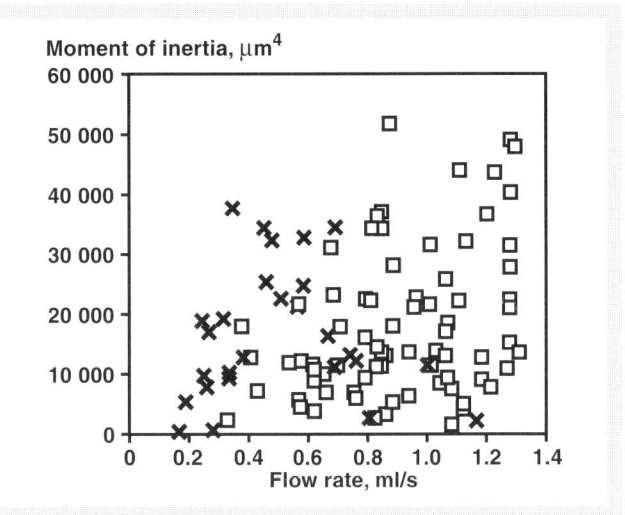

Figure 3. Moment of inertia of fibers as a function of flow rate of coarse (squares) and refined (crosses) fibers. High flow rate = high stiffness in hydrodynamic stiffness measurement[12].

2.2 Pulp fractional composition and fiber properties

Both the fractional composition and the properties of fiber fractions affect the pulp properties. The idea of separately examining the fiber length distribution and the properties of fractions was first introduced by Forgacs in the 1960s[13]. He suggested that these two factors are capable of independent variation due to variations in wood quality or pulping equipment. Properties of fiber fractions were characterized with the specific surface measured using a permeability test described by Robertson and Mason[14]. Figure 4 shows how the specific surface of shorter fractions is higher than that of longer fractions, but the specific surface is independent of the original fiber length of the pulps. The pulps were made of different species. From the ten pulps, three were refiner mechanical pulps and the rest were groundwood pulps. Forgacs further showed that the specific surface measured with the permeability method is closely related to the bonding potential of the mechanical pulp fractions. Based on the data presented in Fig. 4, Forgacs suggested that the relative level of the curves could be defined by the specific surface of the 48/100 fraction. He named it the shape factor for which he used the symbol S. The length factor L was defined as the percentage of the pulp retained on 48 fraction, which means the sum of all the fractions retained on the 48 mesh and coarser screens.

Mannström developed the ideas of Forgacs further, and in place of the specific surface he started to use the specific filtration resistance[15]. Karnis has successfully used both S factors to predict the tear and tensile strength of mechanical pulps[16]. Later Mohlin, Lindholm, and Corson[17-20] have experimentally studied the properties of fiber fractions and their effect on pulp properties. Mohlin mainly investigated TMP pulps and their bonding properties. She showed that the bonding ability of TMP pulps can vary widely (Fig. 5). All the fractions of a poorly bonding pulp were inferior to those of a well bonding pulp. She also showed how the bonding ability of the long fiber fraction improved in the reject refining, which further improved the properties of the whole pulp. Corson concluded that the medium and short fiber fractions had a greater impact on the handsheet properties when the long fiber fraction had a poor bonding ability. Long fibers that had been refined to a higher level of quality were able to consolidate better and, in this case, increasing the proportion of medium or short fibers of the original quality acted to impair the properties of the furnish.

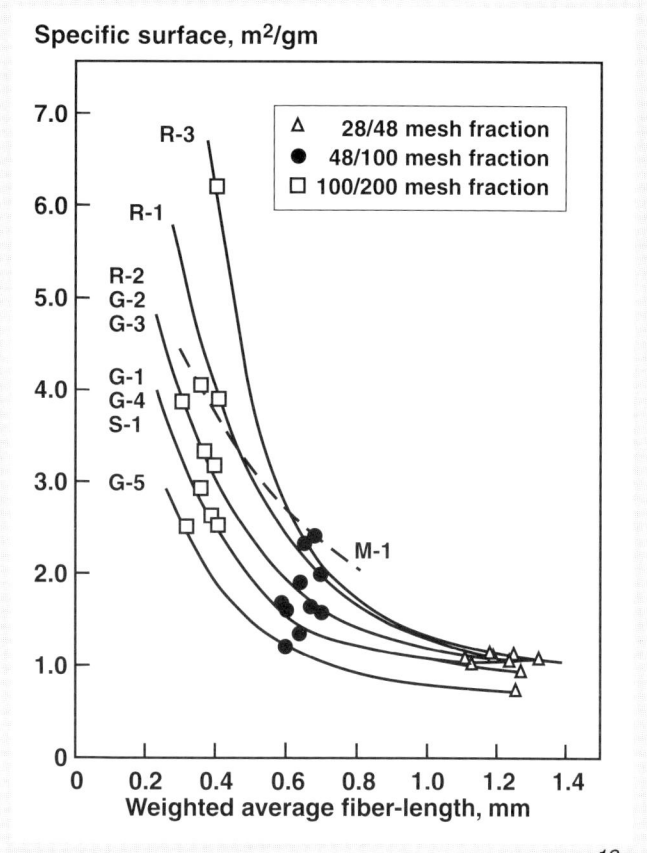

Figure 4. The specific surface of fractions of various mechanical pulps[13].

CHAPTER 15

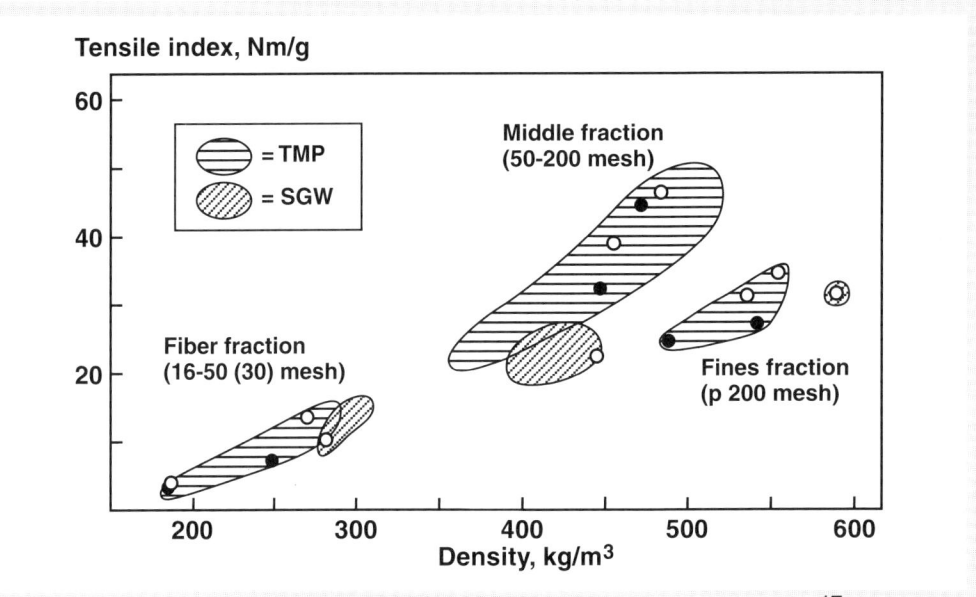

Figure 5. Tensile index vs. density for the coarse fraction, the middle fraction, and the fines fraction[17].

Lindholm tried to distinguish from each other the effect of the fractional composition and the properties of isolated fiber fractions on the handsheet properties. He compared three mechanical pulps – GW, PGW, and TMP – that were prepared from the same lot of raw material, and the GW and PGW were even manufactured using the same equipment. He used Scott bond strength as a measure of the bonding ability. At the same fines content, PGW pulps had the highest Scott bond value and GW the lowest (Fig. 6). On the other hand, the light scattering of all three pulps was similar when compared at the same fines content (Fig. 7).

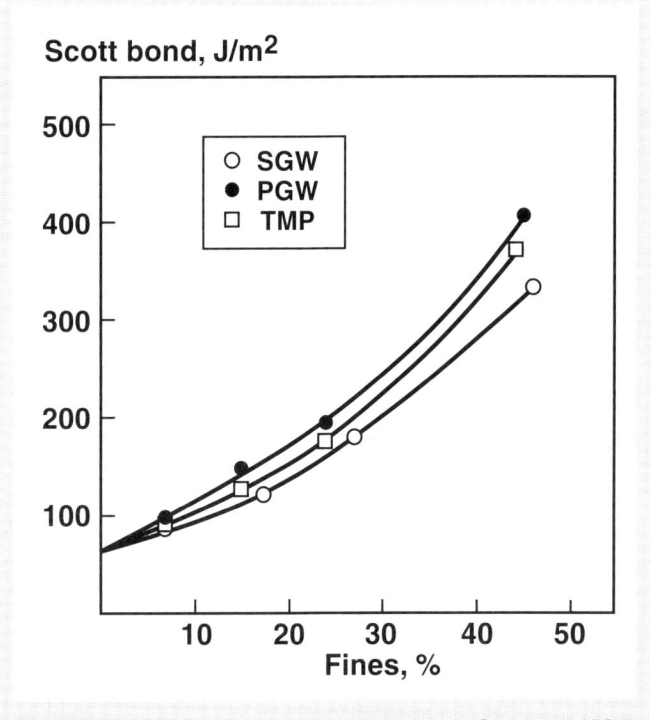

Figure 6. Contribution of different fines fractions on the Scott bond of PGW fiber fraction[19].

Wood and Karnis studied the specific surface of mechanical pulps by fractionating them with the help of a small hydrocyclone[21, 22]. They concentrated upon the linting of newsprint. They slushed a linting paper sample and analyzed its fractions. The results showed that a linting paper sample contained more material with a low specific surface than a nonlinting pulp. They suggest that linting is dependent on the amount of low bonding material rather than the average bonding potential. This means that a method which gives the distribution of the specific surface is required for the prediction of the linting propensity. They recommend the use of a hydrocyclone-type fractionator for that. Later they have also shown that a turbidity measurement can be used to determine the specific surface of each fraction instead of the specific filtration resistance.

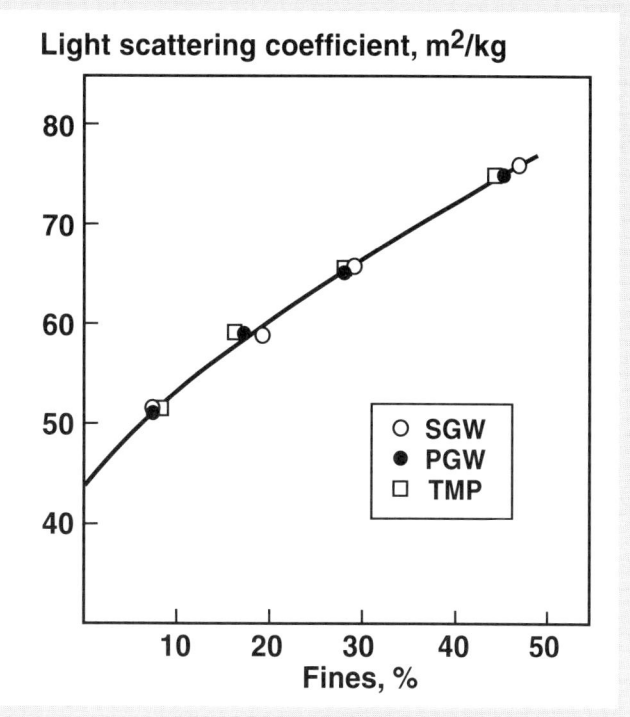

Figure 7. Contribution of different fines fractions on the light scattering of PGW fiber fraction[19].

2.3 Latency in mechanical pulps

When fibers are defiberized at high temperatures and high consistency, they are deformed due to stresses they encounter. According to Htun *et al.*, the polymeric flow enables fibers to be deformed by the stresses in the refiner[23]. Fibers are compressed, twisted, and curled. When cooling down at high consistency, the fibers remain twisted and curly. This phenomena is called latency and it is encountered mainly in refiner mechanical pulps, but similar behavior has also been found in pressure groundwood[24]. Even the thickening of pulps tends to form deformations in fibers. The latency can be removed by agitating the pulp at low concistency and high temperature. After latency removal, a lower freeness and a higher tensile index is obtained. In the industrial processes, latency is removed either in large latency removal chests or in pulpers with powerful mixing right after the refiner discharge. In laboratory procedures, the hot disintegration is a part of routine pulp testing and it is carried out according to a standardized method at the temperature of 80°C–90°C. Htun *et al.* have found that the latency can be removed from CTMP at lower temperatures because of the lower

CHAPTER 15

softening temperatures of sulfonated pulps. They suggest the use of 60°C, which also has an advantageous effect on the brightness. Karnis has reviewed the effect of latency on fiber properties using both published and new results[25]. With latency removal, curls and kinks are removed and both the fiber flexibility and the fiber length are increased because curly fibers behave differently in the fiber length classification. A small increase in true fiber length was concluded to be mainly due to removal of microcompressions. The greatest effects of latency removal were on CSF, tensile strength, tear strength, and modulus of elasticity.

3 Methods used to determine pulp suspension properties

3.1 Method vs. property

The pulp consists of particles of various sizes and of various properties. Both the particle size distribution and the properties of the particles have effects on the behavior of the pulp suspension as well as on the properties of the web forming from the pulp suspension. Pulp properties can be determined using various methods that measure the fiber properties, pulp suspension properties, or the handsheet properties. It has been pointed out that the choice of characterization methods has to be done keeping in mind the objective of the characterization[26].

Examples of fiber properties are the fiber dimensions, structure, and chemical composition of the fiber wall and the surface. For instance, the fiber length can be measured with different methods starting from a microscopic measurement of each fiber to the detection of fibers in a flowing pulp suspension, which is the principle of various measuring devices.

The pulp suspension properties were originally developed for testing the water drainage on a wire section of the paper machine. Drainage properties of pulp suspensions change both when the fiber length distribution is changed and when the properties of fractions are changed. This fact is a basis of various test methods, the best known of which is the Canadian Standard Freeness test. The drainage ability is often inversely correlated with the pulp's bonding ability and the capability to form a smooth and dense sheet, which is why this measurement has become so widely used.

Handsheet properties were developed to simulate the functional paper properties and the runnability both at the paper machine and converting machines. The handsheet preparation and testing are carried out according to standard methods. The retention of fines in handsheet preparation is important, as mechanical pulps are rich in fines. Handsheets can be prepared either from the mechanical pulp alone or from furnishes containing mechanical and chemical pulps and even fillers. To investigate the surface properties, the handsheets are often calendered. Mohlin has investigated the influence of the properties of the mechanical pulp component on mixtures of chemical and mechanical pulps[27, 28]. The blends of mechanical and chemical pulps seemed to behave as independent fiber networks. The response of different types of mechanical pulps to filler addition was also similar, and the authors conclude that the effect of mechanical pulp quality on the strength level of the final paper is essential.

3.2 Principles of drainage measurements

The most common way to characterize the pulp suspension properties is to determine the drainage of the pulp suspension. The drainage properties change both due to the changes in fiber length distribution and in the properties of fibers and fiber fractions. When fibers are refined, the fiber wall becomes thinner, more flexible, and they collapse more easily. At the same time, material is rubbed off from the fiber wall and the proportion of fines is increased.

The most common way to determine the drainage of mechanical pulps is the Canadian Standard Freeness (CSF). The method is described in standards ISO 5267-2 and SCAN-M 4:65. A pulp suspension containing 3.00 ± 0.02 g pulp is diluted to 1000 ml and filtrated through a funnel equipped with an opening both at the bottom and at the side of the funnel. The filtrate flowing through the side opening is collected in a metering glass and the volume is measured. The freeness is the volume of the filtrate expressed in ml.

The Water Retention Value (WRV) is mainly used for chemical pulps but, in some cases, it is also used for mechanical pulps. The principle is that a pad of pulp fibers is formed by dewatering a pulp suspension on a wire screen. The test pad is centrifuged under a specific centrifugal force (SCAN 3000 g, TAPPI 900 g) for a specific time, weighed, dried, and weighed again. The water Retention Value is calculated from the wet mass weight of the centrifugated test pad and the mass weight of the dry test pad.

The specific filtration resistance is determined with a constant rate flow apparatus described by Mannström[15]. The time needed to reach a given pressure difference across the fiber mat is measured. Because of the compressibility of the fiber mat, the specific filtration resistance is a function of the pressure difference. If the compressibility of the pulp is determined separately, the specific volume and the specific surface area of the sample can be determined.

3.3 Principles of fiber classification methods

The fiber length distribution can be determined either by mechanical classification devices or by optical instruments. The most common of the mechanical classification equipment is a McNett- or Clark-type apparatus, in which isolation of fibers is determined by length and the sieves are in vertical position.

The operating principle is given in standards TAPPI T 233 and SCAN-M6. An agitated pulp suspension is fractionated by means of a screening process employing vertical screens with increasing wire sieve cloth numbers. The screens are mounted in tanks arranged in cascade. The mass of the fibers retained by each screen is determined and expressed as a percentage of the oven dry mass of the original sample. Using Clark-type or McNett-type classifiers will produce identical results within the stated precision. Table 1 provides the two series of sieves and the screen openings used:

CHAPTER 15

Table 1. Screen openings.

Tyler series	ASTM E11	Opening, mm
10	12	1.68
12	14	1.41
14	16	1.19
20	20	0.841
28	30	0.595
48	50	0.297
100	100	0.149
200	200	0.0747

Tasman made a review of the length of fiber fractions isolated with a McNett classifier[29]. The average length of the fibers retained on each screen was fairly constant, and he summarized the results to a curve, which can be used to read the average fiber length as a function of sieve opening or the mesh size of the screen. Bentley has determined the fiber lengths of different McNett fractions using a Kajaani FS-200 apparatus. Table 2 shows the results[30].

Table 2. Average fraction fiber lengths for black spruce pulps[30].

	Kraft, mm	TMP, mm	Avg mm
R14	2.66	2.87	2.76
R28	2.01	2.04	2.02
R48	1.33	1.29	1.31
R100	0.75	0.63	0.69
R200	0.32	0.24	0.28
P200	0.16	0.08	0.12

The Brecht-Holl apparatus is a flat screen and its operating principles are given in ZM Merkblatt VI/1. The shives content is determined simultaneously using a slit width of 0.20 mm. For fiber classification, only two sieves are used. In this case, the sieve no. 16 equals the opening size of 0.40 mm and the sieve no. 50 the size 0.125 mm. Results are given as follows:

Shives content (Splitter), %

Long fiber fraction (Faserlangstoff), %

Short fiber fraction (Faserkurzstoff), %

Fines (Feinstoff), %

or

Rs = shive content, g

R16 = fibers retained on the sieve no. 16, g

R50 = fibers retained on the sieve no. 50, g

E = amount of pulp used in the classification.

3.4 Principles of shive content measurements

The shive analyzers based on mechanical screening give the shive content as a weight percentage. Some optical systems like PQM give the number of shives and the shive matrix, in which shives are classified according to their dimensons.

The Somerville-shive content (TAPPI UM 242) is the weight-% of shives retained on a vibrating slotted screen plate. In one screen plate, there are 765 slits each 45 mm long and 150 μm wide. During screening, the pressure of the waterjets above the screen plate is 124 kPa and the flow 8600 ml/min. The size of the specimen is 25 g. The screenings are collected, dried, and weighed. The PFI- minishive content (SCAN-M 13:83) is a weight % of shives not passing an 80 μm slit. The screening effect is based on water flow which is induced by a pulsating diaphragm. Size of the specimen is 2.00 ± 0.02 g. The von Alfthan shive analyzer (TAPPI UM241) has a slit width of 180 μm and the method is based on the blocking of a narrow slit. Size of the specimen is 20 g.

In the Haindl-fractionator (ZM-Merkblatt VI/1.4/86), the slit width is 150 μm. The size of the specimen is 10 g. The Brecht-Holl flat screen (ZM-Merkblatt VI/1) has a slit size of 200 μm x 200 mm. The size of the specimen is 10 g. Pulmac-type screen slit sizes used are at the range 75–375 μm. The test specimen required for screening can be 2–100 g depending on the level of contaminants present.

3.5 Principles of optical fiber analyzers

The principle of optical methods is the measurement of individual fibers from a dilute pulp suspension flowing through a cyvette. The flow is illuminated with a source of light, fibers are detected with a camera, and image analysis is used to reduce the huge amount of information to understandable characteristics.

Kajaani FS-200 (Valmet Automation) is the most widely spread optical fiber measurement device[31]. It gives the fiber length distribution which can be used to calculate the average fiber length and the coarseness of a sample, provided the sample has been weighed accurately beforehand. The method is becoming accepted as a standard laboratory fiber measurement method. The measuring principle is to detect the beginning and the end of a fiber and calculate the fiber length from that information and the flow speed. The FS-200 and the earlier model FS-100 differ in the respect that the whole sample is measured in the FS-100, but the FS-200 measures a side flow. The FS-200 uses polarized light and, because of that, only birefringent fibers are seen. The on-line version of the FS-200 is called FSA.

According to the authors' experience, the fiber length distribution is easily measured and the average values calculated by the device are widely used and understood. Some problems have been encountered because the shives and sometimes even long

CHAPTER 15

fibers can block the capillary, and this can cause difficulties with coarseness measurement. A new development named Fiberlab has been introduced recently. In addition to the earlier measurements, the fiber width and fiber wall thickness index are obtained. It measures the fiber length with the same principle as in FS-200 but, for the measurement of fiber width, a CCD-camera and image analysis is used. In Fiberlab, the whole sample is measured instead of a side flow.

Pulp Quality Monitor (PQM) is a well known on-line apparatus sold by Sunds Defibrator[32]. PQM 1000 is the laboratory model of the pulp quality monitor, and the widely used on-line version is PQM 400. A linear detects the fibers in the flowing suspension. Image analysis is used to measure the size and shape of the fibers and shives. As a result, PQM gives the distributions and average widths and lengths of the fibers and shives as well as the curl index of the fibers. Coarseness can be determined with the laboratory model by weighing the pulp sample before it is loaded into the equipment.

Fiber Quality Analyzer FQA was jointly developed by Paprican, the University of British Columbia, and Op Test[33]. The patented flow cell orients the fibers in the middle of the flow which prevents the fibers from blocking the flow-cell. A ccd-camera and a polarized light source are used. This measurement results in obtaining fiber length distribution, fiber curl, and mean kink angle and kink-index. Coarseness can be calculated in the same way as in the FS-200.

Fiber Master has been developed at the STFI[34]. It gives the fiber length distribution, width, and curl. The width is calculated from the perimeter and the area of a fiber. The apparatus is not yet commercially available.

Pulp Expert is a semiautomatic testing device used at mills. Lately the company has introduced an optional fiber analyzer unit[35]. The principle of the fiber analyzer is similar to the others except that the pulp suspension is flowing between two glass plates. At present, no published information of the operation the fiber analyzer is available.

4 Fiber properties

Fiber properties can be characterized using various methods. Heterogeneity of the material makes it difficult to get the absolute values for certain properties but, in most of the cases, it is enough to see reliably the change in the properties. Table 3 gives examples of the methods used for the measurement or evaluation of a property.

4.1 Flexibility and stiffness of fibers

Two main methods are used to determine the stiffness of fibers. The hydrodynamic method was developed by Tam Doo and Kerekes[36, 37] and the Steadman method[38]. In the hydrodynamic method, the fiber is placed on a groove of a thin (about 1 mm) capillary, which is under water. The fiber is forced to bend by the water flow through the capillary. The fiber stiffness is calculated from the water flow and the fiber width. In the Steadman method, a thin fiber mat is pressed on metallic wires. The length of unbonded fiber segments is measured with the help of a light microscope and an image analyzer. Fiber stiffness is calculated from those unbonded segments and the width of the fibers. Based on that method, a semiautomatic instrument Cyberflex was developed by CyberMetrics.

Table 3. Examples of fiber properties and methods used to analyze them.

Property	Method
Fiber stiffness	Hydrodynamic method Steadman method Density of a handsheet prepared from a fraction
Deformations in fiber wall	Simons staining and light microscopic classification of fibers according to the fiber color Specific filtration resistance
Fiber wall thickness	Light microscope + image analysis - from cross-sections - in transmitted light from longitudinal preparate
Fiber coarseness	Optical measurement methods: FS-200, PQM, Pulp Expert, ... Clark method
Fiber length	Optical measurement methods: FS-200, PQM, Pulp Expert, ...
Fibrillation of fiber surface	Fibrillation index using light microscope + image analyzer Fiber classification according to fibrillation Specific filtration resistance
Chemical composition of fiber surface	FTIR ESCA
Properties of fines	Sedimentation volume Length of the fines particles Proportion of fibrillar material Turbidity

4.2 Simons staining

In mechanical pulping, the structure of the fiber wall is changed and local damages are formed into the fiber wall. This phenomena is called internal fibrillation. One way to characterize this is to use the Simons stain[39, 40]. It was used earlier to characterize chemical pulp refining and microbial damages. In this method, fibers are stained using a 1% solution of a two-component stain, Pontamine fast orange and Pontamine sky blue. Mechanical pulp fibers are stained to dark or light blue or yellow depending on how the particles are able to penetrate into the fiber wall. With refining, the proportion of yellow fibers is increasing.

4.3 Fiber wall thickness

Fiber wall thickness can be measured either from the cross sections of fibers or from normal fiber slides used for the direct light microscopy. The cross sections have been measured, using both a light microscope or electron microscope[8, 9, 12]. The confocal laser microscope enables the determination of the optical cross section of the fibers without physically sectioning the fibers[12]. In direct light microscopy, the fiber wall is measured as it is seen in transmitted light. In both direct microscopy and confocal laser microscopy, the fibers can be either moist or dry, but the measurements of cross

CHAPTER 15

sections are always made of dried samples. Image analyzers are used in all measurements. The microscopic transmission ellipsometry has also been used for the determination of the fiber wall thickness and the S2 fibril angle[41]. The method is based on the retardation of polarized light when light is transmitting the fiber.

4.4 Methods for the characterization of fines

Marton characterized mechanical pulps by determining their settling rate[42]. Dilute pulp suspension is allowed to settle down in a metering glass. The height of the clear water above the pulp is plotted as a function of the elapsed time. The slope of the tangent to the initial part of the curve is called the settling rate. As a result, she showed that the pulps of greater strength settled more slowly than those of lower strength. The settling rate measurements can also be used for measurement of the fines fraction which is not the case with freeness-measurements or the

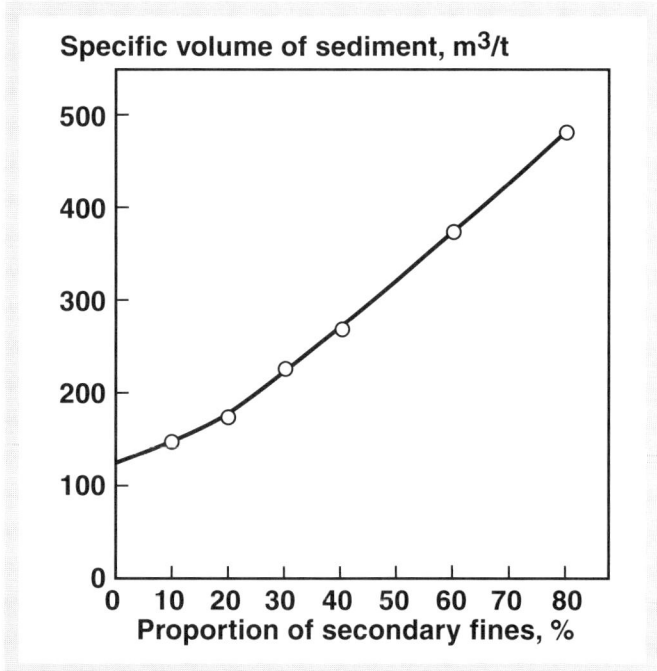

Figure 8. Specific volume of the sediment of a mixture of primary and secondary fines[44].

permeability measurements. Except for the settling rate, Marton also used the terminal volume which is the volume of the turbid suspension after the movement of the particles has stopped. Giese used the sedimentation volume particularly for the characterization of fines fraction[43]. The specific sedimentation volume was the volume of turbid suspension per mass of the fines in the suspension. Giese also used the centrifuge to help the settling of the fines. The method used lately has been the specific sedimentation volume after 24 hours of sedimentation[44]. The results show that the specific sedimentation volume correlates with the length of the fines particles and their bonding ability, which was determined by mixing fines of different types with the same fiber fraction (Fig. 8).

The length of fines particles have been determined with a light microscope and image analyzer[45]. Using this method, the average length of the fines was in the range of 16–25 µm. The distribution of particles is very skewed, the greatest number of particles being short. With increasing energy usage in refining, longer fibrillar particles are formed and the average length of the particles is increasing[44].

Wood *et al.* used the turbidity measurement to determine the specific surface of a fines suspension[46]. The measurement is based on the light scattering of the fines suspension. Luukko has been developing a method to determine the proportions of fibrillar and flake-like material from the fines[47].

5 Handsheet properties and paper testing

5.1 Preparation of handsheets

In spite of the rapid development of devices that determine the fiber and pulp suspension properties, the most common way to characterize the pulp is still to prepare handsheets and use traditional testing methods to test their strength and optical properties. Handsheets are prepared in a sheet mold by filtering a dilute pulp suspension on a wire cloth, couching, wet pressing, and drying them under standardized conditions. The conditions used in the handsheet preparation influence their properties and, because of that, international and national standardized methods are used. Even small deviations in details can cause great differences in testing results. The same handsheet preparation methods are at present standardized for both chemical and mechanical pulps. Compared to the chemical pulps, the mechanical pulps have stiffer fibers and a large proportion of fines. According to our experience, the use of the same handsheet preparation method as for chemical pulps does not give a correct characterization of mechanical pulps. This is why recirculated white water is widely used in a sheet mold when handsheets are prepared of mechanical pulps. The principles of sheet preparation techniques are described in the following standards:

ISO 5269-1, SCAN-M5, TAPPI T205

> A circular rectangular or square sheet is formed in a conventional sheet former from a pulp suspension on a wire screen under suction. The sheet is subjected twice to a pressure of 400 kPa. The sheet is dried in contact with a drying plate in conditioned air. The plate is used to restrict handsheet shrinking.

ISO 5269-2, DIN 54358/1, ZM V/8, V/10

> A circular sheet is formed in a Rapid- Köthen sheet former. The sheet is subjected to the pressure and dried in a dryer with almost complete prevention of shrinkage.

Modified ISO 5269-1, SCAN M5, sheets with recirculated white water.

> Sheets are made according to the standard methods, but recirculated white water is used. In order to ensure the proper retention of fines in the handsheet, the first 10–14 sheets are prepared to obtain a balance between the water and the handsheet.

ISO 3688, SCAN-CM 11

> This is a standard for the preparation of handsheets for brightness measurement.

CHAPTER 15

5.2 Testing of handsheet properties

The volume "Pulp and Paper Testing" in this textbook series describes the testing of hand sheets in more detail. The following list provides only the most common tests and standards:

Standard atmospheres for conditioning and testing

Preferred atmosphere: relative humidity 50±2%, temperature 23±1°C

For tropical countries: relative humidity 65±2%, temperature 20±1°C

Mechanical properties

Basis weight ISO 536, SCAN-P6

Thickness and density ISO 534, SCAN-P7

Tensile strength, stretch, tensile energy absorption, and tensile stiffness ISO 1924-2, SCAN-P38, SCAN-P67

Bursting strength, ISO 2758, SCAN-P24

Tearing strength ISO 1974, SCAN-P11

Air permeance, Bendtsen ISO 5636-3, SCAN-P21

Air permeance, Gurley ISO 5636-5, SCAN-P19

Optical properties

Brightness ISO 2470, SCAN-P3

Y-value, opacity, light-scattering coefficient ISO 2471, SCAN-P8

5.3 Typical sheet properties of mechanical pulps

When compared to chemical pulps, mechanical pulps typically form bulky sheets with high light scattering. Using mechanical pulp, it is thus possible to make paper with acceptable stiffness and opacity at a considerably lower basis weight than using chemical pulp. When compared at the same freeness level, TMP normally has the highest bulk and CTMP the lowest bulk of the mechanical pulps, with PGW and GW placed in between. GW usually has the highest air permeability followed by PGW, TMP, and CTMP in that order when the comparison is performed at the same freeness. CTMP is normally manufactured to quite a high freeness compared to pure mechanical pulps and, at this high freeness, both bulk

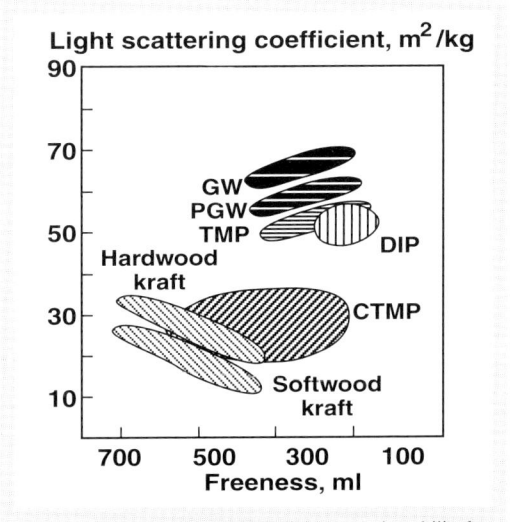

Figure 9. Typical differences in light-scattering ability for various pulps[48].

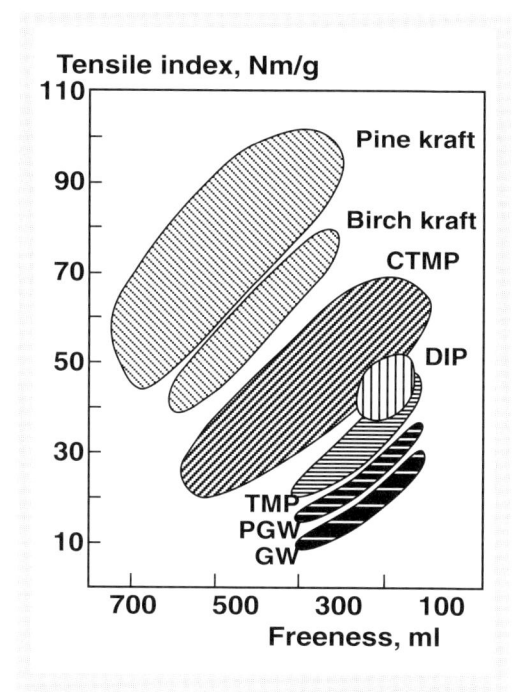

Figure 10. Typical differences in tensile strength of various pulps[48].

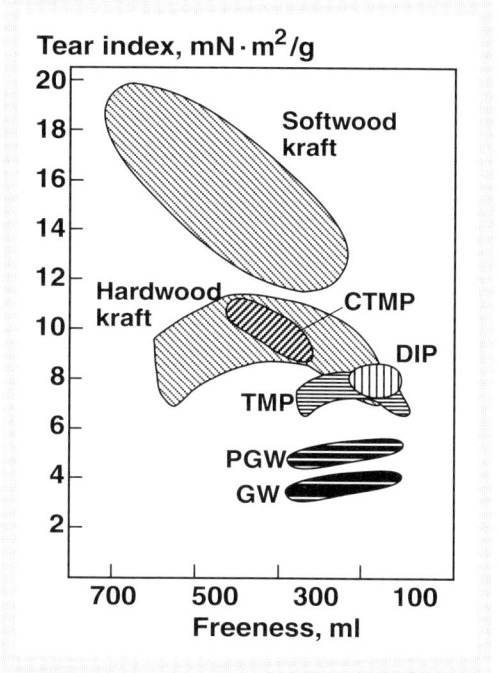

Figure 11. Typical differences in tear strength of various pulps[48].

and air permeability can be quite high.

The groundwood pulps usually exhibit the best optical properties because of the comparatively low process temperatures used. GW has the highest unbleached brightness and light-scattering power, followed closely by PGW. Figure 9 gives normal light-scattering values for some common pulp grades.

The refiner mechanical pulps (TMP and CTMP) exhibit the best strength properties. Because of their good strength properties, they need less reinforcement pulp when manufacturing say LWC paper, which compensates for their lower light-scattering power and makes their use economically favorable. Figures 10 and 11 give the tensile and tear strength for some common pulps.

References

1. Vanninen, M. -P., M.Sc. thesis, Technical University, Helsinki, 1996.
2. Tyrväinen, J., "The influence of wood properties on the quality of TMP made from Norway spruce (Picea Abies) wood from old growth forests, first thinnings and sawmill chips," 1995 International Mechanical Pulping Conference (Ottawa) Proceedings, CPPA, Montreal, p. 23
3. Corson, S. R., Tappi J. 74(11):135 (1991).
4. Hatton J. V., "Newer fiber sources and their effects in papermaking," TAPPI 1993 Pulping Conference Proceedings, TAPPI PRESS, Atlanta, p. 169.
5. Mork, E., Papir Journalen 16(4):44 (1928).
6. Corson, S. R. and Ekstam E., Paperi ja Puu 76(5):334 (1994).
7. Koljonen T. and Heikkurinen A., "Delamination of stiff fibers," 1995 International Mechanical Pulping Conference (Ottawa) Proceedings, CPPA, Montreal, p. 79.
8. Mohlin, U-B., "Fiber development during mechanical pulp refining," 1995 International Mechanical Pulping Conference (Ottawa) Proceedings, CPPA, Montreal, p. 71.
9. Kure, K. A., "The alteration of the wood fibers in refining," 1997 International Mechanical Pulping Conference Proceedings, SPCI, Stockholm, p. 137.
10. Karnis, A., J. Pulp Paper Sci.20(10):J280 (1994).
11. Jang, H.F ., Amiri, R., Seth, R. S., Karnis, A., Tappi J. 79(4):203 (1996).
12. Lammi ,T. and Heikkurinen, A., "Changes in fiber wall structure during defibration," Transactions of the 1997 Fundamental Research Symposium, Pira International, Leatherhead, p. 641.
13. Forgacs, O. L., Pulp Paper Mag. Can. 64(2):T89 (1963).
14. Robertson, A. A. and Mason, S.G., Pulp Paper Mag. Can. 50(13):1 (1949).
15. Mannström, B., "On the characterization of mechanical pulps," 1968 International Mechanical Pulping Conference Proceedings, TAPPI, CPPA, Atlanta, p. 35.
16. Shallhorn, P. and Karnis, A., Pulp Paper Can. 5(4):TR92 (1979).
17. Mohlin, U-B., Svensk Papperstid. 83(16):461 (1980).
18. Mohlin, U-B., "Fiber bonding ability – a key pulp quality parameter for mechanical pulps to be used in printing papers," 1989 International Mechanical Pulping Conference Proceedings, KCL, Helsinki, p. 49.
19. Lindholm, C. A., Paperi ja Puu 62(12):803 (1980).
20. Corson, S., Pulp Paper Can. 81(5):T108 (1980).
21. Wood, J. R. and Karnis, A., Paperi ja Puu 59(10):660 (1977).

22. Wood, J. R., Grondin, M., Karnis, A., J. Pulp Paper Sci.17(1):J1 (1991).
23. Htun, M., Engstrand, P., Salmen, L., J. Pulp Paper Sci. 14:J109 (1988).
24. Karojärvi, R. and Nerg, H., "Latency in pressure groundwood," 1987 International Mechanical Pulping Conference (Vancouver) Proceedings, CPPA, Montreal, p. 25.
25. Karnis, A., Paperi ja Puu 75(7):505 (1993).
26. Heikkurinen, A., Levlin, J-E., Paulapuro, H., Paperi ja Puu 73(5):411 (1991).
27. Mohlin, U-B., Nordic Pulp and Paper Res. J. 1(4):44 (1986).
28. Mohlin, U-B. and Wennberg, K., Tappi J. 67(1):90 (1984).
29. Tasman, J. E., Tappi 55(1):136 (1972).
30. Bentley, R: G., Scudamore, P., Jack, J. S., Pulp Paper Can. 95(4):41(1994).
31. Tiikkaja, E., "Mill benefits from on-line fiber length measurement," TAPPI 1996 Process and Product Quality Conference Proceedings, TAPPI PRESS, Atlanta, p. 7.
32. Sköld, H. and Nilsson, P., "PQM 1000 fiber and shive classifier based on image analysis," 1993 International Mechanical Pulping Conference Proceedings, PTF, Oslo, p. 54.
33. Trepanier, R. .J., "Fiber quality analysis – A new approach.," CPPA Tech. Sect. 1995 Annual Meeting Notes, CPPA, Montreal, p. B55.
34. Karlsson, H. and Fransson, P.I., Svensk Papperstid. 97(10):26 (1994).
35. Pulp Expert leaflet, Pulp Expert Oy, Varkaus, 1994.
36. Tam Doo, P. A. and Kerekes, R. J., Tappi 64(3):113 (1981).
37. Tam Doo, P. A. and Kerekes, R. J., Pulp Paper Can. 83(2):46 (1982).
38. Steadman, R. K. and Luner, P., "The effect of wet fiber flexibility of sheet apparent density," Transactions of the 1985 Fundamental Research Symposium, Mechanical Engineering Publications Limited, London, p. 311.
39. Simons, F. L., Tappi 33(7):312 (1950).
40. Blanchette, A. R., Akhtar, M., Attridge, M. C., Tappi J. 75(11):121 (1992).
41. Ye, C. and Sundström, O., TAPPI J. 80(6):181 (1997).
42. Marton, R. and Robie, J. D, Tappi 52(12):2400 (1969).
43. Giese, E. and Link, D., Zellstoff und Papier 6(9):269 (1957).
44. Heikkurinen, A. and Hattula, T., "Mechanical pulp fines – Characterization and implications for defibration mechanisms," 1993 International Mechanical Pulping Conference Proceedings, Eucepa, PTF, Oslo, p. 294.
45. Pelton, R. H., Jordan, B. D., Allen, L.H., Tappi J. 68(2):91 (1985).
46. Wood, J. and Karnis, A., "The determination of the specific surface of mechanical pulp fines from turbidity measurements," TAPPI 1991 International Paper Physics Conference Proceedings, TAPPI PRESS, Atlanta, p. 11.
47. Luukko, K., Kemppainen-Kajola, P., Paulapuro, H., Appita J. 50(5):387 (1997).
48. Sundholm, J., unpublished KCL data, 1997.

CHAPTER 16

Future outlook

1	**Development of paper demand**	**415**
2	**Paper and mechanical pulp quality development**	**415**
2.1	Newsprint	416
2.2	Uncoated mechanical	416
2.3	Coated mechanical and coated fine papers	416
3	**The high energy consumption – a threat to the future of mechanical printing papers?**	**416**
4	**How to cut the energy consumption?**	**416**
5	**Impact of the increasing use of recycled fibers**	**417**
6	**Can wood supply match the rising demand of mechanical and other pulp grades?**	**418**
7	**Pros and cons for PGW and TMP**	**418**
8	**Further development of today's processes**	**419**
9	**Potential of new technologies**	**419**
	References	421

CHAPTER 16

Jan Sundholm

Future outlook

1 Development of paper demand

Jaakko Pöyry Consulting has estimated the growth in mechanical printing paper demand until year 2005. In Table 1, the paper grades are grouped into three classes: newsprint (standard newsprint and specialities), uncoated mechanical (SC and MF magazine), and coated mechanical grades. The development of newsprint demand varies. Good growth is expected in Europe, but in North America the demand is expected to somewhat decrease. Uncoated mechanicals are expected to grow at a moderate pace while the big expansion is expected in coated mechanicals. Coated mechanicals are expected to grow by 64% in Europe from 1996 to 2005 or in 10 years.

Table 1. The demand of mechanical paper grades in selected regions according to Pöyry[1]

	1996 1000 tons	2000 1000 tons	2005 1000 tons
Newsprint			
Europe	10 089	11 920	14 220
North America	11 920	11 820	11 700
World	34 788	38 140	42 320
Uncoated Mechanical			
Europe	4 917	5 410	6 030
North America	4 413	5 000	5 730
World	12 736	14 350	16 360
Coated Mechanical			
Europe	5 051	6 500	8 300
North America	4 295	5 000	5 790
World	12 583	14 960	17 930

2 Paper and mechanical pulp quality development

According to KCL estimates[2], the surface properties of mechanical printing papers are expected to improve substantially during the period 1996–2005. Some of this development will be due to better forming and calendering techniques, but some also will be due to improved mechanical pulp quality. The need for more smooth and non-roughening paper grades will restrict the possibilities for large energy savings while the processes will become more complicated. The trend toward lower basis weight will continue both for mechanical printing papers and fine papers. This will lead to an increased use of highly bleached mechanical pulp in coated fine papers, and also to increased demands on bulk and opacity for the mechanical pulp.

CHAPTER 16

The freeness value (CSF) will still be useful in estimating the "fineness" of the pulp, even if other and better characterization methods will appear. The developing fractionation and fraction treatment technologies will lead to possibilities to improve the smoothness without also lowering freeness. Nevertheless, it is expected that the mechanical pulp freeness still will continue to decrease.

2.1 Newsprint

Newsprint is more and more printed in four colors. The demands on paper surface are rising and mechanical pulp freeness will decrease from 80–110 ml to 70–90 ml and even somewhat lower for speciality grades.

2.2 Uncoated mechanical

During the recession in the early 1990s, the SC pulp freeness was lowered to 25–30 ml, but has since somewhat increased with the introduction of gapformers to SC making. Due to competition with LWC, the quality demands will continue to rise and the mechanical pulp freeeness will drop to 20–30 ml.

2.3 Coated mechanical and coated fine papers

The need for a less roughening sheet means that long fiber refining technologies will develop further and the freeness will decrease to the same level as for SC pulps (20–30 ml). Mechanical pulp for MFC paper will still have somewhat higher freeness.

3 The high energy consumption – a threat to the future of mechanical printing papers?

The high energy consumption has emerged in recent years as a threat to the long-term future of mechanical printing papers. Production of high-grade thermomechanical pulp, for example, can consume electricity at the rate of more than 3 MWh/t. It has been predicted that the price of electricity could rise substantially in the next few years. An even greater threat could stem from demands from the market, already in evidence, concerning the composition of these papers favoring "low energy" pulps. This might give the electricity consumption of mechanical pulping an intrinsic value of its own that will itself drive the market. A prediction by Jaakko Pöyry in 1977 that energy consumption could be the Achilles' heel of mechanical pulping[3] may, in fact, become reality.

4 How to cut the energy consumption?

It has been estimated that the energy reduction potential of modified and optimized grinding and refining processes might amount to 20%–30% in the specific energy consumption. Theoretically the specific energy consumption might be reduced by up to about 50% by the use of proposed new technologies, such as some sort of compressive shear treatment[4]. Such substantial reductions in electricity consumption cannot be achieved without major long-term research on the mechanisms of mechanical pulping.

Research institutes like KCL began research in this field back in the 1980s and have shown that the energy consumption can be lowered in several ways, e.g., by reducing the grindstone speed in the pressure groundwood process and by raising the intensity

Future outlook

of refining in the TMP process. We have already seen some quite successful implementations of these results, e.g., the RTS-TMP process. However, there are still many technical obstacles in the way for achieving the estimated 50% energy reduction potential while still meeting the rising end-product demands. These obstacles can only be removed through fundamental research on how the wood fiber reacts in physical conditions during fiber removal and fiber treatment. More basic information is also needed on the interactions of different mechanical pulp fractions in the paper sheet and which type of fractions would be most favorable from the energy and quality point of wiev.

5 Impact of the increasing use of recycled fibers

The recovery rates of paper will grow in the future and, according to Jaakko Pöyry, the world average will be 47% in 2010 with well over 50% rates in the more developed areas in the world[5]. Should we expect that an increased use of recycled fibers will fill up the growth in paper demand so that new mechanical pulping capacity will not be needed, or the growth in mechanical pulping capacity will be small, as was largely the case in the beginning of the 1990s? Technically, a substantial part of the mechanical pulp required even for magazine papers could be replaced with fibers from recycled paper, thus also conserving energy because the deinking of recovered paper uses much less energy than mechanical pulping.

Figure 1 shows the Jaakko Pöyry forecast for the total use of papermaking fibers until year 2010[5]. Recycled pulp will be more and more important, but so will most virgin fiber grades, including mechanical pulp. From Fig. 1, we can see that the consumption of mechanical pulp is expected to rise from around 35 million tons in the 1990s to around 45 million tons in year 2010.

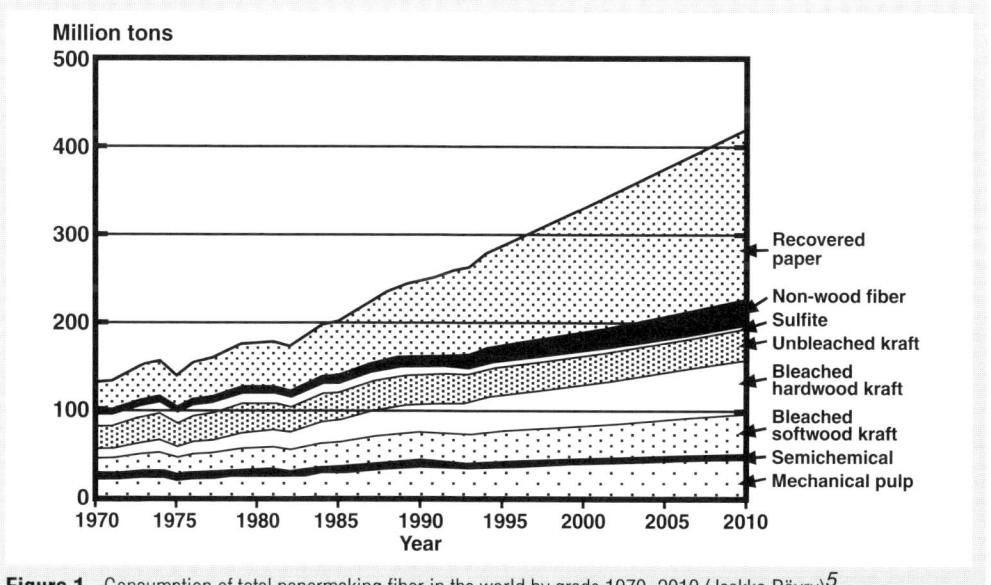

Figure 1. Consumption of total papermaking fiber in the world by grade 1970–2010 (Jaakko Pöyry)[5].

CHAPTER 16

6 Can wood supply match the rising demand of mechanical and other pulp grades?

Acccording to Jaakko Pöyry, the world paper and board consumption will rise from 270 million tons in 1994 to 420 million tons in 2010. Taking into account the increased waste paper recovery rates, this still means an increase in wood demand from 1312 million m^3 in 1994 to 1695 million m^3 in 2010, or an increase of 30%. Gundersby[5] asks if 30% more wood will really be available, and concludes that the industry will have to fight for its wood raw material. Bearing in mind that mechanical pulping needs only half the amount of wood needed for chemical pulping and that the use of mechanical pulps also enables the use of lower basis weights when making printing papers, it seems quite probable that the wood resources required for the increase in mechanical pulping capacity will be found. There might be local restrictions though.

7 Pros and cons for PGW and TMP

When considering the choice of pulp for a new pulping line today, the main modern mechanical processes are TMP, PGW, and CTMP. Of these, CTMP can be considered as a modified TMP, advantageous for some special products like liquid board because of its cleanliness, or for mechanical printing papers because the process enables the use of wood raw materials unsuitable for the pure TMP process. When considering the production of mechanical printing papers using high-quality spruce wood, the best quality can generally be achieved using the pure mechanical processes: either TMP or PGW. The following text discusses some of the pros and cons which have to be considered.

Both PGW and TMP are used for the same kind of products. Acceptable quality and economy of production can be achieved with both, as so many of the factors that affect the comparison, vary from mill to mill. The main differences which affect the choice of pulp are production costs for the pulp making and the quality characteristics of the pulps (for data see chapters "What is mechanical pulp?" and "The character and properties of mechanical pulp"). The differences in production costs come mainly from the energy consumption and investment costs. TMP has a much higher energy consumption than PGW. PGW, on the other hand, has higher investment costs. There are also other cost differences, but they are not decisive.

TMP has excellent strength properties and thus enables a reduction in the use of reinforcement pulp in papermaking, which usually results in a TMP based paper having the lowest production costs at the paper mill, even when taking into account the lower energy consumption of PGW. PGW, on the other hand, offers better optical and paper surface properties, but about the same level of these can be achieved also using TMP when taking into account the lower amount of reinforcement pulp in the final paper sheet. Qualitywise TMP is however a more difficult case than PGW, and the use of high strength TMP in top quality LWC paper is possible only for mills with better then normal know-how and process control.

The general conclusion is that by using PGW for LWC, it is generally easier to achieve the product's quality demands, while by using TMP it is generally easier to achieve good profitability of production. This conclusion holds for today's relationship regarding the price of electric energy on one hand and the price of bleached softwood kraft pulp on the other. If the price of electricity would rise faster than the price of kraft, the situation would be more favorable for PGW.

8 Further development of today's processes

It is quite clear that during the last decade the TMP process has developed at a much faster pace then the PGW process. The reasons lie in the economical profits attainable by developing TMP grades that qualitywise are comparable with groundwood in magazine paper grades. The other factor behind this development is TMP's high energy consumption, which has been regarded (and still is regarded) as a threat to TMP utilization. This has led to the development of several new lower energy TMP processes (see chapter "Thermomechanical pulping").

Technically also the PGW process could be radically modified to produce a superior pulp quality using some 20% less energy than today. This new PGW process is based on a new type of grinder which permits the use of a lower pulpstone speed than today without losing in production rate. There are sound pilot research results that support such a development. The problem is that the development of a new grinder type would be quite expensive.

In the future, less and less of the final pulp quality will be created in the main line refiners or grinders. With improving fractionation methods, an essential part of the pulp manufacturing will consist of separating different fiber fractions from each other for separate refining treatments. In this way, only those fiber fractions that need more mechanical treatment will be further refined. These different fiber fractions will also need different refining conditions.

9 Potential of new technologies

Through the years, many new ways of making mechanical pulp have been proposed. Several of them have developed into laboratory and pilot processes. But still today there has been no real breakthrough, that is to say, no practically and economically viable process yielding good-quality pulp. Some of the most interesting attempts are:

- The reciprocating apparatus[6]
- The Bi-Vis screw extruder[7]
- Fungal or entzymatic treatment[8, 9]
- Explosion pulping[10].

But there have been many more: electron beam treatment[11], fiber separation by cutting[12], axial compression of wood[13], ultrasound treatment[14], and explosion by use of rapid microwave heating[15], just to mention a few. The problem is that most research results seem to be rather unreliable and the scientists' conclusions too optimistic.

The reciprocating laboratory apparatus was developed at Paprican in the early 1960s[6]. Pulp was produced by a specially shaped metallic tool that moved in reciprocating motion over a block of wood. The pulp which was produced between the tool and the wood formed into rolls, which rolled across the surface of the wood and thereby helped to produce more pulp. The pulp had quite high tensile strength (c. 3500 m) and low energy consumption (c. 900 kWh/t), but the amounts of pulp produced were very small. Since then, a number of attempts have been made at building more practical machines on the same operating principle, i.e., rolling fiber bundles. Thus, for example, some researchers from KCL built two different machines in the early 1980s, but they were unable to verify the results of Paprican[16].

The Bi-Vis process was developed by CTP in Grenoble in the early 1980s. The Bi-Vis machine is an extruder with two horizontal screws with a pattern of conveying flights and reversing flights. The reversing flights have openings through which the material is pressed (rather as in a meat grinder). Originally energy reductions of some 15%–30% at the same tensile strength were reported[7]. Since then the process has been further developed both in Europe and in North America. Pilot studies by KCL showed no real energy reduction potential, but it was concluded that further development of the Bi-Vis machine might lead to improved pulp properties[17].

During the 1980s, much research was directed at developing a kind of "biomechanical pulping" by treating wood chips with white-rot fungi[8]. Some energy reductions have been reported, but most research has still been conducted only at the laboratory level and thus the reliability of the results is not clear. This process also has several disadvantages, e.g., long treatment times (days or even weeks), low yield, and low brightness. The process would be technically more attractive if pure enzymes could be used instead of fungi, and a pre-defibered mechanical pulp could be processed instead of chips. Such tests are under way[9], but, so far, no reliable full scale results are available.

Explosion pulping technology, which has attracted some interest in recent years, uses the Stake reactor[10]. Early laboratory studies showed large energy reductions, but most of the research has been directed at the use of hardwoods. Pilot runs by KCL using the institute's refiner and a Stake pilot plant, however, did not show any energy reduction potential with spruce. Stake-treatment at 180°C–200°C did in fact increase the energy consumption compared with TMP and even compared with CTMP at the same chemical addition level. Yield and brightness were also much lower[18].

In the light of the above, the potential of these new processes would not appear very great, at least not in the short term. They might have some potential for development in the long term, particularly if researchers concentrate more than they have so far on influencing specific subprocesses in the transformation process of wood into pulp. One could envisage procedures in several stages, which might include low intensity pre-fiberizing using something like the Bi-Vis machine before the complete defibration to make sure the fiber length is retained, and enzyme treatment of coarse fibers before the final mechanical fibrillation steps. On the other hand, it seems unlikely that anyone will invent an entirely new type of machine which would turn wood into high-quality pulp in just one step.

The "new" technologies will have no significant impact during the next decade but, in the long term, some sort of compressive pretreatment or enzyme treatment could be part of a new energy-efficient technology. New technology is indeed needed if we are ever to be able to cut energy consumption by some 50% as has been proposed[4].

References

1. Jaakko Pöyry Consulting, 1998.

2. Sundholm, J., unpublished KCL report, 1995.

3. Pöyry, J., "Are energy costs the Achilles' heel of mechanical pulp?," 1977 International Mechanical Pulping Conference Proceedings, Finnish Paper Engineers' Association, Helsinki, key-note speech.

4. Sundholm, J., "Can we save energy in mechanical pulping?," 1993 International Mechanical Pulping Conference Proceedings, PTF, Oslo, p. 133.

5. Gundersby, P., Paper Technology 38(5):29 (1997).

6. May, W.D. and Atack, D., Pulp Paper Mag. Can. 66(8):T422 (1965).

7. De Choudens, S. C., Angelier, R., Combette, P., Revue ATIP 38(8):405 (1984).

8. Harpole, G. B., Leatham, G. F., Myers G. C., "Economic assessment of biomechanical pulping," 1989 International Mechanical Pulping Conference Proceedings, KCL, Helsinki, p. 398.

9. Sundholm, J. (Ed.), "KUITU Energy efficient mechanical pulping," Interim report 1988–1990, Ministry of Trade and Industry, Energy Department, Reviews B:97, 1991.

10. Kokta, B. V. and Vit, R., CPPA 1987 Annual Meeting Notes, CPPA, Montreal, p. A143.

11. Rantala, E., "Development of electric defibation process" (in Finnish), Kuitu report 10, KCL, 1990.

12. Ebeling, K., Review of Development Possibilities for New Mechanical Pulping Processes (in Finnish), The Finnish Ministry of Trade and Industry, 1986.

13. Frazier, W., "Reduction of specific energy in mechanical pulping by axial precompression of wood," CPPA 1981 Annual Meeting Notes, CPPA, Montreal, p. A105.

14. Laine, J. E., MacLeod, J. M., Bolker, H. I., Goring, D.A. I., Paperi ja Puu 59(4a):235 (1977).

15. Bergström, J. R., Tiberg, E. B., Swed. pat. nr. 385 027 (27.9.1974).

16. Nyblom, I. and Vaarasalo, J., internal report, KCL, 1982.

17. Heikkurinen, A., internal reports, KCL, 1986 and 1990.

18. Keskinen, A., internal report, KCL, 1989.

Conversion factors

To convert numerical values found in this book in the RECOMMENDED FORM, divide by the indicated number to obtain the values in CUSTOMARY UNITS. This table is an excerpt from TIS 0800-01 ìUnits of measurement and conversion factors.î The complete document containing additional conversion factors and references to appropriate TAPPI Test Methods is available at no charge from TAPPI, Technology Park/Atlanta, P. O. Box 105113, Atlanta GA 30348-5113 (Telephone: +1 770 209-7303, 1-800-332-8686 in the United States, or 1-800-446-9431 in Canada).

Property	To convert values expressed in RECOMMENDED FORM	Divide by	To obtain values expressed In CUSTOMARY UNITS
Area	square centimeters [cm^2]	6.4516	square inches [in^2]
	square meters [m^2]	0.0929030	square feet [ft^2]
	square meters [m^2]	0.8361274	square yards [yd^2]
Burst index	kilopascal sq. meters per gram [kPa C • m 2/g]	0.0980665	grams-force per square centimeter per (gram per square meter) [(gf/cm^2) (g/m^2)]
Density	kilograms per cubic meter [kg/m^3]	16.01846	pounds per cubic foot [lb/ft^3]
	kilograms per cubic meter [kg/m^3]	1000	grams per cubic centimeter [g/cm^3]
Energy	joules [J]	1.35582	foot pounds-force [ft • lbf]
	joules [J]	9.80665	meter kilogams force [m • kgf]
	millijoules [mJ]	0.0980665	centimeter grams force [cm • gf]
	kilojoules [kJ]	1.05506	British thermal units, Int. [Btu]
	megajoules [MJ]	2.68452	horsepower hours [hp • h]
	megajoules [MJ]	3.600	kilowatt hours [kW • h or kWh]
	kilojoules [kJ]	4.1868	kilocalories, Int. Table [kcal]
	joules [J]	1	meter newtons [m • N]
Force	newtons [N]	4.44822	pounds-force [lbf]
	newtons[N]	0.278014	ounces-force [ozf]
	newtons [N]	9.80665	kilograms-force [kgf]
	millinewtons [mN}	0.01	dynes [dynes]
Frequency	hertz [Hz]	1	cycles per second [s-1]
Length	nanometers [nm]	0.1	angstroms[Å]
	micrometers [μm]	1	microns
	millimeters [mm]	0.0254	mils [milor 0.001in]
	millimeters [mm]	25.4	inches [in]
	meters [m]	0.3048	feet [ft]
	kilometers [km]	1.609	miles [mi]
Light absorption coefficient	square meters per kilogram [m^2/kg]	0.1	square centimeters per gram [cm^2/g]
Light scattering coefficient	square meters per kilogram [m^2/kg]	0.1	square centimeters per gram [cm^2/g]

Conversion factors

Property	To convert values expressed in RECOMMENDED FORM	Divide by	To obtain values expressed In CUSTOMARY UNITS
Mass	grams [g]	28.3495	ounces [oz]
	kilograms [kg]	0.453592	pounds [lb]
	metric tons (tonne) [t] (= 1000 kg)	0.907185	tons (= 2000 lb)
Mass per unit area	grams per square meter [g/m^2]	3.7597	pounds per ream, 17 x 22 - 500
	grams per square meter [g/m^2]	1.4801	pounds per ream, 25 x 38 - 500
	grams per square meter [g/m^2]	1.4061	pounds per ream, 25 x 40 - 500
	grams per square meter [g/m^2]	4.8824	pounds per 1000 square feet [lb/1000 ft2]
	grams per square meter [g/m2]	1.6275	pounds per 3000 square feet [lb/3000 ft2]
	grams per square meter [g/m2]	1.6275	pounds per ream, 24 x 36 - 500
Mass per unit volume	grams per liter [g/L]	7.48915	ounces per gallon [oz/gal]
	kilograms per liter [kg/L]	0.119826	pounds per gallon [lb/gal]
	kilograms per cubic meter [kg/m^3]	1	grams per liter [g/L]
	megagrams per cubic meter [Mg/m^3]	27.6799	pounds percubic inch [lb/in3]
	kilograms per cubic meter	16.0184	pounds per cubic foot [lb/ft3]
Power	watts [W]	1.35582	foot pounds-force per second [ft • lbf/s]
	watts [W]	745.700	horsepower [hp]=550 foot pounds force per second
	kilowatts[kW]	0.74570	horse power[hp]
	watts [W]	735.499	metric horsepower
Pressure, stress, force per unit area	kilopascals [kPa]	6.89477	pounds-force per square inch [lbf/in^2 or psi]
	Pascals [Pa]	47.8803	pounds-force per square foot [lbf/ft^2]
	kilopascals [kPa]	2.98898	feet of water (39.2°F) [ft H2O]
	kilopascals [kPa]	0.24884	inches of water (60°F) [in H2O]
	kilopascals [kPa]	3.38638	inches of mercury (32°F) [in Hg]
	kilopascals [kPa]	3.37685	inches of mercury (60°F) [inHg]
	kilopascals [kPa]	0.133322	millimeters of mercury (0°C) [mm Hg]
	megapascals [Mpa]	0.101325	atmospheres [atm]
	Pascals [Pa]	98.0665	grams-force per square centimeter [gf/cm^2]
	Pascals [Pa]	1	newtons per square meter [N/m^2]
	kilopascals [kPa]	100	bars [bar]
Speed	meters per second [m/s]	0.30480	feet per second [ft/s]
	millimeters per second [mm/s]	5.080	feet per minute [ft/min or fpm]
Tear index	millijoules per gram [mJ/g] [mN C•m^2/g]	0.0980665	100 grams-force per gram (gram per square meter [100 gf/(g/m^2)]
Tensile energy absorption (TEA)	joules per square meter [J/m^2]	14.5939	foot pounds-force persquare foot [ft • lbf/ft^2]
	joules per square meter[Jj/m^2]	175.127	inch pounds-force per square inch [in • lbf/in^2]
	joules persquare meter [J/m^2]	9.80665	kilogram-force meters per squaremeter [kgf • m/m^2]
	joules per square meter [J/m^2]	1	joules per square meter [J/m^2]
Tensile index	newton meters per gram [NC•m/g]	1	newton meters per gram [N • m/g]
	newton meters per gram [NC•m/g]	9.80665	kilometers breaking length [km]
Tensile strength	kilonewtons per meter [kN/m]	0.175127	pounds-force per inch [lbf/in]
	kilonewtons per meter [kN/m]	0.29655	pounds-force per 15 millimeter width [lbf/15mm]
	newtons per meter [N/m]	10.945	ounce-force per inch [ozf/in]

Property	To convert values expressed in RECOMMENDED FORM	Divide by	To obtain values expressed In CUSTOMARY UNITS
	kilonewtons per meter [kN/m]	0.65378	kilograms-force per 15 millimeter width [kgf/15mm]
	kilonewtons per meter [kN/m]	0.39227	kilograms-force per 25 millimeter width [kgf/25mm]
	kilonewtons per meter [kN/m]	0.980665	kilograms-force per centimeter [kgf/cm]
	newtons per meter [N/m]	9.80665	grams-force per millimeter [gf/mm]
	newtons per meter [N/m]	66.6667	newtons per 15 millimeter width [N/15mm]
Thickness or caliper	micrometers [µm]	25.4	mils [mil] (or points or thousandths of an inch)
	millimeters [mm]	0.0254	mils [mil] (or 0.001 in.)
	millimeters [mm]	25.4	inches [in]
Volume flow rate	liters per minute [L/min]	3.78541	gallons per minute [gal/min]
	liters per second [L/s]	28.31685	cubic feet per second [ft^3/s]
	cubic meters per second [m^3/s]	0.0283169	cubic feet per second [ft^3/s]
	cubic meters per hour [m^3/h]	1.69901	cubic feet per minute [ft^3/min or cfm]
	cubic meters per second [m^3/s]	0.76455	cubic yards per second [yd^3/s]
	cubic meters per day [m^3/d]	0.00378541	gallons per day [gal/d]
Volume, fluid	milliliters [mL]	29.5735	ounces [oz]
	liters [L]	3.785412	gallons [gal]
Volume, solid	cubic centimeters [cm^3]	16.38706	cubic inches [in^3]
	cubic meters [m^3]	0.0283169	cubic feet [ft^3]
	cubic meters [m^3]	0.764555	cubic yards [yd^3]
	cubic millimeters [mm^3]	1	microliters [µL]
	cubic centimeters [cm^3]	1	milliliters [mL]
	cubic decimeters [dm^3]	1	liters [L]
	cubic meters [m^3]	0.001	liters [L]

Index

A
abrasive 126
acetic acid 384, 386
agitation 349
air emissions 377
alkaline peroxide mechanical pulping 245
amplitude 47
arabinogalactan 382
Asplund 29
axial compression 419
axial load 188

B
bar 188
bark 72
barrier screening 256
basic density 68
basket 255
Bersano 32
Bi-Vis 419
BOD 381
bonding 399
bow screen 290, 348
breaker bar 178
brightening 313
brown groundwood 27
burr 128

C
Canadian standard freeness 32
carbonyl groups 335
cascade control 144
cavitation 189
cell wall thickness 88
centrifugal force 52, 197
centrigrinder 32
chain grinder 121
chip quality 99
chops 252
COD 381
collapsibility 397
condensing 213
conditioning 132
conductometric titration 228
conical disc 160, 174, 185
consistency 200
cyclic stress 46
cyclone 182
DD 179

D
deinked pulp 20
digester 233
disc filter 345
dissolved substances 380
double disc 179
drainage properties 402
DTPA 320

E
earlywood 87
EDTA 320
efficiency 254
elastic 35–37
elastic modulus 35–37
electron beam treatment 419
energy consumption 57, 161
entzymatic treatment 419
epsom salt 322
exchanger 214
explosion pulping 419
external purification 390
extractives 71, 382

F
fatigue 37, 40
Fenerty 23
fiber classification methods 403
fiber coarseness 397
fiber flexibility 340, 397–398, 402, 413
fiber length 395
fiber wall thickness 397
fiber width 397
fibrillation 398

fine cleaner	281
fines	395
flexibility	397
fluff pulp	21
folding boxboard	153, 372
formamidine sulfinic acid	335
FQA	406
fractionation	256
freeness	402
frequency	37, 49, 199
FSA	405

G

galactoglucomannan	382
galacturonic acid	383
Great Northern	28
grinding zone	40
grit	40, 126
groove	131, 188

H

handsheet properties	409
heartwood	71
high speed	205
hole screen	264
hot grinding	27
Hot Loop	118
Hydrodynamic stiffness measurement	398

I

image analyzer	406
impact	197, 316
Impressafiner	168, 247
intensity	54, 183, 196, 300, 355
internal fibrillation	407
interstage sulfonation	245

J

juvenile wood	70

K

Keller	23
knotwood	73

L

latency	401
latewood	44
LE	189

length of fines particles	408
light absorption	313
light scattering	313
light-induced	338
lignans	382
lipophilic extractives	382
liquid packaging board	244
load controls	142
log aligner	136
log feeding	135
LWC	371

M

magnesium sulfate	322
martensitic	189
mature wood	93
mechanical printing paper	21
medium consistency	325, 269
moment of inertia	398
MRSD	170
multistage process	204
MuST	270

N

Nacke refiner	29
newsprint	367

O

ozone	335

P

paper grade	365
particle size distribution	402
pc number	339
pectins	382
peeling off mechanism	397
peracetic acid	335
perhydroxyl ion	319
phenolic	315
photography	56
plate gap	52, 197, 206–209
Pocket grinders	109
power control	144
PQM	405
pre-screen	269
pressurized disc filter	347
PREX	166, 235

Index

primary screen 270
printability 21, 153, 308
PRMP 160
production rate 201
Pulp Expert 406
pulp suspension properties 402

R
reboiler 213
recycled fibers 417
refiner bleaching 334
refining consistency 200
refining intensity 54, 183, 300, 355
removal efficiency 252, 262, 283, 306, 390
residence time 52, 201
resin 385
reversion 337
RGP 171, 174
Rinheat 212
RTS 204

S
sand removal 283
sandstone 27
sapwood 71
SB 169
screw impregnator 167
screw press 168, 290
scrubber 215
SD 160, 169
secondary screen 270
sedimentation volume 408
sharpening 129
shredder 117
silicon carbide 126
Simons staining 407
single disc 160, 169
single stage 193
slightly mechanical 21
slot 266
sodium bisulfite 335
sodium borohydride 330, 335
sodium silicate 322
sodium sulfite 226
softening 35
specific edge load 355
specific filtration resistance 399
specific surface 398

specific surface load 356
speed control 143
SRE 254
steam pressure 202
steaming 202
stiffness 398

T
TDC 172
temperature 202
tertiary screen 270
Thermopulp 32, 205
thickening ratio 252
thinnings 95
TOC 381
toxic 385
turbidity 409
turpentine 218
Twin 177

U
ultrasound treatment 419

V
viscoelastic 35
Voelter 24
Voith 24
volatile organic compounds 212, 387

W
water jet 132
wedge wire 264
WRV 403

X
xylan 382

Y
yellowing 337